Boundary Retracement

Boundary Retracement

Boundary Retracement
Processes and Procedures

Donald A. Wilson

CRC Press
Taylor & Francis Group
Boca Raton London New York

CRC Press is an imprint of the
Taylor & Francis Group, an **Informa** business

CRC Press
Taylor & Francis Group
6000 Broken Sound Parkway NW, Suite 300
Boca Raton, FL 33487-2742

© 2017 by Taylor & Francis Group, LLC
CRC Press is an imprint of Taylor & Francis Group, an Informa business

No claim to original U.S. Government works

Printed on acid-free paper

International Standard Book Number-13: 978-1-4987-2710-5 (Hardback)

This book contains information obtained from authentic and highly regarded sources. Reasonable efforts have been made to publish reliable data and information, but the author and publisher cannot assume responsibility for the validity of all materials or the consequences of their use. The authors and publishers have attempted to trace the copyright holders of all material reproduced in this publication and apologize to copyright holders if permission to publish in this form has not been obtained. If any copyright material has not been acknowledged, please write and let us know so we may rectify in any future reprint.

Except as permitted under U.S. Copyright Law, no part of this book may be reprinted, reproduced, transmitted, or utilized in any form by any electronic, mechanical, or other means, now known or hereafter invented, including photocopying, microfilming, and recording, or in any information storage or retrieval system, without written permission from the publishers.

For permission to photocopy or use material electronically from this work, please access www.copyright.com (http://www.copyright.com/) or contact the Copyright Clearance Center, Inc. (CCC), 222 Rosewood Drive, Danvers, MA 01923, 978-750-8400. CCC is a not-for-profit organization that provides licenses and registration for a variety of users. For organizations that have been granted a photocopy license by the CCC, a separate system of payment has been arranged.

Trademark Notice: Product or corporate names may be trademarks or registered trademarks, and are used only for identification and explanation without intent to infringe.

Library of Congress Cataloging-in-Publication Data

Names: Wilson, Donald A., 1941- author.
Title: Boundary retracement : processes and procedures / Donald A. Wilson.
Description: Boca Raton : Taylor & Francis, a CRC title, part of the Taylor & Francis imprint, a member of the Taylor & Francis Group, the academic division of T&F Informa, plc, [2017] | Includes bibliographical references.
Identifiers: LCCN 2016038571| ISBN 9781498727105 (hardback : alk. paper) | ISBN 9781498727112 (ebook)
Subjects: LCSH: Boundaries (Estates)–United States. | Land titles–United States. | Surveying--Law and legislation--United States.
Classification: LCC KF639 .W55 2017 | DDC 346.7304/38--dc23
LC record available at https://lccn.loc.gov/2016038571

Visit the Taylor & Francis Web site at
http://www.taylorandfrancis.com

and the CRC Press Web site at
http://www.crcpress.com

This work is dedicated to today's surveying and legal scholars, with many of whom I have had the pleasure and honor to collaborate and share thoughts and ideas. Others I have known in passing or have otherwise drawn upon their wisdom and expertise. Taking the risk of omitting someone of importance, but having the compulsion to recognize modern leaders, I wish to recognize the following who have had an impact on my professional life and have influenced my thinking, helping me to become a better retracement surveyor: Hon. Ruggerio Aldisert, R. Ben Buckner, Lane J. Bouman, Jerry Broadus, Curtis Brown, George Butts, Frank Emerson Clark, George Cole, Gary Kent, Kristopher Kline, Wendy Lathrop, Joel Leininger, Jeffrey Lucas, Dr. Alec McEwen, Roy Minnick, Dennis Mouland, Michael Pallamary, Walter G. Robillard, Ray Hamilton Skelton, Herbert Stoughton, and Francois Uzes. Without these dedicated individuals, we would not have progressed as far as we have in both the surveying and legal professions.

I am also indebted to Kris Kline for his reviews and helpful suggestions. He has been a strong supporter of this project since its conception.

Generally speaking, surveyors have no task more challenging than that of boundary retracement. Any given boundary will present the surveyor with dozens, perhaps hundreds, of pieces of evidence leading to an equal number of potential solutions to the problem. The challenge of sifting through the data for relevance and subjecting it to the proper legal and mathematical tests is taxing under the best of circumstances; if the surveyor lacks the correct doctrine, the problem of accurately retracing the boundary is nearly insurmountable.

Joel M. Leininger, PLS
Institutes of Boundary Retracement, Part I

Contents

Preface xv
Author xxi

1. The Concept 1
Principles of Retracement 2
  Apportionment—Reason and Necessity 5
Conditions for the Rule to Apply 6
  Following Footsteps Contrary to Established Rule 34

2. Title Issues 37
Definitions 37
  Definition of Title 37
  Definition of Boundary 37
  The Establishment of Title 38
  Significance of the Original Survey 40
  The Establishment of Title to Land, or Rights in Land 40
  Written Title 41
  Combination 42
  Unwritten Title 42
  Prescription 43
  Missing Title 45
  Notice 50
  Ambiguities 51
  Ambiguity in Field Notes 53
  When There Is No Ambiguity 53
  Multiple Markers 54
  Descriptions by Abutting Tracts 56
  Boundary Line Agreements 60
  Intent of the Parties 60
  Intention 61
  Intent of the Surveyor 63
When There Is Ambiguity 64
The Void Instrument 65

3. Corners, Lines, and Surveys 69
Types of Corners and Their Definitions 70
  Corner, Existent 70
  Corner, Nonexistent 71
  Corner Accessory 71
  Corner Contiguity 71

vii

viii *Contents*

Closing Corner...71
Quarter-Section Corner 71
Section Corner 71
Sixteenth-Section Corner..72
Township Corner..72
Standard Corner 72
Reestablished Corner...72
Lost Corner 72
Obliterated Corner...72
Extinct Corner..72
Meander Corner 73
Half-Mile Post..73
Witness Corner (PLSS)...73
Witness Tree (PLSS)...73
Monument..74
Center of Section 74
Monument Accessory 75
Missing Monument..76
Types of Lines...77
Right of Way Line..78
Fence Line...78
Blazed Line...78
Lot Line...78
Conditional Line..78
Consentable Line..78
Conventional Line..79
Compromise Line...79
Extended Line 79
Ideal Line 79
Lost Line 79
Meander Line..79
Partition Line (Division Line)...80
Property Line 80
Right Line 80
True Line..81
To Range With..81
Presumption as to Straight Lines..81
Continuity of Line..81
Types of Surveys 81
Survey 81
Other Categories of Survey 85
Accurate Survey 85
Original Survey versus First Survey 89
Retracement versus Resurvey 93
Establishing a, or the, Line..94

Independent Survey	96
Tracking a Survey	96
What If There Are No Actual "Footsteps?"	99
Early Court Decision	101
Not Limited to Public Land Survey States	102
Early Guidance in Textbook	103

4. Protracted Boundaries 105

Definition	105
Protraction outside the PLSS	107
What to Do about Protracted Lines	128
Types of Protractions	129
Lines from Other Surveys	129
The Unmarked Line	130
Guidance	130
Protraction in the PLSS	131
Additional Footsteps	139

5. Which Set of Footsteps Is Which? 141

Wrong Set of Footsteps	142
Who Left the Footsteps?	143
Learning from Field Notes	145
Admissibility of Field Notes	146
Discrepancy between Field Notes and Plat	165
Passing Calls	165
Principles Concerning Field Notes	166
Misleading Evidence in Field Notes	167
Ambiguity in the Field Notes	180

6. Figures, Numbers, and Symbols 181

Types of Survey	181
Problems with Directions	184
By the Needle	186
Definition of Course	187
Which Meridian	187
Type of Compass Used	187
Declination	188
Impact on Successful Corner Location	188
Local Attraction	189
Directions Given in Reverse Order	206
Problems with Distances	207
Measurement Allowances	208
Errors in Chaining	210
Errors in Taping	210
Presumption of a Straight Line	211

x   *Contents*

Slope versus Horizontal Measurement.. 218
Actual Survey versus Paper Survey... 219
    Distance Deficiencies 222
    Compare Measurements with Original: The Rule 223
    Problems with Area Recitations 223
Problems with Mathematics.. 224
    Adjustments and Getting Rid of Closure Errors 224
    Adjustments or Not: Shifting Positions 224

7. Resolving/Reconciling Errors... 227
Land Surveying as an Art... 227
Errors in Measurements... 229
Errors 229
Discrepancies.. 229
Errors of Closure 230
Random Errors 230
Index Errors 230
Personal Errors 230
Natural Errors... 231
Systematic Errors 231
Accidental Errors 231
Computational Errors.. 232
Mistakes.. 232
Total Error 233
Reliability 233
Lower Precision Surveys.. 233
Measure of Precision 234
Errors in the Collection of Data 234
Errors in Reporting Data... 235
Words from the Courts... 240
Summary of the View by the Courts on the Subject of Surveys and
    Their Errors 240
The Early Surveys in Hawaii... 242
Course and Distance in Perspective... 243

8. Recognizing What Was Left Behind 245
Evidence.. 246
Tree Names 246
Blazed Tree.. 247
Witness Trees 247
Scribe Marks 248
Stump Holes 248
Tree Rings... 249
Wooden Posts 250
Wooden Stakes and Their Remains.. 250

Contents   xi

Stake and Stones 251
Pits and Mounds 251
Other Witnesses.. 252
    Corner Accessories... 253
        From Manual of Surveying Instructions, 2009 253
Bearing Trees and Bearing Objects 253
Memorials 255
Mounds of Stone 255
Pits.. 256
Accessories to Special-Purpose Monuments 257
    Memorials.. 258
    Fences.. 259

9. Boundary Agreements.. 263
Statute Law in New Hampshire 269
The Meaning of Uncertainty 270
Adverse Possession 272
Acquiescence 273
Establishment of Boundary by Acquiescence........................... 273
Estoppel.. 273
Parol Agreement 274
Effect of Statute of Frauds.. 274
Necessity for Uncertainty or Dispute as to Location of Boundary....... 274
Agreement upon a Definite Line 275
Practical Location of Boundary.. 275
Statute of Frauds 278
Not Meeting the Basic Requirement 280
Consentable Lines.. 282
Conditional Lines... 284
Recognition and Acquiescence 286

10. Proper Procedure 287
Procedure 287
    What Makes an Effective and Successful Retracement
        Surveyor 287
    The Original Creation and the Original Surveyor 287
    In General.. 289
Government Procedure, Manual of Surveying Instructions 296
The Ideal Line... 297
The Corner Never Set or the Line Never Run 298
About Corners and Witness Trees 298
Reversing Course.. 301
Government Surveys.. 303
Locating Unrecorded Maps, Plans, and Sketches 304
The Unrecorded Plat.. 304

xii **Contents**

Probate Records..305
Historical Collections 307

11. The Lost Corner..309
What If the Corner Is Truly Gone? 309
Restoring Lost Corners...309
    Where the Corner Formerly Stood 310
    Lost *v.* Obliterated Corners.. 311
    How Lost Is "Lost"?..312
Procedure 313
Parol Evidence and Reputation 320
Newly Discovered Evidence 324
Not Getting It Right Only Invites Future Problems................................330

12. Nonfederal Rectangular Surveys...333
Beginnings in the New World...334
Division by Half...337
Division by Area 337
Idiosyncrasies within the System 340
Bounding Descriptions...340
Proprietors' Records 341
Other States..343
Kentucky: The Jackson Purchase..343
Tennessee Surveyors Districts...344
The French System of Longlots...345
Louisiana...345
Texas...347
    The State of Texas 347
Longlots in Other Areas 348
Lottings in Georgia 348
Military Grants 348
Special Grants...349
In Summary 349

13. Resolving Overlapping Grants...351
Seniority of Title..351
Original Grants...353
Recent Decisions 359
Opinion..366

14. Forensic Applications 371
Definition 371
The Value of Historical Knowledge and Research...................................372
The Value of a Complete and Thorough Investigation...........................374
Modern Tools..374

Contents

xiii

What Forensic Science Can Do Outside of the Criminal
Environment 374

References 375

**Appendix I: List of Significant Cases on Following
Footsteps, by State** 377

Appendix 2: Lost Corners, by State 405

Final Thoughts 435

Table of Authorities 437

Index 455

Preface

While it would seem that the concept of "following the footsteps" of the original surveyor has been overdone in writings and presentations, it remains a basic legal standard that is not fully understood by many, and as many still, after grasping the concept, lack sufficient training and experience to carry it out properly and completely.

Even so, there are two parts to this concept which have not been adequately addressed. Nowhere has there been a comprehensive review of the court decisions dealing with the subject; and secondly, nowhere has it been described as to how to identify the proper set of footsteps to follow when there is more than one set, especially when two or more conflict with one another.

This book will attempt to accomplish both of the aforementioned shortcomings, at least to an extent. Also, it should be pointed out, at the outset, that the theme of the book is "Following Footsteps," which includes "Following *the* Footsteps," for all too often there is more than just one set that needs to be followed.

Most surveyors have heard or have read the phrase, follow [in] the footsteps of the original surveyor. However, it is questionable as to how many actually do exactly that, or even know what it means to do *exactly* that. Following the footsteps of the original surveyor is a legal standard adopted by the courts in all jurisdictions, and for very good reason.

One of the confusing issues is that the original surveyor's footsteps are no longer visible. Snow has melted; landscapes have changed through erosion and weather exposure, which, along with other factors not the least of which are lost and deteriorated records, have served to obscure original footsteps. Another important issue is that, and several presentations have emphasized, it is impossible to follow exactly in someone else's footsteps since it is impossible to duplicate the exact conditions that existed when the original surveyors did their work. Therefore, human nature convinces us that since we cannot do what is required, we may rely on the next best thing, and, done properly, in theory will yield the same result. This text will demonstrate how and why that is not true. In some situations achieving the same result is not possible, or at least the average surveyor may not be capable of it. This could be due to lack of ability, lack of knowledge, lack of experience, lack of self-confidence, or because of dictates by others, such as lack of time, funds, or resources. Any successful retracement, or any work for that matter, demands three things: sufficient time, sufficient funds, and sufficient resources. Surveyors are often handicapped by the lack of one of more of these essential items.

xv

xvi *Preface*

However, the underlying requirement, dictated by the law, stands. **Follow the footsteps of the original surveyor.** It is a simple phrase, but it is often not so simply accomplished. I would submit that when that phrase was coined, and subsequently used, judges and others knew exactly what they were stating, and understood the reasoning behind it. Equally important are the concepts of long-term stability of property boundaries and sanctity of titles, which will appear throughout this book: Neither is of value to anyone if it is constantly shifting. Above all, people's rights must be stable, and they must be protected. The courts have done a magnificent job of doing that, as can be seen from the multitude of decisions incorporating the subject, and the incredible uniformity with which it has been done. And surveyors, through state statutory law, have been entrusted to do that. Property rights are protected by both state and federal constitutions, and are not to be treated lightly; they are to be acknowledged, protected, and preserved, precisely as they were created.

Thinking about it, it really is a very simple concept. The down side is that it sometimes takes a well-trained, sometimes highly specialized, professional to accomplish this. This is precisely why land surveyors are licensed to practice, regulated, though sometimes not carefully enough, and, in recent years, well-educated, requiring a degree in the subject matter before being allowed to practice and deal with important rights of others. Although this requirement has met with resistance in some circles, it is of great necessity, since land surveying involves many related disciplines: archaeology, astronomy, botany, cartography, dendrology, engineering, forensic science, forestry, geology, history, hydrology, law, and photogrammetry, among others. The average practitioner must at least recognize the aspects inherent in other disciplines, if not possess skill in their use.

The reasoning behind this was summarized as follows for legislative testimony*:

The land surveyor is not an **archaeologist**, but must be able to search for and recover objects placed to mark boundaries hundreds of years ago. He or she must be able to determine if three or four stones under a foot of decayed leaves were placed by the hand of man, or simply left by the melting glacier. He must be able to locate the foundation of the blacksmith shop, which in 1895 "burned 10 years ago."

The land surveyor is not an **astronomer**, but must be able to make observations on stars and planets to position his measurements and relate magnetic observations to the true meridian.

The land surveyor is not a **cartographer**, but must have a broad knowledge of maps and topography to prepare boundary maps and to

---

* Taken in part from, Butts, George F. *What Does a Land Surveyor Do?* The Cornerpost, Vol. XVI, No. 3, pp. 7–14 (September, 1985). Used by permission.

Preface xvii

define those boundaries dependent upon topographical features such as ridges and streams.

The land surveyor is not a **computer specialist**, but must be able to operate computers that perform lengthy calculations required; that direct automatic machines and that do word processing.

The land surveyor is not a **dendrologist**, but must be able to identify a "sugar plum tree" called for in a 1935 deed, or identify a rotten yellow birch stump that has been moldering in the woods for 50 years.

A land surveyor is not a **detective**, but solves problems in a similar matter by assembling piece after piece until the answer appears. The surveyor deals with areas of forensic science on a regular basis, yet is not a **forensic scientist.**

The land surveyor is not an **engineer**, but must have a broad knowledge of such engineering works as highways, power lines, pipe lines, and railroads, because these works often control property boundaries.

The land surveyor is not a **farmer**, but must be able to recognize farming methods in use during since the early settlement of the state. Did the farmer erect that fence to mark the boundary or to keep the cattle out of a swamp? Where was the "north barway"? What was the "hop house"? Where is the 1888 boundary between the "mowing" and the "pasture"?

The land surveyor is not a **forester**, but must be able to separate marks and objects placed by foresters in their management of timber stands, from marks and objects placed to control property boundaries.

The land surveyor is not a **geologist**, but must have a general understanding of the land forms and an ability to identify the various types of stones used for monumentation of land corners.

The land surveyor is not a **handwriting expert**, but must be able to read the writing of the early town clerks. As the incumbent clerk grew older, the penmanship became progressively less readable, until, suddenly a new clerk's handwriting appears. In addition, such early English words as "staddle," "rood," "square perch," and others must be understood.

The land surveyor is not an **historian**, but must have a large, specialized knowledge of the state's early history and the histories of the various towns in which he or she practices. To put together a chain of ownership back to a deed containing a description that is something more than a reference to earlier deeds, the surveyor often has to trace the title back to the early nineteenth, even to the eighteenth, and occasionally into the seventeenth century.

The land surveyor is not a **hydrologist**, but must have knowledge of waters and water courses, for riparian rights such as spring rights, flowage or dam rights, and the problems caused when floods change stream channels; all enter into property boundary determinations.

The land surveyor is not a **lawyer**, but must be able to locate boundaries so that their positions will withstand review by the courts.

The land surveyor is not a **logger**, but must be able to separate the marks left by the cutting foreman from marks made to delineate a property boundary.

The land surveyor is not a **judge**, but must be able to make decisions founded on law concerning the locations of property boundaries. Decisions that will withstand review by the courts.

The land surveyor is not a **juror**, but must be able to come to decisions of fact that will also withstand review by the courts.

The land surveyor is not a **photogrammetrist**, but must be able to make measurements on aerial photographs and to identify objects thereon.

The land surveyor is not a **writer**, but must be able to describe in words the location of boundary lines and give directions for their location. Directions that often will not be used until years in the future when all persons having knowledge of the time of the original survey are dead.

In addition to being an **expert measurer**, the land surveyor must be able to act as an **expert witness**, with special knowledge, wisdom, or information acquired through study, investigation, observation, practice, and experience. One must be able, after reaching the correct decision, to present the evidence and conclusions drawn from the evidence, in a manner that will enable others to reach the same decision. It is not enough to be right; one must also be able to persuade others.

Once making the claim that land surveying is as complex as medicine, the statement was quickly met with the answer that it is often more complex, since it demands both left- and right-brain thinking. The exposure to modern day use of forensic science has resulted in many land surveyors agreeing that that is what they have been doing all along—detective work. And many surveyors will quickly state that that is what they like best about their work. And, rightly so, take great pride in their perseverance and ability to solve a particular problem, especially where others have failed. Law courts do this on a regular basis, reducing arguments and positions to simple legal principles, often through the use of what is known as precedent. Surveyors, in their training, do much the same thing, by learning legal principles, on which they are examined in order to be licensed to practice their professional calling. Legal principles are derived from legal precedents. Following the footsteps is just one of them, but one that has been a long-standing, much used, standard required in retracing an existing boundary, incorporating any and all of the others.

Preface xix

One of the biggest differences between the surveyor relying on principles and courts relying on precedent is that courts continually revisit the reason for the rule, or the decision in the previous case, to insure that it applies, and fits the issue. Surveyors do not always study the reasons behind the rule, or even understand its foundation, frequently relying on what has been presented to them, through a textbook, or a formal presentation, such as a seminar for example. It is as important to understand the reason behind a rule as it is to know the rule. Many rules are incorrectly used, or applied, because they don't properly fit with the issue even though they may appear to at first. In other cases, rules and principles are applied for convenience because without proper understanding, they appear to suffice. Quoting from one important decision, "Blind devotion to a rule may lead to infinite failure."*

One thing that people in general usually do not think about is what an existing boundary actually is. Today's surveyor is burdened with someone else's creation, in some way, at some point back in time, perhaps way back in time, and identified in some manner. It may be poorly described, its description may contain mistakes and errors (more on that difference in a later section), and it may be misleading. Others, making their own interpretations, may further mislead the follower with incorrect statements, conclusions, and interpretations, some of which may even conflict with adopted legal principles. Therein lies a basic problem, which is why the standard requires following the *original* surveyor, not some intermediate surveyor, or other, following surveyor. There are other, better reasons, which will also be explored.

Finding, and locating, an existing boundary, one created at some point in the past, is accomplished through what is known as *retracement*. The term may be defined in several ways: as a procedure, as a surveying method for resurrecting lost property corners, and as a legal standard. Each of these brings us to the same place.

Donald A. Wilson
Newfields, New Hampshire

---

* *Bradford v. Pitts*, 2 Mills. Const. Rep. 115 (South Carolina, 1818). Read this case, and you will not only fully understand the rule, but will also probably never forget it.

Author

Donald A. Wilson, president of Land & Boundary Consultants, has been in practice for over 50 years, consulting to groups throughout the United States and Canada. He is both a licensed land surveyor and professional forester, having conducted more than 500 programs on a variety of topics, including description interpretation, boundary evidence, law, title problems, and forensic procedures.

Mr. Wilson has more than 200 technical publications in several areas, and has been involved with over 60 books, which include titles on Maine history and several books on fishing. Besides being coauthor of *Evidence and Procedures for Boundary Location* and *Boundary Control and Legal Principles*, he is author of *Deed Descriptions I Have Known ... But Could Have Done Without, Easements and Reversions, Interpreting Land Records,* and *Forensic Procedures for Boundary and Title Investigation.* His latest titles are *Easements Relating to Land Surveying and Title Examination* and a co-authorship titled, *Land Tenure, Boundary Surveys and Cadastral Systems.*

Don is an instructor for RedVector's on-line professional courses and, for 38 years, a presenter in the University of New Hampshire's Professional Development Program. He is part owner of and lead instructor in Surveyors Educational Seminars and a regular seminar coordinator for the University of New Hampshire. In his professional practice, Don has testified numerous times, in a variety of courts, on boundary and title matters.

1

The Concept

There is no branch of detective science which is so important and so much neglected as the art of tracing footsteps.

Sherlock Holmes
A Study in Scarlet

One of the more recent writers of legal articles summarized the relationship between title and boundaries in 1960 while contrasting the basic concepts of retracement and apportionment. Robert J. Griffin published *Retracement and Apportionment as Surveying Methods for Resurrecting Lost Property Corners.** In a few pages, concentrating on proper principles, practices, and legal sources, Griffin detailed the legal basis of property corners, descriptions and the location of title lines. He began with a discussion of title, then proceeded to the surveyor's role in the process, finally emphasizing that retracement is paramount, and apportionment a process for use only when the former fails to produce the location of a corner. The remainder of this book will detail many of the concepts with their significance to the boundary location process.

Griffin's opening premise states that "conservation and perpetuity of boundary lines is the primary aim of the law of boundaries. The location upon the ground of such lines is determined by re-survey, and re-surveys may be classified according to method." Two methods exist, retracement and apportionment.

Griffin first outlines the rules and principles of retracement, stating that "Intention is paramount in determining location," and that "the purpose of a re-survey of land is to locate and mark upon the ground the boundaries of the parcel of land evidenced by the description given in a particular deed.† The extent of the parcel actually transferred by the deed is resolved by the intention of the grantor, so far as that intention is effectively expressed in the

---

* *Marquette Law Review*, 43(4), 1960, Article 5, 484–510.
† Descriptions may appear in a variety of instruments in addition to deeds, as discussed in some detail in Chapter 2. Most of the rules apply equally to descriptions found in any legal instrument serving as a conveyance of title, rights or interests, as well as related documents, such as field notes and plats.

deed interpreted in the light of then existing conditions and circumstances."*
He adds, "The expression of this intention in the deed may be incomplete and
ambiguous; but, regardless of how 'bunglingly expressed,' there is a strong
presumption that the grantor intended a certain encompassing boundary to
define the lands granted."

Continuing, Griffin stresses, "the intention of the original grantor, as
expressed and as inferred from the deed, is the paramount consideration
in determining the location of property lines and corners" (citing numerous
decisions).

> When a deed is interpreted in the light of then existing conditions and
> circumstances, the interpreter considers the original survey which
> marked the boundaries. The highest and best proof of intention lies, not
> in the words of expression, but in the work performed upon the ground
> itself. Lines actually run and corners actually established upon the
> ground prior to the conveyance are the most certain evidence of inten-
> tion. It is by the work as executed upon the ground, not as projected
> before execution or represented on a plan afterward, that actual bound-
> aries are determined.
>
> Retracement is a process for gathering evidence. It seeks to accumu-
> late evidence of intended location of property corners when they have
> become lost, obscured, confused or obliterated. The evidentiary suffi-
> ciency of a retracement depends upon the observance of accepted prin-
> ciples which govern the process.

As noted, Griffin's discussion begins with the concept of title, and sets
forth three very important basic and "accepted principles governing the process of retracement," as follows.

Principles of Retracement

**First Principle. Location of a Boundary Line is Determined as of the Time
of Its Creation.** Here, he highlights a very important concept, one which a
lot of people lose sight of:

---

* Griffin references the Maine decision in the case of *Perry v. Buswell*, 113 Maine 399, 94 Atl. 483,
1915, an exceedingly important and significant decision, which states, "The cardinal rule for
the interpretation of deeds and other written instruments is the expressed intention of the
parties gathered from all parts of the instrument, giving each word its due force, and read in
the light of existing conditions and circumstances. *It is the intention effectually expressed, not
merely surmised. This rule controls all others.*" (emphasis supplied).

The Concept

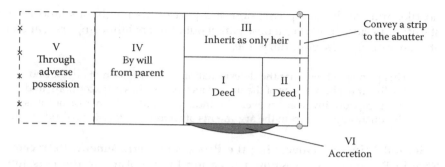

FIGURE 1.1
Example of an ownership made up of several tracts, sources of title, and combination of boundaries.

A boundary line, once established, should remain fixed through any series of mesne conveyances.* (emphasis supplied). A grantee acquiring a particular parcel determines his boundaries as of the time that the particular parcel was carved out of some larger tract.

> He takes to the bounds of the estate of his grantor, who in turn took to the limits of his grantor's estate, etc., to the time of creation of the boundary. A grantee acquiring only a part of the lands of his grantor will determine the common boundaries as of the time of the conveyance, while he will determine boundaries on the perimeter of the grantor's original tract with reference to the time that they were created. Each line of the same parcel must be considered separately, and a determination of the proper surveying method to be used must be made with respect to each line of the parcel.
>
> The time of creation principle is closely related to the original government survey of public lands, since private boundary lines often run along lines established by this first survey or are described and located in relation to the original survey corners† (Figure 1.1).

Griffin added a paragraph to his discussion concerning adverse possession, stating "the doctrine of adverse possession does not defile the principle that location of a boundary line reverts to the time of its creation. When the disseisor owning adjacent property takes and holds a portion of his neighbor's lands, he is not extending or expanding the boundaries of his own property, but rather, is creating an entirely new line dividing his neighbor's land. The boundary of the disseisor's property remains in the position fixed at the time of its creation. By holding actual, open and notorious, exclusive,

---

* Citing one of the most important decisions, written by Justice Thomas M. Cooley, *Diehl v. Zanger*, 39 Mich. 601, 1878.
† At times, if not coincident with original boundaries, they are influenced by original boundaries, having the same direction as perimeter lines (parallel with).

hostile, and continuous possession of a part of his neighbor's land for the statutory period, he carves a new parcel out of the neighboring tract and creates an entirely new boundary line.*

> This principle then, that the determination of the location of any boundary line must be made as of the time of its creation, has no exceptions, and is inescapably involved in every boundary dispute whether or not it is obviously apparent from the arguments of counsel or decisions of judges.

Second Principle Governing the Process of Retracement. Retracement Should Proceed from a Known Location to Hypothecate the Unknown. Retracement is the process of uncovering physical evidence of monuments and corners by intelligent search on the ground for the calls of the description and field notes of the original survey, guided by the controlling influence of known points. It should proceed from a known location to hypothecate the unknown.

Third Principle Governing the Process of Retracement. Retracement Should Apply Rules of Construction to Contradictory Evidences of Intention. "When ambiguities appear in the description, and when discrepancies arise between adjoining descriptions or between the description and the physical evidence of the boundaries as it exists on the ground, rules of construction are applied to determine intention. The rules, which are based on reason, experience and observation, and pertain to the weight of evidence, state the order of preference and relative importance of calls in a grant. This order of priority, in general, is as follows:

1. Lines actually surveyed and marked prior to the original conveyance control over calls for monuments.
2. Calls for fixed monuments control over calls for adjoinders.
3. Calls for adjoinders control over calls for course and distance.
4. Calls for course and distance control over calls for quantity.

The first rule of construction is but an abbreviated phraseology of the principle discussed previously that the highest and best proof of intention lies in the work performed upon the ground. A re-survey should normally apply these accepted rules of construction to contrary evidences of intention, absent any circumstances suggesting their inapplicability."

In summary, Griffin includes the following statements: "In conducting the retracement, care should be exercised to retrace the lines as they were originally run, and not where subsequently and erroneously re-established by

---

* It effectively establishes a new, independent title of a stand-alone parcel with its own boundaries. It is an unwritten transfer of the parcel through possession ripening into title based on the statute of limitations barring an action for recovery after a specified period of time and satisfying other specified requirements.

The Concept

some intermediate owner or holder. When a corner is lost and cannot be re-established from traces of the original monument or its accessories, or from acceptable collateral evidence of the original position, the surveying method of apportionment is considered. The rule of apportionment is applied as a last resort for distributing discrepancies to determine intention in the most equitable way possible."

AUTHOR'S NOTE: Several comments are in order regarding apportionment. Griffin emphasizes that (1) the corner must be lost,[*] and (2) cannot be re-established. When that is the case, apportionment is *considered*.

According to Skelton,[†] there are a number of conditions that must be met in order to apportion excess or deficiency:

1. Where the tract is divided into lots by a plan, and conveyance is made solely by lot number with reference to the plat.
2. Where the tract is divided in proceedings in partition.
3. Where the tract is divided into two or more parts of designated area.
4. Where the whole tract is intended to be conveyed by two or more deeds executed at the same time, and no lines are laid down on the ground, but it is the intention that the tracts shall adjoin.
5. Where the tract consists of a number of surveys lying between lines fixed by the adjoining tracts.
6. Where the tract consists of two or more surveys made into one inclusive survey or made at or about the same time by the same surveyor.
7. Where the lots conveyed are intended to adjoin, but it appears when laid down as described in the conveyances there is a space between them.

Apportionment—Reason and Necessity

Apportionment is applied when retracement fails to yield sufficient evidence of the exact location of a lost property corner. The Wisconsin court in *Jones v. Kimble*[‡] stated the rule as follows:

When the whole length of the line between ascertained corners varies from the record length called for, and intermediate property corners are lost, it must be presumed in the absence of evidence indicating the contrary that the variance arose from an imperfect measurement of the whole line. The variance then is distributed between the several subdivisions of the whole line in proportion to their respective lengths.

---

[*] For a detailed discussion of meaning of the word "lost" and when a corner is truly "lost," see Chapter 11 The Lost Corner.

[†] Skeleton, Ray Hamilton, *Boundaries and Adjacent Properties*. Indianapolis: The Bobbs-Merrill Company. 1930. §§ 216–218.

[‡] 19 Wis. 430, 1865.

Based on this decision, Griffin suggests the reason for the rule: "It must be presumed in the absence of evidence indicating the contrary that the variance arose from an imperfect measurement of the whole line."

Since there are variances in measurements due to a number of factors, apportionment, from a mathematical standpoint, yields the correct position of intermediate points along a line. Stability of location* is demanded in the public interest and can, very often, be obtained only by application of the rule of apportionment.

If a gross blunder in the original survey causes the deficiency or the excess, the variance should not be distributed, but the correction applied at the location where the error occurred, *provided its position can be determined*. However, when there are small discrepancies due to careless surveying and there are no circumstances suggesting the position of the error, the law of probabilities supports the apportionment rule.

Conditions for the Rule to Apply

1. Failure of retracement
2. Predominant intention
3. Parcels created simultaneously

Griffin closes his discussion with the statement, "the proportionment of surplus or shortage over the whole line among the many units comprising the whole is the practical effect of the realization that surveying is the *art* of measurement and not an exact *science*.[†] Changes in nature generally as well as in human nature preclude exact duplication of original measurements, and insignificant unit differences soon accumulate to substantial discrepancies. This practical realization, or some sufficiently expressed intention of the grantor, may indicate that proportionment closely approximates the original work and distributes the excess or deficiency as equitably as possible. The limitations on the surveying method of apportionment are but particular instances of the applicability of the surveying method of retracement. In the

---

* There is a presumption in favor of permanency of natural boundary lines. *Normanoch Ass'n v. Deiser*, 40 N.J. 100, 190 A.2d 845, 1963.

† Courts are more than well aware that there is no such thing as a perfect measurement, and that surveyors are subject to a host of errors with their measurements. "It is a well-known fact that surveyors are apt to differ from each other, and surveyors employed by the United States government are not immune from the frailties of their profession." *Hagerman, et al. v. Thompson, et al.*, 235 P.2d 750, Wyoming, 1951.

See Chapter 8 for a discussion of measurement errors and some suggestions of how to deal with them.

The Concept   7

final analysis, apportionment is but a rule of last resort; it is applied only in absence of any markings upon the ground of the division lines between parcels carved out of the same tract."

In short, apportionment is mostly inappropriate. Regarding the rule in the PLSS employing single and double proportionate measurement for the replacement of lost corners see Chapter 11 concerning this particular topic, those carry distinct requirements and caveats as well, which will be discussed in some detail in that chapter.

Although our decisions in this country originated shortly after the Revolution, there were prior English cases and writings by legal scholars such as Hale, Kent, Coke, and Blackstone. Boundary and title cases were decided in this part of the world early on, some based on decisions in other jurisdictions, then refined and built upon to suit the purpose at hand. Generally the best way to begin is at the beginning, but with the deliberations and refinements that have taken place over a period of more than 200 years, it may make more sense to begin with recent decisions and work our way back. Following that procedure, it would make sense to begin with a review of *Rivers v. Lozeau* (539 So.2d 1147 (Fla. App. 5 Dist. 1989)) then move on to *Tyson v. Edwards*, 433 So.2d 549 (Fla. App. 5 Dist., 1983) and their related cases, ending with earlier decisions which provided a base from which to work. The court system, relying heavily on precedent, the doctrine of *stare decisis*, has crafted a useful and trustworthy set of rules and guidelines. These have proven to be very sound and have stood not only the test of time, but also withstood continual judicial review.

The case of *Rivers v. Lozeau* is by far one of the most important recent decisions, and is a land boundary line dispute case. The controversy in the case involved the correct location of the line between two parcels of land lying within the 40 acre quarter-quarter section described as the Southeast 1/4 of the Southwest 1/4 of Section 15, Township 14 South, Range 24 East, in Marion County, Florida. In 1964 Joseph Rizzo and his wife owned that portion of this quarter-quarter section that is in question. The US Forestry Service owned the land to the north. At that time, the Rizzos retained a surveyor, Moorhead Engineering, to survey their land and to establish certain internal land lines dividing it into parts. Moorhead undertook to locate and monument Rizzos' external boundary lines and corners and to establish and monument the terminal points of certain internal division lines.

> In 1969, the Rizzos conveyed to Marcus E. Brown and wife by deed containing the following land description: The North 400.00 feet of SE 1/4 of SW 1/4 of Section 15, Township 14 South, Range 24 East, Marion County, Florida.
>
> The west, north, and east lines of the Brown parcel followed the outer or external boundary lines of the property owned by the Rizzos. The south line of the Brown parcel did not follow any internal line established by the Moorhead survey. Mr. Rizzo showed Marcus Brown the monuments Moorhead had set as being the north corners of this

quarter-quarter section and certain other Moorhead monuments which the Rizzos told Marcus Brown were 33 feet south of the south line of the parcel the Rizzos conveyed to Brown. Later in 1977 or 1978, Marcus Brown measured 33 feet north of the Moorhead monuments shown him by Mr. Rizzo and placed a metal rod at the point Mr. Rizzo had told him was his south boundary line. Marcus Brown conveyed this property by the same description to George Brown who conveyed by the same description to appellees Raymond S. Lozeau and his wife.

In 1975, the Rizzos conveyed a parcel of their remaining land to Paul W. Adams and wife, which parcel was described by reference to the boundary lines of this quarter-quarter section with the north line of the property conveyed being described as thence N 89 53'01''' E. along a line 400.00 feet south of and parallel to the North line of said SE 1/4 of SW 1/4 a distance of 1327.04 feet to a point on the East line of said SE 1/4 of SW 1/4. Using substantially the same land description, the Adamses conveyed to Daniel E. Reader and wife, who conveyed to appellants Harold J. Rivers and wife.

In 1982 the U.S. Bureau of Land Management did a "dependent resurvey" of the lands of the U.S. Forestry Service which retraced the lines of the original government survey and identified, restored, and remonumented the original position of the corners of the original U.S. government survey.[1] This remonumenting of the original government survey, along with a 1986 survey by Whit Holley Britt, made obvious to all the true location of the north line of this quarter-quarter section on the ground and that the Moorhead monuments intended to denote that line were actually located 28.71 feet north of the true location of that line as it was originally established by the official U.S. government survey and reestablished by the 1982 government "dependent survey."

Appellees Lozeaus brought this action in ejectment and for declaratory judgment against the appellants Riverses who had possession of the south 28.71 feet of the north 400 feet measured from the north line of the quarter-quarter section according to the U.S. government (and Britt) surveys. The Lozeaus argued that they acquired legal title to the disputed land by virtue of the 1969 deed from Rizzo to Marcus Brown and the successive conveyances to them. The Riverses argued that Moorhead was the original surveyor and that his monuments on the ground controlled the location of the land subsequently conveyed by Rizzo, notwithstanding that "later" surveys, i.e., the government survey of 1982 and the 1986 Britt survey, may show the Moorhead monuments to have been in error.[2] After a non-jury trial, the trial court found that the property descriptions of the parties overlapped and ordered that the exact dimensions of the overlap be established and the overlapping property split evenly between the plaintiffs and defendants. The Riverses appeal and the Lozeaus cross-appeal.

Since time immemorial, parcels of land have been identified and described by reference to a series of lines or "calls" or "courses" that connect to completely encircle the perimeter or boundaries of a particular parcel. A particular property description may consist entirely of descriptions of original lines that compose it or it may, in whole or in

The Concept 9

part, refer to other sources which themselves show or describe previously surveyed and existing lines or calls. An individual line or call in a property description usually, but not always,[3] refers to an imaginary straight line customarily described in several ways: (1) by reference to its length, (2) by reference to its terminal points (commonly called "corners" or "angles"), (3) by reference to its angle with regard to true north, magnetic north, or to one or more other lines. A property description composed of descriptions of its constituent boundary lines or calls is known as a "metes and bounds" description. Of the ways that boundary lines are described, the reference to terminal points is the strongest and controls when inconsistent with other references.[4] In effect, real property descriptions are controlled by the descriptions of their boundary lines which are themselves controlled by the terminal points or corners as established on the ground by the original surveyor creating those lines. A property description that refers to, and adopts by reference, the description of a boundary line is DEPENDENT upon the proper location of the adopted line, which is dependent upon the location of the terminal points of the adopted line, which are dependent on their location on the ground as established by the original surveyor creating that adopted line.

Although title attorneys and others who regularly work with them develop expertise as to land descriptions, the only professional authorized to locate land lines on the ground is a registered land surveyor.[5] In fact, the definition of a legally sufficient real property description is one that can be located on the ground by a surveyor. However, in the absence of statute, a surveyor is not an official and has no authority to establish boundaries; like an attorney speaking on a legal question, he can only state or express his professional opinion as to surveying questions. In working for a client, a surveyor basically performs two distinctly different roles or functions:

First, the surveyor can, in the first instance, lay out or establish boundary lines within an original division of a tract of land which has theretofore existed as one unit or parcel. In performing this function, he is known as the "original surveyor" and when his survey results in a property description used by the owner to transfer title to property[6] that survey has a certain special authority in that the monuments set by the original surveyor on the ground control over discrepancies within the total parcel description and, more importantly, control over all subsequent surveys attempting to locate the same line.

Second, a surveyor can be retained to locate on the ground a boundary line which has theretofore been established. When he does this, he "traces the footsteps" of the "original surveyor" in locating existing boundaries. Correctly stated, this is a "retracement" survey, not a resurvey, and in performing this function, the second and each succeeding surveyor is a "following" or "tracing" surveyor and his sole duty, function and power is to locate on the ground the boundaries corners and boundary line or lines established by the original survey; he cannot establish a new corner or new line terminal point, nor may he correct errors of the original surveyor. He must only track the footsteps of the original surveyor. The following surveyor, rather than being the creator of the boundary line, is only its discoverer and is only that when he correctly locates it.[7]

When there is a boundary dispute caused by an ambiguity in the property description in a deed, it is often stated that the courts seek to effectuate the intent of the parties. This is not an accurate notion. The intent of the parties to a contract for the sale and purchase of land, both the buyer and the seller, may be relevant to a dispute concerning that contract, but in a real sense, the grantee in a deed is not a party to the deed, he does not sign it and his intent as to the quality of the legal title he receives and as to the location and extent of the land legally conveyed by the deed is quite immaterial as to those matters. The owner of a parcel of land, being the grantee under a patent or deed, or devisee under a will or the heir of a prior owner, has no authority or power to establish the boundaries of the land he owns; he has only the power to establish the division or boundary line between parcels when he owns the land on both sides of the boundary line he is establishing. In short, an original surveyor can establish an original boundary line only for an owner who owns the land on both sides of the line that is being established and that line becomes an authentic original line only when the owner makes a conveyance based on a description of the surveyed line[8] and has good legal title to the land described in his conveyance.

Subject only to certain rights of individuals under Spanish grants, the United States became the owner of all land now in the State of Florida by virtue of a treaty with Spain dated Feb. 22, 1819 and ratified Feb. 22, 1821 and, as original governmental owner, caused Florida to be surveyed in accordance with a rectangular system of surveys of public lands adopted by Acts of Congress. The permanent seat of government having been established at Tallahassee, an initial point of reference was located nearby through which a north-south guide line was run according to the true meridian and a base (township) line was run east-west on a true parallel of latitude.[9] North-south range lines, six miles apart and parallel to the Tallahassee Principal Meridian, were run throughout the state except where impracticable because of navigable waters, etc. Likewise, East-West township lines, six miles apart and parallel to the base line, were also run throughout the state to form normal townships six miles square each of which were divided into thirty-six square sections, one mile long on each side containing as nearly as may be, 640 acres each. These sections were numbered respectively, beginning with the number one, in the northeast corner and proceeding west (left) and east (right) alternately through the townships with progressive numbers. Sections were divided into squares of quarter sections containing 160 acres. The quarter-quarter section corners are placed on the line connecting the section and quarter-section corners, and midway between them. Although theoretically conceived and invisible, these lines are not merely theoretical concepts but are real lines, actually run and marked on the ground with terminal points monumented by surveyors acting under the authority of the cadastral engineer of the Bureau of Land Management. The approved and accepted boundary lines established by the federal government surveyors are

The Concept

11

unchangeable and control all references in deeds and other documents describing parcels of land by reference to the federal government of sections, townships and ranges.

In establishing the internal lines within Rizzo's subdivision, Moorhead acted as an 'original surveyor' but in attempting to locate and monument Rizzo's external boundary lines which are described by reference to the federal rectangle system of surveying, Moorhead was a 'following surveyor' and not only failed to properly find the northern boundary of this quarter-quarter section where it was located by the original government surveyor (and also re-established by an authorized federal government resurvey) but to evidence his erroneous opinion as to the true line, the Moorhead surveyor placed monuments 28.71 feet north of the true north line of this quarter-quarter section. From the time the federal government granted this quarter-quarter section to the original grantee down to the Rizzos, the title conveyed was to a tract of land located according to the original government survey and by the deed from the Rizzos to Brown, and subsequent deeds, the Lozeaus acquired title to the north 400 feet of this quarter-quarter section according to the true boundary line established by the original government surveyors. This is true regardless of the fact that Mr. Rizzo showed Marcus Brown the erroneous monuments set by the Moorhead surveyors[10] and regardless of where anyone erroneously thought or believed the correct location of this land boundary line to be. Neither the title to land nor the boundaries to a deeded parcel move about from time to time based on where someone, including a particular surveyor, might erroneously believe the correct location of the true boundary line to be. In 1975, the Rizzos conveyed to appellant Rivers' predecessor in title property the northern boundary of which is defined as being 400 feet south of, and parallel to, the north line of this quarter-quarter section. Regardless of any assertion that this conveyance was made relying on the Moorhead survey, the description itself does not describe the line in question by reference to the survey or monuments set by the Moorhead surveyor. On the contrary, that description adopts by reference the true north line of this quarter-quarter section which is necessarily controlled by the location of that line as established by the original government survey. Even if the description in the subsequent deed is considered to overlap the south 28 feet of the property previously conveyed by the Rizzos to Lozeaus' predecessor in title (which it does not), it is quite immaterial because, at the time of the conveyance to Paul W. Adams, Mr. and Mrs. Rizzo did not own that south 28 feet, they having previously conveyed legal title to it to Marcus Brown, Lozeaus' predecessor in title. All else argued in this case is immaterial. The Lozeaus are entitled to prevail in this controversy. All legal theories that could change the result in this case, such as those relating to adverse possession, title by acquiescence, estoppel, lack of legal title, etc., were neither asserted, nor argued, nor material in this case. This case is reversed and remanded with instructions that the trial court enter a judgment in favor of the appellees Raymond S. Lozeau and wife, in accordance with the land description as controlled by the official U.S. government survey.

"NOTES:

[1] This is only a re-establishment of the true position of the original survey by retracement. Clark on Surveying and Boundaries, § 650 Dependent surveys, page 956 (Grimes 4th Ed.1976).

[2] See *Akin v. Godwin*, 49 So.2d 604 (Fla.1950); *Willis v. Campbell*, 500 So.2d 300 (Fla. 1st DCA 1986); *Zwakhals v. Senft*, 206 So.2d 62 (Fla. 4th DCA 1968); City of Pompano *Beach v. Beatty*, 177 So.2d 261 (Fla. 2d DCA 1965) and *Froscher v. Fuchs*, 130 So.2d 300 (Fla. 3d DCA 1961).

[3] Property descriptions sometimes refer to irregular natural lines capable of identification, such as the banks, shores, and high and low marks of bodies of water such as oceans, lakes, rivers and streams, and to the midtread of streams, the face of cliffs, the ridge of mountains, etc.

[4] In a similar manner, when there is an inconsistency between the description of a corner (a line terminal point) in field notes and plats subsequently made and recorded and the original monument evidencing that corner on the ground, the original monument on the ground controls. See *Tyson v. Edwards*, 433 So.2d 549 (Fla. 5th DCA 1983), rev. denied, 441 So.2d 633 (Fla.1983).

[5] See § 472.005(3), Fla. Stat.

[6] This is a most important qualification.

[7] See Clark on Surveying and Boundaries, Chap. 14 Tracking a Survey, pg 339 and generally (Grimes 4th Ed.1976).

[8] Neither the 1969 deed from the Rizzos to Marcus Brown nor the 1975 deed from the Rizzos to Paul W. Adams contains property descriptions of lines bounded by monuments set by surveyor Moorhead in 1964. This would be an entirely different case if the land descriptions in question described lines "commencing at (or running to) a concrete monument set in 1964 by surveyor Moorehead, etc."

[9] See § 258.08, and Fla. Stat. Annot., Vol. 1, page 119, (West 1961). Unfortunately, this helpful material has been omitted from the 1988 edition of this volume of F.S.A.

[10] Notwithstanding that Rizzo and Brown both may have subjectively believed or intended Rizzo's deed to Brown to convey the land between the erroneous Moorhead monuments, because the deed described land by reference to the U.S. government survey it conveyed the legal title to the north 400 feet of this quarter-quarter section as measured from the true location of the original government survey. To the extent that Rizzo's deed conveyed legal title to land Rizzo did not intend to convey, Rizzo's remedy would have been to have brought a reformation suit in equity to have his deed reformed to describe the correct parcel by a correct description.

The Concept   13

Of course, the resulting litigation can be easily visualized: Rizzo would claim that he and his grantee Marcus Brown intended Rizzo's deed to convey land only south to a point 33 feet north of one of Moorhead's monuments and his deed should be reformed accordingly. Brown would admit that was true but would then claim that the parties also obviously intended that Brown was to obtain property 400 feet wide from north to south and that Brown should either keep the 400 feet described in the deed or be entitled to obtain money damages from Rizzo or to rescind the transaction because of Rizzo's misrepresentation that he owned to the errone-ous Moorhead monument located 28.71 feet north of Rizzo's true line and Rizzo did not own that northern 28.71 feet. These con-tentions, which never matured, existed only between the original parties and do not inure to any subsequent good faith purchasers who took legal title to their parcels according to the land descrip-tions contained therein, and the equitable and legal rights between Rizzo and Brown being personal to them are immaterial in litiga-tion between subsequent owners."

There are three very important points in this decision: (1) only a regis-tered or licensed land surveyor may locate land lines on the ground; (2) suffi-ciency of description; and (3) the distinction between an original survey and a retracement survey, followed by the statement that a retracement survey is not a resurvey. First it explains the test of whether a description is sufficient: "the definition of a legally sufficient real property description is one that can be located on the ground by a registered land **surveyor**." Second, this court also emphasized *retracement*, and not *resurvey*. Some recognize this differ-ence although some courts have used the terms interchangeably, calling a retracement survey a resurvey, when technically it is not. This distinction and the reasons for it will be discussed in detail in a later section. And third, reducing boundary surveying to its simplest terms, contrasting the differ-ence between original surveys and retracement surveys, and that the latter simply locates the former.

The *Rivers* case references *Tyson v. Edwards,* another very important deci-sion regarding retracement surveys:

When there is a discrepancy as to the location of the boundary of a sur-veyed lot of land between the monuments placed on the ground by the original surveyor and the written plat of that survey, which controls?

Long ago a large number of government sections of land were sub-divided by the Narcoossee Farm and Townsite Company with each government section of 640 acres generally divided into 64 square lots of about 10 acres each less roadways and with some irregular lots caused by lakes and the prior subdivision of adjoining lands. In this scheme, of course, the section lines and quarter section lines would appear to fall on lines between lots or in the center of roads. However, the plat shows many undimensioned lines and the exact location of the government sections are not shown.

The boundary line dispute here is between the south line of Lots 32 and 33, Section 5, Township 25 South, Range 31 East, according to the New Map of Narcoossee as filed and recorded in Plat Book 1, Pages 73 and 74, Osceola County, Florida (which lots are owned by appellants), and the north line of Lot 24, Section 8, Township 25 South, Range 31 East, according to the same map (which lot is owned by appellees). The east-west boundary line in dispute (between Lots 32 and 33 to the north and Lot 24 to the south) is shown on the New Map of Narcoossee to be an east-west extension of the same boundary line which, across the road and to the east, is the south line of Lot 14 (Section 4) and the north lines of Lots 1 and 2 (Section 9). Lots 1 and 2 (Section 9) as well as the lots south of them (Lots 20 and 21, Section 9) were actually platted and occupied before the New Map of Narcoossee was made. However, the north boundary of Lots 1 and 2 was originally located, occupied and fenced on the ground at a point now established to be 380 feet north of what is now recognized as the true location of the section line between Sections 4 and 5 on the north and Sections 8 and 9 on the south. Of course, according to the plat of the New Map of Narcoossee the north boundary of Lots 1 and 2 would appear to be the section line between these sections. Likewise, according to the same plat, the lot boundary line in dispute in this case would appear to be the same section line. Owners of the next 5 lots north of Lots 1 and 2 as well as other lot owners in the subdivision, including appellees' predecessors in title took occupancy according to the occupancy of Lots 1 and 2, all being 380 feet north of their location as measured from what is now recognized as the true location of the section line. The result of this was that the location of many lots, as established by ground monuments and actual occupancy, was moved north and two odd-shaped double sized lots (Lots 9 and 16, Section 4) located about three-fourths of a mile north of the line disputed in this case and acreage immediately north of lots 32 and 33, was squeezed or shortened by the 380 foot difference between the location of the property lines on the ground and their paper location if located solely according to the New Map of Narcoossee and the now known and accepted location of the government section line.

The difference between the location of the disputed lot boundary as shown by the plat and as it was established on the ground by monuments and occupancy is illustrated by the following sketch (Figure 1.2).

Therefore, it is obvious from the undisputed facts in this case that, as to many lots in this subdivision, including those in question, there is a 380 foot north-south discrepancy between where, from the plat, it is apparent that the original surveyor intended lots to be with relationship to the true location of the government section lines in this area between Sections 4 and 5 to the north and Sections 8 and 9 to the south, and where the original surveyor actually laid out and monumented the boundaries of these lots on the ground. It is further obvious that at least some of the lot owners who subsequently purchased lands by reference to the plat took occupancy according to the monuments on the ground.

The Concept

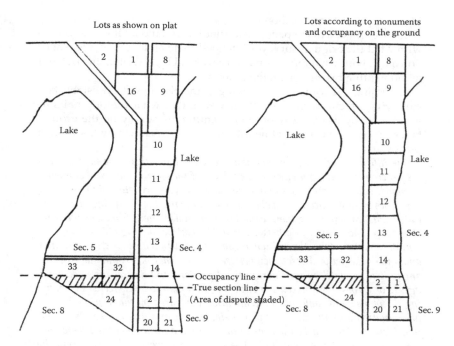

FIGURE 1.2
Plan from *Tyson v. Edwards*.

In this case appellants claim to a lot boundary line that is correct according to the apparent intention of the original surveyor and appellees claim to a lot boundary line that is consistent with lot lines established by the original surveyor on the ground and occupied by owners. Thus, a classic boundary line question is presented. Technically the question is:

Where an original surveyor subdivides and lays out boundaries of parcels in a tract of land which has theretofore existed as a single unit and runs lines and places monuments establishing the location of the subdivided parcels or plots or lots on the ground and the surveyor draws a plat of survey or written map of his work which is recorded and subsequently one or more parcels are conveyed by deed describing the parcels according to such plat of survey and some parcels are sold according to the plat but purchasers take actual possession according to the survey as monumented on the ground and there is a discrepancy and conflict between the location of parcels as located by the original survey on the ground and as they are shown to be located according to the recorded plat, is the correct legal location of a particular parcel as it was actually originally located and possession taken on the ground or is it as can now be located by following only the intent revealed by the recorded plat?

More simply put the question is:

In the event of a discrepancy as to subdivided land lot lines, do you go with what the original surveyor intended to do as shown by the plat or do you go with what the original surveyor did by way of laying out and monumenting his survey on the ground?

Surprisingly, because of surveying principles based on established surveying practices, the correct answer is that what the original surveyor actually did by way of monumenting his survey on the ground takes precedence over what he intended to do as shown by his written plat of survey.

The difficulty with the problem is that the role and practice of the surveyor and his function in solving a surveying problem of the type in this case is misunderstood. Lawyers, architects and design engineers are accustomed to achieving objectives by first conceiving of abstract ideas or plans, then reducing those ideas (intentions) to paper, and then using the written document from which to construct a physical object or otherwise tangibly achieve the original goal as written. When this is done, the written document is always considered authoritative and any deviation or discrepancy between it and what is actually done pursuant to it is resolved by considering the deviations and discrepancies as being defects or errors in the execution of the original plan to be corrected by changing the physical to conform to the intention evidenced by the writing. In only one situation does the surveyor play a similar role and that is when he, in the first instance, lays out boundaries in the original division of a tract which has theretofore existed as a single unit. Thereafter the surveyor's function radically changes. It is not the surveyor's right or responsibility to set up new points and lines establishing boundaries except when he is surveying theretofore unplatted land or subdividing a new tract. Where title to land has been established under a previous survey, the sole duty of all subsequent or following surveyors is to locate the points and lines of the original survey. Later surveyors must only track and "trace the footsteps" of the original surveyor in locating existing boundaries. They cannot establish a new corner or line nor can they correct erroneous surveys of earlier surveyors, even when the earlier surveyor obviously erred in following some apparent original "over-all design" or objective. The reason for this lies in the historic development of the concept of land boundaries and of the profession of surveying. Man set monuments as landmarks before he invented paper and still today the true survey is what the original surveyor did on the ground by way of fixing boundaries by setting monuments and running lines ("metes and bounds"), and the paper "survey" or plat of survey is intended only as a map of what is on the ground. The surveying method is to establish boundaries by running lines and fixing monuments on the ground while making field notes of such acts. From the field notes, plats of survey or "maps" are later drawn to depict that which was done on the ground. In establishing the original boundary on the ground the original surveyor is conclusively presumed to have been correct and if later surveyors find there is error in the locations, measurements or otherwise, such error is the error of the last surveyor. Likewise, boundaries originally located and set (right, wrong, good or bad) are primary and controlling when inconsistent with plats purporting to portray the survey and later notions as to what the original subdivider or surveyor intended to be doing or as to where later surveyors, working, perhaps, under better conditions

The Concept

and more accurately with better equipment, would locate the boundary solely by using the plat as a guide or plan. Written plats are not construction plans to be followed to correctly reestablish monuments and boundaries. They are "as built" drawings of what has already occurred on the ground and are properly used only to the extent they are helpful in finding and retracing the original survey which they are intended to describe; and to the extent that the original surveyor's lines and monuments on the ground are established by other evidence and are inconsistent with the lines on the plat of survey, the plat is to be disregarded. When evidence establishes a discrepancy between the location on the ground of the original boundary survey and the written plat of that survey the discrepancy is always resolved against the plat.

AUTHOR'S NOTE: Italics supplied for emphasis.

From a correct surveying viewpoint it is immaterial that it may be "clear and apparent from an examination of the plat" that the original intent and overall design or plan was to subdivide sections along lines that would have placed the disputed boundary line on the section line between Sections 5 and 8. The lots in question are not government lots laid out by the original government surveyor to subdivide irregular government sections, as is sometimes done.[1] These lots were laid out and created by a private surveyor for the Narcoossee Farm and Townsite Company. The problem here is, perhaps, confused by the fact that rather than giving each lot in the subdivision a different number or, what is more common, dividing groups of lots into lettered or numbered blocks, which permits the duplication of lot numbers in different blocks, here the original surveyor treated each government section as a block and duplicated the lot numbers in each section. This necessitated a reference to the government section in describing the lot by reference to the plat as distinguished from describing each lot by a metes and bounds description or consecutively numbering all lots or using numbered or letter blocks of lots. Nevertheless the lots in question were actually established by their own monuments and occupancy on the ground and the government section line and its location is immaterial to the proper inquiry as is the original government survey and field notes of the government survey locating the section line.

Appellants' position depends on three assertions with which we cannot agree, viz: (1) That when there is a discrepancy between the original survey on the ground and the plat drawn to illustrate that survey, the intent of the surveyor as gleaned from the plat should govern over the boundaries and lines physically located by the surveyor on the ground; (2) that the survey erroneously locating the lot boundary on the ground in this case was made subsequent to the drawn plat; and (3) that the real disputed property boundary line in this case is the government section line between Section 5 and Section 8 and not the privately surveyed boundary line between Lots 32 and 33 to the north and Lot 24 to the south as shown on the New Map of Narcoossee. If the disputed line was truly that between Section 5 and Section 8 then the government survey that first laid out Township 25 South, Range 31 East, into the

36 sections that contain Section 5 and Section 8 would be the original survey in question. However, the boundary line here in dispute is the boundary between lots first laid out by the private surveyor whose survey work on the ground is depicted on the plat known as the New Map of Narcoossee. In this case, the private surveyor, not the government surveyor, laid out lots according to the New Map of Narcoossee, and the private surveyor is the original surveyor whose footsteps on the ground must be followed by all following surveyors. Just as the monuments evidencing the line between Sections 5 and 8 (and Sections 4 and 9) are now there on the ground; likewise, the monuments set by the original survey to evidence the line between Lots 32 and 33 to the north and Lot 24 to the south was, based on evidence presented, found by the trial court to be there on the ground, but at a different location than the monuments establishing the line between Section 5 and Section 8. In this particular case, it is only incidental that the disputed boundary line between Lots 32 and 33 to the north and Lot 24 to the south happens to be shown on the Map of Narcoossee (but not as located on the ground), as being at the same point as the now undisputed boundary line shown on the government survey of Township 25 South, Range 31 East, to be between Section 5 and Section 8. The survey problem in this case, and its proper solution, would be exactly the same as to the discrepancy that also exists between the location on the ground and the location which is shown on the recorded plat of New Map of Narcoossee as the lot boundary lies between Lots 13 and 14, 13 and 12, 12 and 11, 11 and 10, and between the north line of Lot 10 and the south line of Lots 9 and 16, Section 4, New Map of Narcoossee (as shown on the two sketches in Figure 1.2), although according to the New Map of Narcoossee those lot lines are not shown to be coextensive with any government section lines. According to the testimony of Mr. Johnston, the surveying problem caused by this discrepancy between the original survey and the original plat of the "Old" and New Map of Narcoossee stops at this point because occupancy on the ground of acreage north of Lots 32 and 33 has absorbed this 380-foot error, as have Lots 9 and 16 in Section 4. Therefore, there is an easy escape from the horns of the dilemma and of a forced choice between the two alternatives of upholding the trial judge's application of good general law to the facts of this particular case and the specter of this case re-drawing all of the internal east-west subdivision lines of the entire Section 5.[2]

Bishop v. Johnson, 100 So.2d 817 (Fla. 1st DCA 1958), cert. denied, 104 So.2d 596 (Fla.1958), differs from this case in two material particulars. First, Bishop involved the somewhat unusual case where the original government surveyor not only subdivided the township into sections but, further, platted and subdivided the section in question into government lots. See note 1 infra. Therefore, unlike this case, the original government survey establishing the section boundaries was also the original survey of the lot lines involved in Bishop. More importantly, Bishop did not involve a discrepancy between the original survey and the depiction of that survey on the original plat, as is the problem in this case. In Bishop, it was assumed that the original government plat

The Concept

19

of survey correctly portrayed the original survey it purported to depict, and the problem was that a subsequent private survey did not follow the original survey but instead found a peninsula of land appended to a different section than that shown on the prior original government plat of survey. Since there was no suggestion of a discrepancy between the original government survey on the ground and the original government plat of that survey, the court in Bishop used the terms official government "survey" and "plat" as being the same and as being interchangeable. In that circumstance and case and context, and with that understanding, Bishop is exactly consistent with this opinion. We have only a semantical argument with Bishop in that the true survey line is always fixed by the original survey and not by the original plat which only attempts to picture or portray the actual survey work that has already been physically accomplished on the ground. When there is no discrepancy between the two, as in Bishop, the distinction is meaningless; but when there is a discrepancy, as in this case, the difference is the problem and the original survey controls over the original plat. It is also immaterial that a "mistake" was made in originally setting the boundary which is shown as the north lines of Lots 1 and 2 in Section 9 on the New Map of Narcoossee although that fact explains why a similar mistake was originally made in setting the boundary line in question in this case because these two boundary lines are east-west extensions or projections of each other. However, it is likewise immaterial that a "mistake" was made in originally locating the boundary line in question on the ground and how or why that "mistake" was made.[3] In this case the "New" Map of Narcoossee is not the "original survey" for two reasons. The first is, as explained above, no written document is the "original survey" which term properly refers to the original surveying work locating and fixing the monuments and lines on the ground that constituted the original lot boundaries. Secondly, the New Map of Narcoossee, made about 1913, was not even the original plat of the original survey. At least in part it was merely a replatting and extension of an earlier survey, being the "Old" Map of Narcoossee, surveyed and platted about 1887. Similarly, the State Road Department survey of 1944 was merely a retracing survey. Mr. Hanson, appellants' surveyor, is incorrect in suggesting that the State Road Department survey caused the 380-foot "mistake" in locating the north line of Lots 1 and 2 in Section 9. The State Road Department survey correctly "followed" the original survey and, in keeping with proper surveying theory, found, accepted, adopted and re-documented the "mistakes" of the original surveyor in establishing the original boundaries (as shown by fences, possession and other evidence) including the mistake in putting the north line of Lots 1 and 2 at a point 380 feet north of the section line between Sections 4 and 9. Mr. Johnston, appellees' surveyor, is correct in his view that the original survey on the ground must be taken as correct and that the plat (New Map of Narcoossee) is in error and that the later SRD survey is correct in following the original ground survey and not following the original plat.

"As was said in *Akin v. Godwin*, 49 So.2d 604, 607 (Fla.1950):

In making a resurvey, the question is not where an entirely accurate survey would locate the lines, but where did the original survey locate such lines.

This case does not involve boundaries established or affected by adverse possession (the statute of limitations), agreement or acquiescence. Neither does this case involve the entirely different and distinct surveying problem of properly apportioning an excess or deficiency of certain lines or tracts shown by recent measurements as compared with the original measurements.[4]

In a case such as this, the overall legal problem may be to establish and fix the legal boundary between two parcels of land, but the surveying problem is only to locate the boundary established on the ground by the original surveyor and is not to either establish a new boundary line where the plat shows one or to determine if the original surveyor erred in carrying out his work. Since as to contracts, wills, construction plans, etc., the writing is evidence of some preexisting human intent, and legal work is largely concerned with the original intention of the scrivener, it is sometimes difficult for the legally trained mind to think otherwise as to plats and to understand the surveying problem and the proper surveying approach to, and solution of, it.

The result of all of this is that most boundary disputes essentially present a surveying problem and the surveying profession has its own rules, methods and practices for resolving its problems.[5] Neither the legal nor surveying problem in a boundary dispute involves a question of what the original subdivider or surveyor intended or where, on the ground, a boundary should now be established to conform to the plat. The question is where on the ground the original surveyor did in fact fix the particular boundary and not where he intended, or should have, fixed it. Neither at trial nor on appeal can that question be answered by legal deductive reasoning or by analogy to holdings in other law cases involving disputed land boundaries. The legal problem at trial is more simple than the surveying problem and is the common problem of resolving a question of fact as to the location on the ground of the original surveyor's lines and monuments locating the boundary in question. The legal problem on appeal in such a case is even more simple. Where, as here, there is ample substantial competent evidence in the record to support the finding of the trial judge, as the finder of the fact question in the case, his finding comes to the appellate court clothed with a well-deserved presumption of correctness[6] and he should be, and his final judgment in this case is, AFFIRMED.

SHARP, Judge, dissenting.

I dissent in this case because there was no discrepancy between the original survey and the recorded plat as regards the particular lots in question: lots 32 and 33, section 5 and lot 24, section 8. The dispute is illustrated by the following diagram (Figure 1.3):

The plat was a "grid" superimposed on eight sections of land arranged as follows (Figure 1.4):

The Concept

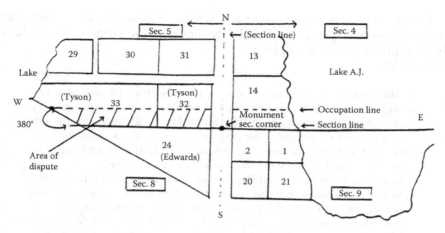

FIGURE 1.3
Diagram from *Tyson v. Edwards*.

Each section was subdivided into 64 equal squares with a dimension of 655.5 feet north to south. Between every two tiers of lots appear platted roads, which coincide with the north-south section lines and quarter section lines. There are also platted roads running east to west along the section and quarter-section lines. In this case there was no question about locating the old section lines, because the section monuments are there on the ground. Although the original surveyor's notes and stakes cannot be found, a mere look at the plat impels the conclusion that this grid or plan was intended to be based on section lines (cf. *Tyner v. McDonald*, 63 So.2d 504, Fla.1953).

Further, it is possible in this case to carry out the plan of the plat because there is enough land in these sections to do so (*Craig v. Russell*, 141 Fla. 105, 192 So.457, 1939). In *Froscher v. Fuchs*, 130 So.2d 300 (Fla. 3d DCA 1961), there was a 50 foot shortage in the plat because the section line fell in the middle of a canal, and there was not enough land left in the section to give each lot the dimensions shown on the plat.

I think this case is analogous to *Akin v. Godwin*, 49 So.2d 604 (Fla.1950). Akin held that the original plat cannot be altered by a later resurvey.

6	5	4	3
7	8	9	10

FIGURE 1.4
Plat as a grid.

The original survey in all cases must, whenever possible, be retraced, since it cannot be disregarded or needlessly altered after property rights have been acquired in reliance upon it.

Id. At 607.

> In this case an error occurred, many years after the plat was recorded, in locating the north boundary of lots 1 and 2 some 380 feet north of the section line. These lots are in the section immediately east of the lots involved in this case, and apparently fences and occupation rights have cemented that error for those lots. There are no similar fence or occupation lines regarding the lots involved in this suit, however.
>
> In addition, the State Road Department map, made long after the plat was recorded, also located the north boundary of lots 1 and 2 380 feet north of the section line. Later surveyors following that map and the fence line on lots 1 and 2, located all the lots in that tier, as well as the lot lines for the tiers involved in this suit 380 feet north of the section and quarter section lines. Mr. Johnston, appellee's expert, testified that the northernmost lots in section 5 had not been pushed north over the county line into Orange County because lots 9 and 16 had been squeezed or shortened by some 380 feet. He agreed a mistake had been made, but he maintained the plat was in error, not the later maps and surveys. The majority opinion in this case agrees with Mr. Johnston.
>
> I cannot accept Mr. Johnston's view. It is a well-established principle in Florida's real estate law that a later survey cannot alter an original recorded plat absent some exceptions which do not apply in this case.

A resurvey that purports to change lines or distances or to otherwise correct inaccuracies and mistakes in an old plat is not competent evidence of the true line fixed by the original plat.

> *Bishop v. Johnson*, 100 So.2d 817, 820 (Fla. 1st DCA 1958), cert. denied, 104 So.2d 596 (Fla.1958). Here that is precisely what has happened. An error in a State Road Department map, and later (post-plat) surveys have altered the straight lines platted for section 5. The plat, as redrawn by this case, will now look something like this (Figure 1.5):
>
> Instead of as it appears of record. An interesting question is how far this wrinkle, erroneously caused by a survey subsequent to the plat, will be allowed to warp the rest of the grid.

"NOTES:

[1] See, for example, the government plat of lots in *Bishop v. Johnson*, 100 So.2d 817 (Fla. 1st DCA 1958), cert. denied, 104 So.2d 596 (Fla.1958).

[2] The suggestion that this case constitutes a wholesale redrawing of all east-west lot lines in Section 5 is an example of the use of muniments ad absurdum or indirect proof which is a special case of the use of the method of conditional proof to test truth value in propositional

The Concept

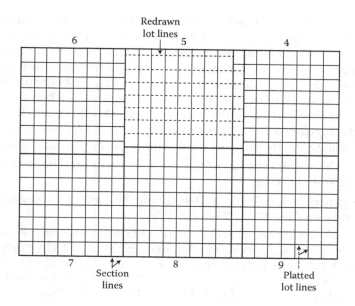

FIGURE 1.5
Plat redrawn.

logic, wherein the negation of a conclusion is asserted as an assumed premise and then an attempt is made to derive a contradiction from it, which in turn is used to derive the conclusion of the argument schema. This deductive proof system does not provide a decision-making procedure here. The majority opinion follows a rule of law to the effect that when there is a discrepancy between a surveyed land boundary line as monumented on the ground and as shown on a plat depicting that survey, the law treats the law on the ground as correct and disregards the contradictory plat once owners have bought and taken possession on the ground in accordance with the property lines as monumented on the ground. The assumption that the application of the rationale of this rule of law will require the "redrawing" of all east-west lot lines in Section 5 is fallacious. Even if it were true, it is far better to redraw lines on a piece of paper to make them consistent with occupancy on the ground than to uproot and move all of the property owners who have in good faith erected homes, fences and other improvements in conformity with monuments on the ground, in order to make their actual occupancy and possession conform to what is erroneously shown on a piece of paper recorded in the courthouse.

[3] Apparently the surveyor in 1913, in staking out on the ground the lots in this area, erroneously assumed that Lots 1 and 2 in Section 4 as monumented, occupied and fenced on the ground were in the

correct location as shown on the earlier (1887) plat; that is, that their north line was at the true location of the section line when in reality the location of their north line was actually 380 feet north of the section line. Therefore, measuring on the ground from those lots the surveyor laid and staked out the boundaries of lots to the north and to the west, including the lots whose boundaries are here in litigation, 380 feet north of where they are depicted on his plat. Therefore, as distinguished from the situation in *Tyner v. McDonald*, 63 So.2d 504 (Fla.1953), here there is a conflict between the plat and the survey.

[4] See J. Grimes, A Treatise on the Law of Surveying and Boundaries §§ 233–256 (4th Ed.1976).

[5] See generally J. Grimes, A Treatise on the Law of Surveying and Boundaries §§ 8–23, §§ 275–300 (4th Ed.1976).

[6] See *Akin v. Godwin*, 49 So.2d 604 (Fla.1950); Trustees of the *Internal Improvement Fund v. Wetstone*, 209 So.2d 698 (Fla. 2d DCA 1968), cert. dism'd 222 So.2d 10 (Fla.1969); *Osborn v. King*, 194 So.2d 912 (Fla. 2d DCA 1967); *Guise v. Shuman*, 188 So.2d 35 (Fla. 3d DCA 1966); *Bittner v. Walsh*, 132 So.2d 799 (Fla. 1st DCA 1961); *Froscher v. Fuchs*, 130 So.2d 300 (Fla. 3d DCA 1961)."

The really important point made in this case is that it distinguishes between a design plan and an as-built plan. The court stated that most people are confused by this (even surveyors) because they do not recognize that there is a difference between the two, believing that a plan is a plan, especially if it has a surveyor's seal on it.

The Tyson case called attention to the problem of misinterpreting the function of a plan: "Written plats are not construction plans to be followed to correctly reestablish monuments and boundaries. They are 'as built' drawings of what has already occurred on the ground and are properly used only to the extent they are helpful in finding and retracing the original survey which they are intended to describe; and to the extent that the original surveyor's lines and monuments on the ground are established by other evidence and are inconsistent with the lines on the plat of survey, the plat is to be disregarded. When evidence establishes a discrepancy between the location on the ground of the original boundary survey and the written plat of that survey the discrepancy is always resolved against the plat."

Therein lies a major problem, especially with laypersons, but also with many professionals and some surveyors. The average landowner pays to have a survey done and receives a plan, believing that the paper is his "survey." As noted in many decisions, the survey is the work upon the ground, while the paper is a "picture of the survey," and is secondary. Too many will attempt to follow the plan, and when physical evidence does not fit exactly what is on the plan/plat, are tempted to, or actually, set new markers purporting to be the correct point, or report discrepancies with the plat, or the resulting

The Concept   25

land description. Too many persons miss this point entirely, or do not have a proper understanding of what a survey actually is, along with its significance.

These two cases, *Rivers* and *Tyson*, both have to do with land in Florida, and both have to do with subdivisions. In *Rivers*, one new lot was created on the perimeter of an existing lot, giving rise to the difference between an existing survey and a new survey, or, put another way, the difference between a retracement survey and an original survey since this case had both. *Tyson* explores the same concepts, although the subdivision consisted of a number of lots rather than just one, and therefore involved a retracement of the perimeter, then creation of interior lines, which later became boundaries. This combination resulted in two sets of footsteps, the original perimeter, and the original lines of the subdivision.

There are two additional Florida cases that are noteworthy, and the four cases should be construed together. The 1961 decision of *Froscher v. Fuchs*[*] stated, "In cases deciding the boundary between two parcels of land, the law is settled that it is the duty of the surveyors to follow the original survey lines under which the property and neighboring properties are held notwithstanding inaccuracies or mistakes in the original survey. The purpose of this rule of law is that stability of boundary lines is more important than minor inaccuracies or mistakes."

A few years prior, the court decided the case of *Wildeboer v. Hack*,[†] which action involved a boundary dispute. "The plaintiff was the owner of lot 3 and the West half of lot 4 of block 10 of Lauderdale Harbors, Section A. The defendants were the owners of lots 1 and 2 of block 1, Samarkand Isles. Samarkand Isles was a resubdivision, including lots 1 and 2 of block 10 of Lauderdale Harbors. The plats show that the east line of lot 2, block 10, Lauderdale Harbors was the same as the east line of lot 1, block 1, Samarkand Isles. The dispute arose over the location of the boundary line between lot 1, block 1, Samarkand Isles, which was the property of the defendants, and lot 3, block 10, Lauderdale Harbors, the property of the plaintiff. Two surveyors testified for the plaintiff and two for the defendants. The plaintiff's witnesses testified that they based their opinions upon stakes and other evidences found on the lot lines of the original survey of Lauderdale Harbors, while the witnesses for the defendants based their testimony on an independent determination from a quarter section corner. Defendants' witnesses located the disputed line approximately 3.48 feet further east than did plaintiff's witnesses.

> The testimony of the witnesses and the decision of the court raised the question of whether or not a junior survey should attempt to correct the descriptions in an older survey, or whether the surveyor, in making the junior survey, should determine where the lines were actually established in the older survey.

---

[*] 130 S.2d 300, Fla., 1961.
[†] 97 So.2d 29, Fla.App. 2 Dist. 1957.

The lower court held that the lines of the old survey should prevail even though a resurvey showed there was error in the dimensions of the lot in favor of the property owner in the new survey over the property owner in the old survey.

It was stated in Florida Supreme Court in *Kelsey v. Lake Childs Co.*, 93 Fla. 743, 112 So. 887, that an original, actual survey of public lands of the federal government, on the faith of which rights have been acquired, control other surveys subsequently made by the government which affect such rights.

In the case of *Akin v. Godwin* Fla.1950, 49 So.2d 604, 607, the Court said "In making a resurvey, the question is not where an entirely accurate survey would locate the lines, but where did the original survey locate such lines." *Clark on Surveying and Boundaries*, 2d Ed., Sec. 411, page 495; *Kahn v. Delaware Securities Corporation*, 114 Fla. 32, 153 So. 308; *LeCompte v. Lueders*, 90 Mich. 495, 51 N.W. 542; *City of Racine v. Emerson*, 85 Wis. 80, 55 N.W. 177; *Dittrich v. Ubl*, 216 Minn. 396, 13 N.W.2d 384. As stated in *8 Am.Jur., Boundaries*, Section 102, page 819: "The object of a resurvey is to furnish proof of the location of the lost lines or monuments, not to dispute the correctness of or to control the original survey. The original survey in all cases, must, whenever possible, be retraced, since it cannot be disregarded or needlessly altered after property rights have been acquired in reliance upon it.' It is generally held, therefore, that a resurvey that changes lines and distances and purports to correct inaccuracies or mistakes in an old plat is not competent evidence of the true line fixed by the original plat. See *Dittrich v. Ubl*, 216 Minn. 396, 13 N.W.2d 384; *Cragin v. Powell* 128 U.S. 691, 9 S.Ct. 203, 32 L.Ed. 566; *City of Racine v. Emerson*, 85 Wis. 80, 55 N.W. 177, 178."

We have examined the record, including the testimony of the witnesses, and conclude there was sufficient testimony for the chancellor below to determine that the professional witnesses of the plaintiff used the correct method in determining the property lines of the parties in this case. Accordingly, the judgment of the lower court is hereby affirmed."

AUTHOR'S NOTE: It may be seen that several decisions and other sources recur in subsequent deliberations of the courts, since they provide insight and guidance, being considered as precedent. These sources may be classified as some of the most important resources for the decision-making process in deciding discrepancies. They are *Clark on Surveying and Boundaries*, *8 Am.Jur., Boundaries*, *Akin v. Godwin (Florida)*, *Cragin v. Powell (U.S., Louisiana)*, *Dittrich v. Ubl (Minnesota)*, and *City of Racine v. Emerson (Wisconsin)*. While these decisions will be used from time to time within the text, there will also be several other significant cases referenced and stressed. The retracement surveyor should make careful note of these resources, and understand their significance.

Another very important decision is the Tennessee case of *Wood v. Starko*,[*] not only for its discussion of the dispute at hand, but, more importantly, for

---

[*] 197 S.W.3d 255, Tenn. App., 2006.

The Concept

its references. This case provides a comprehensive review of decisions dealing with retracement problems throughout the country, as well as at the federal and state levels of jurisdiction. It is of great value for that reason alone.

The court began by emphasizing several important principles, relying on precedent contained in *Staub v. Hampton.** "It is a familiar principle of our system, and one in reason applicable to this species of title, as well as any other, that it is the work on the ground, and not on the diagram returned, which constitutes the survey, the latter being but evidence (and by no means conclusive) of the former ... It is conceded that the patent may be rectified by the return of survey; and why not the return of survey by the lines on the ground, and particularly the numbered tree, which is the foundation of the whole?" In *Dunn v. Ralyea, 6 Watts & S.* (Pa.) 475, Kennedy, J., said: 'That the original lines as found marked on the ground must govern, in determining the location and extent of the survey, is a well-established rule, in general applicable to all cases.' ...'We know, in point of fact, that the marks made on the ground at the time of making the survey are the original, and therefore the best evidence of what is done in making it; that everything that is committed to paper afterwards in relation to it intended and ought to be, as it were, a copy of what was done, and ought to appear on the ground, in the doing of which errors may be committed, which renders it less to be relied on than the work as it appears by the marks made on the ground.'" 46 Pa., [477,] 484, 485 [1864].† *Staub,* 101 S.W. at 784.

> The courses and distances in a deed always give way to the boundaries found upon the ground, or supplied by the proof of their former existence when the marks or monuments are gone. So the return of a survey, even though official, must give way to the location on the ground, while the patent, the final grant of the state, may be corrected by the return of survey, and, if it also differs, both may be rectified by the work on the ground. *Staub,* at 784.

Cited in *Wood v. Starko* are the following with quoted passages:

> The governing rules are near universal and are recited by the Court of Appeals of Ohio in *Sellman v. Schaaf,* 26 Ohio App.2d 35, 269 N.E.2d 60, 66 (1971).

In 11 C.J.S. Boundaries § 3, p. 540, it is said:

> It has been declared that all the rules of law adopted for guidance in locating boundary lines have been to the end that the steps of the surveyor who originally projected the lines on the ground may be retraced

---

* 117 Tenn. 706, 101 S.W. 776, 781, 1907.
† *Lodge v. Barnett.*

as nearly as possible; further, that in determining the location of a survey, the fundamental principle is that it is to be located where the surveyor ran it. Any call, it has been said, may be disregarded, in order to ascertain the footsteps of the surveyor in establishing the boundary of the tract attempted to be marked on the land; and the conditions and circumstances surrounding the location should be taken into consideration to determine the surveyor's intent.

In Clark, Surveying and Boundaries (2d Ed.1939), it is said at page 727, Section 665:

The original survey must govern if it can be retraced. It must not be disregarded. So, too, the places where the corners were located, right or wrong, govern, if they can be found. In that case a hedge planted on the line established by original survey stakes was better evidence of the true line than that shown by a recent survey. In making a resurvey it is the surveyor's duty to relocate the original lines and corners at the places actually established and not to run independent new lines, even though the original lines were full of errors.

In 6 Thompson, Real Property, 594, Description and Boundaries, Section 3047 (1962 replacement), the following is stated.

The line actually run is the true boundary, provided the essential survey can be found and identified as the one called for, and prevails over maps, plats, and field notes. * * * The lines marked on the ground constitute the actual survey and where those lines are located is a matter to be determined by the jury from all the evidence. If the stakes and monuments set at the corners of the parcel in making the survey have disappeared, it is competent to show their location by parol evidence.

At page 599, Section 3049, it is further said that:

Marked corners are conclusive and will control over courses and distances. Although stakes are monuments liable to be displaced or removed, they control so long as it is certain that they mark the corners of the original survey. *Sellman*, 26 Ohio App.2d 35, 269 N.E.2d 60 at 66.

Clearly encompassed in this rule is the fact that it is the monuments laid out by the original surveyor, if they can be located, which govern the boundaries, even if the actual survey used in the plat is in error.

"Moreover, in ascertaining the lines of land or in re-establishing the lines of a survey, the footsteps of the original surveyor, so far as discoverable on the ground by his monuments, should be followed and it is immaterial if the lines actually run by the original surveyor are incorrect. *Vaught*, 116 Mont. at 550, 155 P.2d at 616 (citing *Ayers v. Watson* (1891), 137 U.S. 584, 11 S.Ct. 201, 34 L.Ed. 803; *Galt v. Willingham* (5th Cir.1926), 11 F.2d 757). *See also Buckley v. Laird* (1972), 158 Mont. 483, 491–92, 493 P.2d 1070, 1074–75.

The Concept

29

Olson v. Jude, 316 Mont. 438, 73 P.3d 809, 815 (2003).

In a dispute between adjacent lot owners in a platted subdivision, the Supreme Court of Washington wrestled with differing opinions of professional surveyors and ultimately affirmed the trial court judgment reasoning:

Where a plat contains substantial mathematical errors and discrepancies and with the passing of time questions arise concerning the true boundaries among its component parcels, the question to be answered is not where new and modern survey methods will place the boundaries, but where did the original plat locate them. The main purpose of a resurvey is to rediscover the boundaries according to the plat upon the best evidence obtainable and to retrace the boundary lines laid down in the plat. 12 Am.Jur.2d Boundaries § 61 (1964). Effort should be made to locate the original corners. Despite discrepancies in the original plat, the known monuments and boundaries of the original plat take precedence over other evidence and are of greater weight than other evidence of the boundaries not based on the original monuments and boundaries. Clark, Surveying and Boundaries § 258 (3d ed.1959).

Courts should ascertain and carry out the intention of the original platters. In case of discrepancy, however, between lines actually marked or surveyed on the ground and lines called for by plats, maps or filed notes, the lines marked by survey on the ground prevail (*Stewart v. Hoffman,* 64 Wash.2d 37, 390 P.2d 553 (1964); 11 C.J.S. Boundaries § 49c (1938)).

Staaf v. Bilder, 68 Wash.2d 800, 415 P.2d 650, 652 (1966).

Analogous to the case at bar is *Akin v. Godwin,* 49 So.2d 604 (Fla.1951). In that case, the parties were owners of adjacent lots in a platted subdivision with the plat of record prepared by A.L. Knowlton, engineer. The owner of lot 3 constructed a building supported by three concrete pillars along the boundary line between lots 3 and 4. Plaintiff bought lot 4 and, relying on a sketch prepared well after the subdivision was platted, concluded that the concrete encroached upon lot 4. He sued in ejectment and, following a jury verdict for Plaintiff, the owner of lot 3 appealed. In reversing and remanding, the Supreme Court of Florida held:

Both Lot Three and Lot Four were conveyed to their respective owners with reference to a plat made by A.L. Knowlton, C. E., in the year 1896 duly recorded in the public records of Dade County, Florida. Since this plat is, then, as much a part of the deeds as if copied therein, *see Routh v. Williams,* 141 Fla. 334, 193 So. 71, and *Kahn v. Delaware Securities Corporation, supra,* the real question in this case is: What is the boundary line between Lots Three and Four, *according to the Knowlton plat?*

The plaintiff based her claim of title squarely on the Garris survey of Lot Four; and, as heretofore stated, the verdict and judgment were likewise based thereon. Mr. Garris testified that he made his survey by first ascertaining the block boundaries, as marked by the city monuments establishing the locations and boundaries of the streets on the four sides of Block Fifty; that he then pro rated "any excess of [sic] deficiency in said block," and that the sketch of Lot Four represented the true dimensions of that lot "as pro rated." It appears that the city monuments were established by the so-called Klyce survey undertaken by the City of Miami in 1914 or 1915 for the purpose of establishing the location and marking the boundaries of the city streets, but there is nothing in the record to show what monuments or methods were pursued by Klyce in making his survey.

The plaintiff also adduced the testimony of one Elmore Cormack, a registered civil engineer and surveyor in the employ of the Dade County Surveyor, in an effort to establish the accuracy of the Garris survey. Cormack's testimony showed, however, that he also used the city monuments established by the Klyce survey as the starting point in his survey, and he testified that his survey did not necessarily "go back to the original Knowlton field notes."

It is apparent that, as evidence of the true boundary of Lot Four according to the Knowlton plat, both the Garris and the Cormack surveys suffer from the same infirmity, that is, neither made any effort to retrace the Knowlton survey, nor does it appear that either survey is in any way, either directly or indirectly, related to or tied in with the Knowlton plat, insofar as the exact location, on the ground, of the boundary lines of Lot Four is concerned.

In making a resurvey, the question is not where an entirely accurate survey would locate the lines, but where did the original survey locate such lines. Clark on Surveying and Boundaries, 2d Ed., Sec. 411, page 495; *Kahn v. Delaware Securities Corporation*, 114 Fla. 32, 153 So. 308; *LeCompte v. Lueders*, 90 Mich. 495, 51 N.W. 542; *City of Racine v. Emerson*, 85 Wis. 80, 55 N.W. 177; *Dittrich v. Ubl*, 216 Minn. 396, 13 N.W.2d 384. As stated in 8 Am.Jur., Boundaries, Section 102, page 819: "The object of a resurvey is to furnish proof of the location of the lost lines or monuments, not to dispute the correctness of or to control the original survey. The original survey in all cases must, whenever possible, be retraced, since it cannot be disregarded or needlessly altered after property rights have been acquired in reliance upon it." It is generally held, therefore, that a resurvey that changes lines and distances and purports to correct inaccuracies or mistakes in an old plat is not competent evidence of the true line fixed by the original plat. *See Dittrich v. Ubl*, 216 Minn. 396, 13 N.W.2d 384; *Cragin v. Powell*, 128 U.S. 691, 9 S.Ct. 203, 32 L.Ed. 566; *City of Racine v. Emerson*, 85 Wis. 80, 55 N.W. 177, 178.

As stated in 8 Am.Jur., Boundaries, Sec. 59, page 78: 'Purchasers of town lots generally have the right to locate their lot lines according to the stakes as actually set by the platter of the lots, and no subsequent survey can unsettle such lines. In the event of a subsequent controversy the question becomes not whether the stakes were located with absolute accuracy, but whether the lots were purchased and taken possession of in reliance upon them. If such was the case, the rule appears to be well established that they must govern notwithstanding any errors in locating them. *Akin*, 49 So.2d at 606–07.

While these general rules apparently have their origin in surveys reflecting government grants, such rules are equally applicable to private surveys." *Staub v. Hampton*, 117 Tenn. 706, 101 S.W. 776, 781 (1907).

The Concept

31

AUTHOR'S NOTE: One of the most important statements at the conclusion of this decision is the following: "While these general rules apparently have their origin in surveys reflecting government grants, such rules are equally applicable to private surveys. *Staub v. Hampton*, 117 Tenn. 706, 101 S.W. 776, 781 (1907). Considerable emphasis will be placed on government manuals and procedures along with federal decisions. To the extent that rules and procedures are useful, and appropriate, they may be applied elsewhere.

The decision also provides several additional important sources: *Sellman v. Schaaf (Ohio), Olson v. Jude (Montana) and Staaf v. Bilder (Washington)*.

An additional noteworthy decision from Florida, which is relatively recent, is the case of *Beckman/Tillman v. Bennett*.* The court relied very heavily on *Tyson v. Edwards* which, as noted earlier, is one of the most important recent cases regarding retracement surveys, and the function of survey plats.

> A surveyor cannot set up new points and establish boundary lines unless he is surveying unplatted land or subdividing a new tract. *See Willis v. Campbell*, 500 So.2d 300, 302 (Fla. 1st DCA 1986); *Tyson v. Edwards*, 433 So.2d 549, 552 (Fla. 5th DCA 1983). *Subsequent surveyors may only locate the points and retrace the lines of the original survey; they cannot establish new lines or corners. See Tyson*, 433 So.2d at 552.

Collier, 794 So.2d at 618 (emphasis added).

> After carefully reviewing the deposition testimony of each surveyor, we conclude that only Beckham/Tillman's surveyor conducted a proper retracement utilizing the original monuments. As it was explained in *Tyson v. Edwards*:
>
> The surveying method is to establish boundaries by running lines and fixing monuments on the ground while making field notes of such acts. From the field notes, plats of survey or "maps" are later drawn to depict that which was done on the ground. In establishing the original boundary on the ground the original surveyor is conclusively presumed to have been correct and if later surveyors find there is error in the locations, measurements or otherwise, such error is the error of the last surveyor. Likewise, *boundaries originally located and set (right, wrong, good or bad) are primary and controlling when inconsistent with plats purporting to portray the survey and later notions as to what the original subdivider or surveyor intended to be doing or as to where later surveyors, working, perhaps, under better conditions and more accurately with better equipment, would locate the boundary solely by using the plat as a guide or plan. Written plats are not construction plans to be followed to correctly reestablish monuments and boundaries. They are "as built" drawings of what has already occurred on the ground and are properly used only to the extent they are helpful in finding and retracing the original survey which they are intended to describe; and to the extent that the original surveyor's lines and monuments on the ground are established by other*

---

* 38 Fla. L. Weekly D 1555, 118 So.3d 896, Fla.App. 1 Dist., 2013.

evidence and are inconsistent with the lines on the plat of survey, the plat is to be disregarded. When evidence establishes a discrepancy between the location on the ground of the original boundary survey and the written plat of that survey the discrepancy is always resolved against the plat.

AUTHOR'S NOTE: Italics supplied for emphasis.

In addition, the case of *Rowe v. Kidd*, 249 F. 882 (1916), in discussing a similar situation in Kentucky, another "metes and bounds" state, "In determining the location of an actual survey, the fundamental principle is that it is to be located where the surveyor ran it. As it has been put, the thing to be done is to track the surveyor. This being so, it is where he ran and not where his certificate says he ran that governs. If there is a conflict between where he ran and where he thus says he ran, the latter must yield." In *Dimmitt v. Lashbrook*, 32 Dana (Ky), 1 (1834), Judge Robertson stated: "When a line is actually run, it must be, as so run, the true boundary."

> In reflecting upon this I have been puzzled to reconcile it with the rule that parol evidence is inadmissible to vary a writing. I have reached the conclusion that there is no conflict here, because such a case does not come within the rule. The making of the certificate is not contemporaneous with the making of the survey, though made from notes taken at the time thereof. When it is made the making of the survey is a thing of the past. It is a statement as to how the surveyor actually ran the lines.

It is exceedingly important to understand this difference, which is well entrenched within the law, and recited in numerous decisions. The average persons believe either the deed or the plat (plan) is their survey. However, as continually emphasized through time, the survey is the work on the ground; the deed a written instrument of the transfer of title and the plat a picture of the survey. See further discussion in following chapters, and the rationale of the court system of why this is true.

Discussed later, the Montana case of *Larsen v. Richardson* did not contain an original survey, so there was a question whether there were any actual "footsteps" to follow. The court stated the equivalent of following the original deed writers.

And, in Tyson, the private surveyor, not the government surveyor, laid out lots according to the New Map of Narcoossee, and the private surveyor was the original surveyor whose footsteps on the ground must be followed by all following surveyors.

The Bureau of Land Management, the official organ of US Government for the surveying of public lands, states in its "Manual" by offering the suggestion that the rules therein are founded on good practice: Although it is not statutory authority for the restoration of private property lines and corners, its good sense and technical outline of proper surveying techniques make it an excellent work for general application to private as well as public lands.

The Concept

The courts have recognized the manual as a proper statement of surveying principles.

In the case of *United States v. Doyle** the court stated that "the procedures for restoration of lost or obliterated corners are well established. They are stated by the cases cited [below] and by the supplemental manual on Restoration of Lost or Obliterated Corners and Subdivisions of Sections of the Bureau of Land Management (1963 ed.)." Its footnote states, "This manual is a supplement to the Manual of Survey Instructions (1947) of the Bureau. The courts have recognized the manual as a proper statement of surveying principles. See *Reel v. Walter*, 131 Mont. 382, 309 P.2d 1027."

In 1892, the Oregon court wrote the following in the decision of *Van Dusen v. Shively,†*

> The line as actually run on the ground by Trutch, if it can be ascertained, is the line which must govern in this case, and courses and distances as given in the field-notes must yield thereto. *Goodman v. Myrick*, 5 Or. 65. The location of this line is a question of fact to be ascertained from the evidence. The courses and distances as given in the field-notes are but descriptions which serve to assist in determining where the line was actually run. But where the line can be shown from the marks and blazes on the trees, or other natural monuments or calls, the courses and distances must yield to it. In cases of this kind the object is to follow in the "footsteps of the surveyor" as nearly as possible. No fixed or certain rules can be laid down by which questions of disputed boundaries can be settled, but each case must depend upon its own particular facts. The courses and distances in this case are entitled to but little weight in determining the line in dispute, as they do not correspond with the line as claimed by either party. The evidence is not at all satisfactory, owing in part to the lapse of time since the claim was surveyed, the changes that have taken place in the surface of the country, the destruction of the witness trees at the south-west corner, and the destruction of many of the marked trees along the line, so that we can only hope to approximate a correct result.
>
> It is evident that the surveyor Ashley was not looking for the true government lines, but, as he says, his object was to "pro-rate the surplus." He desired to make the survey conform to a certain line, and he did not desire section 23 to have too much land. But the rule as to restoring lost corners by putting them at an equal distance between two known corners has no application, if the line can be retraced as it was established in the field. The field-notes should be taken, and from the courses and distances, natural monuments or objects, and bearing trees described therein the surveyor should endeavor to fix the line precisely as it is called for by the field-notes. He should endeavor to retrace the steps of the man who made the original survey. If by so doing the line can be located, it must be done, and, when so located, it must control. It is not

---

* 468 F.2d 633, 10th Cir. 1972.

† 22 Or. 64, 29 P. 76, Or., 1892.

34   *Boundary Retracement*

the business of the surveyor to speculate as to whether one government subdivision is short and the other long in acres. He is not authorized to correct what the government has done. The line as surveyed and described in the field-notes is the description by which the government sells its land. If its description makes one section contain three hundred and twenty acres and another nine hundred and sixty acres, the parties must take according to the calls of their patents.

As said in *Kaiser v. Dalto*, 73 P. 828, 140 Cal. 167 (1903), "These surveys may have been correct by courses and distances, but in the course of thirty or forty years a change in the position of the monuments or the surface of the earth might easily cause a variation of four or five inches. The lines as originally located must govern in such cases. The survey as made in the field, and the lines as actually run on the surface of the earth at the time the block were surveyed, and the plats filed must control. The parties who own the property have a right to rely upon such lines and monuments. They must when established control courses and distances. A line, as shown by monuments and as platted by the city authorities, and as acquiesced in for many years, cannot be overturned by measurements alone."*

Following Footsteps Contrary to Established Rule

The Utah case of *Glenn v. Whitney*[†] is an illustration of the rule, its support by the court system, and a lesson to all. As the South Carolina court stated in the case of *Bradford v. Pitts*,[‡] (see Preface for brief discussion), "blind devotion to a rule may lead to infinite error."

After discussing the issue of whether there was acquiescence in a fence line thereby creating an agreed-upon boundary, the court addressed the final issue in the case. "The single remaining question to be determined is whether the court erred in failing to find that plaintiff had established his ownership of the strip of land in question. Plaintiff, on the 29th day of May, 1947, hired W. H. Griffiths, county surveyor for Box Elder County, to make a partial re-survey of Township 14 North Range 5 West in order to locate the true line separating the south half of the north half of section 19 from the south half of the north half of section 20, the boundary line here in dispute. Prior to making that re-survey, Mr. Griffiths obtained the field notes of two government surveys of Township 14. One set were the notes of one Troskolawski who had made the original government survey in 1856. A second set had been made in the year 1887, by one Fitzhugh, also a government surveyor. The notes of each were unusual in that both Troskolawski and Fitzhugh ran their surveys of Township 14 from the west boundary line

---

* *County of Yolo v. Nolan*, 77 P. 1006, 144 Cal. 445, 1904.

† 116 Utah 267, 209 P.2d 257, Utah, 1949.

‡ 2 Mills. Const. Rep. 115, South Carolina, 1818.

The Concept

eastward for the two westernmost tiers of sections and from the east boundary of Township 14 westward for the four remaining tiers of sections. The usual practice, of course, is to run the survey of all sections from the east boundary line westward.

> "Mr. Griffiths commenced his survey by attempting to locate the government monuments establishing the west township boundary line. The reason he gave for running his survey eastward from the west boundary rather than westward from the east boundary, which is the accepted practice, was that this was the boundary from which sections 19 and 20 had been surveyed as shown by the notes of the government surveyors referred to above. Mr. Griffiths further testified that in 1930 he had surveyed the west boundary of this particular township and located the northwest corner monument, the southwest corner monument and also the northwest corner of section 19; that a fence line ran through these three points establishing the west boundary of the township; that although only one of these monuments was to be found when he commenced his survey in 1947, the fence line still existed in a straight line and in the same position as it was in 1930. He admitted, however, that he found no other government monuments. He found a county road, however, running north and south along the east boundaries of sections 20, 17, 20 and 32. He testified that the road was for all practical purposes parallel with the west boundary of the township, that the road had existed there of his own knowledge for about twenty years, and that upon inquiring of local landowners, he learned that the road was considered to be the boundary between the sections on either side of it. He thereupon measured the distances between the road and the west boundary of the township along the north and south lines of sections 19 and 20. He found the distance along the southern boundary of these two sections to be 10,629 feet and that the existing fence line divided that distance unequally giving defendant 5456 feet and plaintiff 5173 feet. Referring to the government notes he found that defendant was entitled to 5326 feet along this boundary and that plaintiff was entitled to 5276 feet. Finding that the total distance between the western boundary of the township and the county road on the east of section 20 was greater than the sum of the two distances called for in the government notes, Mr. Griffiths calculated that the parties were entitled to the distances referred to in the government notes plus a proportional division of the excess. He thereupon determined that plaintiff was entitled to 5310 feet and that defendant was entitled to 5320 feet. Without going into more detail concerning his further calculations, it is sufficient to say that Mr. Griffiths followed the same method to determine where the boundary line should be located as it crossed the northern boundaries of the two sections.
>
> Respondent contends the survey, as made, is legally insufficient for several reasons, the first of which being that under the law as announced in *Henrie v. Hyer*, 92 Utah 530, 70 P. 2d 154, 157, Mr. Griffiths was required to run his survey of all sections from the east boundary of the township

westward and was not at liberty to run it eastward from the west boundary as to sections 19 and 20 as he did. The rule to be found in that case upon which respondent relies is as follows:

Resort should be had, first, to the monuments placed at the various corners when the original government survey of the land was made, provided they are still in existence and can be identified, or can be relocated by the aid of any attainable data. But if this cannot be done and a survey becomes necessary, this must be made from the east, and not from the west, boundary line of the township.

It would appear that reason for the rule requiring a resurvey to be run from the east boundary westward is to establish uniformity in accordance with an established method of survey having been adopted by government surveyors. Such uniformity is desirable in order to preserve, as nearly as possible, the amount of land included within each government patent based upon a government survey. This being true, it is clear in this case, that the rights of the original purchasers of land in sections 19 and 20 were based upon a survey that commenced from the west boundary of the township. In addition, there was at least one monument on the west township line which could be identified. Under circumstances such as these, to give effect to the purpose for which the rule contended for was devised, we hold that the survey of sections 19 and 20 made from the west boundary of the township to the east was proper.

While it is true, as urged by respondent, that because of the curvature of the earth, boundary lines of townships in the northern hemisphere converge slightly to the north and the shortage, if any, in a given township is to be taken up in the westernmost tier of sections, according to established practice, it appears that both government surveyors, in running their surveys of the township here involved took up their shortages where their surveys from the east and west met. Although the rights of the parties, through their predecessors in interest, became established according to an unusual method of surveying, the only way in which to preserve these rights is to follow the survey as originally made. The rule of *Henrie v. Hyer*, supra, is not in conflict with the principle here announced and Mr. Griffiths was not only authorized to commence his survey of sections 19 and 20 from the western boundary of the township, he was required to do so in order to preserve, if possible rights established with reference to the original government survey. *Washington Rock Co. v. Young*, 29 Utah 108, 80 P. 382, 110 Am. St. Rep. 666."

The following quote from an early Texas decision summarizes the foregoing discussed elements nicely:

The actual identification of the survey, the footsteps of the surveyor upon the ground, should always be followed, by whatever rule they may be traced.* This case is discussed in some detail in Chapters 3, 4, & 8.

---

* *Stafford v. King*, 30 Tex. 257, 94 Am.Dec. 304, 1867.

2

Title Issues

Titles to land are not to fail merely because old markers may have disappeared or because it may be difficult to trace footsteps of the surveyor.

Turnbow v. Bland,
149 S.W.2d 604 (Tex.Civ.App., 1941).

Definitions

Definition of Title

The right to or ownership of property. The word is used to designate the means by which an owner of lands has the just possession of his property, the legal evidence of his ownership, or the means by which his right to the property has accrued.[*]

—Patton on Titles, § 1

Definition of Boundary

The invisible line of division between two contiguous parcels of land.

Boundaries may originate, be fixed or be varied by statutory authority, by proved acts of the respective owners (as by plans and deeds, possession, estoppel, or by agreement), or by the courts exercising statutory or inherent jurisdiction.[†]

—Boundaries and Surveys, § 1

The title is the entity that establishes the boundaries; without the former, there can be no latter. Boundaries define the location and extent of the title. Title is a matter of law, while boundaries are a matter of survey, both as

---

[*] Patton, Rufford G. & Carroll G. Patton. *Patton on Titles*. Second Edition. 3 Vols. St. Paul: West Publishing Co. 1957.
[†] Lambden, David W. & Izaak de Rijcke. Boundaries and Surveys. The Carswell Company Limited. 1985.

The Establishment of Title

A Mississippi decision* clearly states the significance of the original survey as a basis of title.

> In the early case of *May v. Baskin*, 12 Smedes & M. 428, this Court said: "The original survey fixed the rights of the parties. The government sold the land according to that survey, and conveyed it to the purchasers. No line could be afterwards run without the consent of all interested, which would vary the rights thus acquired."

A new survey cannot be employed to disturb vested rights acquired in reliance on an earlier survey, nor may such rights be disturbed in violation of a valid agreement between immediately adjacent property owners. Office surveys merely establish boundary lines, and do not determine title to land involved. *Appeal of Moore*, 173 Kan. 820, 252 P.2d 875 (Kan., 1953).

This court went on to say that "were the rule otherwise there could be repeated surveys with the result that each would disturb rights acquired in reliance on a former survey. The very purpose of establishing official permanent boundary lines would be completely defeated."

As early as 1849, the Illinois court stated in the decision of *M'Clintock v. Rogers*,[†] "The rights of the parties in the land in controversy must be determined by the original survey, made under the authority of the government, and returned to the office of the surveyor general; an according to that survey, the disputed township line was a straight one, running through the entire length, from corner to corner; and it cannot now be varied to suit the interests or caprice of purchasers. To permit it to be done would be to annul the authority of all the public surveys; obliterate the lines of demarcation between the property of man and man, and open wide the door to confused, harassing, and endless litigation."

Justice Thomas M. Cooley wrote in the case of *Stewart v. Carleton*,[‡] "the facts [in this case] require some reference to the testimony of the boundary, which seems to us to have been introduced on a somewhat dangerous theory. It appears to have been supposed that surveyors are competent not only to testify to measurements and distances, but also to pass judgment themselves and on information of their own choosing upon the position of lines and starting points. This is not the only case in which we have encountered such evidence on important private rights; and surveyors seem to have the idea

---

* *Boyd v. Durrett*, 216 Miss. 214, 62 So.2d 319, Miss., 1953.
† 11 Ill. 279, Ill., 1849.
‡ 31 Mich. 270, 1875.

Title Issues 39

that they may act entirely on their own judgment in determining important private and public rights.

> This is a very dangerous error. The law recognizes them as useful assistants in doing the mechanical work of measurement and calculation, and it also allows such credit to their judgment as belongs to any experience which may give it value in cases where better means of information do not exist. But the determination of facts belongs exclusively to courts and juries. Where a section line or other starting point actually exists, is always a question of fact, and not of theory, and cannot be left to the opinion of an expert for final decision. And where, as is generally the case in an old community, boundaries and possessions have been fixed by long use and acquiescence, it would be contrary to all reason and justice to have them interfered with on any abstract notion of science. The freaks of opinionated surveyors have led to much needless and vexatious litigation and disturbance, and it is much to be desired that should be confined to their legitimate place as witness on fact, and not on opinion, which lie beyond the domain of science."

He continued in the case of *Diehl v. Zanger** by writing, "This litigation grows out of a new survey recently made by the city surveyor. This officer after searching for the original stakes and finding none has proceeded to take measurements according to the original plat, and to drive stakes of his own. According to this survey the practical location of the whole plat is wrong, and all the lines should be moved between four and five feet to the east. The surveyor testifies with positiveness and apparently without the least hesitation that 'the fences and buildings on all the lots are not correctly located' and there is of course an opportunity for forty-eight suits at law and probably many more than that. When an officer proposes thus dogmatically to unsettle the landmarks of a whole community, it becomes of the highest importance to know what has been the basis of his opinion. The record in this case fails to give any explanation, but the reasonable inference is that the surveyor has reached his conclusion by first satisfying himself what was the initial point of Mr. Crampau's survey, and then proceeding to survey out the plat anew with that as his starting point. Of course by this method if no mistake is made, there is no difficulty in ascertaining with positive certainty where, according to Mr. Crampau's plat, the original street and lot lines ought to have been located; and apparently the surveyor has assumed that that was all he had to do.

> Nothing is better understood than that few of our early plats will stand the test of a careful and accurate survey without disclosing errors. This is as true of the government surveys as of any others, and if all the lines were not subject to correction on new surveys, the confusion of lines land titles that would follow would cause consternation in many

---

* 39 Mich. 601, 1878.

communities. Indeed the mischiefs that must follow would be simply incalculable, and the visitation of the surveyor might well be set down as a great public calamity.

But no law can sanction this course. The surveyor has mistaken entirely the point to which his attention should have been directed. The question is not how an entirely accurate survey would locate these lots, but how the original stakes located them. No rule in real estate law is more inflexible than that monuments control course and distance,—a rule that we have frequent occasion to apply in the case of public surveys, where its propriety, justice and necessity are never questioned. But its application in other cases is quite as proper, and quite as necessary to the protection of substantial rights. The city surveyor should, therefore, have directed his attention to the ascertainment of the actual location of the original landmarks set by Mr. Crampau, and if those were discovered they must govern. If they are no longer discoverable, the question is where they were located; and upon that question the best possible evidence is usually to be found in the practical location of the lines, made at a time when the original monuments were presumably in existence and probably well known. *Stewart v. Carleton*, 31 Mich. 270."

When a section line is discovered to be in error, it does not mean that landowners must readjust their property lines to conform to the resurvey. On the contrary, people relying on the earlier established survey may disregard the resurvey and consider it as having no effect on the ownership of the property. *Affleck v. Morgan*, 12 Utah 2d 200, 364 P.2d 663 (1961) citing *Henrie v. Hyer*, 92 Utah 530, 70 P.2d 154.

Significance of the Original Survey

Once again, relying on the *Rivers* case discussed in Chapter 1, "the surveyor can, in the first instance, lay out or establish boundary lines within an original division of a tract of land which has theretofore existed as one unit or parcel. In performing this function, he is known as the 'original surveyor' and when his survey results in a property description used by the owner to transfer title to property (a very important distinction) that survey has a certain special authority in that the monuments set by the original surveyor on the ground control over discrepancies within the total parcel description and, more importantly, control over all subsequent surveys attempting to locate the same line."

The Establishment of Title to Land, or Rights in Land

Boundaries define the position or location of the title and its extent. Without a title, there are no boundaries.

While the usual consideration is that title is defined, or established, by a written document, the commonly used deed is the first to be considered as

Title Issues   41

the source of the title. However, there are several additional ways that title may be established, not all of which involve a writing. They are categorized as follows.

Written Title

Written title demands some form of writing to complete the process. While documents are not always found in the public records, they may either be located at an appropriate repository or otherwise in proper custody.

1. Patent from the sovereign

   In the public land states, this is usually a writing from the appropriate agency of the United States Government, on file in a designated office. Outside the PLSS, it depends on the nature of the grant. Sovereign grants may be filed in other countries, or in the individual state that was in control when the specific grant was made. Many original records have been preserved in designated offices of state and local government as well as in historical institutions.

2. Private grant, such as a deed

   When filed, such records are generally found in an appropriate registry. The usual consideration is that any deed is filed at the local court house, or similar repository. However, not all deeds and related transfers of title are placed on the public record, so that a "search of the public records," particularly confined to the usual repository, may not uncover them. Title policies will take an exception and not cover "items not on the public record."

3. Will

   Not all wills are filed either, until an estate is probated or the maker of the will has died. When filed, copies are generally kept at offices of probate, frequently located at the local court house building, or other similar institution.

4. Involuntary alienation (bankruptcies & foreclosures)

   While such procedures require writings, such papers are often found in a variety of locations aside from the local court house. Depending on the lien holder, records may be found at a variety of levels, federal, state, county, and local.

5. Eminent domain proceedings

   Eminent domain, being a formal process, involves a series of writings, one or more which contain a description of the taking. They are not always on the public record, though usually on file at the appropriate agency.

Combination

These are categories containing procedures which, depending on the circumstances, can be either written or unwritten.

6. Dedication

   Dedication may be statutory (according to the terms of law) or by common law. It may also be express or implied, as can its often followed procedure, acceptance.

7. Estoppel

   Estoppel may be accomplished in one of three ways, namely by deed, by record, and by matter *in pais*. By deed directly involves a writing of some sort, although it does not necessarily imply a public record. By record also involves some type of writing, and also implies some type of formal preservation in the form of a record, although once again it does not require a public record. Matter *in pais* (or estoppel by conduct) is an equitable estoppel and arises when one, by word of mouth, conduct, or silent acquiescence, reports that a certain state of facts exists, thus inducing another to act in reliance upon the supposed existence of such facts, so that if the party making the representation were not estopped to deny its truth, the party relying thereon would be subjected to loss or injury.

   It is the species of estoppel which equity puts upon a person who has made a false representation or a concealment of material facts, with knowledge of the facts, to a party ignorant of the truth of the matter, with the intention that the other party should act upon it, and with the result that such party is actually induced to act upon it, to his damage.*

   It is more often verbal than by any form of written record. It can also involve silence or inaction in the absence of a responsibility to come forward with the true facts.

Unwritten Title

These are strictly unwritten. While ancillary documentation and information may sometimes be uncovered, the event or procedure itself carries no direct record.

8. Descent

   When a person dies intestate and owning property, legal heirs become the owners. State statutes specify who are the legal heirs. Caution is advised in that timing of the date of death is critical in

---

* *Black's Law Dictionary.*

Title Issues   43

that the statute(s) in most states have changed, often several times, as to specifying the legal heirs.

9. Adverse possession

Adverse possession is the ripening of possession into title through passage of time while certain requirements are being met. It is provided for in the law with a statute of limitations barring an action for recovery after a specified length of time of adverse occupancy by another.

Prescription

Prescription is a form of adverse possession, but with somewhat less stringent requirements of some of the elements. Since it is not possession, but rather arises through use, it results in an easement (known as a prescriptive easement) rather than fee simple title to the property in question.

10. Unwritten agreements

Unwritten agreements come in several forms, the most common of which are parol agreements, whereby two parties verbally agree on something, such as the location of a boundary. A boundary may be established in this manner, depending on certain requirements such as later acquiescence, and so long as such agreement satisfies all of the requirements of agreements in general, it is binding on the parties and their successors in title. See Chapter 10 on boundary agreements.

11. Escheat

When a person dies owning property and neither a will to designate the successor to the property, nor any legal heirs, the property transfers to the state through the process of escheat. Escheat is a common law principle in place to insure that property is not left in limbo upon the death of a person where there is otherwise no procedure for it to transfer. Escheat requires no documentation, and is automatic upon the death of the owner.

12. Accretion

Accretion is the gradual and imperceptible accumulation of new land through the natural action of the elements, water (which is the most common), and wind. While it is said that it effects a change in the boundary, it is more accurately characterized as a receipt of a land title through a natural process. The process may be compared with what is termed avulsion, whereby a sudden, and usually dramatic, change occurs to the landscape through the violent action of water. A common example is the shifting of a river bed, which, if an avulsive act, causes no change to boundaries, even if land should disappear during the process.

Boundary Retracement

13. Parol gift

While the statute of frauds specifically states that, in order to be enforceable, transfers of land must be in writing, it can be seen that under certain circumstances, or with special laws, unwritten transfers may occur. The gift of land by parol, or done verbally, under the right circumstances, is possible, and enforceable.

14. Operation of law

There are several processes that occur automatically as a result of an event. Reversions upon termination of an encumbrance such as an easement, the satisfaction of a condition subsequent, and escheat are common examples. The event takes place, the transfer of land, or rights, occurs, without a written document.

15. Custom

Bouvier's Law Dictionary defines custom as "such a usage as by common consent and uniform practice has become the law of the place, or of the subject matter to which it relates." It creates a right, generally in favor of a group of persons. As a common example, custom differs from prescription, which is personal and is annexed to the person of the owner of a particular estate, while the other is local, and relates to a particular district. The distinction has been thus expressed: "While prescription is the making of a right, custom is the making of a law." Lawson, Usages & Cust. 15, note 2.

16. Prior appropriation

Sir William Blackstone in his "Commentaries on the Laws of England" developed this term as a means of recognizing flowing water as "transient property." While attached to a parcel of upland, the owner of the upland may utilize the water for any and all purposes so long as they do not interfere with the rights of another person, upstream or downstream. The upland owner may divert the water for his use, such as for the operation of a mill.

AUTHOR'S NOTE: It should be emphasized that over half (9 out of 16, with an additional 2 being either) of the procedures do not involve a writing.

Written title documents, of whatever nature and subject matter, contain a description of what is being transferred/conveyed. Such descriptions often leave something to be desired, and are usually less than perfect, but through the use of extrinsic evidence may be adequate for their intended uses.*

This principle has been emphasized by courts of several jurisdictions, one of the most notable of which is the decision of *City of North Mankato v. Carlstrom,*† wherein the Minnesota court stated, "In a deed conveying land,

---

* See *Peacher v. Strauss,* 47 Miss. 353 (1872) in the section on Ambiguities.
† 212 Minn. 32, 1942.

Title Issues 45

the description of the property to be conveyed must be such as to identify it or afford the means of identification, aided by extrinsic evidence." This would seem to apply to any element of transfer of title or rights in real property.

Unwritten transfers contain no such description, and must be surveyed (located and area determined) in order to generate a definition of location or extent. Depending on the nature of the transfer of title or rights, definition may be a complex and time-consuming process, sometimes requiring litigation in order to "quiet the title" in the appropriate parties.

It should be emphasized that nowhere is it stated that a requirement for the transfer is that there must be a confirmation by a court of law. The requirement is already established within, and well documented and explained by, the numerous decisions on the topic. Certainly, if there is a conflict between two sources of title, whether between two deeds, or a deed and one of the other means, it may require a court to make a ruling as to who of the two contenders possesses the better claim, or the better title. Also, there may be additional requirements requiring a ruling to accomplish a particular purpose. One of the most obvious is the requirement of a writing in order for the title to be a marketable title.* Such a requirement falls under the rules for marketability of title in some jurisdictions. But such a requirement does not render the title untransferable.[†]

Missing Title

One of the most recent examples involving lack of title is the Maryland case of *Ski Roundtop Inc. v. Wagerman*,[‡] where the two parties each claimed what they termed an overlap of their respective titles.

This dispute involved the ownership of a tract of land on a mountainside near Emmitsburg, Maryland. Ski Roundtop was the owner of six parcels of land that it contends encompasses the land in question. Four of these parcels of land are on the north side of the disputed realty, and their original title has been traced back to the original patent of a tract of land referred to as "Carolina." The remaining two parcels of land border the south side of the disputed realty and have evolved from an original land patent referred to

---

* A "marketable title" to land is such a title as a court of equity, when asked to decree specific performance of the contract of sale, will compel the vendee to accept as sufficient. It is said to be not merely a defensible title, but a title which is free from plausible or reasonable objections. *Black's Law Dictionary.*

† Title acquired by adverse possession is a new and independent title by operation of law and is not in privity in any way with any former title. Generally it is as effective as a formal conveyance by deed or patent from the government or by deed from the original owner. In fact, it is a good, actual, absolute, complete, and perfect title in fee simple, carrying all of the remedies attached thereto. The title acquired will pass by deed. After the running of the statute, the adverse possessor has an indefeasible title which can only be divested by conveyance of the land to another, or by a subsequent ouster for the statutory limitation period. *3 Am. Jur. 2d, Adverse Possession, § 298.*

‡ 79 Md.App. 357, 556 A.2d 1144, Md.App., 1989.

46   *Boundary Retracement*

as "Nigh Nicking," which itself was the northern tract of a parcel of land referred to in an even earlier patent called "Carricks Chance."

The Brawners (plaintiffs in the case heard by the lower court) contended that a vacancy, winsomely referred to as "Pleasant View," existed between the original patents of Carolina and Nigh Nicking. They contended that they were record title holders or, alternatively, owners by adverse possession of the land within this vacancy. Ski Roundtop contended that the southern boundary of Carolina and the northern boundary of Nigh Nicking are one and the same. The bulk of the evidence introduced at trial by way of testimony by surveyors, deed plottings, and surveys focused on the original surveys of these patents conducted nearly 200 years ago.

The key question at trial was the location of the intersection between the 90th line of the Carolina patent and the 8th line of a patent to the east of Carolina and Nigh Nicking, known as Black Walnut Bottom. If the intersection occurred at the same spot where Nigh Nicking's 29th line intersected at Black Walnut Bottom, the Carolina line and the Nigh Nicking line were one and the same, not merely parallel. At the trial, Ski Roundtop pointed to the fact that the patent for Nigh Nicking referred to the northern boundary of Carricks Chance as Nigh Nicking's 26th through 29th lines. A subsequent survey of Carolina called to the same boundary of Carricks Chance as Carolina's 87th through 90th lines. Thus, relying on this boundary call for the two patents, Ski Roundtop maintained that a common boundary line existed and, therefore, Pleasant View does not.

The Brawners countered that the proper method of plotting the 18th century surveys and, specifically, the 87th through 90th lines of Carolina is to plot, by way of the metes-and-bounds description in the original patent, the eastern boundary of Carolina from the 98th line, where it intersects Black Walnut Bottom at a known point, to determine where the 92nd line of Carolina (also the 11th line of Shield's Delight, a prior patent) intersected the 8th line of Black Walnut Bottom. Once this is determined, the length of the 91st line of Carolina running down part of the 8th line of Black Walnut Bottom can be plotted and this line's intersection with the 90th line of Carolina can be established. Utilizing this method, the Brawners' surveyor determined that the 90th line of Carolina intersected the 8th line of Black Walnut Bottom approximately 10 perches above where Nigh Nicking intersected said line. They contended that their survey indicated that the southern boundary of Carolina and the northern boundary of Nigh Nicking were merely parallel lines, and that a gap existed where Pleasant View is located. Such a survey method of reverse tracing eliminated an apparent foul where Black Walnut Bottom overlaps onto Shield's Delight on Ski Roundtop's survey. (For a view of the disputed lines, see Figure 2.1 at the end of the discussion.)

At the conclusion of the three-day trial, the trial court gave credence to the Brawners' evidence and determined that a gap between Carolina and Nigh Nicking patents existed. The court held that Ski Roundtop's record

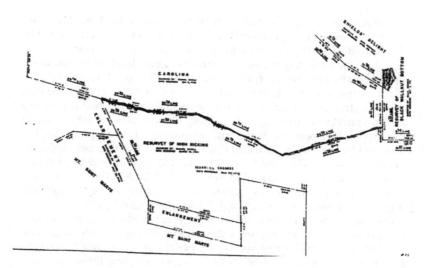

FIGURE 2.1
Diagram from *Ski Roundtop v. Wagerman*.

ownership of the six parcels of land did not include record ownership of the realty located in the gap.

In the court's decision, "Although the case law in this area is sparse, it appears that a requisite for valid title to real property is an original conveyance of public land by the State. See 3 American Law of Property, § 12:16 (1952); 73B C.J.S., Public Lands, § 188 (1983); 2 Patton on Titles, § 281 (2d ed. 1957). Absent such a conveyance, one purporting to transfer an ownership interest in such property transfers nothing, and no quantity of successive transfers by deed nor the mere passage of time will metamorphose good title from void title. Simply put, the Brawners' admission that a patent was never issued to their predecessor vitiates their assertion of record title.

> Logically, it must follow that to the extent the Brawners attempt to rely on calls to Pleasant View in Ski Roundtop's chain of title as establishing record title, any mistaken call to the property of another, the ownership of which is asserted by way of void deed, will not establish the existence of said property.
>
> To the extent the Brawners rely on such calls in cross-appeal Issue II as evidence that a gap exists between Ski Roundtop's properties, we hold that the sole controlling focus should be whether the boundaries of the original patents establish the existence of Pleasant View. These patents precede any of the deeds referred to by the Brawners. Any discussion of subsequent deeds is irrelevant. In the absence of facts giving rise to an estoppel, we decline to establish a rule of law that binds successors to real property to all descriptions of property made by their predecessors in prior deeds, particularly where the original patent contradicts such deeds. Moreover, one purporting to be an adjoining land-owner should

48   *Boundary Retracement*

not be allowed to capitalize on such mistakes where the boundaries are correctly established by even earlier deeds or, in this case, earlier patents.

Effectuation of the intent of the original parties, agreement, or surveyor, as the case may be, is of paramount consideration in boundary dispute cases. See *Wood v. Hildebrand*, 185 Md. 56, 60, 42 A.2d 919 (1945); *Md. Construction Co. v. Kuper*, 90 Md. 529, 548, 45 A. 197 (1900). The trial court's determination came down to an examination of two conflicting modern surveys of the original patents. Determination of which one of the two surveys best effects the true boundaries of the disputed land as intended by the original surveyor is a question of fact. *Zawatsky Constr. Co. v. Feldman Development Corp.*, 203 Md. 182, 187, 100 A.2d 269 (1953); *Dundalk Holding Co. v. Easter*, 195 Md. 488, 73 A.2d 877 (1950). The court below found as a matter of fact that the intent of Samuel Duvall, the original surveyor, was 'to join Nigh Nicking and Carolina.'"

Once the intent of the original surveyor is established, absent a finding of a patent mistake on the part of the original parties of such magnitude to vitiate the original intent, see e.g., *Laflin Borough v. Yatesville Borough*, 54 Pa.Cmwlth. 566, 422 A.2d 1186 (1980) (surveyor's original intent not effectuated due to an 'erroneous assumption that Yatesville then extended to that dividing line between lots'), the trial court is bound by its factfinding to apply general rules of boundary law to effectuate the original intent as found by the court. Of course, most boundary disputes evolve from surveying mistakes or ambiguous deeds. The fact that one surveyor's interpretation of the original survey results in a tidier or neater package, however, does not suffice, of itself, to override the intent of the original surveyor. A mistake of the magnitude that would justify the trial court's non-implementation of the original surveyor's intent was not found here.

Citing *Maryland Coal and Realty Co. v. Eckhart*,[*] "an albatross in the wake of every title searcher is the ominous question of whether he has gone back far enough in the chain of title."

This case emphasizes the importance of a title search back to the sovereign, in order to determine the rights the parties began with. Even though the Wagerman's chain of record title was from 1812, the court found that the title was void because an original land patent was lacking.

The New York case of *Dolphin Lane Associates, Ltd. v. Town of Southampton*[†] adds further insight as to the retracement of early descriptions and understanding of the conditions of earlier times.

To understand the title to the land(s) in question, the court reviewed the early history of the area from the British Crown through the Dutch and British claims and eventually to proprietors and their patents. These patents were confirmed by the act of the Colonial Assembly of May 6, 1691 and by the State Constitution of 1777 (Bradford's Laws, 1694, p. 6; Colonial Laws of NY, I, 224–225). The court emphasized, "This action, by its very nature, involves

---

[*] 25 Md.App. 605, 337 A.2d 150, 1975.

[†] 339 N.Y.S.2d 966, 72 Misc.2d 868, 1971.

Title Issues 49

the tracing of chains of title going back to the earliest settlement of the Town of Southampton. Thus, the history of the early settlement and development of Southampton must be examined. The past must be explored in order to understand the present."

Two recent cases have emphasized the importance and significance of original creation(s) of titles.

On the subject of "ancient rights," the case of *Bruker v. Burgess and Town Council of Carlisle** has for its subject the title to a town square created over 200 years prior. The court stated in its decision, "there is no doubt that the mere fact of the use of [the Square] by the public for now more than 200 years is sufficient to raise a conclusive presumption of an original grant for the purpose of a public square; such is an ancient and well established principle of the law. Nor can it be denied that, where such a dedication has been established and the public has accepted it, there cannot be any diversion of such use from a public to a private purpose, and it is also true that, where a dedication is for a limited or restricted use, and diversion therefrom to some purpose other than the one designated is likewise forbidden."

In the dissenting opinion of two of the judges, "Evidence of dedications and titles based upon ancient documents or events arising out of antiquity cannot possibly be as strong or clear as would be required in matters arising or titles created in modern times; and we must not lightly strike down rights or public uses which have existed for more than a century. Counsel for the Borough frankly admits that if the Borough can change or destroy this market place it can also change or destroy the church which was built on this same public square and like the market place has been used for more than 200 years. This church and this market place were dedicated by Thomas Penn in 1751; and this dedication and use as a market place were thereafter frequently ratified. The Borough merely contends that there exists today no clear evidence that this square was the 'spot' which was dedicated for these purposes.

> Title to this market place or square is not in the Borough of Carlisle; it is in the Commonwealth of Pennsylvania, with a reversionary interest in Penn's heirs. The Borough has no title or estate in this property; it has only a right to control and regulate, in the interest of the public, the public use of the 'market place' for the uses and purposes to which it was dedicated.
>
> In my judgment, the ancient documents, plans, letters and other evidence produced, plus the public use of this part of the Square as a market place for over 200 years, were sufficient to establish a dedication for that purpose or use, and consequently the Borough can and should be enjoined from changing or destroying this use."

Relating more to the usual dilemma of the average practicing retracement surveyor is the Tennessee case of *Burton v. Duncan*,[†] wherein the parties

---

* 376 Pa. 330; 102 A.2d 418, Pa., 1954.
† M2009-00569-COA-R3-CV (TNCIV).

50   *Boundary Retracement*

disputed the boundary between them. The court found the boundaries clear from one description, but not for the other, relying on their surveyor for the location of the "disputed boundary." In determining the two boundaries, the court found that there existed a sliver of land between the two ownerships.

The court stated in its decision, "We recognize that our holding creates a so-called 'no-man's land' concerning the sliver of property at issue in this action. We regret that fact, but it was the duty of the court to determine the boundary line of the Duncans' property and the boundary line of Mr. Burton's property. We have done that. The fact that the disputed property does not lie within the litigants' respective boundaries is problematic. The record suggests there may be others who own or have a claim to the disputed property, but the record is not sufficient to permit us to determine who owns the disputed property. Moreover, the resolution of that issue may require that others, who are not parties to this action, are essential parties to such a determination; accordingly, they may be indispensable parties, which prevents this court from making that determination. Accordingly, we make no decision concerning who may own the disputed property. Our decision is limited to the determination of the respective boundary lines of Mr. Burton's and the Duncans' property."

Until fairly recently, it was not necessary for either a deed or a plat (plan) to be placed on record in order to be valid, or effective. Any plat referenced in a deed is part of the deed, whether or not it is on the public record. In some jurisdictions, and in some instances, either or both must be recorded, according to statute law. An obvious example is a subdivision plat and the conveyances based on it.

A number of jurisdictions have, in the past, honored two basic principles:

A deed or plat need not be on record to be valid
A deed is valid against all who have notice

Notice

Until recently, when title standards were made more comprehensive, and a few State Supreme Court decisions had been handed down, there were commonly known to be two kinds of notice: actual and constructive. Actual notice may be defined as knowledge of the contents of a document or of other facts which may affect title to an interest in real property. If a title examiner has actual notice of a recorded document, it is immaterial whether the document appears in the record chain of title. If a title examiner has actual notice of an unrecorded document, it is immaterial that the document is not recorded. A title examiner must consider the effect of any document of which he or she has actual notice in the preparation of his or her title opinion.[*]

Constructive notice may be defined as being charged by law with notice of the effect on title to an interest in real property of the contents of a document

---

[*] Colorado Title Standards, Section 1.2.2.

Title Issues 51

or of other facts without knowledge of the document itself or the facts themselves. A document recorded in the real property records in the office of the county clerk and recorder is constructive notice of its existence and of its contents to all persons subsequently acquiring an interest in the real property affected by that document even if the document is not properly indexed or copied in the records by the clerk and recorder. While the recording of a document is constructive notice to the persons subsequently acquiring an interest in the real property affected by that document, a title examiner is only responsible for analyzing the effect on title of those recorded documents which would be revealed by a properly conducted search of the real property records by the title examiner or which are contained in the abstract of title examined by the title examiner.*

While these definitions owe their origins to the Colorado Title Standards, they have been the general rule for some time, applicable to both title examiners and other searchers of items affecting title to, or an interest in, real property. As stated prior, retracement surveyors do not do a title search, per se, but they do search the title.

The third and more recent type of notice is termed Inquiry Notice. Relying on the same set of definitions, inquiry notice may be defined as being charged by law with notice of the effect on title of facts that would have been revealed by an inquiry if known facts would cause a reasonable person to inquire. If a person acquiring an interest in real property has knowledge of facts which, in the exercise of common reason and prudence, ought to put him or her upon particular inquiry as to the effect of such facts on the title to such real property, he or she will be presumed to have made the inquiry and will be charged with notice of every fact which would in all probability have been revealed had a reasonably diligent inquiry been undertaken. Whether the known facts are sufficient to charge such person with inquiry notice will depend upon the circumstances of each case.

If, in the course of a title examination, a title examiner discovers a document which is not in the record chain of title but which sets forth or refers to facts (other than the existence of an unrecorded document) that would cause a reasonable person to inquire about the effect of such facts on the title being examined, the title examiner should disclose such facts in the title opinion so that the person for whom the title opinion is written may determine whether to undertake an inquiry.[†]

Ambiguities

Invariably, where descriptions are concerned, they frequently contain ambiguity. Ambiguity may be defined as doubtfulness, doubleness of meaning; indistinctness or uncertainty of meaning of an expression used in a written

---

* Colorado Title Standards, Section 1.2.3.
† Colorado Title Standards, Section 1.2.4.

instrument. It may be latent or patent. It is the former, where the language employed is clear and intelligible and suggests but a single meaning, but some extrinsic fact or extraneous evidence creates a necessity for interpretation or a choice among two or more possible meanings. A patent ambiguity, on the other hand, is that which appears on the face of the instrument, and arises from the defective, obscure, or insensible language used.*

As the Mississippi Court stated in the case of *Peacher v. Strauss*,† "no deed or conveyance of land was ever made, however minute and specific the description, that did not require extrinsic evidence to ascertain its location; and this is so whether the description be by metes-and-bounds, reference to other deeds, to adjoining owners, watercourses, or other description of whatever character. Every contract or deed for the conveyance of land must define its identity and fix its locality, or there must be such a description of the land as, by the aid of parol evidence, will readily point to its locality and boundaries. Looking through the authorities, it will be seen that, under every conceivable state of facts, and in every imaginable circumstance, the cases may be counted by hundreds, if not by thousands, where contracts, wills, and deeds are made effective by the identification, by extrinsic evidence, of the person or subject intended, yet in no wise violating the rule, that such evidence 'cannot be admitted to contradict, add to, subtract from, or vary the terms of a written instrument. A simpler rule, perhaps, in most cases, is this: that evidence may explain but cannot contradict written language.'"

While in the past courts frequently separated latent ambiguities from patent ambiguities, and prescribed rules for each confining extrinsic evidence for the resolution of patent ambiguities within the document, today's prevailing practice is to treat all ambiguities the same, and apply extrinsic evidence wherever necessary in order to uphold the written instrument rather than declaring it void. It is not the preferred practice to declare an instrument void if it can be prevented, or avoided.

As previously discussed, Robert Griffin's Rule # 3 stated *Retracement Should Apply Rules of Construction to Contradictory Evidences of Intention.* "When ambiguities appear in the description, and when discrepancies arise between adjoining descriptions or between the description and the physical evidence of the boundaries as it exists on the ground, rules of construction are applied to determine intention. The rules, which are based on reason, experience and observation, and pertain to the weight of evidence, state the order of preference and relative importance of calls in a grant and provide for a selection of the best solution to a question of meaning."

Some rules of construction are discussed later in this chapter.

The modern tendency is to disregard technicalities and to treat all uncertainties in a conveyance as ambiguities to be clarified by resort to the intention of the parties as gathered from the instrument itself, the circumstances

---

* *Black's Law Dictionary.*

† 47 Miss. 353, 1872.

Title Issues

attending and leading up to its execution, and the subject matter and the situation of the parties as of that time.*

Ambiguity in Field Notes

Rules of construction announced by the courts, and familiar to all surveyors, are to be resorted to only when latent ambiguities in the field notes are developed from an accurate ground resurvey. If, perchance, the metes and the bounds should coincide on a resurvey, no latent ambiguity would exist, and there would be no need for resorting to rules of construction or the introduction of parol evidence.

Judge Smedley, in *Bond v. Middleton*,[†] said: "The decision in the Dow case, as is apparent from the opinion, is the application of the well settled rule that if ambiguity arises when the description or field notes contained in a deed are applied to the ground, extrinsic evidence may be admitted to interpret properly the description and thus to give effect to the true intention of the parties."

In the case of *United States v. Champion Papers, Inc.*,[‡] (See Chapter 5 for details of this case) it was "readily apparent that [there was] an ambiguity in the description contained within [the surveyor's] field notes, since the survey failed to close as it should and did not coincide with the sketch on the original field notes." The various subsequent resurveys did not coincide with either the original field note description or the field note sketch. It was "therefore necessary to apply the various rules of construction and admit parol evidence to resolve the conflicts and to establish the lines with were actually run by the surveyor."

When There Is No Ambiguity

When there is no ambiguity, the deed says what it says, and there is no need to go outside to seek extrinsic evidence. This is a very important concept, since there is frequently a temptation to consider any and all evidence relating to the records and on the ground in the vicinity of the subject property. For example, a fence nearby a record boundary may have some relation to the boundary, but if it is not recited with the description of the property, then it is not part of the record title. While it may relate to a boundary, it is a separate matter besides the interpretation and location of the deed, or other land description.

It is proper practice to locate the fence, or any other evidence, so as to consider whether it has any relation to the property, so long as the fact that it is not part of the record title is not lost sight of.

---

* See 17 Am. Jur. 2d Contracts, § 240; 30 Am. Jur. 2d Evidence, § 1065.

† 137 Tex. 550, 155 S.W.2d 789, 1941.

‡ 361 F.Supp. 481, 1973.

In 1987, the South Carolina court stated in the case of *Bellamy v. Bellamy,** "Since there is no ambiguity in the description of realty conveyed by the deed, the grantor's intent must be determined from within the four corners of the deed; and he is deemed to have intended to convey land which was accurately and well described within the deed."

Multiple Markers

There are legitimate situations where multiple markers exist, and that is where they both, or all, mark true corners (Figures 2.2 and 2.3).

The problem arises when several surveyors, for a variety of reasons, each set their own marker all of which purport to mark the same corner (e.g., one true corner, but several markers, each reported as marking it). Common causes are the difference between surveys due to different procedures used, different beginning points of surveys, different sets of legitimate errors, use of different adjustment routines, among others. In most cases, there can only be one true corner, and the result of the subsequent surveys is nothing short of confusion, for there can only be one correct marker, if that. Frequently all of the markers are incorrect, and the true corner exists at a place that either no surveyor found, if it was marked, or no surveyor was capable of determining the location of the true corner in the absence of the marker.

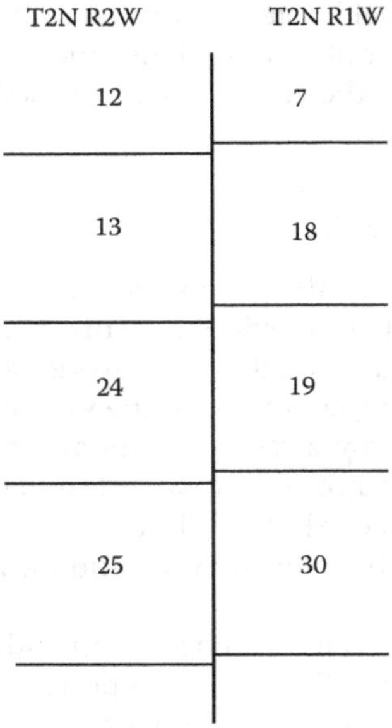

FIGURE 2.2
Double corners in the PLSS.

_____

* 292 S.C. 107, 355 S.E.2d 1, 1987.

Title Issues

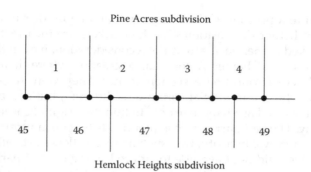

FIGURE 2.3
Double corners in the private sector.

Two decisions relate to this problem. In *Hagerman v. Thompson*,* the Wyoming court addressed three mineral surveys, and was unable to decide which of the three was the correct retracement of the original survey. The court said, "The purpose of a resurvey is to ascertain lines of original survey and original boundaries and monuments as established and laid out by survey under which parties take title to land, and they cannot be bound by an resurvey not based on survey as originally made and monuments erected." In this case, "all the three surveys in question here were resurveys, binding on no one, unless one of these perchance should ultimately in a proper proceeding be found to be correct. We have no way of knowing which one of them is correct. It is a well-known fact that surveyors are apt to differ from each other, and surveyors employed by the United States government are not immune from the frailties of their profession. Which one of these resurveys is correct is a question of fact."

In the case of *Johnson v. Westrick*,† the Wisconsin court, in addition to stating that incorrect retracement surveys are troublesome, extended their feelings by sending a message. "It was held in this case that the east line of the street was where the original surveyor placed it, not where it should be according to resurveys or subsequent surveys; that subsequent surveys are worse than useless; they only serve to confuse, unless they agree with the original survey."

Retracement people are generally quite proficient at finding and subsequently locating markers. Whether the marker(s) has any significance is quite another matter. Remember that a corner exists, with or without a marker, and where the problem arises is when the retracement person is incapable of determining the corner. Some do not even consider the corner as a separate entity, and accept whatever marker they find that seems to fit the purpose. Notwithstanding the fact, however, that they are inclined to

---

* 235 P.2d 750, 1951.
† 200 Wis. 405, 1930.

call attention to a problem if there is more than one marker in the vicinity of a corner. Whether the position(s) of the markers are incorrect in relation to the unmarked corner generally is not a consideration, if even thought of.

The discussion in Chapter 11 concerns a situation where there were four private markers surrounding a section corner. They were recovered by a BLM official, who set an official corner at a different, but nearby, location. This official returned at a later date only to find the original corner—at some distance away. The bad part of this is that deeds had been written based on the incorrect surveys. In a situation such as this, sellers have either tried to convey land they did not have title to, or are leaving behind parts that they do have a right to convey. Either way, the record is severely encumbered. If discovered early enough, there may be a chance to correct the problem, especially if the affected parties are available to execute corrective deeds. In the extreme case, picture such a problem being discovered 50 or more years after the fact, when none of the parties are still alive. At the least, there will be claims of adverse possession, acquiescence, and estoppel. Place whatever label on it you wish, the fact of the matter remains that there is a marked difference between what is on the ground and what is in the record. A careful surveyor will eventually uncover these problems since they will not go away by themselves. It is always a surveyor's nightmare, one that is probably unpredictable, has increased the cost of the work several fold, and will likely result in some sort of litigation, either a simple quiet title action, or a major boundary dispute.

The problem of newly discovered evidence will also be discussed in some detail in Chapter 11.

Descriptions by Abutting Tracts

This category of description often leads people to believe an ambiguity exists when in reality it does not. The seeming ambiguity arises when the parcel description is compared with evidence on the ground, the information does not appear to be harmonious. A somewhat vague description, such as one by abutting tracts and without calls for monuments or measurements, does not provide the retracement surveyor much to work with. However, from a title point of view, when parcels call for one another, the descriptions do not leave gaps or overlaps, and by law, called-for abutters, if existing at the time and identifiable, are considered monuments. Therefore, actually, such a parcel is bounded by four monuments, items of the highest dignity.*

---

* Calls in a deed for an adjoining tract of land are calls for a monument, and where the location of such adjoining tract of land is certain, it becomes a monument of the highest dignity. Where the description in a deed conveying a tract of land calls for the line of an adjoining tract, the location of which is undisputed or clearly established, it will control in locating such tract of land. *Vandal v. Casto*, 81 W.Va. 76, 93 S.E. 1044, 1917.

Title Issues

In *United Fuel Gas Co. v. Snyder,*[*] the court had this to say, "The general rule in determining what is included in a conveyance is that general calls for quantity must yield to more certain and locative lines of the adjoining owners which are or can be made certain." 9 C.J. 224. The statement in a conveyance of the amount of land to be conveyed is not a factor in establishing boundaries, unless the more particular description expressed is indefinite and uncertain. It is very generally considered that quantity is the least certain element of description and the last to be resorted to in fixing boundaries. Ordinarily all other elements of description must lose their superior value through ambiguities and uncertainties before resort can be had to quantity. *South Penn Oil Co. v. Knox,* 68 W.Va. 362, 69 S.E. 1020; *State v. King,* 64 W.Va. 546, 63 S.E. 468.

> Even *the unmarked line* of an adjacent tract, if it is well established and if its position can be ascertained with accuracy, will control a call for courses and distances. In the application of this rule such a line may be given the dignity of an artificial object.[†]

With the case of *Matador Land & Cattle Co. v. Cassidy-Southwestern Commission Co.,*[‡] the Texas court speaks to this issue, saying: "One of the rules adopted by the courts in establishing boundaries is that those calls shall be adopted which are more certain, avoid conflicts in surveys, etc., and harmonize with the evident purpose of the state in making the grant, even though other calls less certain, even though a natural object, must be disregarded. *J.W. Robinson v. Elias Mosson et al.,* 26 Tex. 249; *Bass v. Mitchell,* 22 Tex. 285; *Jones v. Burgett,* 46 Tex. 284; *Linney v. Wood,* 66 Tex. 22, 17 S.W. 244. In the leading case of *Maddox Bros. & Anderson v. Fenner,* 79 Tex. 279, 15 S.W. 237, it is said that:

> When unmarked lines of adjacent surveys are called for, and when from the other calls of such adjacent surveys the position of such unmarked lines can be ascertained with accuracy, and when, in the absence of all evidence as to how the survey was actually made, there arises a controversy as to whether course and distance or the unmarked line of another survey shall prevail, we see no good reason why the survey line should not be given the dignity of an "artificial object" and prevail over course and distance."

Bounding Descriptions

Those descriptions categorized as "bounding," or "abounding," or "bounded," are those which call for abutting parcels on all sides, or just some of the sides. They are commonly denoted as "descriptions by adjoiners." When there are only four calls, these descriptions can easily lead one

---

[*] 102 W.Va. 75, 135 S.E. 164, W.Va., 1926.
[†] 11 C.J.S. Boundaries, § 53(1).
[‡] 207 S.W. 430, Tex.Civ.App., 1918.

to believe that the resulting parcel is either rectangular or square, thereby creating a mindset which may be misleading and difficult to overcome. For example, the following description.

Bounded south by the highway, east by [land of] Brown, north by [land of] Smith, and west by [land of] Jones. The description may or may not recite an area of the parcel. The reader may easily be misled by this type of description, in that, in reality, (1) it is probably not truly rectangular, and (2) the area figure, if stated, may be merely an estimate. These are usually undeserving to be categorized as rectangular parcels, for most times they are not. Tracing the title back to its origin frequently results in a better description, ideally one with measurements based on an original survey.

Typical of descriptions of this nature is that the original description, often a metes-and-bounds description, has been dropped in favor of the several adjoiners. Adjoining descriptions can be terribly misleading, since brevity has overtaken accuracy somewhere along the way, and left the reader with more than one, sometimes several, interpretations. Figure 2.4 is an example of a modern description traced back in the records to its original, retraceable, description. The retracement surveyor should always be on guard against this situation, since interpretations of modern descriptions can often be incorrect, and additional evidence, often not called for in the description, can easily lead to an incorrect conclusion.

One of the biggest problems existing today is when an individual, or a retracement surveyor, has only the bounding description to work from and, on the ground, a scattering of fences. While this work is not a discussion of fences, one which could easily take up an entire volume by itself, a few

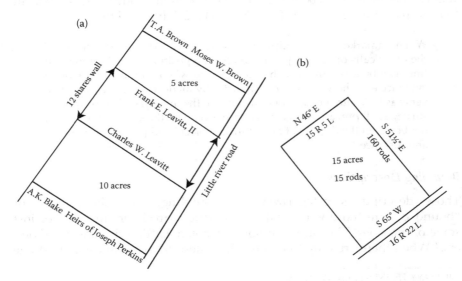

FIGURE 2.4
Current deed description (a) and 1832 deed description, based on a survey (b).

Title Issues

points are in order to illustrate why a poor practice may be uncovered, which is likely incorrect. Temptation, in the absence of better information, comes into play and the tendency is to seek the combination of fences, relying more heavily on the oldest ones, which come closest to the recited area figure. While they almost never satisfy the area call exactly, the difference, in the minds of some, is easily explained away in that measurements, especially those of area, are "more or less." First of all, fences not called for in a deed are not part of the description* so they cannot control the elements of the description (the monuments[†]) and second, area is being used as a controlling factor, when it should be considered the last, in the absence of better, controlling, information. Frequently, the reasoning behind accepting the [uncalled-for] fences is based on the theory of acquiescence, whereby parties have occupied to the fence and (it is believed) they have created this as their boundary through its acceptance over time.

Before deciding whether to consider accepting a fence as a boundary through acquiescence of the parties, one should read and understand the court's thinking in the case of *Pilgrim v. Kuipers*,[‡] which is discussed in its appropriate context in Chapter 9 on boundary agreements.

> There is a critical distinction between a fence which establishes a boundary line, and a fence that merely separates one side of the fence from the other. The former is a monument as well as a fence, while the latter is merely a fence.
>
> Nor does a fence establish a boundary line when it does not conform to the true line, even though the property owners thought it was the boundary.
>
> Where two adjoining owners are divided by a fence, which both owners suppose to be on the line, such fence is a division fence, as between them, until the true line is ascertained, when they must conform to the true line.

Again, there is widespread variation in the requirements for and results of occupation among the jurisdictions. Since there is no clear set of guidelines concerning this issue that will apply in every case, a study must be made not only of the circumstances surrounding the situation, but should include an understanding of the court's determinations for the particular jurisdiction.

Acquiescence is a type of agreement, and, referring to Chapter 9 on boundary agreements, may suffice, depending on the jurisdiction and the circumstances, only when certain conditions are met, and certain requirements satisfied.

---

* Fences not referred to in a deed cannot control the distances stated in the deed. *Kashman v. Parsons*, 39 A. 179, 70 Conn. 295, 1898.

    A line called for is quite as controlling as any natural or artificial boundary. *Parran v. Wilson*, 160 Md. 604, 154 A. 449; *Ramsay v. Butler, Purdum & Co.*, 148 Md. 438, 129 A. 650.

[†] 209 Mont. 177, 679 P.2d 787, 1984.

[‡] 209 Mont. 177, 679 P.2d 787, 1984.

Because title examinations are usually limited in scope, searches usually do not go very far back in time. The retracement surveyor is working to a different standard, and not doing a so-called "title search" although he or she is actually *searching the title*. The retracement surveyor is seeking the original survey, which may be coincident with the creation of title, usually found prior, sometimes many years prior, to the resting deed in a title search. As earlier stated, frequently the best information, sometimes the original information, is found in the description contained in the creating document(s).

Boundary Line Agreements

Boundary line agreements can present a title problem when done inappropriately. They may be encountered during the title searching process, whether for boundary or title purposes, and the tendency is to assume that this is the final answer to the identification and location of that particular boundary. However, recent studies attempting to locate properties as well as to solve other title problems have uncovered evidence, which either brings the agreement into question or demonstrates that the true boundary is actually identifiable, and located elsewhere.

A study of the court decisions demonstrates that many so-called boundary line agreements violate the statute of frauds since, perhaps inadvertently, they identify a line different from the actual boundary, thereby creating a [sometimes small or a sliver] parcel of land between the two lines resulting in an attempt to transfer title without a writing. Those statutes that provide for boundary line agreements are specific in nature, and contain certain requirements. All of the types of agreements require certain conditions to be met, one of which is that the true boundary is uncertain, unknown, or unascertainable. Frequently this is not really the case, thereby making the agreement inappropriate, and likely contrary to law. The ultimate result is that it may be void, but in the interim, especially if there has been a reliance on it as a locator of the boundary, it may cause one or more title problems. In the extreme, should the agreement be found to be void, it may serve to void the subsequent instruments relying on it or incorporating it.

Chapter 9 is devoted to boundary agreements, and details their requirements, shortcomings, and consequences.

Intent of the Parties

Many state court decisions state and re-state the importance of the intent of the parties [to the transaction]. There is probably no concept in title and boundary law that is more misunderstood than this term. Intent of the parties and the intention of the parties is not always the same thing. One source had stated that there are three potential truths when it comes to conveyancing and surveying: (1) that which was intended to be surveyed; (2) that which was actually surveyed; and (3) that which was reported to have been

Title Issues

surveyed. The comparison of these items shows up frequently in courts' analyses in various decisions.

It seems to arise all too frequently where someone is equating what the parties meant to do with their intent, and therefore treating them as equivalent. One of the most glaring examples is where a party intended to convey a series of lots of even width and the lots are marked or monumented at variance to that planned width, even if by a small amount, yet resulting plans, and subsequent deeds, report the "planned width." According to the universal rule of law, the lots are as surveyed and marked on the ground. Many people have a difficult time digesting the rule of law, since they know the parties' intent is to govern, and the parties intended to lay out lots of the specified width, but was actually accomplished on the ground is at variance.

Second guessing from the wording in the description as to what parties meant to do, or what they wished to do, only invites a deviation from reality. Different readers may interpret a given description different ways, and therefore reach different conclusions. The truth is, the parties to the transaction agreed on a particular description of premises—the grantor signed it and presumably read it and the grantee placed it on record presumably accepting it, and courts have laid down myriad rules for their interpretation when there is ambiguity or confusion. What parties wanted to do, or had planned to do, is insignificant, as demonstrated by the decisions presented below.

Selected decisions illustrating this concept follow. This confusion has apparently arisen frequently, so that the courts have taken it as a serious issue.

The basics of the word "intention" is summarized in the following library reference. Simply, it is about what the parties put in writing in the document, not what they said, not what they contemplated, not what they thought, perhaps not even what they planned to do. But intention is what they put in the document.

Intention

"Intention" as applied to the construction of a deed is a term of art, and signifies a meaning of the writing. 26 C.J.S., § 83.

Numerous court decisions from a variety of jurisdictions have addressed this issue. A selection of some of the more illustrative ones follow.

While in a case of ambiguity "……...it is the duty of the court to place itself as nearly as possible in the situation of the parties at the time the instrument was made, that it may gather their intention from the language used, viewed in the light of the surrounding circumstances," yet the rule "that the construction of a written document is the ascertainment of the intention of the parties……...does not authorize the use of evidence of matter not proper for consideration in the interpretation of a writing."

What is to be determined is the meaning of the deed, and not the parties' understanding of the meaning. What they in fact intended cannot control or affect the language used when its meaning is ascertained. The process of construction builds upon the language to develop the intention, and not upon the intention to interpret the language. *Smart v. Huckins.**

The intention sought in the construction of a deed is that expressed in the deed, and not some secret, unexpressed intention, even though such secret intention be that actually in mind at the time of execution. *Sullivan v. Rhode Island Hospital Trust Co.*, 185 A. 148, 56 R.I. 253 (1936).

It is not what the parties meant to say, but the meaning of what they did say that is controlling. *Urban v. Urban*, 18 Conn. Sup. 83 (1952).

Intent of the grantor as spelled out in the deed itself must be interpreted, not the grantor's intent in general, or even what he may have intended. *Wilson v. DeGenaro*, 415 A.2d 1334, 36 Conn. Sup. 200 (1970).

As early as 1866, the New Hampshire court spoke about intention in the case of *Wells v. Jackson Iron Mfg. Co.†*: The actual intention of the parties, or the surveyor, is not admissible to affect the construction of the deed.

In ascertaining the intent of the grantor of a deed, it is not the function of the court to attempt to ascertain and declare what he meant to say. And in doing that it is the duty of the court to presume that he knew the legal meaning of the terms which he employed in the deed. The application of that rule may sometimes appear to work a harsh result. But even so, it is a safer and wiser rule of decision which rests titles to property on the basis of the somewhat definite and ascertainable legal meaning of the words which the grantor used rather than upon the always uncertain and imponderable basis of what some court might decide that the grantor meant to say. *Holloway's Unknown Heirs v. Whatley*, 104 S.W.2d 646 (Texas, 1937).

Nothing passes by a deed except what is described in it regardless of the intention of the parties. *Tompkins v. American Republics Corp.*, 248 S.W.2d 1001 (Texas, 1952).

Buie v. Miller, et al., 216 S.W. 630 (Texas, 1919).

In re Lynch's Estate; In re Duffy, 272 N.Y.S. 79 (1934).

National City v. California Water and Telephone Company, 22 C. 560 (California, 1962).

> it is not enough that the parties had * * * [a particular] intention in fact [that is, as a unanimous state of mind], unless they have expressed it in some way in * * * [the] deed. The question is not what did the parties actually mean to say, but, what is the meaning of what they have said. *Bartholomew v. Murry*, 61 Conn. 387, 23 A. 604 (1891); *Lampson Lumber Co. v. Caporale*, 140 Conn. 679, 102 A.2d 875 (1954).

It is the expressed, rather than the unexpressed, intent.

In construing a grant, the intention of the parties is ascertained by giving suitable effect to all the words of the grant, reading them in light of the

---

* 82 N.H. 342, 1926.
† 47 N.H. 235, N.H., 1866.

Title Issues 63

circumstances attending the transaction; but the supposed intention of the parties, however fortified by circumstances, cannot be permitted to overcome the effect of the express language of the grant, taken as a whole, and properly construed. *Whitmore v. Brown & Gilley; Smallidge, et al. v. Same, 100 Me. 410* (1905).

A decision which adds perspective to the phrase "the intent of the parties" is *Roth v. Halberstadt*[*]: "The primary function of a court faced with a boundary dispute is to ascertain and effectuate the intent of the parties at the time of the original subdivision."

In this case the disagreement involved the ownership of a spring mentioned in the deed. However, since the spring was found not to have existed when the original subdivision took place, it was not part of the original creation of the dividing line and therefore not a controlling call. The rules of construction had to be applied to the original description at the time of the subdivision.

The original subdivision is that point where a parcel was severed from a parent tract, and created as a single entity for the very first time. That does not necessarily insure that the description creating that parcel also creates its boundaries, for one or more of the boundaries may be coincident with boundaries of the parent tract, created at an earlier point in time.

Intent of the Surveyor

Referring again to the Texas case of *United States v. Champion Paper, Inc.*,[†] the court elaborated on conclusions of law, saying, at first, "the purpose of the inquiry in a boundary dispute action is to locate and follow the footsteps of the original surveyor. Various rules of construction for purposes of ascertaining boundaries have been adopted by the courts to aid in following the surveyor's footsteps. In accordance with these rules the priority of the calls which are found in the original surveyor's field notes is as follows: (1) natural objects; (2) artificial objects; (3) courses; (4) distances; and (5) quantities.

> The purpose of the rules of priority of calls in an original survey is to aid the Court in finding the best evidence of what the original surveyor actually did on the ground. In the event the footsteps of the original surveyor can be more accurately ascertained by following a call of lower priority are inapplicable.
>
> These rules of construction are designed to carry out the intention of the parties. The intention of the parties is considered to be essentially the same as that of the surveyor. The surveyor's intention is to be ascertained by scrutinizing what he actually did in making the survey as reflected by his field notes and the attending totality of circumstances of the survey."

---

[*] 392 A.2d 855, Pa. Super., 1978.
[†] 361 F.Supp. 481, 1973.

In determining location of boundary line, it is not where surveyor intended to run a boundary or should have run it but it is where boundary line was actually run which controls.* In ascertaining the surveyor's intention, the inquiry is not the intention which exists in his mind, but the intention which may be deduced from what he did in making the survey as reflected by his field notes.[†] Moreover, the court is not concerned with some secret intention of the surveyor not expressed in his field notes and in his acts, but it is concerned with his purpose, to be gathered from what he did in making his survey, what he called for in his field notes, and all of the attending circumstances.[‡]

Furthermore, it is a fundamental principle of law that boundaries are to be located on a resurvey where the original surveyor ran the lines and called for them to be located in his field notes.[§]

When There Is Ambiguity

As Griffin[¶] detailed, rules of construction are applied to determine intention when:

Ambiguities appear in the description.

Discrepancies arise between adjoining descriptions.

Discrepancies arise between the description and the physical evidence as it exists on the ground.

Most courts have stated that when there are ambiguities within the description, rules of construction should be relied upon to ascertain the intent of the parties. Griffin expanded upon the view, stating that rules of construction are appropriate when problems arise between adjoining descriptions, and when discrepancies arise when attempting to apply the description to the ground, as emphasized in the New Hampshire case of *Smart v. Huckins.*[**]

Carrying that directive even further, the Pennsylvania court stated that "the object of *all rules* for the establishment of boundaries is to ascertain the actual location of the boundary as made at the time. The important and controlling consideration, where there is a conflict as to a boundary, is the parties' intention, *whether expressed or shown by surrounding circumstances*" (emphasis supplied).

---

* *Stanolind Oil & Gas Co. et al. v. Wheeler et al.*, 247 S.W.2d 187, Tex., 1952.

[†] *Finberg v. Gilbert*, 104 Tex. 539, 141 S.W. 82, 1911.

[‡] *Masterson v. Ribble*, 78 S.W. 358, Tex. Civ. App., 1904.

[§] *Thatcher v. Matthews*, 101 Tex. 122, 105 S.W. 317, 1907; *Bolton v. Lann*, 16 Tex. 96, 1856; *Falby v. Booth*, 16 Tex. 564, 1856.

[¶] *Marquette Law Review*, 43(4), Spring, 1960, Article 5, 484–510.

[**] 82 N.H. 342, 1926. The application of the deed to the land and not the land to the deed.

Title Issues

Referring again to *Smart v. Huckins*, the New Hampshire court stated, "while in a case of ambiguity, it is the duty of the court to place itself as nearly as possible in the situation of the parties at the time the instrument was made, that it may gather their intention from the language used, viewed in the light of the surrounding circumstances."

Referencing the Pennsylvania decision in the *Dallas Borough Annexation Case*, 169 Pa. Super. 126, 82 A.2d 676 (1951) noted above, it states:

> The object of all rules for the establishment of boundaries is to ascertain the actual location of the boundary as made at the time. The important and controlling consideration, where there is a conflict as to a boundary, is the parties' intention, whether express or shown by surrounding circumstances; * * *. 11 C.J.S., Boundaries, § 3, pp. 538, 539.
>
> [t]he original survey in all cases must, whenever possible, be retraced, since it cannot be disregarded or needlessly altered after property rights have been acquired in reliance upon it. *Id.* At 652, 493 N.W.2d at 283 (quoting *Boundaries*, 12 Am. Jur. 2d 462–63 § 57 (1997)) (emphasis omitted).

The Void Instrument

As a general rule, a deed or conveyance will not be declared void for uncertainty in description if it is possible by any reasonable rules of construction to ascertain from the description, aided by extrinsic evidence, what property is intended to be conveyed; and it is sufficient if the description in the deed or conveyance furnishes a means of identification of the land or one by which the property conveyed can be located.[*]

Courts prefer to uphold an instrument whenever possible, regardless of how incomplete, varying in form, or, as Griffin noted, "words bunglingly expressed," rather than render it void. An instrument may be found void if does not properly identify the subject matter of the conveyance. In *Peacher v. Strauss*,[†] the Mississippi court stated "No deed or conveyance of land was ever made, however minute and specific the description, that did not require extrinsic evidence to ascertain its location; and this is so whether the description be by metes and bounds, reference to other deeds, to adjoining owners, watercourses, or other description of whatever character.

> Every contract or deed for the conveyance of land must define its identity and fix its locality, or there must be such a description of the land as, by the aid of parol evidence, will readily point to its locality and boundaries.

---

[*] 12 Am. Jur. 2d. Boundaries, § 104.
[†] 47 Miss. 353, 1872.

Looking through the authorities, it will be seen that, under every conceivable state of facts, and in every imaginable circumstance, the cases may be counted by hundreds, if not by thousands, where contracts, wills and deeds are made effective by the identification, by extrinsic evidence, of the person or subject intended, yet in no wise violating the rule, that such evidence cannot be admitted to contradict, add to, subtract from, or vary the terms of a written instrument. A simpler rule, perhaps, in most cases, is this: that evidence may explain but cannot contradict written language."

The Mississippi decision in the case of *Goff v. Avent*** summarizes a number of decisions from several sources and jurisdictions regarding the uncertainty of a description and its effect on the validity of an instrument.

A deed is not void for uncertainty because there may be errors or an inconsistency in some of the particulars. If a surveyor by applying the rules of surveying can locate the land, the description is sufficient, and generally the rule may be stated to be that the deed will be sustained if it is possible from the whole description to ascertain and identify the land intended to be conveyed. 2 Devlin on Deeds, section 1012, page 320.

Where the tract of land conveyed is described only by the name of the township or the subdivision of the township, such a tract is a subdivision according to the United States survey, the deed is considered as referring to the line of the survey made by the United States and the monuments then erected. 2 Devlin on Deeds, section 1032 page 349.

Generally therefore any description is sufficient by which the identity of the premises can be established. A conveyance is also good, if the description can be made certain within the terms of the instrument for the maxim, id certum, est quod certum, reddi potest applies, extrinsic facts pointed out in the description may also be resorted to to ascertain the land conveyed, and the property may be identified by extrinsic evidence, as in the case of records of the county, where the land is situated. 13 Cyc., page 544.

"If a surveyor can locate the land from the description, it is sufficient." *Campbell v. Carruth*, 32 Fla. 264, 13 So. 432; *Smiley v. Fries*, 104 Ill. 416; *Pennington v. Flock*, 93 Ind. 378; *Oxford v. White*, 95 N.C. 525. "If the description affords sufficient means of ascertaining and identifying the land intended to be conveyed, it is sufficient." *Armijo v. New Mexico Town Co.*, 3 N. M. 244 5 P. 709.

Uncertainty is immaterial if the premises can be identified, by means of the description in connection with other conveyances, plats, lines, or records well known in the neighborhood or of file in public offices. *Pittsburgh etc. R. C. O. v. Beck*, 1521 Ind. 421, 53 N.E. 439.

---

* 122 Miss. 86, 84 So. 134, 1920.

Title Issues

Where description of territory incorporated is not so uncertain as to render determination of territory impossible, the act is not void for uncertainty of description. The recognized rule in this jurisdiction is that where the description of territory incorporated within municipal limits by a special law does not utterly fail to inclose or cover some area, and the description is not so uncertain as to make it impossible to determine the territory intended to be included in the municipality, the law is not void for uncertainty of description. *State v. City of Sarasota*, 92 Fla. 563, 109 So. 473 (Fla. 1926).

In some instances, under some circumstances, courts have declared a description, and therefore the instrument, void for uncertainty. For example, North Carolina courts have been known to take a significantly restrictive approach to the description issue, holding that a grant or reservation of easement is void for uncertainty if it does not adequately describe the location of the easement.*

The Supreme Court of Appeals of West Virginia has also adopted the approach that an easement is void if it is not adequately described in the instrument seeking to create it.†

A few other courts have expressed a comparatively strict standard regarding the designation of an easement's location, requiring that the easement description meet the general conveyancing standard for identifying a parcel of land. Others apply the "surveyor sufficiency" test: certainty such that a surveyor can go upon the land and locate the easement from the description as described in *Rivers*.

---

* See *Allen v. Duvall*, 311 N.C. 245, 316 S.E.2d 267 (1984) and a line of additional cases.
† *Highway Properties v. Dollar Sav. Bank*, 189 W.Va. 301, 431 S.E.2d 95 (1993) quoting *Allen v. Duvall* with approval.

3

Corners, Lines, and Surveys

A boundary line is located by survey, either original or retracement. A boundary line is defined by two boundary corners. The word "boundary" has also been defined as "a line or object indicating the limit or furthest extent of a tract of land or territory; a separating or dividing line between countries, states, districts of territory, or tracts of land."[*]

A **corner** has been defined as the intersection of two converging lines or surfaces; an angle, whether internal or external; as the "corner" of a building, the four "corners" of a square, the "corner" of two streets. A mere variation in a line does not constitute a "corner."[†]

Not all corners are monumented. But they still exist, as a corner, by definition. Notwithstanding the Colorado decision in the case of *Lugon v. Closier*[‡] wherein the court stated that a corner that has never been set is a myth.

Some corners have never been monumented, others have been monumented and the monuments have become obscured, or destroyed, or have otherwise disappeared.

AUTHOR'S NOTE: There is a tremendous amount of confusion between a corner and a monument. When asked about the existence of a corner, a person will often immediately think of monument, and respond accordingly. A monument is evidence of the location of a corner, and, if existing, may or may not mark the corner's precise location. Considerations include whether the marker has been disturbed, whether it was set in the proper location, or whether it was set as a witness marker to the corner.

At the intersecting points of boundary lines are corners, which are designated by certain marks. The same law of the U.S. Congress that established the Public Land Survey System (PLSS) declared that these corners "shall be established as the proper corners of the sections or subdivisions of sections which they were intended to designate." All the land has been sold subject to the provisions of this law, and all titles run directly or indirectly to those boundary lines and corners established by the United States survey.[§]

---

[*] 11 *C.J.S. Boundaries,* § 1 Definitions.
[†] *Christian v. Gernt. et al.,* Tenn. Ch., 64 S.W. 399, 1900.
[‡] 78 Colo. 141, 240 P. 462, Colo., 1925.
[§] Hodgman, Francis. *About Corners.* Paper read before the Michigan Association of Surveyors and Engineers, and published in the *Proceedings of the Second Annual Meeting,* Lansing, January 11–13, 1881.

Surveyors in all jurisdictions should study and be familiar with this entire procedure since it is a formalization of refined practices from earlier jurisdictions. Early instructions and earlier laws were similar, and accomplished the same result. Everything began with a sovereign, surveyors were sent out with instructions, measured lines, set corners, drew maps, and made returns of their work. As Hodgman continued to state, "the same principles which apply to the corners and lines of the United States survey, I understand also to apply to the corners and lines of all original surveys, such as for village plats, etc., after the land has been sold by them. The fundamental points and lines having been thus legally determined when the first stakes were set, the surveyor who is now called on in the practice of his profession has a two-fold duty to perform. *First*, He must take, sift and weigh the evidence which tends to show where the original corners were located, and must determine at what precise point the preponderance of evidence indicates them to be. An argument based on false premises is likely to come to a false conclusion. A survey based on a wrong starting point is certain to come to a false ending if it be otherwise accurately performed. Hence the necessity for a correct beginning."

> When the lines and corners have been once established, the surveyor's function changes. He has to determine the location of these original corners before he can even begin the mechanical part of his labors. This is essentially a judicial proceeding which is often very difficult, and requires a tact and skill, in procuring and interpreting evidence bearing on the point, only acquired by long experience and practical work in the field.
>
> The corners of the United States survey, so far as I have observed, were marked by planting a post or stake in the ground. These stakes had notches cut in them, were squared at the top, and set in certain regular positions in the ground. These marks tended to distinguish them from other stakes that might chance to be driven in the ground for any purpose. When trees stood conveniently near, two of them were marked and their directions and distances from the corner were given in the field notes. When no trees were near, a mound was sometimes raised about the post.

Types of Corners and Their Definitions

Corner, Existent

One whose position is identifiable by evidence of monument, its accessories, or description in field notes, or can be located by acceptable supplemental survey record, some physical evidence, or testimony (*Reid v. Dunn, 20 Cal. Rptr. 273, 1962*).

Even though its physical evidence may have entirely disappeared, a corner will not be regarded as lost if its position can be recovered through the

Corners, Lines, and Surveys

testimony of one or more witnesses who have a dependable knowledge of the original location (*ASCE Definitions*).

Corner, Nonexistent

A corner which has never existed cannot be said to be lost or obliterated and established under the rules relating to the establishment of lost or obliterated corners, but should be established at the place where the original surveyor should have put it (*Lugon v. Crosier, 240 P. 462, 78 Colo. 141*).

Corner Accessory

A physical object adjacent to a corner, to which the corner is referred for future identification or restoration. Accessories include bearing trees, mounds, pits, ledges, rocks, and other natural features to which distances or directions, or both, from the corner or monument are known. Accessories are part of the monument (*ASCE Definitions*). When still in existence, in the absence of the monument, may be the most useful information for the replacement, or location, of the corner. Accessories are often considered as part of a corner monument itself rather than as "aiding in a secondary way."

Corner Contiguity

When parcels of land or mining claims have angle points (corners) in common—though they do not share a common boundary line—they are said to have "corner contiguity." (*Glossaries of BLM Surveying and Mapping Terms*).

Closing Corner

A corner (usually marked by a monument) that indicates where the new line intersects a previously established land boundary. Closing corner monuments are not considered as fixed in position, but may be adjusted by a later surveyor to the line closed on. They are an exception to the rule that "wherever an original monument is set, its position is unalterable" (*Brown's Boundary Control and Legal Principles*).

Quarter-Section Corner

Also known as quarter corner. A corner at an extremity of a boundary of a quarter section, midpoint between or 40 chains from the controlling section corners, depending on location within the township (*ASCE Definitions*).

Section Corner

A corner at the extremity of a section boundary (*ASCE Definitions*).

Sixteenth-Section Corner

Also known as sixteenth corner. A corner at an extremity of a boundary of a quarter-quarter section; midpoint between the controlling corners on the section or township boundaries (*ASCE Definitions*).

Township Corner

A corner at the extremity of a township boundary (*ASCE Definitions*).

Standard Corner

A corner on a standard parallel or base line (*ASCE Definitions*).

Reestablished Corner

One that has been relocated and replaced through the application of appropriate rules.

Lost Corner

A point of a survey whose position cannot be determined, beyond reasonable doubt, either from traces of the original marks or from acceptable evidence or testimony that bears upon the original position, and whose location can be restored only by reference to one or more interdependent corners (*Manual of Instructions Bureau of Land Management 1973*).

Obliterated Corner

One at whose point there are no remaining traces of the monument or its accessories, but whose location has been perpetuated, or the point for which may be recovered beyond reasonable doubt by the acts of testimony of the witnesses, or by some acceptable record evidence (*Manual of Instructions Bureau of Land Management 1973*).

Extinct Corner

> There are two senses in which the word extinct may be used in this connection. One, the sense of physical disappearance; the other, the sense of loss of all reliable evidence (*Chief Justice Thomas M. Cooley, The Judicial Function of Surveyors*).

If a corner monument is obliterated, it must be reestablished where the evidence shows the original surveyor actually located it, regardless of where it should have been located. "And it is immaterial if the lines actually run by the original surveyor are incorrect." *Ayers v. Watson*, 137 U.S. 584, 11 S.Ct. 201,

Corners, Lines, and Surveys

34 L.Ed. 803; *Galt v. Willingham*, 11 F.2d 757. Or as the court stated in *Tyson*, "boundaries originally located and set (right, wrong, good, or bad) are primary and controlling when inconsistent with plats purporting to portray the survey and later notions as to what the original subdivider or surveyor intended to be doing."

There can, in fact, be such a thing as a lost corner, although it is likely quite a rare situation, considering what a corner is by definition, and whether it had ever been marked. Corners exist, regardless of whether they were ever marked. But to be considered lost means there is absolutely no hope of anyone ever determining its location. Chapter 11 is devoted to the lost corner, how to determine if it is, in fact, lost and how to determine a reasonable location for it when it is truly lost.

Meander Corner

A corner marking the intersection of a township or section boundary and the mean high-water line of a body of water. Also a corner on a meander line (*ASCE Definitions*).

Half-Mile Post

In early survey practice, in parts of Alabama and Florida, so-called "half-mile posts" were established. In some cases the "half-mile post" was not at mid-point on the section line. In other cases the "half-mile post" was in the true position for the quarter-section corner. In still other instances the "half-mile post" was not on a true line nor a mid-point on the line. Each set of field notes regarding "half-mile posts" requires individual consideration, as the survey practices were not uniform even within the same surveying district (*Glossaries of BLM Surveying and Mapping Terms*).

Witness Corner (PLSS)

A monumented survey point usually on the line of survey near a corner established as a reference mark when the corner is so situated as to render its monumentation or ready use impracticable.

Witness Tree (PLSS)

According to the General Instructions of 1846, and other instructions prior to that year, "*Witness trees signalized and marked as [bearing trees], but the course and distance to them, as well as the small chop, are omitted.*" Later, all trees used as corner accessories were marked as bearing trees, and the distance and bearing from the corner was recorded (*Glossaries of BLM Surveying and Mapping Terms*).

Monument

A "monument" is a natural or artificial object on the ground which helps to establish a line; "natural monuments" are such things as trees, rivers, stone outcroppings, creeks, and land features; "artificial monuments" are such things as fences, stakes, roads, and things placed by human hand.*

A monument when used in describing land can be defined as any physical object on the ground which helps to establish the location of the line called for. It may be either natural or artificial and may be a tree, stone, stake, pipe or the like (*Parran v. Wilson*, 160 Md. 604, 154 A. 449).

Center of Section

The center of a section is not a physical government monument, but it is a point capable of mathematical ascertainment, "thus constituting it, in a legal sense, a monument call of the description."

The Court's rationale was thus:

In *Fagan v. Walters*, 115 Wash. 454, 197 P. 635, construing a call of a deed which read: "Thence west along the north boundary of land sold and conveyed by William D. Simpson and Annie Simpson, his wife, to Isabelle Gibb on January 4, 1889," we said: "A reading of the description contained in the deed fairly and clearly fixes the north boundary of the tract conveyed as the north boundary of the land sold and conveyed by William D. Simpson, and the Simpson boundary thus becomes a monument which controls the distance, and the distance 90.7 feet must give way to such monument. Authorities are numerous which hold that, where the call for a boundary in another deed or for the boundary of another tract is expressed as 'by such a line' or 'by the north line of such a tract,' the line or boundary referred to is locative and fixes the boundary definitely, and is a monument"—citing a number of authorities. In our early decision in *Shelton Logging Co. v. Gosser*, 26 Wash. 126, 66 P. 151, remarks were made in harmony with this view, wherein a government fractional lot of a section was referred to as "a monument" for purpose of description of adjoining tideland. Manifestly, such a lot would be no less a monument in that description if any physical evidence of its boundary calls were obliterated because they could be mathematically established upon the ground, just as the center of this Section 34 can be and was later correctly established upon the ground (*Matthews v. Parker*, 163 Wash. 10, 299 P. 354, 1931).

A physical object is not a monument unless the deed's description makes a reference to it for that purpose. *Proctor v. Hinkley*, 462 A.2d 465, 469 (Me.1983). In this case, the referee found that "the most important monument" was the pin located near the hemlock tree on top of the bank in the northern line of the Hinkley lot. The appeals court stated that the referee was clearly in error

---

* *Sowerwine v. Neilson*, 671 P.2d 295, Wyo, 1983.

Corners, Lines, and Surveys

since neither the hemlock tree nor the pin was referred to, even indirectly, in the description of the Hinkleys' lot, so that neither could serve as a "monument." A physical object is not a monument for the purpose of locating a boundary unless the description makes reference to it for that purpose.

For a discussion concerning monuments and wood evidence, see Chapter 8.

Monument Accessory

The Texas court in the case of *Stadin v. Helin*[*] had this to say about witness trees: "A witness tree is not an established corner, but merely an object by means of which, in connection with the field notes, if correct, the corner may be found. The course and distance from a witness tree given in the field notes are just as liable to be erroneous as any others. In fact, it is the common experience of surveyors that the course, or, as the witnesses call it, the 'angle' from the witness tree to the corner designated in the field notes is very often erroneous. When an established monument is wanting, so that resort must be had to the field notes in order to ascertain where it was located, and the courses and distances contained in the field notes are inconsistent, and cannot be reconciled, there is no universal rule which requires that certain ones should be preferred to the others. Such a case is very much like one where living witnesses contradict each other. We have to accept as true the testimony of those who, under all the circumstances, are most entitled to credit, and whose testimony is therefore more likely to be in accordance with the actual facts" (Citing *Loring v. Norton*, 8 Me. 61; *Jones v. Burgett*, 46 Tex. 284).

AUTHOR'S NOTE: The rule states that the monument must be "lost." Refer to Chapter 11 for a discussion on what constitutes a "lost" corner.

The decision of *Lugon v. Crosier*[†] is of interest at this point: "The plaintiffs in error say, however, that if the monument is rejected, the rule of the General Land Office as to restoring lost monuments must be followed. The rule invoked is Gen. L. O. Reg. 47:

> A lost or obliterated closing corner from which a standard parallel has been initiated or to which it has been directed will be reestablished in its original place by proportionate measurement from the corner used in the original survey to determine its position.

The corner in question is a closing corner, but not one from which a standard parallel has been initiated nor one to which a standard parallel has been directed; we do not see, therefore, that the rule relates to this case, but if it did we doubt that the corner can be regarded as lost or obliterated. It never existed, and so cannot, strictly speaking, be said to be lost or obliterated. If the monument were lost or obliterated, there would be some reason

---

[*] 76 Minn. 496, 79 N.W. 537, Minn., 1899.
[†] 78 Colo. 141, 240 P. 462, Colo., 1925.

to attempt to relocate it, and perhaps the method prescribed in rule 47 is as good a way as any other, but when it is a myth, never on the ground, the natural, straightforward, and sensible way is to establish the corner at the place where the original surveyor ought to have put it, and that is where the north course of the east line of the section meets the correction line at right angles, and that is where the report puts it. Everybody knows that that is where the section line ought to have closed, and where the original surveyor, honest or dishonest, meant to close it; that his duty required him to close it there, so that the inclosure of his lines might be a rectangle or nearly so. Why should courts be less reasonable than reasonable men?"

The court maintained the same philosophy with a line which had never been run. In *Ralston v. M'Clurg,** the court described the problem as "this is not a question of tracing an actual boundary, or of discovering a lost one, or one which may be presumed to have been completed; but of constructing a survey by adding two lines which were never actually run. And the cardinal object is to ascertain what the surveyor would have done if he had gone on to complete the work. *Beckley vs Bryan &c. 2 Bibb 493.* This is to be ascertained, not by vague conjecture, but by rational deductions from his report, as compared with the existing facts."

Only through thorough understanding of how a particular surveyor did his work is it possible to speculate on what he would have done—where he would set a corner, or where and how he would have run a line. Chapter 8 presents a discussion of understanding the surveyor who went before.

Missing Monument

The Maine case of *Lloyd v. Benson*[†] addressed this situation very well.

> When a monument referenced in a deed is missing, it does not lose its significance as a monument if its original location can be determined. We have previously expressed this principle in both positive and negative terms. For example, in *Theriault v. Murray*, 588 A.2d 720, 722 (Me. 1991), we stated: "The physical disappearance of a monument does not end its use in defining a boundary if its former location can be ascertained."
>
> In contrast, in *Milligan v. Milligan*, 624 A.2d 474, 478 (Me. 1993), we stated, "[t]he physical disappearance of a monument terminates its status as a boundary marker *unless* its former location can be ascertained through extrinsic evidence." (Emphasis added.) We concluded in *Milligan* that "[b]ecause the unrebutted testimony in this case was that no pin... was ever located at [the terminus described in the deed], the pin could no longer be considered a monument." *Id.*
>
> Regardless of whether expressed in the positive or negative, the principle remains the same: The location of a monument that is described

---

* 39 Ky (9 Dana) 338, 1840.
† 2006 ME 129, 910 A.2d 1048, Me., 2006.

Corners, Lines, and Surveys

in a deed, but is missing from the face of the earth, can be established through extrinsic evidence. *See Hennessy*, 2002 Me 76, 796 A.2d at 48.

Once so established, the monument has the same legal significance as if it were not missing. *Theriault*, 588 A.2d at 722 (stating that if the locations of missing monuments can be determined, the "monuments as a matter of law must prevail over the deed's course and distance calls").

The physical disappearance of a monument terminates its status as a boundary marker unless its former location can be ascertained through extrinsic evidence. *Ricci v. Godin*, 523 A.2d 589, 592 (Me.1987) (citing *Bailey v. Look*, 432 A.2d 1271, 1274 (Me.1981)). Because the unrebutted testimony in this case was that no pin (other than that placed by the defendants' surveyor) was ever located at 450 feet along the shore, the pin could no longer be considered a monument.

As was said by Macfarlane, J., in Whitehead v. Atchison, *supra*: "One who purchases a surveyed lot, or tract of land, without notice of the actual boundary, or corners, has the right to rely upon what appears from the original survey, or plat thereof, and is not bound by monuments which do not appear therefrom to have been placed upon the land." But the rule is otherwise, if the plat or survey calls for monuments, for then the plat will be governed by them, if their location can be definitely ascertained, although they may have been removed, destroyed or decayed. McKinney v. Doane, 155 Mo. 287, 56 S.W. 304 (1900). Ref. Whitehead v. Atchinson, 136 Mo. 485.

Types of Lines

The definition of a **line**, in surveying and dividing grounds means prima facie, a mathematical line, without breadth;[*] yet this theoretic idea of a line may be explained, by the facts referred to and connected with the division, to mean a wall, a ditch, a crooked fence, or a hedge, that is, a line having breadth.[†]

A **boundary line** is a line along which two areas meet. The term "boundary line" is usually applied to boundaries between political territories, as "state boundary line," between two states. A boundary line between privately owned parcels of land is termed a property line by preference, or if a line of the United States public land surveys, is given the particular designation of that survey system, as section line, township line, etc.[‡]

A **property line** is generally considered to be a boundary between two parcels of real property, or between two titles, or interests. Like a boundary line, it need not define the division between two parcels in fee simple, since

---

[*] *Baker v. Talbott*, 6 T.B. Mon. 179.
[†] *Ibid.*
[‡] ASCE, *Definitions of Surveying, Mapping, and Related Terms.*

it may be a dividing line between a fee simple ownership and an easement interest, or the like.

Right of Way Line

A line defining the edge or limits of a right of way (easement, or fee), such as commonly along a road or highway, and occasionally along a path or a trail. Commonly used in the case of railroad lines, where the line itself is frequently called the right of way of the railroad.

Fence Line

A line defined by a fence, which may be straight or nearly so, or meandering. Fence lines are sometimes evidence of occupation and boundaries, but more often do not mark a property boundary.

Blazed Line

In wooded country, especially where there are no fences, boundaries are frequently designated by a line of blazes, marks or hacks on trees, sometimes painted. Since trees generally do not grow in a straight line along the location of a boundary, blazed lines tend to meander, and are usually on approximately the location of a boundary.

Lot Line

The line defining the outward limit of a lot, or small division of land.

Conditional Line

A conditional line in eastern Kentucky is a line made by agreement of parties, generally without the aid of a surveyor. *Hoskins Heirs v. Boggs*, 242 S.W.3d 320 (Ky, 2007). See Chapter 9 on boundary agreements for further discussion on this case and on conditional lines in general.

Consentable Line

Fences may be recognized by respective property owners as consentable boundary lines (*DiVirgilio v. Ettore*, 149 A.2d 153, 188 Pa. Super. 526).

A consentable line is the term used by Pennsylvania courts to signify an agreed upon boundary by adjoining owners. Discussions throughout the cases and in the treatises contain the required elements consistent with other jurisdictions. See Chapter 9 on boundary agreements for a detailed discussion of this area.

Corners, Lines, and Surveys

Conventional Line

See conditional line.

Compromise Line

See conditional line.

Extended Line

Extended line is a produced line.*

Ideal Line

By ideal line is meant not necessarily one that was not run, but one that is not actual. It is an open line as contrasted with one not marked. Of course a line that was not run is always an ideal line (*Rowe v. Kidd*, 249 F. 882 1916).

> There is a difference, in this respect, a palpable and essential difference betwixt actual and an ideal line, or a marked and open line. (*Mercer v. Bate*, 4 J.J. Marsh. 334, Ky).

AUTHOR'S NOTE: Refer to Chapter 10 for the proper procedure as to how to correctly work with the situation of an ideal line.

Lost Line

A boundary which has lost its distinctive character as such by removal, displacement, decay, or change, so that it no longer answers the purpose of a bound in defining the true line between the tracts.[†]

Meander Line

The line along a watercourse showing the place of the watercourse and its sinuosities, courses, and distances.[‡]

In *Seabrook v. Coos Bay Ice Co.*,[§] the Oregon court stated that in the interpretation of the deed before them, "the only reference is 'along the meander line,' which can only mean the actual meander line of the bay, because the meander line of the public survey is not a permanent visible or ascertained boundary. In *Railroad Co. v. Schurmeier*, 7 Wall. (U.S.) 272, 19 L.Ed. 74, it is held: 'Meander lines are run in surveying fractional portions of the public lands

---

* *McAndrews, etc. Co. v. Camden Nat. Bank*, 87 N.J.Law 231, 94 A. 627.

† *Perry v. Pratt*, 31 Conn. 433.

‡ *Schurmeier v. St. Paul, etc., R. Co.*, 10 Minn. 82, 88 Am.D. 59.

§ 49 Or. 237, 89 P. 417, Or., 1907.

80 *Boundary Retracement*

bordering upon navigable rivers, not as boundaries of the tract, but for the purpose of defining the sinuosities of the banks of the stream;' and this is the rule generally, both in the state and federal courts."

Partition Line (Division Line)

One or more lines dividing a parcel jointly owned into individual shares. It is usually done among heirs to an estate and through a judicial proceeding. The process of division itself is generally termed a partition.

Property Line

"Property line" is a division between two parcels of land,* or, less frequently, between two title entities.

Right Line

A straight line from one point to another point.

A mathematical line is defined to be length without breadth: it exists only in the mind. A surveyor's line has a local habitation: it consists of a series of marked or established points on the ground, approaching a right line, according to the skill and correctness of the surveyor and perfection of his instruments. An actual township line consists of a series of section corners, never in point of fact falling in a right line from township corner to township corner. When these corners are gone, we resort to the next best evidence to ascertain where they were originally placed (*M'Clintock v. Rogers*, 11 Ill. 279, Ill. 1849).

Where the line of a survey was not in fact marked, or the evidence of where it was actually run has become extinct a right line from corner to corner must govern.

But a line proved, although deviating from a right line, must be the boundary (*Lyon v. Ross*, 4 Ky. 466, Ky.App. 1809).

A line or corner established by a surveyor, in making a survey upon which a grant has issued, cannot be altered because the line is longer or shorter than the distance specified, or because the relative bearings between the abuttals vary from the course named in the plat and certificate of survey. So, if the line run by the surveyor be not a right line, as supposed from his description, but be found by tracing it to be a curved line, yet the actual line must govern; the visible actual boundary, the thing described, and not the ideal boundary and imperfect description, is to be the guide and rule of property. These principles are recognized in Beckley v. Bryan, prin. dec. 107, and Litt. Sel. Cas. 91; Morrison v. Coghill, prin. dec. 382; *Lyon v. Ross, 1 Bibb* 467; *Cowan v. Fauntleroy, 2 Bibb* 261; *Shaw v. Clement, 1 Call.* 438, 3rd point; *Herbert v. Wise,*

---

* *Ujka v. Sturdevant*, 65 N.W.2d 292, N.D., 1954.

Corners, Lines, and Surveys 81

3 *Call.* 239; *Baker v. Glasscocke*, 1 *Hen. & Munf.* 177; *Helm v. Small, Hard.* 369. *Baxter v. Evett's Lessee*, 23 Ky. 329 (Ky.App. 1828).

True Line

True line is a straight line.[*]

To Range With

A line "to range with" another, from the end of which it begins, must follow the path of the other line when extended and a continuation of it.[†]

Presumption as to Straight Lines

Where a line is described as running from one point to another, it is presumed, unless a different line is described in the instrument, or marked on the ground, to be a straight line, so that by ascertaining the points at the angles of a parcel of land the boundary lines can at once be determined.[‡] The rule of surveying, as well as of law, is to reach the point of determination by the line of shortest distance.[§]

Continuity of Line

Ordinarily, a boundary line which is marked for a part of the distance should be continued in the same direction for the full distance.[¶]

Unusual and troublesome calls have been, and may be, encountered. They must be evaluated on a case-by-case basis. Examples include "a line to be later determined" and "a line to be agreed on."

Types of Surveys

Survey

The definition of a survey is not always well understood, though most people, especially surveyors, believe they know of what it consists. However, a

---

[*] *Lillis v. Urrutia*, 9 Cal. App. 557, 99 P. 992.
[†] *Lilly v. Marcum*, 214 Ky. 514; 283 S.W. 1059.
[‡] *Halstead v. Aliff*, 78 W.Va. 480, 89 S.E. 721.
[§] *Bartlett Land, etc. Co. v. Saunders*, N.H. 103 U.S. 316, 26 L.Ed. 546.
[¶] *Banks v. Talley*, 194 A. 362 (Del.Super. 1937).

review of definitions and court decisions involving the point indicates otherwise. Simply stated, a survey is a procedure which locates a parcel of land, or a right in land, on the surface of the earth. It defines its position, and may define its extent. The Iowa case of *Kerr v. Fee** had as its main issue what is meant to do a survey, and stated concisely, "to survey land means to ascertain corners, boundaries, divisions, with distances and directions, and not necessarily to compute areas included within defined boundaries; such computation being merely a matter of mathematics."

The Georgia court stated that "a 'survey' is a process by which a parcel of land is measured and its contents ascertained and is also a statement of or a paper showing the result of the survey with the courses and distances and quantity of the land.†" This court focused on a "parcel of land," whereas the Georgia licensing statute specifically mentions "property rights," as well as "any estate, or interest therein," being more extensive.‡

More importantly, for purposes of the theme of this book, is the Texas case of *Outlaw v. Gulf Oil Corporation*,§ which appropriately stated, "'Survey' is the substance and consists of the actual acts of the surveyor." The court elaborated on the confusion between types of evidence of survey, by stating that "a 'map' is a picture of a survey, 'field notes' constitute a description thereof, and the 'survey' is the substance and consists of the actual acts of the surveyor, and, if existing established monuments are on the ground evidencing such acts, such monuments control because they are the best evidence of what the surveyor actually did in making the survey and are part at least of what the surveyor did."

Predating this case, although not referenced or cited as precedent, is the Maine case of *Bean v. Bachelder*.¶ The case had to do with a description from a plan of the original layout of a town, worded as "Lot No. 5, in the third range in Greenfield, according to Herrick's plan." Herrick had surveyed the south half of the town into lots and ranges, the north half having been previously surveyed into lots and ranges by another surveyor. The jury was instructed, in effect, that the lines run by Herrick upon the surface of the earth, as and for the boundaries of lot 5, would still be the boundaries of that lot, if their

---

* 161 N.W. 1079, 179 Iowa 545, 1917.

† *Overstreet v. Dixon*, 131 S.E.2d 580, 107 Ga.App. 835, 1963.

‡ "Land surveying" means any service, work, or practice, the adequate performance of which requires the application of special knowledge of the principles of mathematics, the related physical and applied sciences, and the requirements of relevant law in the evaluation and location of property rights, as applied to

(A) Measuring and locating lines, angles, elevations, natural and manmade features in the air, on the surface of the earth, in underground works, and on the beds of bodies of water, for the purpose of determining and reporting positions, topography, areas, and volumes.

(B) Establishing or reestablishing, locating or relocating, or setting or resetting of monumentation for any property, easement, or right of way boundaries, or the boundary of any estate or interest therein; as well as other functions. § 43-15-2. **Definitions, paragraph 6.**

§ 137 S.W.2d 787, (Tex.Civ.App., 1940.)

¶ 78 Me. 184, 3 A. 279, Me., 1886.

Corners, Lines, and Surveys

locality could be found. The plan was merely a picture. The survey was the substance. The plan was not made to show where the lots were to be hereafter located, or how they were to be hereafter bounded. It was made as evidence of where they had before been located and bounded. The lot actually surveyed, bounded by the lines actually run, was the lot intended to be conveyed. The plan was named in the deed, rather as a picture, indicating the location and lines of the lot. Still the actual boundaries, rather than the pictured boundaries, were to be sought for. The picture might not be wholly accurate.

Justice Joseph Lumpkin, II wrote in the case of *Thompson v. Hill*,[*] "A plat is not a mark on the land, but a representation of the land on paper, appealing to the eye by means of lines and memoranda rather than by words alone."

More recently the Florida case of *Tyson v. Edwards*[†] adds perspective with additional discussion.

In this case the appellants claimed to a lot boundary line that is correct according to the apparent intention of the original surveyor and the appellees' claim to a lot boundary line that is consistent with lot lines established by the original surveyor on the ground and occupied by owners. Thus, a classic boundary line question is presented. Technically the question is: Where an original surveyor subdivides and lays out boundaries of parcels in a tract of land which has theretofore existed as a single unit and runs lines and places monuments establishing the location of the subdivided parcels or plots or lots on the ground and the surveyor draws a plat of survey or written map of his work which is recorded and subsequently one or more parcels are conveyed by deed describing the parcels according to such plat of survey and some parcels are sold according to the plat but purchasers take actual possession according to the survey as monumented on the ground and there is a discrepancy and conflict between the location of parcels as located by the original survey on the ground and as they are shown to be located according to the recorded plat, is the correct legal location of a particular parcel as it was actually originally located and possession taken on the ground or is it as can now be located by following only the intent revealed by the recorded plat?

More simply put the question is: In the event of a discrepancy as to subdivided land lot lines, do you go with what the original surveyor intended to do as shown by the plat or do you go with what the original surveyor did by way of laying out and monumenting his survey on the ground?

Surprisingly, because of surveying principles based on established surveying practices, the correct answer is that what the original surveyor actually did by way of monumenting his survey on the ground takes precedence over what he intended to do as shown by his written plat of survey.

---

[*] 137 Ga. 308, 73 S.E. 640, 1912.
[†] 433 So.2d 549, Fla.App. 5 Dist. 1983.

84   *Boundary Retracement*

The difficulty with the problem is that the role and practice of the surveyor and his function in solving a surveying problem of the type in this case is misunderstood. Lawyers, architects, and design engineers are accustomed to achieving objectives by first conceiving of abstract ideas or plans, then reducing those ideas (intentions) to paper, and then using the written document from which to construct a physical object or otherwise tangibly achieve the original goal as written. When this is done, the written document is always considered authoritative and any deviation or discrepancy between it and what is actually done pursuant to it is resolved by considering the deviations and discrepancies as being defects or errors in the execution of the original plan to be corrected by changing the physical to conform to the intention evidenced by the writing. In only one situation does the surveyor play a similar role and that is when he, in the first instance, lays out boundaries in the original division of a tract which has theretofore existed as a single unit. Thereafter the surveyor's function radically changes. It is not the surveyor's right or responsibility to set up new points and lines establishing boundaries except when he is surveying theretofore unplatted land or subdividing a new tract. Where the title to land has been established under a previous survey, the sole duty of all subsequent or following surveyors is to locate the points and lines of the original survey. Later surveyors must only track and "trace the footsteps" of the original surveyor in locating existing boundaries. They cannot establish a new corner or line nor can they correct erroneous surveys of earlier surveyors, even when the earlier surveyor obviously erred in following some apparent original "over-all design" or objective. The reason for this lies in the historic development of the concept of land boundaries and of the profession of surveying. Man set monuments as landmarks before he invented paper and even today the true survey is what the original surveyor did on the ground by way of fixing boundaries by setting monuments and running lines ("metes and bounds"), and the paper "survey" or plat of survey is intended only as a map of what is on the ground. The surveying method is to establish boundaries by running lines and fixing monuments on the ground while making field notes of such acts. From the field notes, plats of survey or "maps" are later drawn to depict that which was done on the ground. In establishing the original boundary on the ground the original surveyor is conclusively presumed to have been correct and if later surveyors find there is error in the locations, measurements or otherwise, such error is the error of the last surveyor. Likewise, boundaries originally located and set (right, wrong, good, or bad) are primary and controlling when inconsistent with plats purporting to portray the survey and later notions as to what the original subdivider or surveyor intended to be doing or as to where later surveyors, working, perhaps, under better conditions and more accurately with better equipment, would locate the boundary solely by using the plat as a guide or plan. Written plats are not construction plans to be followed to correctly reestablish monuments and boundaries. They are "as built" drawings of what has already occurred on the ground and are properly used only

Corners, Lines, and Surveys

to the extent they are helpful in finding and retracing the original survey which they are intended to describe; and to the extent that the original surveyor's lines and monuments on the ground are established by other evidence and are inconsistent with the lines on the plat of survey, the plat is to be disregarded. When evidence establishes a discrepancy between the location on the ground of the original boundary survey and the written plat of that survey, the discrepancy is always resolved against the plat.

Other Categories of Survey

Accurate Survey

The concept of an accurate survey can mean different things to different people, and how it might be used. However, the West Virginia court, out of necessity, came up with a definition. In the case of *CSX HOTELS, INC., dba the Greenbrier Resort, a West Virginia corporation, v. CITY OF WHITE SULPHUR SPRINGS, West Virginia*, et al., Frederick W. Kretzer, et al., v. City of White Sulphur Springs, West Virginia, et al.,* differences of opinion were the result of what constitutes an accurate survey. "The circuit court found that a map of the territory (adjacent, unincorporated land) to be annexed that was attached to the petition was not an 'accurate survey map' as required by the West Virginia Code because no physical, on-the-ground survey had been conducted of the territory. Because the map was prepared using previously-conducted surveys, property descriptions contained in publicly-filed deeds, and other documents, the circuit court determined that the annexation petition was 'fatally flawed.'"

> Using the procedure set forth in *W.Va.Code*, a petition was filed with the City…setting forth the change proposed in the metes and bounds of the municipality and asking that a vote be taken upon the proposed change. According to the Code, the petition shall be verified and shall be accompanied by an accurate survey map showing the territory to be annexed to the corporate limits by the proposed change. The metes and bounds description in the petition of the territory to be annexed, and the map of the territory accompanying the petition, were prepared by a registered professional engineer using the calls and distances culled from preexisting surveys, public records and from the deeds describing the boundaries of the various properties in the area to be annexed. These deeds were recorded by the property owners with the Greenbrier County Clerk. No on-the-ground examination or measurement of the proposed new City boundary lines was made.

---

* 217 W.Va. 238, 617 S.E.2d 785, W.Va. 2005.

One of the grounds upon which the appellees sought an injunction was that the City had not performed an actual, on-the-ground survey of their property and the other properties encompassed by the annexation petition, and had therefore not prepared an 'accurate survey map' of the territory to be annexed into the City. The City, however, took the position that as long as the map of the territory to be annexed was reasonably accurate, and residents and landowners could determine whether they or their properties were affected by the annexation, then the map was an "accurate survey map" that complied with the Code.

Early in the litigation, comments by the circuit court indicated the court's belief that the Code required a detailed, physical, on-the-ground survey. Therefore, before engaging in protracted, expensive litigation over the factual accuracy of the description and map of the territory to be annexed, the parties agreed to submit to the circuit court the narrow legal question of whether an on-the-ground survey was required or not.

On February 9, 2004, the circuit court entered a "Final Order Granting Permanent Injunction" against the City. Applying several dictionary meanings, the circuit court found, as a matter of law, that the City's annexation petition was not accompanied by an 'accurate survey map' as required by the Code (a) because the map was not "accurate" and "free from error or defect...careful or meticulous;" and was not a "survey" because it did not involve 'the process by which a parcel of land is measured and its boundaries and contents ascertained.' The circuit court found the City's January 15, 2003 petition for annexation to be 'fatally flawed,' and permanently enjoined the City from taking any action pursuant to the petition.

The City then appealed the circuit court's order.

The appellant City challenged the circuit court's interpretation of the phrase 'accurate survey map' in the Code. The appellant argued that the statute requires the production of a reasonably accurate map so as to permit residents and landowners, of the municipality and of the territory to be annexed, to quickly determine whether they are or are not affected by an annexation petition. The appellant also argues that the annexation survey map need only be accurate enough to determine which properties are within the City's boundaries for purposes of voting, taxation, and the provision of services. The appellant takes the position that the statute does not require the production of a precise, on-the-ground survey sufficient to resolve boundary disputes between property owners. The appellant contends that a map prepared using previously prepared surveys or using property descriptions contained in other documents, such as the publicly-filed deeds to the land being annexed, are sufficient.

The appellees counter that the Legislature chose the phrase 'accurate survey map' to indicate the map had to be (1) precise and (2) prepared from measurements done in an actual survey performed by a person qualified to perform surveys. The appellees argue that the Code (a) must be read in conjunction with the statutes pertaining to the licensing and

Corners, Lines, and Surveys

practice of land surveyors, which direct that all "survey" maps must be prepared by a licensed land surveyor in a manner conforming to specific technical standards.

We can find no definition of 'accurate survey map' in the Code regarding surveying, or any other portion of the *Code* pertaining to the annexation process. '[W]here there is some ambiguity in the statute or some uncertainty as to the meaning intended...resort may be had to rules of construction of statutes.' *Crockett v. Andrews*, 153 W.Va. 714, 718, 172 S.E.2d 384, 386-87 (1970) (*quoting* 17 M.J., Statutes, § 31). "In the absence of any definition of the intended meaning of words or terms used in a legislative enactment, they will, in the interpretation of the act, be given their common, ordinary and accepted meaning in the connection in which they are used." Syllabus Point 1, *Miners in General Group v. Hix*, 123 W.Va. 637, 17 S.E.2d 810 (1941). *In accord*, Syllabus Point 6, *State ex rel. Cohen v. Manchin*, 175 W.Va. 525, 336 S.E.2d 171 (1984). ("Undefined words and terms used in a legislative enactment will be given their common, ordinary and accepted meaning.")

Black's Law Dictionary (5th Ed.1979) defines a "survey" as a 'process by which a parcel of land is measured and its boundaries and contents ascertained; also a map, plat, or statement of the result of such survey, with the courses and distances and the quantity of the land.' Nowhere in the definition of 'survey' do we find that a survey is exclusively a process by which land is physically measured 'on the ground' and boundaries established by direct observation. Instead, it is merely *any* process by which an area of land is measured, and its boundary determined—even if done by indirect means.

The appellees contend that, in order for the measurement of land boundaries to be accurate, the measurement must be done by some physical examination of the land. The appellees contend—again referring to the various statutory requirements regulating the profession of land surveying—that a proper, accurate survey is one that establishes boundary monuments, and clearly measures distances and angles using conventional survey equipment or using global positioning system equipment.

The City, however, asserts that the direct-observation survey methods proposed by the appellees would be cost prohibitive, making the annexation process nearly impossible for most municipalities. Furthermore, because of the size of the tracts of land owned by the appellees—apparently encompassing several thousand acres—a physical, on-the-ground survey would be highly impractical. That impracticability becomes even more apparent, according to the City, because a physical survey of the boundaries of the appellees' tracts would require a surveyor to trespass onto the appellees' lands. Finally, the City contends that while such a survey would result in a detailed report sufficient to resolve boundary disputes between property owners, such elegant detail is not necessary for purposes of putting the citizenry on notice regarding who will be impacted by the annexation.

The appellants argue that the measurement and determination of land boundaries can be accurately accomplished for purposes of annexation by compiling the distances and courses contained in preexisting surveys, in the boundary descriptions within the deeds of the properties that are to be annexed into a municipality, or in other records. We agree.

While we do not believe that the statutes pertaining to the licensing and practice of land surveyors are controlling on the issue of annexation, the language contained within those statutes guides us in accepting the correctness of the appellant's argument.

For example, *W.Va.Code*, 30-13A-3(hh) [2004] defines the practice of land surveying as including the act of determining the configuration or contour of the Earth's surface, or the position of fixed objects thereon, 'using such sciences as...photogrammetry, and...making geometric measurements and gathering related information pertaining to the physical or legal features of the earth, improvements on the earth, the space above, on or below the earth[.]' In other words, it appears that even licensed land surveyors may determine a point, line, object or area on the surface of the Earth by using geometric measurements taken from sources other than direct, physical, on-the-ground observations. The statute recognizes those sources can include aerial photography, or sources identifying "improvements" and 'legal features of the earth.'

Furthermore, the circuit court concluded that an 'accurate' survey is one which is 'free from error or defect...precise; careful or meticulous.' However, *W.Va.Code*, 30-13A-26(g) [2004]—which sets forth 'minimum technical criteria' governing surveys of property boundaries—states that while distance must be 'reported in feet or meters, or parts thereof,' and angles or directions 'reported in degrees or parts thereof,' these observations must only be 'measured to a precision that will produce the desired level of accuracy.' Further, the quantity or area of land within the boundary must only be 'measured and reported to a precision consistent with the purpose of the survey.' In sum, even a boundary survey of a single tract of land by a licensed land surveyor need not be precise and free from any error or defect; such a survey must only reach the desired level of accuracy consistent with the purposes of the survey.

We therefore hold that in an annexation proceeding initiated by a municipality under *W.Va.Code*, 8-6-2(a), an 'accurate survey map' is a map reflecting the course and distance measurements, boundaries, and contents of the territory that is proposed to be incorporated into the municipality's limits. The map must reach the desired level of precision consistent with the purposes of the survey, namely to provide notice to residents and freeholders of the municipality and the territory encompassed by the annexation petition of a potential change regarding who will vote in municipal elections; taxation and revenues; and the provision of services.

We believe that the circuit court erred by concluding that the survey map attached to the City's annexation petition, as a matter of law, was required to be absolutely free from error or defect and could not be assembled from preexisting surveys or boundary descriptions contained in deeds filed with the county clerk. A municipality may prepare an annexation survey map by whatever means it chooses, so long as the map is of reasonable accuracy such that residents and landowners in

Corners, Lines, and Surveys

the City and in the territory to be annexed can easily understand which parcels of land are being incorporated into the City's boundaries.

This case is an excellent example of the difference of opinion concerning what a survey actually is. The court sought appropriate definitions and reasoning to determine the answer, and is guidance not only to the definition of a survey, but also the procedure for determining the answer to a legitimate question.

Original Survey versus First Survey

An original survey is just what it says, the survey originally done when a subject tract was separated from its source parcel, or the survey that formed the basis for the first subdivision. In contrast, the first survey is the very first survey done after the separation has been made. It is possible, obviously, to convey a tract or parcel of land without having a survey made. Sometime later, perhaps days or even years after the parcel is created, it is actually "surveyed." This is the first survey, meaning it represents the first time the parcel is actually located on the ground and defined with measurements, or surveyed. However it cannot be an original survey since it was not done at the time of severance of the new tract, or at the origin of the title to the parcel. At that time, the severance and its resulting description somewhat and somehow defined the parcel and its location. Those tracts acquired through any one of several unwritten means provide examples. For instance, a tract added to an existing parcel through accretion is a separate parcel, unsurveyed, although it may attach to a surveyed parcel. Unless, and until, someone actually measures and locates it, it does not become a surveyed tract. There is no original survey, the parcel exists and has a title history, and if it becomes surveyed, that survey is the first survey.

Because the first survey occurs sometime after a parcel is created, it is the product of that first surveyor, and therefore *his opinion* as to the location and extent of the parcel created at some time in the past. A first survey is not given the dignity of an original survey, since it is not "original" and is merely an opinion based on what the surveyor knew, or could find in terms of evidence, at the time of his survey.

The South Dakota case of *Titus v. Chapman** distinguished between an original survey and a first survey. The parties were adjoining owners in the Northeast Quarter of the Southeast Quarter, Section 12, Township One North, Range Six East of the Black Hills Meridian, the 1/16th line being the boundary separating the two ownerships. The U.S. Forest Service originally surveyed the area in question in April of 1879 and August of 1886. This survey resulted in a plat of Section 12, Township One North, Range Six East of the Black Hills Meridian. The general practice at the time was to locate

---

* 687 N.W.2d 918, S.D., 2004.

90   *Boundary Retracement*

artificial or natural monuments to indicate section corners and quarter section corners.*

Section 12 was next surveyed in 1946 for the purpose of subdividing the Northwest Quarter of the Southeast Quarter of Section 12, together with other property. This survey platted Lot A of the Northwest Quarter of Southeast Quarter of Section 12 with an eastern boundary collinear with the 1/16th line of Section 12. The eastern boundary of Lot A was not monumented. However, the survey indicated an artificial monument, an iron pin, on the southern portion of the 1/16th line of Section 12 at the southern quarter corner between the Southwest Quarter of the Southeast Quarter and the Southeast Quarter of the Southeast Quarter. The record is unclear as to who set the iron pin, or if it was merely located when the survey was done.

A third survey of the area was conducted in October of 1970 for the purpose of further subdividing the Northeast Quarter of the Southeast Quarter of Section 12 into Tracts A, B, and C. The Chapman lot was not platted at the time, but was eventually to be further subdivided in 1983 from Tract B. The western boundary of Tract B was intended to terminate at the 1/16th section line according to this survey.

However, this third surveyor did not follow the original U.S. Forest Service Black Hills monuments to determine the original location of the 1/16th line. Instead, he located an iron pin or pipe without a surveyor's cap at the southwest corner of Tract C. He assumed the iron pipe was the same pin indicated on the second survey as the 1/16th line. There were no facts to indicate the origins of the iron pin, or to identify it as denoting the original 1/16th section as located by the U.S. Forest Service survey. Despite the absence of clear and convincing evidence as to the identity of the pin, this third surveyor accepted it as the 1/16th line and conducted measurements to locate the western edge of Tract B, which would eventually become the western edge of Lot 2. He indicated he marked the other three corners of Tract B with 5/8th inch rebar. The end result was the addition of approximately 34 feet to Tract B on its western boundary.

In 1983, a plat subdividing Tract B into lots, including Lot 2 which was eventually purchased by Chapman, was filed by Surveyor Number 4. He relied on the third survey and set the western edge of Lot 2 at the 1/16th section line. However, the 1/16th section line was platted at the same location as the third survey rather than the original location as designated by the

---

* In surveying and subdividing townships, government surveyors were required to establish section corners by "a mound, four pits and a stake; and a quarter section corner by a mound, two pits and a stake or post; the stakes or posts to be properly marked, to indicate the corner represented." *Randall v. Burk Tp. of Minnehaha County*, 4 S.D. 337, 57 N.W. 4, 9, 1893.

"Visible marks or indications left on natural or other objects indicating the lines and boundaries of a survey, are monuments. They include posts, pillers, [sic] stone markers, cairns, fixed natural objects[,] blazed trees and watercourses. Any natural or artificial physical object on the ground which helps establish the location of a line is a monument." *Block v. Howell*, 346 N.W.2d 441, S.D., 1984 (citing *Black's Law Dictionary* 909, 1979).

Corners, Lines, and Surveys

U.S. Forest Service survey. Surveyor 4 then marked the western boundary corners of Lot 2 with a 5/8th inch rebar.

Following their purchase of Lot A in 2001, the Tituses commissioned a survey of their property prior to beginning construction of a new home. The survey, conducted by Surveyor 5, retraced the original footsteps of the U.S. Forest Service survey and called upon the three remaining original corner stones placed around the perimeter of Section 12 by the U.S. Forest Service in 1879 and 1886, which monumented the southeast, southwest, and northwest corners of Section 12. Instead of using the iron pin of unknown origin as the location of the 1/16th section line as had Surveyors 3 and 4 before him, Surveyor 5 attempted to locate the missing northeast corner stone in order to properly divide Section 12 in its entirety by aliquot portions to arrive at the original location of the 1/16th section line as required by the Manual of Instructions for the Survey of Public Lands of the United States.

When Surveyor 5 was unable to ascertain the exact location of the northeast corner due to the obliteration of the monument, he extended his survey out one additional mile north from the vicinity of the missing northeast corner stone to a replacement monument located at the northeast corner of adjoining Section 1 placed by the U.S. Forest Service surveyors. Using methods prescribed by the Manual of Instructions for the Survey of the Public Lands of the United States, the location of the northeast corner of Section 12 was computed. Once the complete exterior boundary of Section 12 was determined, Surveyor 5 properly computed the aliquot divisions to arrive at the original location of the 16th section line as intended by the U.S. Forest Survey. The Tituses filed suit, moving for summary judgment, and the trial court established the 1/16th line as originally set by the U.S. Forest Service as the property line common to the Tituses and Chapman.

The court, in its decision, stated that government surveys, not surveys conducted by private individuals, create, rather than merely identify, boundaries.[*] The term "original survey" refers to the official government survey performed under the laws of the federal government by its official agency.[†] A subsequent survey by a private individual or non-government entity is more accurately described as a retracing or resurvey.[‡] In a retracing or resurvey, a surveyor must "take care to observe and follow the boundaries and monuments as run and marked by the original survey."[§] Boundaries as established by original government surveys are unchangeable and must control disputes.[¶]

Original monuments, those located by the original surveyor, mark true corners.[**] "Where the location of the original monument can be found, or

---

[*] *Cox v. Hart*, 260 U.S. 427, 436, 43 S.Ct. 154, 157, 67 L.Ed. 332, 337, 1922.

[†] *See Id.; Block v. Howell*, 346 N.W.2d 441, 444-45, S.D., 1984; Walter G. Robillard & Lane J. Bouman, Clark on Surveying and Boundaries § 4.12, 5th ed., 1976.

[‡] *Block*, 346 N.W.2d at 444; *Randall v. Burk Tp.*, 4 S.D. 337, 347, 57 N.W. 4, 10, 1893.

[§] *Id.*

[¶] *Christianson v. Daneville Tp.*, 61 S.D. 55, 58, 246 N.W. 101, 102, 1932.

[**] *Lawson v. Viola Tp.*, 50 S.D. 555, 557-558, 210 N.W. 979, 980, 1926.

can be established by evidence, such location shall be held to be the true corner, regardless of the fact that resurveys may show that it should have been located elsewhere."[*] Where the original monument is obliterated, that is it cannot be located nor established by evidence, then a corner can be established by a new survey.[†] Only upon obliteration of an original corner may a new survey be made from points that can be determined in accordance with the original surveyor's field notes.[‡] However, if the point at which an original monument was located can be ascertained by the court, the line as indicated by the government survey prevails.[§]

SDCL 43-18-7 provides:

In retracing lines or making the survey, the surveyor shall take care to observe and follow the boundaries and monuments as run and marked by the original survey, but shall not give undue weight to partial and doubtful evidence or appearances of monuments, the recognition of which shall require the presumption of marked errors in the original survey, and he shall note an exact description of such apparent monuments.

The 1970 survey No. 2 located an iron pipe or pin without a survey cap. The surveyor assumed the pin indicated the 1/16th section line as intended by the U.S. Forest Service. That survey failed to "walk in the footsteps" of the original surveyor in order to obtain clear and convincing evidence of the identity of the iron pipe as the 1/16th line based on its location in relation to the true section corners established by the original government survey. Instead, Surveyor 2 conducted some retracing of the original corners of Section 12 in the field, but the majority of the measurements were calculated on paper and based on the location of the iron pin or pipe which he assumed denoted the location of the 1/16th section line. The subsequent surveys relied upon by Chapman and her predecessors in interest all relied on the same erroneous assumption utilized by Surveyor 2 in 1970.

The Tituses' survey conducted in 2002 did "walk in the footsteps" of the original government surveyor. The Tituses' surveyor, Surveyor 5, located three original corner monuments, and retraced the original U.S. Forest Service survey in order to ascertain the location of the final missing corner monument and accurately determine the exterior perimeter of Section 12. He then properly computed the aliquot divisions, including the 1/16th section line, in compliance with the Manual of Instructions for the Survey of the Public Lands of the United States.

---

[*] *Id.* (citing *Byrne v. McKeachie*, 34 S.D. 589, 149 N.W. 552, 1914; *Hoekman v. Iowa Civil Township*, 28 S.D. 206, 132 N.W. 1004, 1911; *Randall*, 4 S.D. 337, 57 N.W. 4; *Beardsley v. Crane*, 52 Minn. 537, 54 N.W. 740, 1893; *Ogilvie v. Copeland*, 145 Ill. 98, 33 N.E. 1085, 1893; *Nesselroad v. Parrish*, 59 Iowa 570, 13 N.W. 746, 1882).

[†] *Lawson*, 50 S.D. at 558, 210 N.W. at 980 (citing *Randall*, 4 S.D. at 355, 57 N.W. at 10; *Washington Rock Co. v. Young*, 29 Utah 108, 80 P. 382, 1905).

[‡] *Id.*

[§] *Dowdle v. Cornue*, 9 S.D. 126, 127, 68 N.W. 194, 194–195, 1896.

Corners, Lines, and Surveys

Chapman contends that the Tituses were required to use the Number 2 survey in order to determine the boundary line. Their argument relies on the term "original survey" in SDCL 43-18-7 as indicating the first survey to plat the tracts from which Chapman's Lot 2 was subdivided. Reliance on the 1970 Surveyor 2 retracing is fatal to Chapman's claim. The "original survey" in SDCL 43-18-7 refers to the original U.S. Forest Service survey conducted in the 1800s, not the first private survey to plat the subdivision of the tracts in question.

The record indicates that the Tituses' survey complied with the requirements of SDCL 43-18-7 in that it retraced the footsteps of the original government surveyor. Conversely, the record indicates that Chapman's surveys did not follow the requirements for retracing an original survey.

This case provides an excellent example of the necessity of reviewing any and all surveying work that bears a relationship to the situation at hand. Only two of the five surveys were useful, or acceptable: The original survey by the U.S. Forest Service, and the retracement of that survey by Surveyor No. 5.

Retracement versus Resurvey

People in general, surveyors included, often confuse retracements and resurveys. Even many court decisions use the terms interchangeably, designating resurvey, when they are actually discussing a retracement survey. It is therefore exceedingly important to understand what the two terms are and not to confuse them, as they mean entirely different things.

The Florida decision of *Rivers v. Lozeau** is helpful to an extent. It discusses what an original survey is, as opposed to a retracement survey. Even though it makes a point of distinguishing between a retracement survey and a resurvey, it does not define or describe a resurvey.

> In working for a client, a surveyor basically performs two distinctly different roles of functions:
>
> **First**, the surveyor can lay out or establish boundary lines within an original division of a tract of land which has existed as one unit or parcel. In performing this function, he is known as the "original surveyor" and *when his survey results in a property description used by the owner to transfer title to property*[†] that survey has a certain special authority in that the monuments set by the original surveyor on the ground control over discrepancies within the total parcel description and more importantly, control over all subsequent surveys attempting to locate the same line.[‡]

---

* Fla.App. 5 Dist., 539 So.2d 1147, 1989.
† This is a most important qualification.
‡ Clothed with this authority, it is incumbent on the original surveyor to perform the best work possible under the circumstances, and to leave appropriate "footsteps" for succeeding following surveyors.

94 *Boundary Retracement*

> **Second**, a surveyor can be retained to locate a boundary line on the ground which has been established. When he does this, he "traces the footsteps" of the "original surveyor" in locating existing boundaries. Correctly stated, this is a "retracement" survey, not a resurvey, and in performing this function, the second and each succeeding surveyor is a "following" or "tracing" surveyor. The "following" surveyor's sole duty, function and power is to locate the boundary corners and boundary lines established by the original survey. He cannot establish a new corner or new line terminal point, nor may he correct errors of the original surveyor. The following surveyor, rather than being the creator of the boundary line, is only its discoverer and is **only that when he correctly locates it.**[*]

The court went on to say, "The intent of the parties to a contract for the sale and purchase of land may be relevant to a dispute concerning the contract, but in a real sense, the grantee in a deed is not a party to the deed. The grantee does not sign it. His intent as to the quality of the legal title he receives, and as to the location and extent of the land legally conveyed by the deed, is quite immaterial as to those matters. The owner of a parcel of land, being the grantee under a patent or deed, or devisee under a will or the heir of a prior owner, has no authority or power to establish the boundaries of the land he owns; he has **only** the power to establish the division or boundary line between parcels when he owns the land on both sides of the boundary line he is establishing. In short, an original surveyor can establish an original boundary line only for any owner who owns the land on both sides of the line being established. The line becomes an authentic original line *only when the owner makes a conveyance based on a description of the surveyed line* and has good legal title to the land described in his conveyance."

Establishing a, or the, Line

Another incorrect statement that finds its way into survey discussions and court decisions, surveyors' testimony, clients' direction, lawyers' request, and courts' reporting is that "the line was established by the surveyor." Only when an original survey is performed is a line established, as in the case of a new subdivision, and then only when there has been a transfer of title, but never with a retracement. In a retracement, the line being retraced had already been established at some time in the past, at the time it was created followed by a transfer of title to that line.[†] It is the "sole duty, function and power," as the court said in *Rivers*,[‡] of the following or tracing surveyor to locate the line at the same place where it was originally established. This

---

[*] *See Clark on Surveying and Boundaries*, Chap. 14 Tracking a Survey, pg 339 and generally, Grimes, 4th Ed., 1976.

[†] This is the most important qualification *Rivers*, Ibid.

[‡] Ibid.

Corners, Lines, and Surveys

court also noted that "the following surveyor, rather than being the creator of the boundary line, is only its discoverer and is only that when he correctly locates it."* Discovery comes as the result of some form of investigation.

It is important to realize, and not forget, that the court emphasized "sole duty, function and power." There is no doubt, as we examine the numerous decisions that follow, that this is as accurately and succinctly as it could be stated.

A related decision and one worth knowing is the early decision in *Pereles v. Gross*,† which held, "in resurveying a tract of land according to a former plat or survey, the surveyor's only function or right is to relocate, upon the best evidence obtainable, the corners and lines at the same places where originally located by the first surveyor on the ground. Any departure from such purpose and effort is unprofessional, and, so far as any effect is claimed for it, unlawful."

A later case from the same jurisdiction was that of *Johnson v. Westrick*,‡ which states "to fix lines variant from the originals and according merely to his notion of a desirable arrangement of lots and streets leads naturally to confusion of claims among lot owners, and, when done by a city surveyor as a basis for occupation of land for streets, is attempted confiscation."

> It was held that the east line of the street was where the original surveyor placed it, not where it should be according to resurveys or subsequent surveys; that subsequent surveys are worse than useless; they only serve to confuse, unless they agree with the original survey.

The words chosen for these two decisions by the Wisconsin court are strong and important words, as they define a surveyor's duty, and the consequences for failing to fulfill that duty.

Interestingly, both of these cases had to do with street locations, the former a right of way line, the latter an intersection of two streets. As the Ohio court stated in *Manufacturer's National Bank of Detroit v. Erie County Road Commission*,§ where an argument in the case was by a landowner claiming that requiring a land user to determine exactly where rights-of-way lines are located places too great a burden on the land user. The court replied "a landowner or occupier is under an obligation to know the boundaries of the property. The border of the right-of-way is a boundary line like any other."

All boundaries, whether real property, easements, locations of improvements, or property rights, even districts such as for schools or for zoning, should be treated the same, using the same procedures under the same requirements of law and survey standards.

---

* Ibid.
† 126 Wis. 217, 1905.
‡ 200 Wis. 405, 1930.
§ 63 Ohio St.3d 318; 587 N.E.2d 819, 1992.

Independent Survey

This survey does not appear to have been made from any known government corner and is called by one of the witnesses "an independent survey"—that is to say a survey not dependent upon any known government corner. *Bentley v. Jenne*, 33 Wyo. 1, 236 P. 509 (Wyo. 1925).

Tracking a Survey

Clark on Surveying and Boundaries* titles Chapter 13 just that: Tracking a Survey. The first paragraph provides a nice summary: "The cardinal principle guiding a surveyor who is running the lines of a previous survey is to follow in the footsteps of the previous surveyor. Where a survey is once made and parties have acted on the strength of the surveyor's lines, property rights have arisen which cannot be taken away without the consent of the owners, regardless of the errors committed by the original surveyor. It is the extensive duty of the retracing surveyor to see what the first surveyor did, not what he should have done. No matter how inaccurate the original survey may have been, it will be conclusively presumed to be correct, and that if there be error in the measurements or otherwise, such error is the error of the last surveyor. Hence, the surveyor will, at all times, keep in his mind this presumption and conform his acts thereto."

These words are very much like Robert Griffin's: A boundary line, once fixed, should remain in its original position through any series of mesne conveyances.

Chapter 10 summarizes in detail the proper procedure(s) in conducting a retracement survey.

In surveying a tract of land according to a former plat or survey, the surveyor's only duty is to relocate, upon the best evidence obtainable, the courses and lines at the same place where originally located by the first surveyor on the ground. In making the resurvey, he has the right to use the field notes of the original survey. The object of a resurvey is to furnish proof of the location of the lost lines or monuments, not to dispute the correctness of or to control the original survey. The original survey in all cases must, whenever possible, be retraced, since it cannot be disregarded or needlessly altered after property rights have been acquired in reliance upon it. On a resurvey to establish lost boundaries, if the original corners can be found, the places where they were originally established are conclusive without regard to whether they were in fact correctly located, in the respect it has been stated that the rule is based on the premise that the stability of boundary lines is more important than minor inaccuracies or mistakes. But it has also been said that great caution must be used in reference to resurveys, since surveys made by different

---

* Frank Emerson Clark, *A Treatise on the Law of Surveying and Boundaries*, Third Edition by John S. Grimes; Indianapolis: the New Bobbs-Merrill Company, Inc., 1959. This treatise is frequently relied upon by the court system.

Corners, Lines, and Surveys

surveyors seldom wholly agree. A resurvey not shown to have been based upon the original survey is inconclusive in determining boundaries as will ordinarily yield to a resurvey based upon known monuments and boundaries of the original survey.*

The lines of a survey marked on the ground are controlling as to boundaries fixed with reference to such survey, and such lines when ascertained constitute the true lines, or, as the rule is sometimes expressed, the tracks of the surveyor, so far as discoverable on the ground with reasonable certainty, should be followed. Nevertheless, this principle has no application unless the footsteps of the surveyor and the actual survey called for in the grant or description can be found and identified; and further, lines and boundaries cannot be constructed with reference to objects which may be found on the ground as indicated by the footsteps of the surveyors where there are no calls in the grant for such objects.

The footsteps of the surveyor must be traced before course and distance may be ignored. In the absence of monuments and marks, or in case of repugnancy between them, the determination of what are the true lines called for in an entry, grant, or conveyance is a pure question of construction, dependent on the intention of the parties, as shown by the instrument and the circumstances of the transaction.†

In relocating or reestablishing the lost lines of an old survey, the tracks of the original survey should be followed so far as it is possible to discover them, and the purpose of a resurvey is to find where the original lines ran. All locations, calls, and distances must, if found, be followed.‡

Resurvey. A resurvey may be defined as correcting and redoing an original survey, to make it more nearly in conformance with its intended location.

Clark,§ as it is commonly known, provides a detailed discussion of Resurveys. In § 163, it is stated, "There are two types of government resurveys, the dependent and the independent survey.

When the retracement shows that the principal resurvey problem is one of obliteration, with comparative absence of large discrepancies, that is,—that the early survey had been made faithfully, then that official survey can be reconstructed or restored as it was in the beginning; the methods applied are termed a 'dependent resurvey.' Conversely, if the retracements show the existence of intolerable discrepancies, that is,—that the early survey had not been faithfully executed, usually including the fact that some of the lines had not been established, having no actual existence, and therefore incapable of reconstruction or restoration to conform with a fictitious record, entirely different methods can and must be applied to the public-land area termed an 'independent resurvey.' In the latter circumstance, the boundaries of the

---

* 12 Am Jur 2d, Boundaries, §61, Resurveys.
† 11 C.J.S. Boundaries, §14, Lines and Their Location.
‡ 11 C.J.S. Boundaries, §15, Relocation of Lost Line.
§ Ibid., § 258.

patented lands are given special treatment for their protection, as nearly in harmony as possible with what could be afforded by local court decree."

Perhaps the discussion in the case of *Cragin v. Powell** best serves to illustrate the view of the court system concerning resurveys. In this case, in 1881 a surveyor had been appointed by the court to survey an original township, and he discovered what he termed to be gross errors in the original survey done in 1837. He stated that the original survey "was so incorrect, and the traces of its lines and corners so difficult to identify, that he was unable to locate any proper line between the lands in question, except upon the bases of a resurvey of the entire township, in accordance with certain corrective resurveys of adjoining townships." He proceeded to "correct" those errors, producing a new survey and plat despite the fact that titles had been created and lands transferred based on the original survey and plat. In his report and as oral evidence, he stated that "this governmental survey is incorrect; some of it more incorrect than the rest, but especially erroneous in the length of its lines and [a location]." "The plat," he reported, "is totally inconsistent with that of the governmental survey, and should have been rejected." The court noted that whether the official survey made by [the original surveyor] is erroneous, or should give way to the extent of its discrepancies to the survey reported is a question which was not within the province of previous courts or of this one. "The mistakes and abuses which have crept into the official surveys of the public domain form a fruitful theme of complaint in the political branches of the government. The correction of these mistakes and abuses has not been delegated to the judiciary," except for some listed exceptions.

In the past, some townships were designated by the surveyor general to be resurveyed, but not this one. In a letter to the surveyor general, the Commissioner of the General Land Office soon put an end to the practice of undertaking resurveys. He stated, "the making of resurveys or corrective surveys of townships once proclaimed for sale is always at the hazard of interfering with private rights, and thereby introducing new complications. A resurvey, properly considered, is but a retracing, with a view to determine and establish lines and boundaries of an original survey ..... but the principle of retracing has been frequently departed from, where a resurvey (so called) has been made and new lines and boundaries have often been introduced, mischievously conflicting with the old, and thereby affecting the areas of tracts which the United States had previously sold and otherwise disposed of."

The court went on to state that "the power to make and correct surveys of the public lands belongs to the political department of the government and that, whilst the lands are subject to the supervision of the General Land Office, the decisions of that bureau in all such cases, like that of other special tribunals upon matters within their exclusive jurisdiction, are unassailable

---

* 128 U.S. 691, La., 1888.

Corners, Lines, and Surveys 99

by the courts, except by a direct proceeding; and that the latter have no concurrent or original power to make similar corrections, if not an elementary principle of our land law, is settled by such a mass of decisions of this court that its mere statement is sufficient." (listing several citations).

Unfortunately, as good a work as it is, Francis Hodgman, in *A Manual of Land Surveying*,* confuses the terms, including a complete chapter, titled Resurveys,† while providing discussion and a wonderful collection of court decisions, on retracement surveys. He begins the chapter by saying "In an old settled country, the principal work of the surveyor is to retrace old boundary lines, find old corners, and relocate them when lost."

He continues with "In making resurveys the surveyor is called upon—1. To construe descriptions in deeds; 2. To find the location of corners and boundary lines; 3. To renew corner monuments and to mark anew boundary lines. All this, technically, and legally, has to do with retracement, not resurvey."

What If There Are No Actual "Footsteps?"

The foregoing discussion is not more than an elaboration of concepts, precepts, and resolution and guidance from legal principles derived from manuals of instruction and court decisions. Relatively recent on the scene is the answer to a long-existing dilemma: what to do if there is no original survey, or no actual "footsteps" to follow? For the answer to this question, reference is made to the Montana case of *Larsen v. Richardson.*‡ The case had to do with a difference between two parties, both of which employed a surveyor, concerning the location of a boundary.

The court noted that "surveyors retracing a metes-and-bounds description are guided by a hierarchy or "priority of calls" in the event of a conflict between elements within the land description. Monuments generally control over inconsistent courses and areas (i.e., number of acres), the rationale being that the latter depend for their correctness on a variety of circumstances and, therefore, are generally less reliable than a monument established on the ground at the time of the survey. *See* Walter G. Robillard & Lane J. Bouman, *Clark on Surveying and Boundaries* § 14:21, 397 (7th ed., Lexis Law 1997), § 15:08A, 147 (LexisNexis Matthew Bender Supp. 2010); Walter G. Robillard & Donald A. Wilson, *Brown's Boundary Control and Legal Principles* 121-23 (6th ed., John Wiley & Sons 2009)." Note here that even though Montana is considered a Public Lands state, this court is discussing a metes-and-bounds problem.

---

* The F. Hodgman Co., Climax, Michigan, 1913.
† Chapter X.
‡ 361 Mont. 344, 260 P.3d 103, Mont., 2011.

100   *Boundary Retracement*

Both surveyors began their respective analyses by, first, obtaining the deeds in the Larsens' chain of title and the chains of title of the adjacent landowners. This is standard practice in conducting a retracement survey, as both surveyors testified. *See also* Robillard & Bouman, *Clark on Surveying and Boundaries* § 14:21, 397 (1997). ("In ascertaining the true and correct boundaries of a parcel, the surveyor is obligated to consider any and all evidence. This rule is inflexible.")

The court then noted that "the object of all rules for the establishment of boundaries is to ascertain the actual location of the boundary as made at the time (citing precedent). As to boundary disputes, the primary purpose is to track the footsteps of the original surveyor, to locate the survey as it was intended to be located on the ground by him (again citing precedent). Although, in the present case, the deeds to the properties at issue here were prepared without the benefit of a proper field survey, we conclude that the foregoing principles nevertheless apply in determining the locations of the boundaries described in those deeds. In other words, the duty is to track the courses laid out by the deed writers in the boundary descriptions." This will be a key issue in a later discussion of protraction, or protracted boundaries, in Chapter 4.

This should come as no surprise, since the definition and location of the parcels came at the time of their creation. Both scenarios, survey versus non-survey, are based on the same premise: the title has been established, and the resulting parcel(s), along with their boundaries, are now fixed.*

Before concluding, the court added the following: *See* Robillard & Bouman, *Clark on Surveying and Boundaries* § 15:08, 124 (Supp. 2010) ("When a surveyor is unable to follow the precise 'footsteps' of his or her predecessor, then a surveyor must attempt to track the original surveyor's work using whatever recoverable evidence that exists."); Curtis M. Brown, Walter G. Robillard, & Donald A. Wilson, *Evidence and Procedures for Boundary Location* 42 (2d ed., John Wiley & Sons 1981). ("Discovery of the original monument itself is not a necessity, since many types of evidence can be resorted to that will suffice as proof of the original location.")

Continuing, citing authority, the court stated that one party's approach is contrary to the law of surveying and without any evidentiary foundation. "If there were any principle that perhaps is misinterpreted and possibly misapplied by surveyors, attorneys, and courts, it is the priority of calls." Robillard & Wilson, *Brown's Boundary Control and Legal Principles*, 92. The priority of calls was developed through case law in the 1800s and is also governed by statutory law. The general hierarchy is as follows: lines actually run on the ground by the creating surveyor prevail over natural monuments (e.g., a tree), which prevail over artificial monuments (e.g., surveyor's stakes), which prevail over references to adjoining boundaries (e.g., "to Hunter's property

---

* Later discussion will compare and contrast parcel boundaries vis-à-vis ownership boundaries. Griffin, also discussed in a later section, makes a very strong point of this.

Corners, Lines, and Surveys 101

line"), which prevail over directions (e.g., northwest), which prevail over distances (e.g., 30 feet), which prevail over area (e.g., 5 acres), which prevails over place names (e.g., "the Quinn farm"). Robillard & Bouman, *Clark on Surveying and Boundaries* § 14:21, 397 (1997), § 15:08A, 147 (Supp. 2010); Robillard & Wilson, *Brown's Boundary Control and Legal Principles* 121–23; adding precedent.* These rules "gr[ew] out of the peculiar exigencies of the country, and were moulded by experience, to meet the demands of justice." *Riley*, 16 Ga. At 148; *see also Booth v. Upshur*, 26 Tex. 64, 70 (1861) (the rules are "founded on reason, experience and observation" and "pertain[ ], not to the admissibility, but to the weight of evidence"). The rationale is that the lower-ranked calls are generally less reliable than the higher-ranked calls. As explained by the court in *Riley*, 16 Ga. At 148, "any natural object, when called for distinctly, and satisfactorily proved—and the more prominent and permanent the object, the more controlling as a locator—becomes a landmark not to be rejected, because the certainty which it affords, excludes the probability of mistake, 'whereas course and distance, depending, for their correctness, on a great variety of circumstances, are constantly liable to be incorrect. Difference in the instrument used, and in the care of surveyors and their assistants, lead to different results.'" *See also McCullough v. Absecon Beach Co.*, 48 N.J. Eq. 170, 21 A. 481, 487 (1891).

Early Court Decision

One could quickly assume, maybe even be led to believe, that the directive **Following the Footsteps of the Original Surveyor** originated from one of the states in PLSS, since discussion is so widespread in decisions and standards of practice in those states. But it must be remembered that, at the outset, all states, as well as other land definitions within and without the United States of America, had public land, whether it was administered by a king, a group formed by a higher power, or by some other governmental body of the country or region.

It would appear that the first time the concept becomes apparent in American law is in what we would today call a non-Public Land state, and considered to be part of the metes-and-bounds system. The earliest case law seems to be found in the Texas courts, a state that has produced many sound and useful decisions as well as a treatise on the very subject.[†]

---

* *Pollard v. Shively*, 5 Colo. 309, 313, 1880; *Riley v. Griffin*, 16 Ga. 141, 147–148, 1854; *M'Clintock v. Rogers*, 11 Ill. 279, 296–97, 1849; *Tewksbury v. French*, 44 Mich. 100, 6 N.W. 218, 218–219, 1880; *Hoffman v. Beecher*, 12 Mont. 489, 502, 31 P. 92, 96, 1892; *Lodge v. Barnett*, 46 Pa. 477, 484–485, 1864.

† Orn, C. L., Vanishing footsteps of the original surveyor. *Baylor Law Review*, IV(3), 1952.

The Texas court, citing earlier decisions, stated in 1941,* "In all of the cases, from *Stafford v. King*, 30 Tex. 257, 94 Am. Dec. 304 (1867),† on down, the law has been that search must be made for the footsteps of the surveyor, and that, when found, the case is solved." As will be discussed in some detail in Chapter 5, if the location of the corner‡ can be found, the inquiry is finished; if not, then the problems have only begun. In the latter situation, the investigator must determine who established the corner and by what methodology, then reconcile their errors. That can become a forensic study of the highest order.

In *Stafford*, Judge Smith included in his lengthy and very instructive decision, "The identification of the actual survey, as made by the surveyor, is the *desideratum* of all [these] rules. The footsteps of the surveyor must be followed, and the above rules are found to afford the best and most unerring guides to enable one to do so." There it is, so stated in simple, concise terms.

Not Limited to Public Land Survey States

At least one other state court has referred to the terminology and guidance from Public Land states. In its analysis of a survey problem, the Tennessee (considered to be a metes-and-bounds state, although containing a small area of rectangular surveys) court relied on a number of decisions from Public Land states (Ohio, Montana, Washington, Florida, Wisconsin, Minnesota, Louisiana, as well as Pennsylvania, and the United States Supreme Court§ along with important references, both surveying and legal).¶

It went on to say "While these general rules apparently have their origin in surveys reflecting government grants, such rules are equally applicable to private surveys," citing *Staub v. Hampton*, 117 Tenn. 706, 101 S.W. 776 (1907). It is noteworthy that more than one hundred years ago the Tennessee court was relying on, and giving guidance to, principles and directions from, instructions having to do with Public Land Surveys.

---

* *Hart v. Greis et al.*, 155 S.W.2d 997.

† 1867.

‡ Definition of corner; not marker, but corner. See the early part of this chapter for definitions of types of corners.

§ Each of these referenced decisions will be discussed with the following text material. Some will be noted as significant in the state where they were decided.

¶ *Wood v. Starko*, 197 S.W.3d 255, Tenn.App., 2006.

Corners, Lines, and Surveys 103

Early Guidance in Textbook

The procedure appears in the supplement to BLM Manual, *Restoration of Lost or Obliterated Corners & Subdivision of Sections*. The rules for the restoration of lost corners have remained substantially the same since 1883, when first published as such.*

---

* Restoration of Lost or Obliterated Corners, and Subdivision of Sections, March 13, 1883, 1 L.D. 339; 2d edition 1 L.D. 671; revised October 16, 1896, 23 L.D. 361; revised June 1, 1909, 38 L.D. 1; reprinted in 1916 and 1936; revised April 5, 1939; reissued May 8, 1952; reprinted with corrections July 11, 1955; reprinted in 1960; revised June 3, 1963; reprinted in 1965 and 1968.

4

Protracted Boundaries

Where lines are in fact run by the surveyor, his real location will always be followed; but where he does not in fact run a line, there is no location by him to be followed.

Kentucky Union Co. v. Hevner, et al., 210 Ky 121,
275 S.W. 513 (Ky.App. 1924).

Definition

Parcels of land or lots drawn on a subdivision map but not surveyed or monumented on the ground by an original survey are said to be created by protraction.

The Texas treatise on following the footsteps* begins with its opening paragraph stating "When the public domain of Texas was surveyed in the early days two kinds of surveys were made—one, a survey on the ground, and the other a survey in an office." Even when surveys were made on the ground, not all lines were actually run and monumented, but many were projected. The projected lines thus, in effect, became office calls.[†]

Protraction is the method whereby boundary lines are created without being run on the ground. They may be drawn on a plat, or merely written out in a land description. It is not unusual, in some areas and points in time, to find them in probate files where lands are divided among heirs, but they are also found as extensions of bona fide surveys and conveyances. Wherever found, they can create a title since they are usually used for description purposes, thereby leaving the parcel(s) so created as an unsurveyed tract awaiting the *first surveyor* to actually locate it on the ground and mark it. They are troublesome at the least, and although not recommended, may be encountered in retracement. They are found commonly in the mid-Atlantic states, where they have been the subject of considerable litigation.

---

* Orn, C. L., Vanishing footsteps of the original surveyor, *Baylor Law Review*, IV(3), 1952.
† Citing *Humble Oil & Ref. Co. v. State*, 162 S.W.2d 119, Tex. Civ. App., 1942.

106   *Boundary Retracement*

The earliest decisions dealing with protracted lines are found in the eastern states, where some of the very early surveys were made. The West Virginia court analyzed the problem very candidly:

The basic problem with protracted surveys was reported in a West Virginia case,* wherein the court stated:

> The result of this loose, cheap, and unguarded system of disposing of public lands was that in less than 20 years after the adoption of the system nearly all of them were granted, the most part to mere adventurers, in large tracts or bodies, containing not only thousands, but in many cases hundreds of thousands, of acres in one tract. Often the grantees were nonresidents, and few of them ever saw their lands or expected to improve or use them for purposes other than speculation. The entries and surveys under warrants so cheaply and easily obtained were often made without reference to prior grants, thus creating interlocks, thereby covering land previously granted, so that in many instances the same land was granted to two or more different persons. Sometimes upon one survey actually located others were laid down by protraction—constructed on paper by the surveyors, without ever going upon the lands, thus creating on paper blocks of surveys containing thousands of acres, none of which were ever surveyed or identified by any marks or natural monuments. Thus, while the state was rapidly disposing of her large domain at 2 cents per acre, it was not attaining the main objects in view—the settlement of the country and revenue from the owners of lands. It was found that the grantees not only failed to settle upon and improve their lands, but in most instances they wholly neglected to pay the taxes due thereon, whereby revenue failed and the improvement of the territory was embarrassed and retarded. Nonoccupation of the lands and nonresidence of the owners made a resort to the lands themselves the only fund for delinquent taxes.

Protracted surveys are the subject of great detail in the case of *Rowe v. Kidd.*[†] It states that a protracted line is one that is only run mentally, and therefore not marked. See further discussion later in the chapter.

AUTHOR'S NOTE: Since following the boundary lines of the original creators, whether a surveyor or a scrivener, is the basic rule, it is necessary to know by what method the title, and ultimately its boundaries, was created. See Chapter 2 for a detailed discussion of the creation of title.

Lines of rectangular lots, both inside and outside the PLSS, may have been created by protraction, that is drawn on paper, but not marked on the ground. Or, in the alternative, as has been noted in a number of decisions, once surveyed and marked, but now unable to be identified on the ground. Either way, the procedure for resolution is the same. However, a word of

---

* *Fay v. Crozer*, 156 F. 486, S.D.W. Va., 1907 citing Hutchinson on Land Titles, pp. 1,2.

† 249 F. 882, E.D. Ky, 1916.

Protracted Boundaries 107

caution is in order. Before a conclusion is drawn that such lines have not been surveyed (and therefore no actual "footsteps"), all possibilities must be explored, including original town layouts, original subdivision layouts, proprietors' records and the like, otherwise there is no proof that there was no original survey.

Protraction often leads to multiple markers set for the same corner, known widely as "pincushions," due to different surveyors' opinions, procedures, or legitimate measurement error. These possibilities are discussed in Chapter 8.

Protraction outside the PLSS

Protraction may also be encountered outside of the PLSS, although it may not be as common. It is more likely to be found with original grants and patents, from which cases are not infrequent. This presents a special concern for the retracement surveyor, since any line of a subsequent tract which is part of an original grant line, must be retraced in its own context, and may require a retracement of the entire original grant line and, rarely, the entire grant. More complex than that, in cases of gross acreage conflicts and with remainder parcels, especially if they are ill-described or contain ambiguities, sometimes may only be defined by locating the original source tract, and taking out each subsequent tract in the order they were created to determine what remains.

Protracted lines can be used to subdivide parcels of land, although most jurisdictions today prevent that, requiring an actual survey. However, the problems will likely arise from divisions made prior to zoning and subdivision regulations being adopted. Problems with modern divisions will likely be of a different nature, although they too occur. Protracted lines may also be found creating new lines for either adding to an existing parcel, or creating a new parcel next to, and probably adjoining, an existing parcel. Either way, if followed by a conveyance, a title will have been created, and its boundaries established (Figures 4.1 and 4.2).

One of the more recent decisions involving protracted lines is a case from the Maine court* illustrating an extreme example as to what a protracted situation may lead to. It also contains an interesting twist recognized by the court of appeals, but not by the lower court. As the Supreme Court noted, the lower court came to the correct conclusion but for the wrong legal theory.

> On appeal, Wilson contends that the court erred in its determination of the location of the boundary line between his lot and the lot belonging to Steinherz. The court located the line according to a survey made after the conveyances to Wilson and Steinherz. Wilson contends that the

---

* *Steinherz v. Wilson*, 1998 ME 22, 705 A.2d 710, Me. 1998.

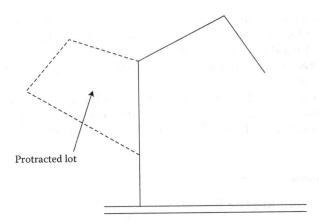

FIGURE 4.1
Protracted lot from a known perimeter.

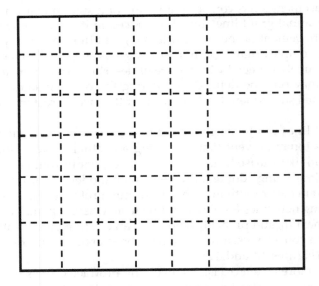

FIGURE 4.2
Protracted subdivided lots within a known perimeter.

language in his deed requires that the boundary line must be located on the face of the earth by reference to the recorded plan of the Brooks Farm Subdivision. Because the boundary recognized by the court was one created by parol agreement, and was binding on the parties, we affirm the judgment, although on grounds different than those relied on by the Superior Court.

Wilson and Steinherz both purchased their adjacent lots from Jonathan Milligan. Wilson owns lot 3, and Steinherz is the owner of lot

Protracted Boundaries

3B as identified on a subdivision plan. The perimeter of the tract was surveyed by Roland Libby in 1905 using boundary monuments. Its 40 individual lots, however, were superimposed on the tract between 1905 and 1961 by an unknown drafter, apparently without being surveyed. In 1961 Milligan hired a surveyor to prepare a recordable plan of the Brooks Farm Subdivision. Without the benefit of a field survey, the surveyor transposed a map that had distance and course data onto a map that had only lot lines. The resulting subdivision plan was recorded in 1961. During this process and in the surveyor's presence, Milligan drew a line on the map through original lot 3, without using a stated length, compass course, or survey markers. This purported to create the lots in dispute: 3 and 3B.

No actual survey of any individual lot was done until its sale, resulting in what is known as a 'paper' or 'protracted' subdivision.* Milligan hired his surveyor beginning in 1962 to survey a number of the lots sold in conjunction with the sales.

As he surveyed lots, the surveyor found numerous inconsistencies between the angles, equations, and distances shown on the Brooks Farm plan and the location of landmarks on the face of the earth. He reconciled these inconsistencies using a "best-fit" approach, retaining consistent data while revising inconsistent data in order to fit each lot within the perimeter boundaries of the Brooks Farm plan.

Wilson purchased lot 3 from Milligan in 1980 by a deed that referenced the Brooks Farm subdivision plan.† Wilson looked at the land with Milligan for fifteen to twenty minutes prior to purchase. Milligan did not testify at trial, but Wilson testified that Milligan showed him a high point of land with an old foundation, and told him that the foundation area was on his side of the boundary. Wilson testified that he 'didn't know exactly where [the boundaries] were.' The Brooks Farm subdivision plan referenced by Wilson's deed also appeared to depict a structure on lot 3. Milligan and Wilson agreed before the conveyance that Milligan would have his surveyor mark the boundaries of lot 3 on the face of the earth, including the boundary line between lots 3 and 3B. He did not mark that boundary line immediately after the Wilson purchase, however. It was not until 1982, in conjunction with Steinherz's purchase of her lot, that he surveyed and marked that boundary line.

Steinherz purchased lot 3B from Milligan in 1982. Her deed description also referred to the recorded plan. Steinherz made placement of precise boundary markers at Milligan's expense prior to closing an express condition of the contract. The surveyor surveyed lot 3B for Milligan in

---

* This method contrasts with the layout of a subdivision after a survey is completed and boundary markers have been laid down, which allows the lots to be conveyed with metes and bounds descriptions.

† The Milligan to Wilson deed contained the following description:

A certain lot or parcel of land with any improvements thereon, as more fully delineated on "Plan of the Brooks Farm at Cape Porpoise," dated 1905, made by R.W. Libby, Engineer, Saco, Maine, and recorded in the York Registry of Deeds, Plan Book 33, Page 7, said lot being LOT NO. 3, as shown on said Plan, to which Plan reference may be had for a more complete description.

the summer of 1982. That survey located the boundary line between 3B and 3. Steinherz then hired an architect to begin the building process. When the architect found discrepancies between the Brooks Farm subdivision plan and the surveyed boundary lines, the surveyor did a second survey. His second survey confirmed to him that Steinherz's land included much of the disputed high point of land and the old foundation. On that basis Steinherz had her house built near the old foundation, which her contractor filled and graded. No improvement has ever been made to Wilson's lot.

Shortly after Steinherz's house was completed in May 1983, Wilson, who is a resident of Massachusetts, visited his lot. Upon seeing the house, he expressed to Steinherz that he thought his boundary was further up the knoll, and asked her for a plan for her lot. Wilson later requested information from Milligan, who verified that the boundary was essentially as it had been surveyed, and as evidenced by the distinct tree line that the surveyor cut and that Steinherz has maintained since 1983.

Wilson took no action until 1988, when he learned that Middle Branch Engineering (MBE) had done a systematic survey of the entire subdivision to resolve numerous inconsistencies between the Brook Farm subdivision plan and the surveys of individual lots that had been done "piecemeal" over the years. The MBE survey, commissioned by prospective developers of some of the lots, was a mathematical construction of how the 1961 Brooks Farm subdivision plan would appear if projected on the ground. The MBE survey concluded that the Steinherz house had been built on lot 3, Wilson's lot. Wilson purchased the MBE survey and recorded it. Steinherz then brought this action to quiet title and, because of the recording of the MBE survey, for slander of title. Wilson counterclaimed for trespass and sought an injunction to remove Steinherz's house. Noting the fact that the description in Wilson's deed does not 'precisely locate the boundary on the face of the earth,' and that Wilson at least implicitly agreed to have the surveyor work the boundary between lots 3 and 3A, the court concluded that the boundary worked by the surveyor in 1982 became the legal boundary. The court entered judgment for Steinherz on her quiet title action and for Wilson on Steinherz's claim for slander of title. It decided in favor of Steinherz on Wilson's counterclaim. Wilson then filed this appeal.

Wilson contends that the issue is one of law. He notes that his deed refers to a recorded subdivision plan with specific dimensions and landmarks, and that his deed identifies his land by lot number. He contends that even though the boundary line between lots 3 and 3B on the plan shows no course distance or monument, nevertheless that line can be located on the face of the earth, the MBE survey determines its true location, and he cannot be divested of the land within that boundary. We disagree.

The court recognized the boundary established by the surveyor as being the true boundary between Wilson's lot 3 and Steinherz's lot 3B. The court explicitly found that Wilson and Milligan agreed that the surveyor would survey the boundary between lots 3 and 3B, and that there was a similar agreement between Milligan and Steinherz. Those findings

Protracted Boundaries

111

are amply supported in the record. That surveyor in fact conducted such a survey and located that boundary, which had theretofore been both unsurveyed and ambiguous. Accordingly, pursuant to the agreement Wilson had with Milligan, albeit delayed until the Steinherz purchase, the surveyor's survey established that boundary. Once established, the legal effect of the boundary so established was not diminished by the later MBE survey. Although the trial court did not rely precisely on the doctrine of the establishment of a boundary by parol agreement, nevertheless its findings that Wilson agreed to have the surveyor survey and determine the boundary between the lots lead to the application of the doctrine.

We have previously recognized boundaries by parol agreement. *Bemis v. Bradley,* 126 Me. 462, 464, 139 A. 593 (1927) ('It is a familiar and well settled principle of law that a boundary line may, under certain circumstances be established by parol agreement of adjoining owners.'). See also *Ames v. Hilton,* 70 Me. 36, 46 (1879).

Although Wilson had only an oral agreement with Milligan that his surveyor would survey the boundary, a written agreement is not necessary.

A contract between owners of adjoining tracts of land fixing a dividing boundary is within the Statute of Frauds but if the location of the boundary was honestly disputed the contract becomes enforceable notwithstanding the Statute when the agreed boundary has been marked or has been recognized in the subsequent use of the tracts. Restatement (Second) of Contracts § 128(1).

The reason that a parol agreement regarding an unascertained, uncertain or disputed boundary falls out of the Statute of Frauds is that "no estate is created [and] ... the coterminous proprietors hold up to it by virtue of their title deeds, and not by virtue of a parol transfer of title." *Osteen v. Wynn,* 131 Ga. 209, 62 S.E. 37, 39 (1908). Cf. *Piotrowski v. Parks,* 39 Wash.App. 37, 691 P.2d 591, 596 (1984) (oral agreement valid if parties designate a visible boundary); *Foster v. Duval County Ranch Co.,* 260 S.W.2d 103, 109-10 (Tex.Civ.App.-San Antonio 1953) ('where parties are in doubt as to where the true division line between them of their lands may be, they may fix it by parol agreement which would be mutually binding on them, even though they were mistaken as to its true locality'); *Gulf Oil Corp. v. Marathon Oil Co.,* 137 Tex. 59, 152 S.W.2d 711, 714 (1941) (agreement may be proven by acts and conduct falling short of express statement or acquiescence).

In conjunction with Wilson's purchase of lot 3, he and Milligan orally agreed that Milligan's surveyor would survey the boundary. The trial court found that, although after the boundary was marked Wilson did not explicitly agree to it, and that although he began to inquire into the boundary when he first saw the house, he had 'requested and expected' that it would be marked by the surveyor. In short, Milligan and Wilson made a parol agreement that Milligan's surveyor would mark the boundary, and Steinherz is the beneficiary of that agreement. While Wilson's subsequent behavior does not demonstrate an uncontradicted agreement as to the location of the actual line marked, he must be held to

112   *Boundary Retracement*

the agreement itself and thus to its result. If a valid parol agreement is found, the agreement can determine an uncertain boundary. See, e.g., *Emery v. Fowler,* 38 Me. 99, 102 (1854) (boundary marked by mutual agreement of parties with intent to conform to a deed already given controls notwithstanding inconsistency with the deed).

Although the conclusion of the trial court that the survey established the boundary between lots 3 and 3B was arrived at by reason of a different legal theory, nevertheless the court's conclusion was correct, and its judgment must be affirmed. See *Baybutt Construction Corp. v. Commercial Union Ins* (where trial court's ultimate conclusion is correct in law, it must be sustained on appeal, although its conclusion may have been reached on incorrect legal theory).

The New Jersey case of *Niscia v. Cohen** is an example where protraction failed because there was no monument, or permanent point, from which to make the protraction. The parties to this action entered into a written contract which, by its terms, required the plaintiff in error to convey and the defendant in error to purchase, a lot of land, 'No. twenty-eight (28) on a map of the town of Hudson, made by Clerk & Bacot, city surveyors, and duly filed in the office of the clerk (now register) of the county of Hudson.' When the contract was signed the defendant in error paid $500 on account of the purchase price. It is admitted that the defendant in error tendered, as required by the contract, a deed which described the property as in the contract, and that it was refused by the purchaser upon the sole ground that the vendor "could not convey by reason of encroachment of the real estate upon other real property."

After such refusal the purchaser brought his suit to recover the amount paid on the purchase price, the cause of action being, as set out in the declaration, that "the said defendant then and there was unable to deliver to said plaintiffs the deed of warranty free from encumbrance of said premises, part of the same on the east and south being occupied and possessed by the adjoining property owners." At the trial the defendant in error recovered a verdict upon which the judgment now under review was entered.

The plaintiff in error, defendant below, moved, at the close of the plaintiff's case, for a nonsuit, and also at the close of the whole case for a direction in his favor, and both motions being refused, exceptions were sealed and error assigned thereon. We think the defendant below was entitled to a favorable ruling on both of these motions, and that the refusal requires a reversal of the judgment assailed. There is no testimony in the case which justifies the inference that the land described in the deed was 'occupied and possessed by the adjoining property owners,' and therefore nothing was shown to sustain the issue raised which could be properly submitted to a jury.

The plaintiff below called but a single witness to support his claim that a part of the lot as described in the deed was in possession of the adjoining landowners, and this witness, who testified that he was a surveyor, producing a sketch showing the result of what he claimed to be a survey

---

* 84 N.J.L. 351, 86 A. 395, N.J.L. 1913.

Protracted Boundaries

of the lot, said that it "was a practical location of lot twenty-eight," but he admitted that the sketch did not agree with the map mentioned in the contract and deed, and that it was not made from the map, but from measurements which he had made without regard to it. What this witness did, as appears from his testimony, was to measure one hundred and twenty-five feet easterly from a highway called Smith street, and establish that as the westerly corner, on Newark avenue on which it fronts, of lot No. 28, while no such street appears on the original map, and there is nothing in the testimony of this witness which shows any basis for the assumption, made by him, that the westerly line of the lot is where he protracted it on his sketch. The testimony of this witness is rested upon presumptions without facts to support them. No deed of the lot on the west is produced, or fact testified to, which fixes the westerly line of this lot as beginning one hundred and twenty-five feet easterly from Smith street, nor is the true line of that street located by any survey produced, or other competent testimony. If a survey is based on a measurement from a given point the correctness of that point, as a monument, must be shown, and the reason for the use of the distance measured should appear in the testimony, for we cannot accept as correct a survey showing the "practical location" of a lot, made without regard to the filed map on which it is laid down, or without proof of the location of the adjoining lots as protracted on the original map along the lot in question. Whether the lot sought to be conveyed has been encroached upon and is in the possession of other owners, depends upon the true location of the beginning point adopted by the person making the survey, and before the distance called for as the width of the lot can be said to extend so far in a given direction as to disclose that a portion of it has been encroached upon by an adjoining owner, the beginning point should be clearly established, and that manifestly was not done in this case according to the proofs offered at the trial.

The plaintiff in this case failing to prove, by competent evidence, the conditions stated in his declaration as his reason for refusing to carry out his contract, he has not made out the basis of his action, and in such case the defendant was entitled to a direction in his favor, no other right to recovery appearing.

The court had a problem with the fact that the surveyor did not protract from a verified corner. Note that in PLSS requirements, and in the subsequent court decisions, protraction must be based on an accepted government corner. Without such a basis, a line is just "hanging out there" tied to nothing.

While it is an accepted practice to rely on courses and distances to locate a position where a monument is lost, the measurements must be tied to something.

A discussion of original grants and patents is within Chapters 12 and 13, however the following case is illustrative of several courts' reasoning as to how to locate an original tract, part of which was defined by protracted lines. It is early enough to be a ground-breaking case, although several of the courts relied on precedent they had at their disposal. The parcel is a large

114   *Boundary Retracement*

tract, but such a case may be very useful when the retracing surveyor must deal with an original grant line.

In this case,* Davis claimed title under a patent issued September 25, 1845, on a survey dated March 6, 1845. The defendant, Bramblet, claimed under a junior patent issued January 28, 1846, on a survey dated March 6, 1845. The first patent, calling for 86,000 acres of land, was issued to Ledford, Skidmore, and Smith, and will be referred to as the Ledford patent. The second, calling for 9500 acres, was issued to Moses Cawood. The controversy grows out of an alleged conflict in the boundaries of the two patents, and involves the true location of the Ledford patent; that of the Cawood patent being undisputed. Located in one way, the Ledford patent embraces the entire land called for by the Cawood patent. Located in another, the conflict is but slight. The description of the land granted by the Ledford patent is as follows:

> A certain tract or parcel of land, containing 86,000 acres, by survey, bearing date the third day of March, 1845, lying and being in the county of Harlan, and bounded as followeth, to wit: Beginning on Crank's creek, on two beeches and two sugar trees, beginning corner to said Smith's 1,500-acre survey; thence S. 70 degrees W. 664 poles, to three beeches, beginning corner to Smith's 600-acre survey; thence S. 28 degrees W. 400 poles, to a stake on the top of Cumberland Mountain; then S. 60 degrees W. 8,320 poles, to a stake near Cumberland Gap; thence No. 15 degrees E. 3,200 poles, to a stake; thence N. 55 degrees E. 8,820 poles, to a stake; thence S. 5 degrees W. 3,150 poles, to the beginning, with its appurtenances.

The order of the Harlan county court for the survey was made in March 3, 1845, and on the same day the certificate of survey, on which the patent was issued, was made. This survey describes the property precisely as it is described in the patent. Accompanying it was a plat in the following form (Figure 4.3).

Laid out by the courses and distances given above, the boundary of the survey closes, defining a tract of land of about the shape shown above, running in its longest direction from northeast to southwest, the southeast side of which (made up of the first, second, and third lines) is approximately 30 miles long, the opposite or northwest side (bounded by the fifth line) 27 1/2 miles long, and the ends (being the fourth and sixth lines) about 10 miles long. It contains in round numbers 300 square miles, or 192,000 acres, over 100,000 acres more than called for.

Coming to the location of the patent, it will be observed that the first and second lines call for known natural objects as corners—the first corner being "on Crank's creek, on two beeches and two sugar trees, beginning corner to said Smith's 1,500-acre survey"; the second corner, "three beeches, beginning corner to Smith's 600-acre survey"; and the third corner, "a stake on the top of Cumberland Mountain." By their use, the first line, with a slightly changed

---

* *Bramblet v. Davis*, 141 F. 776, 6th Cir. 1905.

Protracted Boundaries

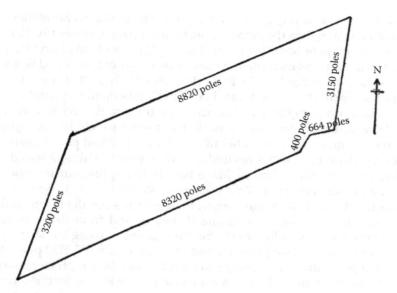

FIGURE 4.3
Plat in *Bramblet v. Davis*.

course, was found to be 911.5 poles, instead of 664 poles, in length, and the second line 463.4 poles, instead of 400 poles, in length. The increase in length of these two lines, amounting to 310.7 poles, and the slight change in their bearing, would necessarily produce a slight variation at the point of closure, but so slight in so great a perimeter (almost 80 miles in length) that the survey, with the first and second lines thus located, may be said practically to close. The first and second lines, thus defined by natural objects, are the only ones about whose location there is no dispute. The second ends at "a stake on the top of Cumberland Mountain." The state line between Kentucky and Virginia in this locality runs along the top or highest point of Cumberland Mountain, so that the third corner is on the boundary which marks the limit of the territorial jurisdiction of Kentucky.

Right here is where the difficulty in locating the patent begins, for the third line, which is defined as "then S. 60 degrees W. 8,320 poles, to a stake near Cumberland Gap," if run from the third corner by this course and distance, crosses the state line, passes through the wedge-like point of Virginia between Kentucky and Tennessee, and ends in Tennessee about four miles southeast of Cumberland Gap. Taking the point thus fixed by course and distance as the fourth corner, and running the remaining lines (the fourth, fifth, and sixth) by the courses and distances given, the boundary practically closes. But because Kentucky had no authority to run its survey into other states, and no power to issue a patent for land located in other states, doubt was naturally thrown about the true location of the patent, and the matter has been before the courts, both state and federal, a number of times.

116   *Boundary Retracement*

The first suit was brought by Maria Mott Davis, the grandmother of the complainant, and then the owner of the Ledford patent, against W. C. Farmer and others, to quiet her title to a tract of 12,900 acres claimed under a junior patent. This suit was instituted in the Circuit Court of the United States, and was decided by Judge Barr in 1894. 141 F. 703. He held that the fourth corner of the patent should be located at or in Cumberland Gap, and the third line should run with the state line from the third to the fourth corner thus ascertained. A few years later, a similar suit was brought by the same person in the same court to quiet her title under the Ledford patent against one Hinckley, claiming 67,000 acres under a junior patent. This suit was decided by Judge Evans, who followed Judge Barr in fixing the fourth corner at or in Cumberland Gap. 141 F. 708. Taking the fourth corner as being at or in Cumberland Gap, both Judge Barr and Judge Evans ran the fourth and fifth lines according to their courses and distances, and from the sixth corner, thus ascertained, ran a line to the beginning, thus closing the survey. As a result, the sixth or closing line, instead of running S. 5° W. 3150 poles, ran S. 5° E. 4560 poles, the course being changed 10°, and the length increased 1410 poles, or nearly 5 miles. This is the location contended for by the complainant below.

While the present case was pending in the court below, a suit involving the location of this patent was begun in the state court and decided by the Court of Appeals of Kentucky. This was the case of Creech *v.* Johnson, decided in October 14, 1903, and reported in 116 Ky. 441, 76 S.W. 185. The highest court of Kentucky there held that, having located the first and second lines, about which there seemed to be no dispute, and thus reached the state line at the third corner, the way to deal with the difficulty presented was to go back to the beginning, reverse the courses, and run the lines according to the calls of the patent, until the state line should again be reached. From this point, where the fourth line reversed should intersect the state line, run with the state line to the third corner, itself on the state line. The location thus fixed as compared with that made by Judges Barr and Evans is roughly shown by a diagram printed in the opinion. 116 Ky. 447, 76 S.W. 186.

The court below had before it all of the opinions when it rendered its decision, but concurred in none of them. It declined to follow the decision of the Court of Appeals of Kentucky as one in a matter of local law, and in a most learned and elaborate opinion determined on a location of its own, different from any one of the others. Having run the first and second lines, as practically agreed upon, to the top of Cumberland Mountain, being the state line, it ran the third line, rejecting its course, with the state line, the distance called for in the patent, namely, 8320 poles. Here it stopped and located the fourth corner, being a point 720.4 poles (about two and one-fifth miles) from Cumberland Gap. From the fourth corner thus fixed it ran the fourth line according to the course and distance given in the patent, thus fixing the fifth corner. It then went back to the beginning corner, reversed the course of the last line and ran it the distance of the call, thus fixing the sixth corner. It then

Protracted Boundaries

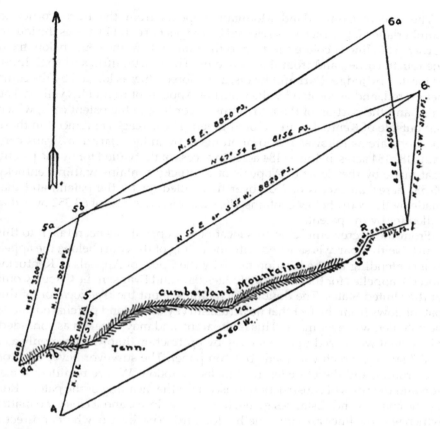

FIGURE 4.4
Illustration from *Bramblet v. Davis* of locations by different courts.

closed the survey by running a line from the fifth corner thus fixed to the sixth corner thus fixed, without regard to either the course or distance of the patent; the result being that the fifth line, instead of running north 55° E. 8320 poles, is made to run north 67.54° E. 8156 poles. The locations made by the different courts are shown in Figure 4.4.*

---

* Owing to the difficulty of imprinting one color upon another, as would be necessary if this cut was prepared as indicated by the court, the locations made by the different courts have been indicated as follows:

The lines according to the calls of the patent are indicated by 1, 2, 3, 4, 4c, 5, 6, 1. That portion which falls outside of Kentucky is represented by the lines south of the Cumberland Mountains, being 3 to 4 and 4 to 4c.

The lines as located by Judges Barr and Evans are indicated by numbers 1, 2, 3, 4c, 4b, 4a, 5a, 6l, 1.

The lines as located by the court below are indicated by 1, 2, 3, 4c, 4b, 5b, 6, 1.

The lines as located by the Court of Appeals of Kentucky are indicated by 1, 2, 3, 4c, 5, 6, 1. Editor of case volume.

118   *Boundary Retracement*

The line of Cumberland Mountain appears from the third corner to Cumberland Gap, which was selected by Judges Barr and Evans as the fourth corner. This line is colored red up to the point where the court below fixed the fourth corner, and from there on, blue. The fourth, fifth, and sixth lines, as located by Judges Barr and Evans, are colored blue; as located by the court below, red; and as located by the Court of Appeals of Kentucky, yellow. The third and the portion of the fourth line, according to the patent calls, which fall outside of Kentucky, are colored green. The acreages embraced in these locations are as follows: The location made by Judges Barr and Evans contains 182,184 acres, being 96,184 acres in excess of that called for by the patent; that made by the Court of Appeals of Kentucky contains within Kentucky 93,552 acres, an excess of 7522 over that called for by the patent; and that made by the court below contains 149,352 acres, an excess of 63,352 over that called for by the patent.

From the decree enforcing this location, an appeal has been taken to this court. Neither party is satisfied with the ruling of the court below; the appellant contending for the location made by the Court of Appeals of Kentucky, and the appellee for that made by Judges Barr and Evans in the Circuit Court of the United States. The doubt which hangs about the true location of this patent flows from the fact that no actual survey of the land purchased and to be patented was ever made. This is apparent, and may be taken as conceded. The patent was based upon a survey by protraction, known commonly as a "call" survey, which was made only on paper. The surveyor never went on the ground, and therefore left no "tracks" to follow. We are remitted to reason and the rules of construction to ascertain the meaning of the patent. Run by the courses and distances given, the survey closes, and it closes no matter which way the lines are run, whether forward to backward, whether directly or by the reverse method. The first, second, and third corners being fixed by natural objects, and the others described by courses and distances, there is no difficulty in locating the entire survey according to the strict terms of the patent. But it appears, when thus located, that part of it falls outside of Kentucky. Well, what of it? If the patentee saw fit to place part of his survey outside of Kentucky, ought the court therefore to be expected to relocate the patent and place all of it inside of Kentucky? We think not. If part of it clearly falls outside of Kentucky, that part fails, just as does part embraced in a prior patent.

The duty of the court in this contingency is not to strain the law and dislocate the survey, in order to bring all its lines within Kentucky, but simply ascertain and define the portion of the survey which does really lie within Kentucky. This can best be done by first locating the lines which lie in Kentucky, and locating them as nearly as possible according to the calls of the patent. This is what common sense suggests and the settled rules of construction in such cases require. Let us quote the principles laid down in the leading case of *Pearson v. Baker,* 34 Ky. 321, all of which apply to the present situation:

Protracted Boundaries

First. In the general, distance yields to course, or, in the absence of any circumstance bringing the mind to a contrary conclusion, the courses shall be first pursued, contracting or extending the distances, as the case may require, to make the survey close. (*Bryan v. Beckley*) Litt. Sel. Cas. 91.

Second. The beginning corner in the plat or certificate or survey is of no higher dignity or importance than any other corner of the survey. *Beckley v. Bryan, Ky. Dec.* 91; 1 Pirtle's Digest, 114.

Third. The order in which the surveyor gives the lines and corners in his certificate of survey is of no importance to find the true position of the survey. Reversing the courses is as lawful and persuasive as following the order of the certificate. (*Thornberry v. Churchil*) 4 T.B.Mon. 32 (16 Am.Dec. 125).

Fourth. That construction is to prevail which is most against the party claiming under an uncertain survey. It is his duty to show and establish his corners. *Preston's Heirs v. Bowmar, 2 Bibb,* 493. From which it will follow that he who sets up and relies on an outstanding claim must show that it embraces the land in contest, and should not succeed by using it, when it is uncertain whether it embraces it or not.

When, then, having located the first and second lines we come to the beginning of the third, and find that, by running it according to the course and distance given in the patent, it crosses into Virginia and ultimately ends in Tennessee, a difficulty is presented which in our opinion entirely justifies us in going back to the beginning, reversing the calls, and tracing the lines the other way, so as first of all to locate the lines which do lie within Kentucky, and ascertain the boundaries of the granted land which is located there. There is no trouble in doing this, and having run the sixth and fifth lines reversed, the fourth line reversed brings us again to the top of Cumberland Mountain, from which, if run out on the same course, the line would, at the distance given in the patent, strike the end of the third line run out according to the patent call. But since it is a vain thing to run the lines of a Kentucky survey outside of Kentucky, and since we have reached the boundary of the state at two widely separated points, one the beginning of the third line, and the other, practically, the beginning of the fourth, so that between the lines thus run and the boundary itself all the Kentucky land granted by the patent is really embraced, we may reject the third line and the part of the fourth outside of Kentucky, and adopt a new third line, which runs with the state line along the top of Cumberland Mountain between the two points mentioned. In this way all the lines are located as called for in the patent, except the third and part of the fourth, for which a portion of the state line is substituted ex necessitate, thus closing the survey and giving the patentee every acre of land in Kentucky which he is entitled to on that side. *Bruce v. Taylor, 2 J.J.Marsh.* 160.

The method adopted by Judges Barr and Evans, and by the court below, after ascertaining that the third line, run by course and distance would leave Kentucky, cross the wedge-like point of Virginia, and land in Tennessee, was to make a new third line with a new fourth corner, all located in Kentucky. To do this, Judge Barr and Evans rejected both the course and distance given in the patent, fixed the fourth corner at or in Cumberland Gap, and ran the line by the meanders of the top of the mountain to it, while the court below, rejecting the course of the patent, ran the line by the meanders of the top of the mountain the distance of the call, and stopped, fixing his fourth corner there. We agree with the court below that Judges Barr and Evans were in error in fixing the fourth corner "at or in Cumberland Gap," merely because the patent says that the third line, after being run S. 60° W. 8320 poles, would end "at a stake near Cumberland Gap." Under such a call, how "near" the stake, marking the terminus of the third line, would be to Cumberland Gap must be determined by running out the line according to its course and distance. *Harry v. Graham*, 18 N.C. 76, 27 Am.Dec. 226. And the terminus of a line run a distance of 26 miles might well be said to be near a place so well known as Cumberland Gap if it should be said to be near a place so well known as Cumberland Gap if it should turn out to be about 4 miles from it. Rejecting Cumberland Gap as the fourth corner, and disregarding the course of the call, the court below ran the third line by the meanders of the top of Cumberland Mountain the full distance called for in the patent, and stopped there, although still 2 miles or more from Cumberland Gap. The fourth corner, thus fixed, was regarded near enough to come within the description of the patent.

In our opinion, the court below avoided one error to fall into another. It, too, created a corner not contemplated in the patent, but in a different way. Clearly the surveyor never intended to run the third line with the meanders of the state line either to Cumberland Gap or the distance of the call, or he would have said so. He intended to run the line S. 60° W. 8320 poles, and thus fix the fourth corner, because the survey was a paper one, and only from the fourth corner thus fixed would the remaining lines of the call, when traced, close the survey. He probably supposed that a line run from the third corner S. 60° W. would stay in Kentucky; but he made a mistake, and the patentees took the chance. The fault we find with Judges Barr and Evans and the court below is that they made a new line and created a new corner before they were compelled to. Obviously, the running of the third and part of the fourth lines outside of Kentucky necessitated a new boundary on that side, which, of course, would be the state line; but it seems to us the proper way to find how much of the survey is bounded by the state line, is to go back to the beginning, reverse the courses, and run the calls until the state line is again reached. This we think is what would naturally be done if a possible purchaser under a junior patent wished to find out the boundaries of the Ledford patent on the northwest side.

The creation of a new fourth corner made more trouble. The fourth, fifth, and sixth lines, beginning at the new fourth corner and run according to the patent, would not close the survey by many miles. Various expedients for closing it were suggested and different ones adopted. Judges Barr and Evans

Protracted Boundaries 121

ran the fourth and fifth lines according to the calls, and from the sixth corner thus fixed ran to the beginning. This changed the course of the sixth line from S. 5° W. to S. 5° E., and increased its length from 3150 poles to 4650 poles, a distance of almost 5 miles. If they had run the fifth line by its course until it intersected this sixth line, reversed and prolonged, the fifth line would have been 9468 poles, instead of 8820 poles, in length, an increase of 948 poles, or about 3 miles, and the sixth line 5518 poles, instead of 3150 poles in length, an increase of 2368 poles, or over 7 miles. To obviate the alteration in the course and distance of the sixth line, or the large increase in the distance of both the fifth and sixth lines, if run according to their courses, the court below, after running the fourth line the course and distance of the call, and thus fixing the fifth corner, went back to the beginning, reversed the call, ran the sixth line the course and distance of the patent, and fixed the sixth corner. The fifth and sixth corners thus fixed, it joined by a line of its own, having neither the course nor distance of the patent, the course being N. 67° 54' E., instead of N. 55° E., a change of 12° 54', and the distance being 8156 poles, instead of 8820 poles, a shortage of 664 poles, or about 2 miles.

We have already given the areas of the different locations. On the one side, it is contended that a location containing 96,184 acres in excess of that called for by the patent (as is the case with the location of Judges Barr and Evans) is by that fact alone put under suspicion; while, on the other hand, it is insisted that a grant of 93,552 acres (the area covered by the location of the Court of Appeals of Kentucky), although 7552 acres in excess of the amount called for by the patent, will be reduced by existing prior grants, far below the 86,000 acres paid for, and to which the patentees were justly entitled. It appears in the record that the appellee's surveyor made a map showing the overlaps upon the location made by Judges Barr and Evans, and estimated they amounted to 105,000 acres, leaving about 87,000 acres. This estimate was not, however, supported by the production of the map, although it was said to be in existence, and the court refused afterwards to reopen the case for the purpose of going carefully into the overlaps, holding the inquiry not to be material at that stage of the case. We must therefore regard the testimony upon this point as unsatisfactory. Whatever might be shown, our conclusion would remain unaffected. If the patentees got more than 86,000 acres within the granted boundaries, it could not be said that they have been defrauded because of overlaps reducing the patentable amount below 86,000 acres. They took the risk of the overlaps when they laid down the survey where they did.

There was much discussion as to the effect which should be given the decision of the Court of Appeals of Kentucky in the case of *Creech v. Johnson*, 116 Ky. 441, 76 S.W. 185. The appellant relies largely upon the holding in that case. The appellee suggests that the case was between parties having but a slight interest in the main controversy; that it was prearranged, while the present case was in the lower court, to secure a location by the Court of Appeals of Kentucky, which would be controlling; and that that court did not have the advantage of a full presentation of their side by the present owners of the

Ledford patent. The matter of the location of the patent was brought to the attention of the Court of Appeals in *Asher v. Howard,* 70 S.W. 277, 24 Ky.Law Rep. 961; Id., 72 S.W. 1105, 24 Ky.Law Rep. 2118, where the court refused a rehearing and declined to locate the patent because the necessary data were not then before it. The court evidently appreciated at that time the large interests involved, for it says so. The motion to advance and the briefs in the case of Creech *v.* Johnson further advised it of the importance of the main question. Under the circumstances, we must hold that whether the location of the Ledford patent, involved in Creech *v.* Johnson, ought to have been heard and adjudicated, was one for the Court of Appeals to pass upon, and, since it entertained jurisdiction and delivered a considered opinion, which appears in the Reports of the court prepared for publication, we must accept its conclusions as its deliberate judgment upon the location of the patent, entitled to the weight such judgments usually are. *Adams Express Co. v. Ohio State Auditor,* 165 U.S. 194, 219, 17 Sup.Ct. 305, 41 L.Ed. 683.

We have been referred to a number of cases bearing upon the effect which should be given by a federal court to a decision of the court of a state upon a question of local law. *Preston v. Bowmar, 6 Wheat.* 580, 5 L.Ed. 336; *Williamson v. Berry, 8 How.* 495, 12 L.Ed. 1170; *Suydam v. Williamson,* 24 How. 427, 16 L.Ed. 742. But it has not become necessary for us to determine nicely the applicable rule, because, treating the location as an original matter, we have reached the same conclusion as the Court of Appeals of Kentucky did.

In conclusion: The present owners of the patent stand in the shoes of the patentees. The deed from the patentees, under which they hold, warned them of a possible mistake in the patent by expressly excepting "any part of the 86,000-acre tract which may be found to lay outside of the state of Kentucky." If they lose some land by this location, it was the patentee's fault, for which they must stand good. The patentees selected, and through the surveyor located, the land, and described it in the patent issued, which by the settled law of Kentucky is to be construed most strongly against the patentees. If they saw fit not to have the land actually surveyed, but a mere "call" survey by protraction made, they took the risk. As one of the witnesses, a surveyor of the early days, referring to such surveys, said:

If you hit the land, you save it; and if you miss it, you lose it.

It closed and bounded a tract of land, but it was laid down in the wrong place. There was an overlap into Virginia and Tennessee. It cannot be presumed that Kentucky intended to grant land outside its boundaries, thus covered by its patent; nor can it be presumed that it intended to make good such a mistake by conveying land inside its boundaries not covered by its patent, as the patentees located it. All that Kentucky intended to grant was the land inside of Kentucky embraced within the boundaries of the patent, and such a grant is effected by locating the patent the way in which, concurring with the Court of Appeals of Kentucky, we have located it.

Protracted Boundaries

The judgment is reversed, and the case remanded for further proceedings in conformity with this opinion.

In the Kentucky case of *Givens v. U.S. Trust Co.** referenced in 11 C.J.S. § 50 earlier, the court stated as follows: In the case of *Albertson v. Chicago Veneer Co.*, 177 Ky. 285, 197 S.W. 831, 835, the court there confronted with a similar contention, when commenting on the rules for ascertaining the boundary of a tract of land as here contended for by appellant, said:

> The foregoing rules, however, have for their prime purpose the ascertainment of the true boundaries of lands, and are not designed to be used for the purpose of imposing boundaries other than the real ones, and the rule that courses and distances must yield to natural objects and established boundary lines, in fixing the boundaries of lands, does not apply, when it is evident that the call for a natural object, or an established boundary line was made, under the mistaken belief that it existed at the point where the surveyor reported it to be, when in fact the natural object and boundary line was not at that point.

Also, in the case of *Bryant v. Strunk*, 151 Ky. 97, 151 S.W. 381, 383, we said:

> While lines of a patent will be extended in order to reach a designated object, this will not be done where it is evident from all the facts that the surveyor simply made a mistake as to the location of the object; he supposing it to be at one place, when, in fact, it was at another.

We there held that a patent will not be run in order to reach lines of other surveys called for in the patent, when to do so would give the patent a different shape from that shown by the plat made by the surveyor and on which the patent issued, and when it is shown that the surveyor did not in fact run the lines, and was mistaken as to the location of the surveys referred to. In such a state of case the patent will be located according to the calls of the patent, so far as this may be done without interfering with older surveys. The same principles underlie the opinion of the court in *Kentucky River Timber & Coal Co. v. Morgan*, 210 Ky. 113, 275 S.W. 12.

It is evident that the surveyor here, in making the call of this seventh line and calling for its terminus in the line of the distant David Turner patent, proceeded under a mistaken belief either as to the distance of such survey or as to its proper location in that his call of 30 poles therefore was less than one-tenth of the distance of 1 mile between the seventh corner of this survey and the terminal line called for as 30 poles distant. This line is not a marked or run line, but was, we are convinced, simply mistakenly protracted by the surveyor.

In *Albertson*, this same court also had this to report:

> It is not to be presumed that the surveyor knowingly made a description for the lands, which it is impossible to run out, according to any recognized

---

* 86 S.W.2d 986, 260 Ky. 762, Ky. App., 1935.

rules of surveying. It is true that the well-established doctrine is to the effect that as a general rule when there is a conflict between courses and distances and recognized objects, which are called for as establishing the boundary lines of a survey, the courses and distances must yield, and natural objects and established boundaries of other tracts called for and designated known points therein must be accepted as the true boundary of the land, and lines actually run and marked by the surveyor control the fixing of the boundary. If a prior survey is called for in a patent as its boundary, and the lines of the prior survey are established lines, then they control the boundary lines of the subsequent survey, which calls for them. *Jones v. Hamilton,* 137 Ky. 256, 125 S.W. 695; *Wallace v. Maxwell, 1 J. J. Marsh.* 447; *Baxter v. Evett's Lessee,* 7 T. B. Mon. 329; *Ewing's Heirs v. Savary, 3 Bibb,* 236; *Bruce v. Taylor, 2 J. J. Marsh.* 160; *Trustees, etc., v. Wagnon, 2 A. K. Marsh.* 379; *Bruce v. Morgan, 1 B. Mon.* 26; *Pitman v. Nunnelly,* 32 S.W. 606, 17 Ky. Law Rep. 793; *Brockman v. Rose,* 90 S.W. 539, 28 Ky. Law Rep. 673; *Morgan v. Renfro,* 124 Ky. 314, 99 S.W. 311, 30 Ky. Law Rep. 553; *Alexander v. Hill,* 108 S.W. 225, 32 Ky. Law Rep. 1147; *Rock Property Co. v. Hill,* 162 Ky. 324, 172 S.W. 671; *Brashears v. Joseph,* 108 S.W. 307, 32 Ky. Law Rep. 1139.

One of the reasons of this rule is that a mistake might creep into the description of the survey, as to course or distance, and the survey is the line where the surveyor actually ran regardless of any improper description. In the last-named opinion the following rule was laid down to be followed in ascertaining the boundary of a tract of land:

That in determining boundaries marked corners are the most satisfactory evidence, then natural objects, such as streams, ridges, and cliffs, then calls for the lines of other patents, which are of record and susceptible of definite location, then courses, and lastly, distances.

The foregoing rules, however, have for their prime purpose the ascertainment of the true boundaries of lands, and are not designed to be used for the purpose of imposing boundaries other than the real ones, and the rule that courses and distances must yield to natural objects and established boundary lines, in fixing the boundaries of lands, does not apply, when it is evident that the call for a natural object or an established boundary line was made, under the mistaken belief that it existed at the point where the surveyor reported it to be, when in fact the natural object and boundary line was not at that point. *Bryant v. Strunk,* 151 Ky. 97, 151 S.W. 381. This exception to the general rule has a peculiar potency, when, as was said in *Ralston v. McClurg, 9 Dana* (Ky), 338.
Where the purpose:

is not to ascertain the position of lines and corners once actually run and established, but to construct a survey by making two lines never run, these lines should be fixed where the surveyor would have made them if he had run them out. *Mercer v. Bate, 4 J. J. Marsh.* 334.

Protracted Boundaries

In *Bryant*, this court emphasized,

> While lines of a patent will be extended in order to reach a designated object, this will not be done where it is evident from all the facts that the surveyor simply made a mistake as to the location of the object; he supposing it to be at one place, when, in fact, it was at another. To run out this survey so as to reach on the east the lands [called for] and on the west the survey [called for], we have not only to extend a line of the survey beyond any reasonable contemplation of mistake, but we have to put in a number of other lines on both sides of the survey. No rules of construction will justify such a departure from the calls of the patent and the plot accompanying the survey, where the lines were not in fact located on the ground, and the calls of the patent must control.

The Texas case of *State v. Coleman-Fulton Pasture Co.** was one of first impression, and dealt with what appeared to be excess land left between established surveys. The court discussed it this way and laid down a sensible rule:

> The intention of the surveyor reading from the testimony of the witnesses and the data in the land office show the very apparent purpose of the locator to acquire the land within the description of the field notes, and patent appropriating thereby the land in controversy. We see no reason, fact, or argument that can justify the conclusion that the surveyors in this case in locating lines would have ever left a space between surveys of such conformation as the state claims was made between all such surveys adjacent and called for in the field notes. The contention of the state is in exact opposition to the correct theory that obtains in locating and reconciling surveys. The intention of the surveyor as expressed in his work in making office surveys must have an almost controlling effect in construing his work, in the absence of a better reason to ignore it. The intention would not control if the footsteps of the surveyor was actually made and could be found and identified on the ground by fixed and identified calls for location objects. *State v. Palacios*, 150 S.W. 229; *Robinson v. Doss*, 53 Tex. 496; *Coleman County v. Stewart*, 65 S.W. 383.
>
> We have not had our attention called to any case as authority for a suit like this: that is, where all the field notes in a grant call for adjacent surveys as this, and represent no vacant land apparent from all the files, surveys, field notes, maps, and plats in the General Land Office, and finally are embraced in a patent, that the state may declare a vacancy between surveys fitting on to each other, and make new field notes, tearing apart and separating the contiguous and companion surveys, thereby wedging in a vacancy, and call it unappropriated public domain, in order to sue for it as public domain. This is not the method of reforming a grant so as to recover an excess.

---

* 230 S.W. 850, Tex. Civ. App., 1921.

126   *Boundary Retracement*

If the theory of the state is to give the Barrett its full quantity of land, and take the excess, if any, on the north by either tearing them apart and slide some further north, or slide the Barrett down, to accomplish such a result, as shown by the "Blucher State Map," we must hold with the trial court there is no vacancy.

> The amount or excess in this survey is but an evidence against appellant's contention, as said by Mr. Jutice Henry in *Maddox Bros. & Anderson v. Fenner*, 79 Tex. 292, 15 S.W. 239, that-
> 'Such excess, however, has never been held by this court a ground for disregarding surveys actually made. In this case the greatness of the excess is not without force as an argument indicating that the surveyor actually intended to make the Stevens embrace all of the land left by the surrounding surveys. It is apparent that by an actual survey he would have been able to locate a survey of 800 acres in the vacancy beginning at the Ryan southwest corner in some regular shape, and that the irregular and peculiar shape given to the Stevens, as is evidenced by the calls of the patent, was adopted because he was acquainted with the boundaries of the surrounding surveys, and intended to make a survey conforming accurately to them as they then appeared.'

This decision referred to *State v. Palacios*,* which, explaining an interesting twist to a common problem, stated, "Where there is an actual survey, and the field notes of such survey are expressed in the grant, it will be conclusively presumed that the grantor intended to convey the land embraced in such survey; and hence the primary rule that in such cases we should follow the footsteps of the original surveyor."

> When a surveyor has made a mistake in some of his calls, how are we to know which of his calls correctly describe his footsteps? In all cases where two calls, descriptive of a line or corner, lead to different results, as where the call is to run from a given point a certain course and distance to another point, and such other point will not be reached by running said course and distance, one or the other of said calls was inserted by mistake, and the mistaken call should be disregarded. The purpose of all rules in reference to the comparative dignity of calls is to ascertain which of contradictory calls are the mistaken ones.
> In this grant there are no field notes, in the sense in which that term is ordinarily used, viz., notes made by the surveyor in the field while making the survey, describing by course and distance, and by natural or artificial marks found or made by him, where he ran the lines and made the corners. We have only a map made by the surveyor and his memoranda explanatory of said map. If we take said map as indicating that he ran from the established southwest corner of the San Anacuas grant south 200 cordels (10,000 varas), thence west 27 cordels (1350 varas),

---

* 150 S.W. 229, Tex. Civ. App., 1912.

Protracted Boundaries

thence south 125 cordels (6250 varas), and thence east, north, east, north, and west to the beginning, then he did not include the land in controversy in said survey. On the other hand, if said survey was made so as to coincide with the boundaries of the San Andres and San Pedro de Charco Redondo grants, as shown by the surveyor's notes explanatory of said map, it did include the land in controversy.

As to courses and distances, it will be seen from the field notes in the La Huerta, as patented by the state, that while the course and distance from the southeast corner of Las Anacuas (south 10,000 varas) is the same as is shown upon the map attached to the La Huerta grant, and the second call (west 1355 varas) is practically the same, the third call in the patent is south 6944 varas, instead of 6250 varas, as shown by said map, the fourth call in the patent is south 80 east 6355 varas, in lieu of east 6500 varas in the grant. The fifth call in the patent is north 10 degrees east 5000 varas, in lieu of north 6250 varas. The sixth call in the patent is south 80 east 1770 varas, in lieu of east 2350 varas, and the seventh call is north 13,428 varas, in lieu of north 10,000 varas. The eighth call in the patent is the same as that in the grant, viz., west 7500 varas. It is evident from the calls and descriptive matter in the field notes of the patent that the same were made from an actual survey. Why the departure on four of the six lines from the calls in the grant? Evidently, in order to make said survey extend to and coincide with the lines and corners of the surveys called for in the grant on the south and east. If there are mistakes in the map as to both course and distance to lines and corners of the adjacent grants on the south and east, does it not render it probable that calls for distance on the west were also mistakes, when it is found that such calls will not coincide with the boundaries of the surveys called for on these lines?

The map shows Arroyo Agua Poquita (creek of Little Water) near the corner 10,000 varas south of the southwest corner of Las Anacuas, and this same creek near the next corner 1350 varas west. The field notes of the patent do not call for any creek near these corners.

The weight to be given to a call for the line or corner of a previous survey depends very much upon whether or not the lines or corners of such survey were probably known to the surveyor at the time he made his survey. Could he have identified them by natural or artificial calls on the ground, or by running from their known corners? The San Pedro de Charco Redondo (St. Peter of the Round Water Hole), as patented by the state, calls to begin "in the middle of Charco Redondo, a round water hole in Palo Blanco creek." Presumably this round water hole was of some notoriety, as it gave its name to the Ramirez grant. The field notes call to run from the beginning corner up the Palo Blanco 5,290 varas to a rock in the bed of said creek; thence north 1 degree 15 west 22,587 varas, "to original rock corner; thence east 6,400 varas to Los Olmos creek, which is shown to be another name for Agua Poquita creek; at 6,500 varas, a pile of rock." It thus appears that lines and corners of the

128   *Boundary Retracement*

Charco Redondo survey could easily have been found and identified by the surveyor who surveyed La Huerta, and who probably made both of said surveys, if La Huerta was actually surveyed on the ground, and that the northeast corner of Charco Redondo is within 100 varas of Agua Poquita. The corresponding corner of La Huerta, marked on the map, should be near said creek.

The other referenced case, *Coleman County v. Stewart*,* added additional guidance:

It is apparent from the statement [made] that either the calls for distance or the calls for the river surveys was mistakenly made, for there is no theory that arises from the facts that would reconcile the two. One or the other must be of controlling influence, and in determining which of the two must yield two rules should be observed. The first is that the survey must be constructed by reference to the calls in the grant, and such calls cannot be aided by the lines and calls of other surveys not mentioned in the field notes. The second is that where the lines and corners were not established on the ground, as is the case here, we must remove the ambiguity or uncertainty arising from the inconsistent calls by ascertaining and giving effect to the intention of the surveyor at the time of making the location.

An important principle is mentioned here, that surveys and calls not mentioned in field notes cannot serve to control the calls made.

What to Do about Protracted Lines

Again, relying on *Rowe v. Kidd* where the court discussed the various possibilities and the reasons for them, the following options exist, depending on the circumstances:

Lines which have been run by protraction only, that is, not actually run and marked, will be located by calls for course and distance where it appears that the latter calls locate the land more nearly in conformity with the plat and the correct acreage,† or that the call for the object was made under a mistaken belief as to its location‡; but if the surveyor was not mistaken as to the location of the visible thing called for, course and distance will yield thereto, whether the line of the survey was actually run or merely protracted.§¶

---

* 65 S.W. 383, Tex. Civ. App., 1901.

† *Givens v. U.S. Trust Co.*, 86 S.W.2d 986, 260 Ky. 762, Ky. App., 1935; *Swift Coal & Timber Co. v. Sturgill*, 223 S.W. 1090, 188 Ky. 694.

‡ *Albertson v. Chicago Veneer Co.*, 197 S.W. 831, 177 Ky. 285.

§ *Rowe v. Kidd*, D.C. Ky., 249 F. 882, affirmed 259 F. 127, 170 C.C.A. 195.

¶ 11 C.J.S. § 50(a) *Lines run by protraction*.

Types of Protractions

There are two types of protraction: (1) straight protraction; (2) existing perimeter, interior lines by protraction. In the first instance, a protracted lot is added as an extension to an existing lot, using one or more lines as an anchor, and drawing the remaining lines on paper, deriving their courses and distances from (usually) an unknown source. In the second instance, an existing perimeter is subdivided into a chosen number of smaller lots. See Chapter 12 for examples of subdivided rectangular grids within a surveyed perimeter. Many of the early rectangular subdivisions were laid out on the ground by survey; however many have been protracted. This is commonly found in the PLSS, where exterior corners were located and markers set, but interior corners left as protracted according to the designs per instructions.

Lines from Other Surveys

Another type of protracted survey is one founded on a constructive or paper survey in which its metes and bounds are taken and copied from, the calls of lines of earlier surveyors forming the exterior boundaries. Such procedure results in a combination, or rather a conglomeration, of a variety of procedures and errors, each of which presents a stand-alone situation that cannot be reconciled as a single unit, but rather as however many units as there are independent surveys defining the perimeter of the subject parcel. The West Virginia case of *Hansford v. Chesapeake Coal Co.** is an early example of such a situation.

While this dispute was due to a contract issue, whereby after survey there was significant overage in the total area, the court nevertheless recognized that the estimate of area was not based on a survey prior to or at the time of the transaction, but occurred a year later by, and for the benefit of, the grantees. The court recited "the vendors knew the original grant for the land had been founded on a survey located by protraction and not by actual survey, that is, it was founded on a constructive or paper survey, its metes and bounds having been copied from the calls of lines of earlier surveys forming its exterior boundaries, and that no actual survey had at any time since been made of the land; none was demanded or made at the time of the sale or afterwards until [the following year] in no way connected with the purchase."

Discussed in Chapter 3 under the section on surveys was the West Virginia case of CSX Hotels located a change of a municipal boundary in this fashion, and while the actual location of the new line was not the issue in the case, what was at issue was the fact that the location had not been accomplished

---

* 22 W. Va. 70, W. Va., 1883.

130   *Boundary Retracement*

by an actual instrument survey on the ground. Since the requirement was that the line be based on an accurate survey, after considerable research and deliberation, the court concluded that an "accurate survey map" is a map reflecting the course and distance measurements, boundaries, and contents of the territory that is proposed to be incorporated into the municipality's limits. The map must reach the desired level of precision consistent with the purposes of the survey, namely to provide notice to residents and free-holders of the municipality and the territory encompassed by the annexation petition of a potential change regarding who will vote in municipal elections; taxation and revenues; and the provision of services.

The Unmarked Line

Lines not marked, such as from protracted surveys, were sometimes called for.

The rule for an unmarked line of an adjacent survey is superior to call for course and distance is not a rule of absolute application, and a call for the line of an adjoining survey should not prevail over a call for distance, unless such line can be located with reasonable certainty and accuracy.*

Guidance

The Kentucky court has had extensive experience with protracted surveys, and its decisions offer guidance in dealing with them. One of the decisions discussed earlier, that of *Bryant v. Strunk*,[†] sets out the reasoning behind their applications in a straightforward manner:

> It is true that the rule is that the calls of a patent for course and distance must give way to known or established objects found on the ground. But, after all, the rules that have been laid down on this subject are for the purpose of establishing the actual location of the lines and corners of the original survey, and they have little application where the lines were not run out in the original survey, but were simply laid down by the surveyor by protraction as was evidently the case in the patent before use. When the lines were not in fact run, we have little to guide us except the calls of the patent and the plot of the land accompanying the original survey. The plot accompanying the original survey is potent evidence in the determination of the general shape of the tract of land intended to be patented. To follow

---

* *State v. Stanolind Oil & Gas Co.*, 96 S.W.2d 297, Tex. Civ. App., 1936.
† 151 Ky. 97, 151 S.W. 381, Ky. App., 1912.

Protracted Boundaries

the lines of the other surveys in the case before us on the east to John Murphy's survey and on the west to King's survey, and then on around with still other surveys would be to make this tract include five times as much land as the grantee paid for and give the tract an entirely different shape from that which was evidently contemplated in the grant.

While lines of a patent will be extended in order to reach a designated object, this will not be done where it is evident from all the facts that the surveyor simply made a mistake as to the location of the object; he supposing it to be at one place, when, in fact, it was at another. To run out this survey so as to reach on the east the lands of John Murphy and on the west the Enos King survey, we have not only to extend a line of the survey beyond any reasonable contemplation of mistake, but we have to put in a number of other lines on both sides of the survey. No rules of construction will justify such a departure from the calls of the patent and the plot accompanying the survey, where the lines were not in fact located on the ground, and the calls of the patent must control.

Protracted lines also create an exception to the long-standing rule that natural monuments are controlling. In the Kentucky case of *Duff v. Fordson Coal Co.*,* the court elaborated on two exceptions to the rule, quoting from American Jurisprudence, Vol. 8, Sect. 54, Page 785:

> Whenever natural monuments or objects, which include mountains, rivers, creeks, and rocks, are distinctly called for and satisfactorily proved, they become landmarks to which preference must be given, because the certainty which they afford excludes the probability of mistake. Ordinarily, a preference is given to natural objects over artificial monuments in determining boundaries, but natural objects cannot prevail when they are doubtful, and in that case recourse is had to artificial marks or monuments or other calls of an inferior degree of accuracy.

There are two recognized exceptions to the rule giving preference to natural objects, namely, as stated by appellant: (1) Where the object called for was located, or rather designated by the surveyor by protraction, and not located on the ground by actual survey; and (2) where there is evidence which makes it appear that the surveyor was mistaken as to the position of the object called for. *Asher v. Fordson Coal Co.*, 249 Ky. 496, 61 S.W.2d 20; *Fordson Coal Co. v. Napier*, 261 Ky. 776, 88 S.W.2d 985.

Protraction in the PLSS

Typically, protraction is relatively common in the Public Lands states, which would normally be expected, since not all lines have been run, nor all corners

---

* 298 Ky 411, 182 S.W.2d 955, Ky. App., 1944.

132   *Boundary Retracement*

monumented, yet titles have been created, and lands have been conveyed. It would seem that sufficient rules for guidance had been established in the eastern states as previously discussed, however following are some selected examples involving PLSS situations.

The Idaho case of *Sala v. Crane** is a good illustration of the necessity of following the rules when dealing with a protraction.

> C. Where public lands conveyed by patent are described by legal subdivisions and lot numbers, "according to the official plat of the survey of said lands returned to the General Land Office by the Surveyor General," and there is a discrepancy in such plat between the lines subdividing the section and the government corners as they exist upon the ground and are shown on the plat and described in the accompanying field notes, the monumented corners will prevail as against a hypothetical dividing line protracted by the draftsman upon the plat in the surveyor-general's office.

Interestingly, the court opened its decision with the following:

The lower court has found that the actual monuments in the field established at the time of the survey are still in existence; those monuments are controlling. (Sec. 4804, U.S. Comp. Stats. 1916; Land Department Regulations of June 1, 1909, secs. 16–20, 25.)

The Supreme Court of the United States has adopted the same view as the *Land Department. (St. Paul & P. R. R. Co. v. Schurmeier, 7 Wall. (U.S.)* 272, 19 L.Ed. 74; *Higuera's Heirs v. United States, 5 Wall. (U.S.)* 827, 18 L.Ed. 469; *Ayers v. Watson,* 137 U.S. 584, 11 S.Ct. 201, 34 L.Ed. 803; *M'Ivers Lessee v. Walker,* 4 Wheat. (U.S.) 444, 4 L.Ed. 611.)

The courts of the states recognize the same rule. (*State v. Ball,* 90 Neb. 307, 133 N.W. 412; *Galbraith v. Parker,* 17 Ariz. 369, 153 P. 283; *Bayhouse v. Urquides,* 17 Idaho 286, 105 P. 1066; *Grand Cent. M. Co. v. Mammoth M. Co.,* 36 Utah 364, 104 P. 573; *Harrington v. Boehmer,* 134 Cal. 196, 66 P. 214; *Keyser v. Sutherland,* 59 Mich. 455, 26 N.W. 865; *Woods v. West,* 40 Neb. 307, 58 N.W. 938; *Bullard v. Kempff,* 119 Cal. 9, 50 P. 780; *Seabrook v. Coos Bay Ice Co.,* 49 Ore. 237, 89 P. 417; *Whiting v. Gardner,* 80 Cal. 78, 22 P. 71; *Foss v. Johnstone,* 158 Cal. 119, 110 P. 294; Silver King Coalition Mines Co. v. Conkling M. Co., 255 U.S. 150, 41 S.Ct. 310, 65 L.Ed. 561.)

What is interesting is the widespread acceptance of the prevailing rule, both in and outside the PLSS.

In *Sala,* the case involved a boundary line between the lands of appellant and respondent in Sec. 6, T. 47 N., R. 3 W., B. M.

The United States patented to respondent, with other lands, Lots 5 and 6, and to appellant, with other lands, Lot 4 in said section. The boundary line in dispute is the east and west line between Lots 4 and 5, which subdivisions

---

* 38 Idaho 402, 221 P. 556, Idaho, 1923.

Protracted Boundaries   133

comprise the west half of the northwest quarter of Sec. 6, Lot 5 being the southwest quarter of the northwest quarter of Sec. 6.

The section was officially surveyed in 1901. The survey began at the township corner between township 47 and 48 north and ranges 3 and 4 west, being also the northwest corner of 6 and the northeast corner of 1 in the said respective townships. A stone was set to mark this township corner, and the survey ran thence south on a line between said townships, the west line of 6 being coterminous with the east line of 1. The official field notes of this survey show that 40 chains south of this township and section corner the surveyor:

> Set a basalt stone, 18 × 15 × 5 ins., 12 ins. in the ground, for 1/4 sec. cor., marked 1/4 on W. face, from which
>
>> A pine, 20 ins. diam., bears S. 89 1/2 [degree] E., 161 links dist., marked 1/4 S 6 B T.
>> A pine, 18 ins. diam., bears N. 36 1/2 [degree] W., 187 links dist., marked 1/4 S 1 B T.

The quarter corner on the west line of Sec. 6, as established by the survey in the field, is thus marked by a stone monument and witness tree in place upon the ground, and its location 40 chains south of the northwest corner of the section and township is not controverted. The quarter quarter corner on this west line between Lots 4 and 5 is not marked.

The plat of the section prepared in the office of the surveyor-general shows the east-and-west half-section line protracted from the monumented corner on the east line of Sec. 6, which is equidistant from the northeast and southeast corners of the section, to a point on the west line 2.02 chains north of this west quarter corner monumented on the ground and described in the field notes, and the boundary line between Lots 4 and 5 as intersecting the west line of the section at a point 17.98 chains south of the northwest corner of this section and township.

The appellant contends that the southwest corner of Lot 5 is at the stone monument located in the field as the west quarter corner of Sec. 6, and that the northwest corner of Lot 5 and the southwest corner of Lot 4 should be equidistant between the west quarter corner and the northwest corner of the section, or 20 chains from either of these monumented corners.

The respondent bases her right of recovery to the land in question upon the fact that the patent, after describing the lands patented to her as the SE.1/4 of the NW.1/4, the NE.1/4 of the SW.1/4, and Lots 5 and 6, in Sec. 6, contains the following: "According to the official plat of the survey of said lands returned to the General Land Office by the Surveyor-General," and that this qualifying clause means according to the boundary lines of the plat subdividing the section as such lines were protracted in the surveyor-general's office, and that the monumented corners, field notes and all other landmarks appearing as a part of the description in the official plat must be governed solely by this protracted line in determining the true boundary between Lots 4 and 5.

134   *Boundary Retracement*

The case was before this court on a former appeal (*Sala v. Crane*, 31 Idaho 191, 170 P. 92), wherein it is held that:

"Where a patent conveys land according to the official plat of the survey returned by the surveyor-general, the plat becomes an integral part of the description of the land."

The cause being reversed, a new trial was ordered and further testimony taken. The trial court made its findings in accordance with the foregoing facts, and as a conclusion of law therefrom held that:

By reason of the decision of the Supreme Court (in the former appeal) this court is compelled to find and does find, that the southwest corner of said Lot 5 is 2.02 chains north of said quarter corner of said section, monumented by the survey and shown on the ground, and the northwest corner thereof is 17.98 chains south of the northwest corner of the township.

The court then entered judgment upon such findings and conclusions awarding to respondent the premises in controversy. From this judgment this appeal is taken, and presents for determination the question as to which of these two conflicting descriptions contained in the official plat shall control.

Sec. 4804, U.S. Comp. Stats. (sec. 2396, U.S. Rev. Stats., 8 F. Stats. Ann., 2d ed., p. 669), approved Feb. 11, 1805, is as follows:

BOUNDARIES AND CONTENTS OF PUBLIC LANDS, HOW ASCERTAINED. The boundaries and contents of the several sections, half-sections and quarter-sections of the public lands shall be ascertained in conformity with the following principles:

First. All the corners marked in the surveys, returned by the surveyor-general, shall be established as the proper corners of sections, or subdivisions of sections, which they were intended to designate; and the corners of half and quarter sections, not marked on the surveys, shall be placed as nearly as possible equidistant from two corners which stand on the same line.

Second. The boundary-lines, actually run and marked in the surveys returned by the surveyor-general, shall be established as the proper boundary lines of the sections, or subdivisions, for which they were intended, and the length of such lines, as returned, shall be held and considered as the true length thereof. And the boundary-lines which have not been actually run and marked shall be ascertained, by running straight lines from the established corners to the opposite corresponding corners; but in those portions of the fractional townships where no such opposite corresponding corners have been or can be fixed, the boundary lines shall be ascertained by running from the established corners due north and south or east and west lines, as the case may be, to the water-course, Indian boundary-line, or other external boundary of such fractional township.

Third. Each section or subdivision of section, the contents whereof have been returned by the surveyor-general, shall be held and considered as containing the exact quantity expressed in such return; and the

Protracted Boundaries

half-sections and quarter-sections, the contents whereof shall not have been thus returned, shall be held and considered as containing the one-half or one-fourth part, respectively, of the returned contents of the section of which they may make part."

The Land Department has construed this section, and in its Revised Rules of June 1, 1909, approved by the Secretary of the Interior, we find the following:

2nd. That the original township, section and quarter section corners established by the government surveyor must stand as the true corners which they were intended to represent, whether the corners be in the place shown by the field notes or not.

3rd. That the quarter quarter corners not established by the government surveyor shall be placed on the straight lines joining the section and quarter section corners, and midway between them, except on the last half-mile of the section lines closing on the north and west boundaries of the township or on other lines between fractional sections.

4th. That all subdivisional lines of a section running between corners established in the original survey of a township must be straight lines running from the proper corner in one section line to its opposite corresponding corner in the opposite section line. (See secs. 75–82.)

All lines are supposed to be actually surveyed, and the intention of the grant is to convey the land according to the actual survey; consequently, distances must be lengthened or shortened and courses varied so as to conform to the natural objects called for. (*M'Iver's Lessee v. Walker*, 17 U.S. 444, 4 Wheat. (U.S.) 444, 4 L.Ed. 611.)

Provision was made by the act of Feb. 11, 1805 (now U.S. Comp. Stats., sec. 4804) that townships should be 'subdivided into sections, by running straight lines from the mile corners, marked as therein required, to the opposite corresponding corners, and by marking on each of the said lines intermediate corners, as nearly as possible equidistant from the corners of the sections on the same.' Corners thus marked in the surveys, are to be regarded as the proper corners of sections, and the provision is, that the corners of half and quarter sections, not actually run and marked on the surveys, shall be placed, as nearly as possible, equidistant from the two corners standing on the same line.... Lines intended as boundaries, but which were not actually run and marked, must be ascertained by running straight lines from the established corners to the opposite corresponding corners. (*St. Paul & P. Ry. Co. v. Schurmier*, 74 U.S. 272, 7 Wall. (U.S.) 272, 19 L.Ed. 74.)

But ordinarily, surveys are so loosely made, and so liable to be inaccurate, especially when made in rough or uneven land or forests, that the courses and distances given in the instrument are regarded as more or less uncertain, and always give place, in questions of doubt or discrepancy, to known monuments and boundaries referred to as identifying the land. (*Higuera's Heirs v. United States*, 72 U.S. 827, 5 Wall. (U.S.) 827, 18 L.Ed. 469.)

136   *Boundary Retracement*

> In ascertaining the lines of land, the tracks of the surveyor, so far as discoverable on the ground with reasonable certainty, should be followed; and marked trees, designating a corner or a line on the ground, should control both courses and distances. (*Ayers v. Watson*, 137 U.S. 584, 11 S.Ct. 201, 34 L.Ed. 803.)

To the same effect, see also *State v. Ball*, 90 Neb. 307, 133 N.W. 412; *Halley v. Harriman*, 106 Neb. 377, 183 N.W. 665; *Galbraith v. Parker*, 17 Ariz. 369, 153 P. 283.

In *Keyser v. Sutherland*, 59 Mich. 455, 26 N.W. 865, the court said:

> The quarter lines are not run upon the ground, but they exist, by law, the same as the section lines. When the township and section lines are run, and the corners marked according to law, the quarter-section lines are ascertained on the plat by protracting lines across the section north and south and east and west from the opposite quarter-section posts set in the exterior lines of the section by the government surveyor, and smaller subdivisions are protracted so as to make one-half and one-fourth of a quarter section.

The cases principally relied upon in support of respondent's contention are *Gazzam v. Lessee of Phillips*, 61 U.S. 372, 15 L.Ed. 958, and *Cragin v. Powell*, 128 U.S. 691, 9 S.Ct. 203, 32 L.Ed. 566. *Gazzam v. Phillips* expressly disapproves the holding in *Lessee of Brown v. Clements*, 44 U.S. 650, 3 HOW 650, 11 L.Ed. 767. The parties in both cases rest their claims upon the patents issued to James Ethridge and William D. Stone, respectively. A review of all the facts and questions of law considered in these cases cannot be made within the proper limits of a judicial opinion. The Gazzam-Phillips case, when considered in connection with the holding in *Brown v. Clements*, which it overrules, appears to lend support to respondent's contention, although in a number of essentials we think that both *Gazzam v. Phillips* and *Cragin v. Powell* present a question materially different from the case at bar. The Gazzam case was one in which the patent expressly stated the acreage conveyed and the precise compensation to be paid to the government for such acreage. The description in the Ethridge patent was "for the southwest quarter of section 22 ... . containing ninety-two acres and sixty-seven hundredths of an acre, according to the official plat of the survey, etc." The description in the Stone patent was "for the south subdivision of fractional section 22 ... . containing one hundred ten acres and fifty-one hundredths of an acre, according to the official plat of the survey, etc."

In the case at bar, the exterior lines of Sec. 6, as run by the survey, are 80 chains in length, except that the northeast and southwest quarter quarters are fractional because of adjoining lakes. The section lines are each parallel to their opposite lines, and the northwest quarter of this section, in which the land in controversy is situated, when its area is computed from the monumented corners, contains the government allotment of 160 acres, each of the

Protracted Boundaries

exterior lines being 40 chains in length. A number of the subdivisions in the south half of the section, as shown by the plat, contain an excess acreage, and it appears from the plat that the only reason the two section corners, that is, the northeast and southwest, were not monumented on the ground and now shown to be in place, a mile from each opposite section corner, is because of the bodies of water mentioned.

In *Chapman & Dewey Lbr. Co. v. St. Francis Levee Dist.*, 232 U.S. 186, 34 S.Ct. 297, 58 L.Ed. 564 at page 196 of the official reports, it is held that any part of the description in the plat must be read in light of the others, for it is a familiar rule that where lands are patented according to such a plat, the notes, lines, landmarks and other particulars appearing thereon become as much a part of the patent and are as much to be considered in determining what it is intended to include as if they were set forth in the patent, citing *Cragin v. Powell, supra*, and *Jefferis v. East Omaha Land Co.*, 134 U.S. 178, 10 S.Ct. 518, 33 L.Ed. 872.

We therefore conclude, and so hold, that the monumented corners shown on the plat should prevail as against this dividing line protracted by the draftsman to a point 2.02 chains north of the quarter corner, both being shown by and being a part of the official plat.

The respondent earnestly contends that the decision in *Sala v. Crane, supra*, is the law of this case, and must control this decision. The doctrine of the law of the case has generally been followed by this court. However, where this court is not a court of final conclusion in the determination of the question presented, and error may be taken to the federal supreme court, the rule of the law of the case contended for is not applicable. (*A. B. Moss & Bro. v. Ramey*, 25 Idaho 1, 136 P. 608.) All courts of intermediate jurisdiction are controlled and bound by the decisions of courts of ultimate resort. (*State v. Moore*, 36 Idaho 565, 212 P. 349.) It is settled that Congress has plenary power to dispose of public lands, and that the federal supreme court is the court of final resort with regard to questions of the character here presented. (*United States v. Gratiot*, 39 U.S. 526, 14 Pet. (U.S.) 526, 10 L.Ed. 573; *California v. Deseret Water, Oil & Irr. Co.*, 243 U.S. 415, 37 S.Ct. 394, 61 L.Ed. 821; *Ruddy v. Rossi*, 248 U.S. 104, 8 A. L. R. 843, 39 S.Ct. 46, 63 L.Ed. 148.) The rule of the law of the case is not an inflexible rule. (*City of Hastings v. Foxworthy*, 45 Neb. 676, 63 N.W. 955, 34 L. R. A. 321; *M. K. & T. Ry. Co. v. Merrill*, 65 Kan. 436, 93 Am. St. 287, 70 P. 358, 59 L. R. A. 711; 2 R. C. L., sec. 188, p. 226, "Appeal and Error.")

And also, where the facts on the second appeal are materially different from the facts in the case previously passed upon, the rule may not always be applicable. The county surveyor, Phinney, who was the only witness who testified at the first trial, stated that the east quarter corner of said Sec. 1 is a stone, while the west quarter corner of said Sec. 6 was not established on the ground, and that he established it 2.02 chains north of the east quarter corner of Sec. 1, because of instructions from the surveyor-general. Upon the second trial, the one from which this appeal is taken, the same witness testified in effect that the west quarter corner of Sec. 6, as monumented on the ground,

138   *Boundary Retracement*

is identical with the east quarter corner of Sec. 1, and the two engineers, Eddelblute and Edwards, who were not witnesses at the first trial, testified positively upon the second hearing that the quarter corner on the west line of Sec. 6 was monumented on the ground and at the place shown by the official plat, and the trial court so found.

It is true that upon a rehearing in the former appeal, the opinion appears to assume that the west quarter corner of Sec. 6 was monumented on the ground, but the testimony at that trial does not show this to have been the fact, while upon the second hearing it is conclusively shown by all the engineers who were called upon to testify. There being this apparent conflict in the official plat itself, and there being no controversy about the location of this quarter corner of Sec. 6 as it was monumented in the field and has ever since been plainly visible, under the foregoing rules and the federal statute above referred to, and the construction and application given to it by the land department in its regulations above referred to, which appear to be in harmony with the decisions of the federal supreme court, particularly as stated in the excerpt from *St. P. & P. Ry. Co. v. Schurmier, supra,* we feel compelled to hold that the proper interpretation of the descriptive clause contained in the deed requires the true quarter corner on the west line of Sec. 6 to be as monumented on the ground, and the quarter quarter corner between Lots 4 and 5 to be equidistant between the quarter corner and the northwest corner of the section.

The cause is reversed and remanded, with instructions to enter a judgment and decree according to the views herein expressed.

Like the South Dakota case of *Titus v. Chapman,** the original monuments, or the places where they stood, govern over unknown markers with no known basis and official maps with protracted, unsurveyed, lines. The map is nothing more than a picture by a draftsperson and the protracted lines are not part of the survey.

In the Florida decision of *Hardee v. Horton,*† the court made the point that:

> a map or plat which represents no survey, but is prepared by projecting lines of a prior erroneous government survey on paper over a space representing a large area of unsurveyed lands, purporting to represent section, township, and range lines according to the rectangular method of surveying, although adopted and referred to in deeds of conveyance as the official map of the grantor, when shown by competent testimony to be inaccurate and unreliable as an aid to locate the unsurveyed lands which are conveyed by description according to the rectangular method of describing lands, is insufficient as a survey of said lands.
>
> A complete title to unsurveyed public lands does not vest in the grantee until the lands conveyed have been identified by an authorized survey; and where unsurveyed lands are conveyed by description according

---

* *Supra.*
† 90 Fla. 452, 108 So. 189, Fla., 1925.

Protracted Boundaries

to the rectangular method of describing lands, although the deed be a grant *in praesenti*, the title vests in the grantee upon delivery of the deed subject to the right and duty of the political authorities of the state to identify and separate by a survey the lands conveyed from the unsurveyed lands within which they are included.

Where an erroneous map, which represents no survey, is referred to in a deed conveying unsurveyed lands by a more particular description, and there is conflict between the map and the more particular description, the lands should be located by the more certain and definite description, and the erroneous map may be treated as surplusage.

Notwithstanding the Montana case of *Larsen v. Richardson*,* discussed in Chapter 1 and subsequently referred to, wherein the court stated that when there are no actual "footsteps," the surveyors duty is to follow what the deed writers said, the Texas court added another perspective. In *Gilbert v. Finberg*,† the Texas court stated, "where there are no footsteps of the surveyor to follow, the rules relating to actual surveys are frequently not applicable." That seems to make perfect sense, since survey rules apply to surveys, but not when there is no actual survey.

All rules relating to the comparative dignity of calls are designed to aid in determining which call or calls of a survey were made by mistake; but in passing upon an office survey, where there are no footsteps of the surveyor to follow, the rules relating to actual surveys are frequently not applicable. In the case of *Boon v. Hunter*,‡ the court said: "It is, however, not believed that the same rules in regard to the lines or corners of other surveys called for in a patent can be applied, when it clearly appears that no actual survey was ever made, and in such case it becomes necessary to look to all matters of description contained in the patent, in order to determine what particular land was conveyed and intended by the state and the grantee to be conveyed by the patent."

Additional Footsteps

Whenever there are protracted lines, there are more than just one set of footsteps. The original line(s) used as anchor(s) are subject to one set of survey techniques with its inherent errors and adjustments, while the protracted lines are a different situation entirely. Without benefit of survey, they may have been estimated, scaled, or merely sketched on paper. Frequently, the procedure was done to depict a certain acreage for transfer. Usually, there

---

* 361 Mont. 344, 260 P.3d 103, Mont., 2011.
† 156 S.W. 507, Tex. Civ. App., 1913.
‡ 62 Tex. 588.

are no markers on the ground, unless someone after the fact decided to make an attempt of marking it. As such, it defies retracement surveyors in their attempts to accomplish a reasonable fit.

Someone's attempt at setting markers merely adds another set of problems to the mix. In other words, here is yet another set of footsteps.

5

Which Set of Footsteps Is Which?

If you walk the footsteps of a stranger, you'll learn things you never knew you never knew.

Pocahontas

Frequently, the retracement surveyor will find that there is more than one set of footsteps. Several scenarios may present themselves: one is original, one is original and perpetuated either in part or in total by previous retracement surveyors, one is new based on educated guesswork, or some theory frequently based on some pure mathematics either with or without an acceptable beginning point, and there are possibly others. When confronted with such a dilemma, it is the retracement surveyor's "sole duty, function and power" as stated in the *Rivers* case, to retrace the footsteps and the line(s) of the original surveyor. Frequently what is found is that one or more retracement surveyor has followed the footsteps of not the original, but a later retracement surveyor, and has accepted the work of that surveyor or others. The temptation to accept that work on face value can be great, especially if the previous surveyor is known, or has a reputation for doing good work. Anyone can make a mistake, take shortcuts, or otherwise fall short of the duty to follow the original, and without a comparison of later work with the original, is to take a calculated risk. The *Ivalis* case following is an example of how a number of surveyors can reach the wrong conclusion.

Finding the original set of footsteps among the others is often not a simple task. Those before, who failed to fulfill their duty, present this retracement person with difficulty and confusing issues. If such previous followers are still available for consultation, or if their plats, notes, and records are clear and complete, the difficult task becomes easier, but that is seldom the case. Some of the following examples will illustrate this confusion and its disservice to other following surveyors, land owners, and the public at large.

Once the original set of footsteps has been found, and located, there may exist yet another set of footsteps, perhaps even more than one. In such cases, there is defined therein an identifiable tract of land between the sets of footsteps, and someone will have title to it. Quite often this is a title dispute waiting to be tried in a court of law. While the determination of title is not the responsibility of the surveyor, it is the duty of the surveyor to acknowledge abutters, recognize obvious title conflicts and overlaps, and report to the appropriate people. Considering contracts, to survey someone's land

141

142

implies surveying their ownership; locating someone's deed on the ground is quite another matter, and may not equate to locating their title, or what they believe they own. As previously discussed, a deed is but one of several means of acquiring, or transferring, title.

Wrong Set of Footsteps

Chapter 1 included discussion of two decisions wherein people followed the wrong set of footsteps. Both had to do with title to a parcel of land claimed by both parties to the action. *Ski Roundtop* is an example where the investigation of the record title did not trace back to an original source, thereby not accounting for an additional title besides the two litigants. The court's analysis of *Burton v. Duncan* resulted in finding a sliver of land between the two litigants, to which neither party had a claim.

The case of *Ivalis v. Curtis v. Harding** demonstrates a situation whereby a number of surveyors, over a period of time, followed the wrong set of footsteps. Between 1859 and 1863, the external boundaries of Section 35 were established by government surveyors who set only the section corners and quarter-corners. The first subdivision of the interior of Section 35, including Government Lots 8 and 9, was accomplished by the Vilas county surveyor, Graham, in 1915, and recorded in 1918. This survey erroneously located the north–south quarter-line (serving as the boundary between Lots 8 and 9), thereby placing the lands conveyed and occupied by the defendants in Lot 8 rather than Lot 9. Later, surveyors used Graham's monuments to further subdivide the surrounding area. Graham's error in establishing the line between Lots 8 and 9 was perpetuated by Harding in 1971, when he conducted surveys.

Another surveyor, Inman, conducted his survey in 1989 and established what the court found to be the "true" north–south quarter-line. This established that the disputed property lay not in Government Lot 8 but in Lot 9. Inman later testified that Harding failed to use the level of care and skill required of a reasonable registered land surveyor in 1971, and thus negligently prepared the surveys of the land sold to the defendants. In his defense, Harding pointed out that other surveyors commonly relied upon Graham's monuments, including Inman, on other occasions. Harding was found negligent for failing to determine the correct line between the two lots.

Even though a number of surveyors had followed Graham's footsteps and accepted his monuments, when tried in court it was determined that they had not followed original footsteps. They had, instead, followed an incorrect set left by a retracement surveyor.

---

* 173 Wis.2d 751, 496 N.W.2d 690, Wis. App., 1993.

Who Left the Footsteps?

With government surveys, it is usually a relatively simple matter of determining who left the *original* footsteps. There should be some sort of record as to who set original corners, and who created original lines. Field notes and plats are generally filed with the appropriate agency.

Unlike government surveys, private surveys, until recently, but even then only in some instances, and in some states, do not share a like requirement. Subdivision surveys in many states now must be recorded, but that is a relatively recent requirement. Other than with surveys directed by the local government (states and state-directed entities), there is no such requirement, and likewise no requirement for the filing of field notes. Fortunately, some of these have been preserved in a variety of locations and repositories, but, unfortunately, not enough.

However, colonial surveyors also had their instructions, and were sworn to undertake their duty in a faithful manner, though it may have varied from town to town and from client to client. Many of those regulations may be found, and, in the absence of information to the contrary, it is presumed that a surveyor faithfully did his duty.

For any government survey (BLM, USFS, etc.), the agency and its surveyors should have followed the appropriate instructions, and courts presume that they did.* Even earlier courts in the metes-and-bounds states discuss presumptions of surveyor being on the land and physically doing the survey:

It the case of official surveys, it will always be presumed that the surveyor did his duty,[†] and that his work was accurate.[‡]

The presumption is that surveys were made as stated in the field notes,[§] and, in the absence of proof to the contrary, the lines of a survey will be presumed to have been run and the corners to have been established as returned.[¶]

To learn about a particular surveyor's habits and procedures, one can follow any survey done by that surveyor. Not being confined to the locus in question, a retracement surveyor can follow surveys at other locations, including in other towns.

Another source of information about a previous surveyor is whether he ever testified in court. If so, there should be a record of testimony, and perhaps a file of exhibits, including field notes and maps.

---

* The original surveys made by the United States government are not to be taken as conclusive presumptions of law; they may be rebutted and impeached as to their correctness; but, prima facie, they are to be presumed to be correct until their accuracy has been properly impeached. State ex rel. Brayton v. Merriman, 6 Wis. 14. (1858).
† *Thatcher v. Matthews*, 101 Tex. 122, 105 S.W. 317, 1907.
‡ *Franklin v. Texas Sav., etc., Assoc.*, 119 S.W. 1166, Tex. Civ. App., 1909.
§ *Lamb v. Bonds & Dillard Drilling Corporation*, 107 S.W.2d 500, Tex. Civ. App., 1935.
¶ *Dolphin v. Klann*, 246 Mo. 477, 151 S.W. 956, 1912.

144 Boundary Retracement

The following testimony is an example given to the court of how a particular surveyor performed his work.

A part of the testimony for the plaintiff was that of C. E. Uren, as follows:

"I am a surveyor and civil engineer. I know the lines in controversy. I went upon the ground this spring at the request of plaintiff to make a survey to establish the corner. I sought for the missing corner, and could not find it. The north and south section corners are in place, and are easily discoverable without the aid of field-notes. Not being able to find the corner, I established it as a lost corner by dividing the distance between the two section corners equally. The line is over a mile long, while the survey calls for a mile long only. I examined carefully all the marks that would aid me. I discovered a line of blazed trees from the south stake, which were originally line-trees. I followed along the same, and at half a mile set a stake, near which and about right distance and direction was the remains of a yellow pine stump badly burned, but which I believe was the bearing-tree."

"The only pine-tree which is now standing has two blazes on the west side, which are old, and made quartering just like line blazes. This tree could not be a witness-tree. I examined it very carefully. It has a large burned place on the side defendant claims the stake to be, and facing stake. There is no mark outside the burned place. Surveyors do not make bearing-marks on dead trees or burned wood."

"About two hundred and fifty feet south of the pine is an oak eighteen feet east of my line with old blaze on the west side. If the line ran where defendant claims, it would be on the east side of this oak-tree. I found line trees all along my line. I also found line-trees on the north half of the line. Section lines are always originally run from the south toward the north. I ran the line through from section-post to section-post, and placed a stake exactly midway; the distance from said post to the north and south stakes being exactly 40.875 chains in each instance. The fence of defendant is 57 3/4 feet north of the stake so placed, and incloses 2.66 acres of plaintiff's land."

"I was absolutely unable to find or establish the original one-quarter section stake, and there is nothing on the ground by which it or a bearing-tree can be established. I examined very carefully."

Higgins v. Ragsdale, 23 P. 316, 83 Cal. 219 (1890).

Studying a particular surveyor to learn as much as possible about him involves reading his journals, studying his field notes, any correspondence and possibly researching the family genealogy. An individual's training, apprenticeship, and education can offer insight as to a surveyor's methods and procedures.

Frequently, if a local surveyor is well known, there will be notations and discussions about him in local and county histories. A search of local libraries, schools, historical societies, and archives of a variety of sorts may yield records and equipment.

Which Set of Footsteps Is Which?   145

Any maps, descriptions in a variety of title documents (deeds, etc.), can lend insight as to measurements, and ultimately to equipment. Knowing the equipment can lead to how it may have been used. Notations are often found in probate records, not only the individual's, but also other landowners who may have relied on the individual's work, or services. Wills, probate partitions and divisions, returns of expenses for services to either the decedent or the estate may be insightful.

Learning from Field Notes

The term "field notes" in its ordinary sense means "notes made by the surveyor in the field while making the survey, describing by course and distance, and by natural or artificial marks found or made by him, where he ran the lines and made the corners."* The law presumes that surveys were made as stated in the field notes approved by the General Land Office.†

The importance and significance of field notes may be shown by its place in the hierarchy of conflicting calls.

Corners and marked lines

Natural monuments

Artificial monuments

Maps, plats, and field notes

Monuments are facts; the field notes and plats indicating courses, distances, and quantities are but descriptions which serve to assist in ascertaining those facts. *Myrick v. Peet*, 180 P. 574 (1919) citing *Martin v. Carlin*, 19 Wis. 454, 88 Am. Dec. 696 (1865).

Field notes, along with maps and plats, are high in the list of priority of calls when ambiguity exists within a description. As many courts have noted, maps and plats, and field notes, go hand in hand. In the process, the surveyor locates and sets corners and marks and notes lines utilizing natural features where available, and placing artificial markers where necessary. While doing, so, notes are made in the field (usually, but sometimes soon after) which constitute "a written description of the survey,"‡ from which maps and plats are later compiled utilizing those notes, observations, and recollections. Therefore, in the overall scheme of available information, field notes can be very important and useful. They are, however, not without their shortcomings.

---

* *State v. Palacios*, 150 S.W. 229, Tex. Civ. App., 1912.
† *Nanny v. Vaughn*, 60 Tex. Civ. App. 290, 187 S.W. 499, 1916.
‡ *Outlaw v. Gulf Oil Corporation*, 137 S.W.2d 787, Tex. Civ. App., 1940.

146 *Boundary Retracement*

Notations and abbreviations are often made, and not all are consistent or adhere to a standard pattern. Sometimes it is a matter of deciphering the notations made by a particular individual.

Admissibility of Field Notes

The New York decision of *Wightman v. Campbell** stated that "field book entries made by a deceased surveyor for the purpose of a survey on which he was professionally employed, are admissible in evidence as being made in the discharge of his professional duty." (Stephen's Digest of the Law of Evidence, arts. 25, 27; *Price v. Earl of Torrington, [1703] 1 Salk.* 285; *Mellor v. Walmesley, L. R.* [1905] 2 Ch. 164; *Walker v. Curtis,* 116 Mass. 98; *People v. Holmes,* 166 N.Y. 540.) If the proper foundation had been laid for the introduction of the notes in evidence, as easily might have been done by showing that they were made within the scope of professional employment, and the notes had been put in evidence, the witness, himself a surveyor and competent to interpret them, could have testified therefrom as to the location of the boundary lines of the Taylor farm, and the evidence would have been entirely proper.

In the decision of *Kennedy v. Oleson,*[†] the issue of admissibility of field notes arose. The court reported, "Defendants contend Exhibits P-1 and P-3 contain opinions and conclusions that make them inadmissible and the opinions of experts based thereon inadmissible. We think not. Perhaps a resurvey would always contain conclusions of the surveyor as to the original line but it is based upon experience and professional knowledge, and the actual location of the original line is a fact. Here the surveyor's notes refer to original gas pipe stakes. The evidence shows he was in a position to know this. He had been a surveyor for 50 years and was acquainted with the surveyor who laid out the original Auditor's Plat. His notes also show a pin referred to as the Mahone pin and state he considers it in error. This pin was discussed by the deceased surveyor with the defendant William Oleson, the plaintiff's husband, and the surveyor's assistant at the time of the survey. And the notes show the distances between the lines found by the surveyor and a different distance for the Mahone pin. If this is an opinion and conclusion, it is an apparent conclusion inherent in the facts of the location of the line and the Mahone pin. The basis of the conclusion is apparent and inheres in it. These are not such conclusions or opinions as will exclude the field notes or render inadmissible the opinion of a qualified expert based thereon. *Hodges v. Sanderson,* 213 Ala. 563, 105 So. 652; *Allely v. Fickel,* 243

---

* 217 N.Y. 479, 1916.
[†] 251 Iowa 418, 100 N.W.2d 894, Iowa, 1960.

Which Set of Footsteps Is Which?

Iowa 105, 49 N.W.2d 544; *Wightman v. Campbell*, 217 N.Y. 479, 112 N.E. 184; *Warcynski v. Barnycz*, 208 Md. 222, 117 A.2d 573, and *In re Estate of Scanlan*, 246 Iowa 52, 54, 67 N.W.2d 5, 6."

The decision in *Daniel v. Florida Industrial Co.** analyzed the survey field notes.

> "This is an action, instituted by a notice of motion for judgment, brought by the Florida Industrial Company, a corporation, against W. F. Daniel to recover the balance due on three notes for $8,797.19 each, which were given that company by Daniel for the deferred installments of the purchase price of fractional section 1, township 43 south, range 29 east, and lots three and four in fractional section 6, township 43 south, range 29 east, in Hendry county, Florida which were conveyed by it to Daniel by a deed dated September 14, 1925."

Even though the subject of the note is Florida land, the action was brought in Virginia.

> "Daniel's defense is a plea of set-off in which he alleges that the land shown on the diagram, which we have inserted, as the area K X Y C, containing seventy-eight acres, is in section 1, township 43 south, range 29 east; that Florida Industrial Company had lost title to this seventy-eight acres through the adverse possession of other persons; and that he is entitled to have the value thereof at the time of his purchase of section 1 set-off against the demand of the plaintiff."
>
> "Space forbids the insertion of a plat covering all the territory to which reference is made in the evidence; and the diagram inserted shows in their entirety only sections 1 and 12. On this diagram the section lines called for by the Government map and field notes are shown as solid lines. The north and south lines and the east and west lines, shown as broken lines, denote lines which the defendant claims were run on the ground by the surveyor for section lines, though the locations thereof do not accord with his field notes."
>
> "The reply of Florida Industrial Company is that the area K X Y C lies in section 12, not in section 1, and was not conveyed, or intended to be conveyed, in the deed from it to Daniel."
>
> "These Florida lands were a part of the public lands of the United States, and were subdivided and disposed of by it to private persons. The United States statutes prescribing rules for surveying and for ascertaining the boundaries and contents of subdivisions of the public lands have remained unchanged in any particular here material from long prior to 1872 to the present time. See sections 2395–2396, Revised Statutes U.S. 1878; U.S. Code Ann. (1928), Title 43, Public Lands, sections 751–752. The pertinent parts of sections 751–752, U.S.C.A., Title 43, Public Lands, are quoted in the footnote.† We note below the rules of law which have been

---

* 159 Va. 472, 166 S.E. 712, Va., 1932.
† 13 Section 751, U.S.C.A., Title 43, Public Lands, paragraphs 1, 3, and 5:

laid down by the courts relating to what constitute Government subdivisions of the public lands, and to ascertaining the boundaries thereof."

"A survey of the public lands of the United States does not describe boundaries, it creates them. *Cox v. Hart* (Cal. 1922), 260 U.S. 427, 43 S.Ct. 154, 67 L.Ed. 332. A 'Government section' or other 'Government subdivision' is the land lying within the lines of that section or subdivision as surveyed and marked upon the ground by the government surveyor. This is true, even though due to some mistake made by him the surveyor ran and marked the lines on the ground otherwise than as was required by the rules and regulations of the government, or as called for in his field notes. It the point at which a section corner was placed by the government surveyor or the line marked by him as a section line is satisfactorily established, it is conclusive as to the location of the corner or line, though the location of the corner or line does not accord with that corner or line as shown on the government map of the subdivision, or with that called for by the field notes of the surveyor who made the survey on which the

---

The public lands shall be divided by north and south lines run according to the true meridian, and by others crossing them at right angles, so as to form townships of six miles square, unless where the line of an Indian reservation, or of tracts of land surveyed or patented prior to May 18, 1796, or the course of navigable rivers, may render this impracticable; and in that case this rule must be departed from no further than such particular circumstances require.

Third. The township shall be subdivided into sections, containing, as nearly as may be, six hundred and forty acres each, by running through the same, each way, parallel lines at the end of every two miles; and by making a corner on each of such lines, at the end of every mile. The sections shall be numbered, respectively, beginning with the number one in the northeast section and proceeding west and east alternately through the township with progressive numbers till the thirty-six be completed.

Fifth. Where the exterior lines of the townships which may be subdivided into sections or half-sections exceed, or do not extend six miles, the excess of deficiency shall be specially noted, and added to or deducted from the western and northern ranges of sections or half-sections in such township, according as the error may be in running the lines from east to west, or from north to south; the sections and half-sections bounded on the northern and western lines of such townships shall be sold as containing only the quantity expressed in the returns and plats, respectively, and all others as containing the complete legal quantity.

Section 752, U.S.C.A., Title 43, Public Lands (omitting par. 3):

The boundaries and contents of the several sections, half-sections, and quarter-sections of the public lands shall be ascertained in conformity with the following principles:

First. All the corners marked in the surveys, returned by the Field Surveying Service, shall be established as the proper corners of sections, or subdivisions of sections, which they were intended to designate; and the corners of half and quarter sections, not marked on the surveys, shall be placed as nearly as possible equidistant from two corners which stand on the same line.

Second. The boundary lines, actually run and marked in the surveys returned by the Field Surveying Service, shall be established as the proper boundary lines of the sections, or subdivisions, for which they were intended, and the length of such lines, as returned, shall be held and considered as the true length thereof. And the boundary lines which have not been actually run and marked shall be ascertained, by running straight lines from the established corners to the opposite corresponding corners; but in those portions of the fractional townships where no such opposite corresponding corners have been or can be fixed, the boundary lines shall be ascertained by running from the established corners due north and south or east and west lines, as the case may be, to the watercourse, Indian boundary line, or other external boundary of such fractional township.

Which Set of Footsteps Is Which?

government map is based. *Watrous v. Morrison,* 33 Fla. 261, 14 So. 805,39 Am.St.Rep. 139; *Kirch v. Persinger,* 87 Fla. 364, 100 So. 166; *Lawler v. Rice & Goodhue Counties,* 147 Minn. 234,278 N.W. 317,180 N.W. 37; *Nesselrode v. Parrish,* 59 Iowa 570, 13 N.W. 746; *Climer v. Wallace,* 28 Mo. 556, 75 Am.Dec. 135; *Hess v. Meyer,* 73 Mich. 259, 41 N.W. 422; *Puget Mill Co. v. North Seattle Imp. Co.,* 120 Wash. 198, 206 P. 954; *Beardsley v. Crane,* 52 Minn. 537, 54 N.W. 740; *Anderson v. Johanesen,* 155 Minn. 485, 193 N.W. 730; *Galbraith v. Parker,* 17 Ariz. 369, 153 P. 283, on rehearing *Ivy v. Parker,* 18 Ariz. 503, 163 P. 258; *State v. Ball,* 90 Neb. 307, 133 N.W. 412; *Ogilvie v. Copeland,* 145 Ill. 98, 33 N.E. 1085; *Yolo County v. Nolan,* 144 Cal. 445, 77 P. 1006."

"'In the sale of lands in sections, or subdivisions thereof, including lots, according to the government survey, the survey as actually made controls.' *Miller v. White,* 23 Fla. 301, 2 So. 614; *Liddon v. Hodnett,* 22 Fla. 442. It is the survey as it was actually run on the ground that governs, if the monuments, corners, or lines actually established can be located or proved. Courses and distances yield to such corners and lines, so long as the latter can be located, and for the reason that the latter are the fact or truth of the survey as it was actually made while the former are but descriptions of the act done, and when inaccurate they cannot change the fact.'*Watrous v. Morrison,* 33 Fla. 261, 14 So. 805, 807,39 Am.St.Rep. 139.'"

"But the field notes and government plats made by the government surveyors at the time public lands were originally surveyed are presumed to be correct; and they are determinative of the true location on the ground of the corners and lines therein called for, until it is shown by clear and convincing evidence that the corner or line in question was in fact established on the ground by the government surveyor at a place other than that called for by the field notes and plat of the original survey. Especially is this true where, as in the case at bar, the field notes call for the location of the corners and lines substantially as the law directs that they shall be established. *Langle v. Brauch,* 193 Iowa 140, 185 N.W. 28; 9 C.J. (Boundaries), page 215, section 130 *et seq.*"

"In a deed between private parties where the land conveyed was a part of the public lands of the United States and is described and designated only as a certain section, or other government subdivision in a designated township and range, without any other description, it means that 'government section' or other 'government subdivision,' and in determining the lines of the land thereby conveyed the same rules apply that apply in determining, as against the government, the lines of a section, or subdivision thereof, which has been conveyed by it. The lines of the subdivision as marked on the ground at the time of the original government survey, when satisfactorily established, determine the lines of the land conveyed, whether they accord with the government map or the field notes of the survey or not. See cases above cited; *Kelsey v. Lake Childs Co.,* 93 Fla. 743, 112 So. 887; *Town v. Greer,* 53 Wash. 350, 102 P. 239; *Trinwith v. Smith,* 42 Or. 239, 70 P. 816; *Fellows v. Willett,* 98 Okla. 248, 224 P. 298; *Desha v. Erwin,* 168 Ark. 555, 270 S.W. 965. As between the parties to a deed this rule is subject to some exceptions, as for instance, where it is shown that there was a mutual mistake of the parties as to the location of the lines of the government subdivision, and that it was their mutual

150   *Boundary Retracement*

intention that the deed should convey a definite piece of land. *Town v. Greer, supra.* But the case at bar does not fall within such exceptions."

"The official government map of township 43 south, range 29 east, shows section 1 as a fractional section bounded by four straight lines. Its north line is a part of the north line of the township, but neither its course nor length is noted on the map. Its east line is a part of the east line of the township and is noted as being 23.85 chains long. Its south line runs from a point on the east line of the township (corner to section 12) S 89 degrees 47' W 80.17 chains, and is a part of a straight line which is shown as the northern boundary of full sections 12, 11 10 and 9. Its west line is 25.53 chains long, and is the northern segment of a straight north and south line running from the southern to the northern line of the township, which is the western boundary of full sections 36, 25, 24, 13, 12 and of fractional section 1. The section is divided on the map into four lots by north and south lines, and the aggregate of the acreages noted as being contained in these four lots is 197.52 acres."

"The fact that the government map shows that the northern boundary of section 9 and the southern boundary of section 1 are parts of the same straight line, running substantially due east and west, is the key to the way this controversy arose."

"The courthouse square at La Belle, the county seat of Hendry county, is in section 9, and in that community the northwest corner of the square is reputed to be the northwest corner of section 9. From the northwest corner of the courthouse square State highway No. 25 runs for two miles on what appears to be a straight line due east. It then bends to the north through a small, but apparent, angle, and runs on a straight line along the new course for a little more than a mile to the point marked A on the diagram heretofore inserted. At this point it is deflected to what is substantially a due east line and continued eastward along the courses shown on the diagram."

"When Daniel was considering the purchase of section 1, he procured maps of the township which purported to be copies of the official government map. He learned that State highway No. 25 was supposed to run along the northern line of section 9, and driving along the highway he noted the bends in it at the points approximately two and three miles from La Belle. He observed that, if the line along which the highway ran for two miles east of the courthouse at La Belle was projected eastward, it would pass some distance south of the line which runs from A to B on the diagram."

"From this information he concluded that the line of the highway from A to B ran through section 1, and at least several hundred feet north of the south line of the section. He purchased the land in this belief, and according to his testimony, the fact that he thought the section included the land south of the road very materially influenced his decision to purchase section 1. But, there is no evidence which shows, or tends to show, that Florida Industrial Company, or any of its agents, was in any way responsible for Daniel's having thought that this was true, or had any reason to believe that he thought so."

"As has been before stated, the government map notes the acreage in section 1, township 43 south, range 29 east, as being 197.52 acres; and

Which Set of Footsteps Is Which?

it appears from the record that lots 3 and 4 in section 6, township 43 south, range 30 east, contain, or are supposed to contain, in the aggregate eighty-four acres. The total of these two acreages is substantially 281 1/2 acres."

"Soon after Daniel received his deed he heard that the land abutting on the south side of State highway No. 25 from about K to B on the diagram was claimed by some other person, but he made no investigation of the matter until after this suite was brought in January, 1931. He then had an investigation made and ascertained that Mrs. Carrie Byrd Shealy and Mrs. Susan G. Hutchinson and her grantee, Mrs. Lee, were claiming to own all the land lying adjacent to and south of the line A C shown on the diagram; and that they claimed under patents issued by the United States for the northwest quarter and the southwest quarter of section 12, township 43 south, range 29 east, which had been issued on homestead entries made in 1909 and 1910."

"He then had Henry S. Gove, a civil engineer, to make a survey of the eastern tier of sections in township 43 south, range 29 east, for the purpose of locating the south line of section 1 as it had been run on the ground at the time of the original survey. It is chiefly upon his testimony that Daniel relies to support his contention that the true south line of section 1 is the line X Y shown on the diagram."

"Gove introduced with his testimony the government map above mentioned, the field notes of W. L. Apthorp who in 1872 surveyed and established for the government the southern and eastern lines of the township and set the corner posts thereon for the sections bounded thereby, and the field notes of Samuel Hamblin, the government surveyor who in 1872 surveyed the inside section lines of the township and set thereon the corner posts for the sections and quarter sections. His testimony with reference to the survey which he made for Daniel presents two phases."

"The first phase relates to the reproduction on the ground of the eastern tier of sections in the township according to the calls of the field notes of Apthorp and Hamblin, with which the government map accords."

"The second phase of his testimony relates to evidence found by him on the ground, upon which he based the opinion expressed by him that Hamblin did not run the west and south lines of section 1 in accordance with their locations as called for by his field notes, and that he actually ran the west line of the section 417 feet east of the line called for by his field notes, and the south line 711 feet further south than called for by his field notes. That is, that the true west and south lines of the section as actually run by Hamblin on the ground are the lines X F and X Y shown on the diagram."

"When Apthorp made his survey the west line of township 43 south, range 29 east, had been theretofore surveyed and the southwest corner of the township established. He ran the east line of township 44 south, range 29 east, and set a corner post for the corner between townships 44 and 43, or as he expressed it: 'Set post corner TP. 43 & 44 Rgs. 29 & 30 from which a pine bears N 38 deg. E 23 1ks., from which a pine bears N 45 deg. W 5 1ks., from which a pine bears S 20 deg. W 41 1ks., from which a pine bears S 12 deg. E 87 1ks.' This point we shall call M."

152   *Boundary Retracement*

"From this point he ran west on a 'random' or trial line for the south line of township 43. This trial line intersected the west boundary of the township six chains north of the corner theretofore established as the southwest corner of the township. So he moved down to the corner which had theretofore been established as the southwest corner of the township, and ran a 'true' line between the corners established as the southwestern and the southeastern corners thereof, setting posts for the section corners as he went. His notes on this line say that at eighty chains from the southeast corner of the township he 'set post corner sections 1, 2, 35 and 36 from which a pine bears N 11 deg. E 50 1ks., a pine bears S 89 deg. W 47 1ks., no trees convenient in sections 1 and 35.' This point we shall call N. He ran thence to the southeast corner which he had established, and from thence ran north along the east line of the township, setting posts for the section and quarter-sections as he went. Omitting the descriptive matter given between the mile posts, except his calls for quarter-section corners which were positively identified by the witness Cove, his notes read as follows:

Thence north on east boundary sec. 36 T 43 R 29 S & E

Affirmed

(M 1)

First Mile.

80:00 set post corner sections 25, 30, 31, and 36 No bearing trees

Second Mile. North on east boundary section 25

(M 2)

80:00 set post corner sections 19, 24, 25, 30 from which

A pine bears N 78 deg. E 27 lks.

A pine bears S 62 deg. W 75 1ks.

A pine bears S 78 deg. E 34 1ks.

No tree convenient in section 24

(M 2 1/2)

Third Mile. North on east boundary section 24

40:00 set 1/2 mi. post from which

A pine bears N 6 deg. E 18 lks.

A pine bears N 48 deg. W 20 lks.

(M 3)

80:00 set post in edge of road to Ft. Thompson N.W. corner sections 13, 18, 19, and 24 from which

A pine bears N 79 deg. E 84 1ks.

A pine bears N 30 deg. W 98 1ks.

A pine bears S 51 deg. W 107 1ks.

Which Set of Footsteps Is Which?

A pine bears S 44 deg. E 47 1ks.

Fourth Mile. North on east boundary section 13

(M 3 1/2)

40:00 set 1/2 mile post from which

A pine bears N 5 1/2 deg. E 20 1ks.

A pine bears N 77 deg. W 30 1ks.

(M 4)

80:00 set post corner sections 7, 12, 13, and 18 from which

A pine bears N 2 deg. E 77 1ks.

A pine bears N 56 deg. W 48 1ks.

A pine bears S 64 deg. E 98 1ks.

No tree convenient in section 13

Fifth Mile. North on east boundary section 12.

40:00 set 1/2 mi. post from which

A pine bears N 22 deg. W 17 1ks.

No tree convenient in section 7

(M 5)

80:00 set post corner sections 1, 6, 7, and 12 from which

A pine bears N 60 deg. E 19 1ks.

A pine bears N 48 deg. W 12 1ks.

A pine bears S 47 deg. W 36 1ks.

A pine bears S 4 deg. E 37 1ks.

Sixth Mile. North on east boundary section 1.

23.85 interest Jacksons line, south boundary section 36 T 42 S; R 29 E 24; 75 ch. west of his post the SE corner of said Tp.

Set post at intersection from which a pine bears

South 20 1ks. No other trees convenient."

The notations M, M 1, M 2, etc. in the left-hand margin are not parts of Apthorp's notes, but are placed by us for purposes of reference.

> "Gove testifies that the point N, the corner to sections 35 and 36, is a definitely known point. At this point he found a government stake, or post, with the proper government marks on it for this corner; and he was able also to verify this post by one of the two witness trees called for by Apthorp and the stub of the other. He testifies that in his whole survey this was the only stake he found which he would say was an original government stake, the others were, 'I would say possibly twenty or twenty-five years old, possibly more.' He cut into the witness tree which

154 Boundary Retracement

is standing, and found at fifty-eight annual rings from its outer surface evidence that it had been blazed."

"When, beginning at N, Gove retraced the remainder of the south line and the east line of the township in accordance with the calls of Apthorp's notes, he found and positively identified the corners established by Apthorp at points M, M 1, M 2, M 2 1/2, M 3, M 3 1/2, M 4, and M 5. He identified them by the witness trees called for by Apthorp, or the stumps thereof. They were all located on the true township line within a fraction of a chain of the distances from each other for which Apthorp calls in his field notes. The point (M 5) at which Apthorp set the post for the corner to sections 1 and 12 is the point C shown on the diagram, and was positively identified by him by finding one of the witness trees and the stumps of two others called for by Apthorp. He cut into the witness tree that was standing, an 18 inch pine, and at fifty-seven annual rings from its outer surface found evidence that it had been blazed."

"Hamblin began his subdivision of the township at N, the corner of sections 35 and 36 on the south line of the township." Omitting certain calls along the lines of sections 36 and 25, which are not here material, his notes read:

From the corner to sections 1, 2, 35, and 36 on the south boundary of this township I run north between sections 35 and 36.

80:00 set post for corner to sections 25, 26, 35, and 36 from which

A pine 14 in. diam. bears N 16 deg. W 50 1ks.

A pine 15 in. diam. bears S 22 deg. W 43 1ks.

A pine 13 in. diam. bears S 33 deg. E 102 1ks.

A pine 10 in. diam. bears N 81 deg. E 71 1ks.

East on a random line between sections 35 and 36.

80:23 intersected east boundary of township 21 1ks.

North of the post corner to sections 25 and 36 from which I ran N 89 deg. 51 min. W on true line between sections 23 and 36.

40:12 set post for 1/4 section corner

80:23 to corner sections 25 and 26–35 and 36

North between sections 25 and 26

80:00 set post for corner sections 23, 24, 25, 26 from which

A pine 13 in. diam. bears S 13 deg. W 29 1ks.

A pine 15 in. diam. bears N 40 deg. W 92 1ks.

A pine 20 in. diam. bears S 74 deg. E 98 1ks.

3d. rate pine and palmetto land.

East on random line between sections 24 and 25

40:00 set post for temp. 1/4 section corner

Which Set of Footsteps Is Which?

80:19 intersected east boundary of township at post corner to sections 24 and 25 from which I run west on true line between sections 24 and 25.

40:09 set post for 1/4 section corner from which

A pine 7 in. in diam. bears N 75 deg. W 87 1ks.

A pine 10 in. in diam. bears S 32 deg. E 92 1ks.

80:19 the corner sections 23-24-25-26

3d. rate pine and palmetto.

North between sections 23 and 24

32:00 to grass pond

40:00 set post in pond for 1/4 section corner. Could not raise mound.

46:00 cross pond to pine

80:00 set post for corner to sections 13, 14, 23 and 24 from which

A pine 10 in. diam. bears N 72 deg. W 43 1ks.

A pine 8 in. diam. bears S 15 deg. W 80 1ks.

A pine 10 in. diam. bears S 44 deg. E 31 1ks.

A pine 10 in. diam. bears N 34 deg. E 43 1ks. 3d. rate pine and palmetto land

East on random line between sections 13 and 24

4:00 to prairie

6:00 to pine land

15:00 to prairie

18:00 to scrub and palmetto

32:00 to pine

40:00 set post for temp. 1/4 section corner

80:15 intersected east boundary of township 23 1ks.

South of post corner to sections 13 and 24 from which I run

S 89 deg. 50 min. W on true line between sections 13 and 24

40:07 set post for 1/4 section corner, from which

A pine 8 in. diam. bears N 66 deg. E 28 1ks.

A pine 10 in. diam. bears S 4 deg. E 10 1ks.

80:15 to corner sections 13, 14, 23, and 24

3d. rate pine and palmetto

North between sections 13 and 14

40:00 set post for 1/4 section corner from which

A pine 10 in. diam. bears N 45 deg. W 66 1ks.

A pine 10 in. diam. bears S 82 deg. E 100 1ks.

80:00 set post in edge of wet prairie for the corner to sections 11, 12, 13, and 14 from which

A pine 10 in. diam. bears N 35 deg. W 56 1ks.

A pine 10 in. diam. bears S 51 deg. W 70 1ks.

A pine 8 in. diam. bears S 64 deg. E 65 1ks.

A pine 6 in. diam. bears N 24 deg. E 30 1ks.

3d. rate pine and palmetto land.

East on random line between sections 12 and 13

3:00 leave wet prairie to pine land

27:00 to prairie

40:00 set post for temp. 1/4 section corner

57:00 cross to pine land

79:78 intersected the east boundary of the township 42 links south of the post corner to sections 12 and 13 from which corner I run S 89 deg. 42 min. west on true line between sections 12 and 13

39:89 set post for 1/4 section corner in mud pits 10 1ks. east and west

79:78 the corner to sections 11, 12, 13, and 14

3d. rate pine and wet prairie land.

North between sections 11 and 12

9:00 cross wet prairie to pine and palmetto

40:00 set 1/4 section post from which

A pine 15 in. diam. bears N 88 deg. W 82 1ks.

A pine 11 in. diam. bears N 32 deg. E 27 1ks.

66:00 leave pine and palmetto to open bottom land

80:00 set post corner to sections 1, 2, 11, and 12 from which

A cabbage bears S 32 deg. W 123 1ks.

A cabbage bears S 29 deg. E 128 1ks.

A cabbage bears N 51 deg. E 138 1ks.

2nd. rate pine and palmetto land—January 29, 1873.

East on random line between sections 1 and 12

7:00 to pine and scattering cabbage

40:00 set post for temp. 1/4 section corner

80:17 intersected east boundary of township 30 1ks. south of the post corner to sections 1 and 12 from which I run S 89 deg. 47 min. W. on true line between sections 1 and 12.

40:08 set post for 1/4 section corner from which

A pine 10 in. diam. bears N 47 deg. W 53 1ks.

Which Set of Footsteps Is Which?

A pine 9 in. diam. bears S 28 deg. E 78 1ks.

80:17 the corner sections 1, 2, 11, and 122nd rate pine and cabbage land

Knowing that I shall intersect the south boundary of township 42 S of R 29 E in less than 80 chains I run north on a true line between sections 1 and 2.

25:53 intersected the south boundary of T 42 S R 29 E and set post for corner to fractional sections 1 and 2 in mound Pitts ten links east and west.

"1st rate rich bottom land subject to overflow."

"It is plain that the calls in Hamblin's notes for the 'post corner to sections 25 and 26,' the 'post corner to sections 13 and 24,' the 'post corner to sections 12 and 13,' and 'the post corner to sections 1 and 12' were intended by Hamblin as calls for the respective corner posts set for these section corners by Apthorp the year before.

"Gove began at the corner of sections 36 and 35 on the south line of the township (point N), and reproduced on the ground the section lines of the eastern tier of sections in accordance with the calls of Hamblin's field notes in the same order in which Hamblin says he ran them. He testifies that when he did so, he found that the south line of section 1 ran from the point C on the eastern line of the township to B and thence along State highway No. 25 from B to H; and that the point C is the known and positively identified point (M 5) at which Apthorp set a post for the corner to sections 1 and 12. He also testifies that when the lines of section 1 are run in accordance with Hamblin's field notes, it contains substantially 197 1/2 acres."

"The depositions of E. E. Goodno, Mrs. Carrie Byrd Shealy, Drew Hampton and S. L. Stewart were taken by the plaintiff, but were read in evidence by the defendant."

"Goodno testified that he had lived at Fort Thompson, which is about two and one-half miles from section 1, for twenty-eight years, and had for twenty-eight years owned Sections 2 and 11 in township 43, and also the section in the next township which lies east of section 12 of township 43. When he purchased these lands there was a post at the northwest corner of section 12. It was 'about three feet high and six inches square; the top was hewed.'"* * *

"'It was in an open place. There was saw palmetto and chaparral around it, but it was an open place I would say sixty or seventy feet across.' * * * 'Everybody knew that corner because it was right side of a cow path were we were through every day nearly.' Soon after he bought sections 11 and 2 he had them surveyed by Mr. Hampton, Drew Hampton's father, and he located that stake as the northwest corner of section 12. It was there before Hampton made this survey and he left it right where it was. That stake was 'in the center of the road', State highway No. 25. Mr. Stewart buried the stake there. 'It is everybody's belief and understanding that that (State highway No. 25) is the true line, everybody in the country.'"

158   *Boundary Retracement*

"Mrs. Carrie Byrd Shealy is the widow of James M. Byrd, who made a homestead entry on the northwest one-fourth of section 12 in 1909, and the person who secured a patent therefor in 1917. James M. Byrd died in 1914. She testifies that Mr. Hampton, who was county surveyor, made a survey of this quarter section for her husband soon after he made his homestead entry, and pointed out to her the section corner—'The northwest corner Post was shown to me right where the hard road is now.' At that time there was only a county road along there. She and her husband were then living in a house he had built which was located 'within a few feet of the north line, the northwest corner of that tract of land.' This house was standing when she moved away from the land about eight or ten years before her deposition was taken, but is not the house that is there now. At one time she had a fence on the road that ran along her north line, which was part of a fence enclosing about fifteen acres. A part of this fence along the road was there when she moved away from the land."

"Drew Hampton, a civil engineer, who is thirty-six years of age, and has done much surveying in this county and in this township, testifies that when he was a boy he helped his father survey for Mr. E. E. Goodno, in 1909 or 1910, a road from La Belle eastward beyond the east line of township 43 south, range 29 east. Before that time his father had made a survey of the northwest one-fourth of section 12 for James M. Byrd. What is now State highway No. 25 runs along the line of the road surveyed by them for Goodno. At that time there was a government stake in place for the northwest corner of section 12. 'It was a wood stake, lightwood stake, a square stake' and had 'government scribe marks on it.' This corner stake was located 'about the center of the road' as it now exists. He and his father followed out the north line of section 12 and found a line of blazed trees all the way across. This line corresponded with the stake mentioned, and 'extended along the road, right on out, prolonged.' The blazed marks on the trees had the appearance of being original government marks. Mr. Goodno built the first road along this line. In 1917, he (Drew Hampton) helped Stewart resurvey this road as a county road, and 'we ran right along by that corner.'"

"S. L. Stewart was county engineer of Lee county from about 1916 to 1923. When Hendry county was cut off from Lee in 1923 he became county engineer for Hendry county. In 1916 or 1917 he surveyed for Lee county what is now State highway No. 25 from what is now the west line of Hendry county to the east line of the county. He began at the west end and worked east. He testifies in part as follows: When, in 1916 or 1917, he ran the survey for what is now State highway No. 25, he found at the northwest corner of section 12 'an old stake which is probably now buried in the middle of that road. We first ran the road and took the section line for the center of our road, and when we found the stake we were going to put the grades down and drove that stake in the ground and set off-set stakes twenty-five feet on each side.' There was no road at that time east of that stake. 'It was one of these old fat lightered stakes. I don't remember the marks on it, but I do know that we had no doubt about it being the government corner.' For the purpose of checking the line

Which Set of Footsteps Is Which?   159

between sections 1 and 12 he 'went through and found the northwest corner' of section 1. From this stake he ran the road along the south line of section 1 for something less than half a mile and then turned it northeast to get on the north line of the township. The location of State highway No. 25 is the same as the location of the road run by him in 1916 or 1917."

"It also appears from the evidence that Mrs. Hutchinson and Mrs. Lee have occupied a house built a short distance east of the point B and less than 300 feet south of the line B C for many years, and that along part of the line B C there is and has been for some years a fence which encloses Mrs. Lee's premises."

"None of these witnesses specifically fixes the exact point along the road at which the stake to which they refer as the northwest corner of section 12 stood; but the inference to be drawn from their testimony is that it stood about at the point A, at which the road makes a bend."

"The testimony of Gove, John H. Caldwell, Jr., Hampton and Stewart establishes the following facts: The north lines of sections 9, 10 and 11, which are shown on the government map as lying in the same north and south tier of section as section 12, have been generally accepted for many years as running along a line which lies south of the east and west line called for in Hamblin's field notes as the north line of section 12. The road shown near the bottom of the diagram, running from east of S to R, is supposed to run along the south line of section 11; and someone at some time has marked as the south line of section 12 a line which corresponds with the line A R. This line runs 669 feet south of the line called for in Hamblin's notes as the south line of section 12."

"Gove also testifies that at 80.50 chains south of the line just mentioned as having been marked for the south line of section 12, he found a line which at some time had been marked by somebody as the north line of section 13, and that the west end of this line is 581 feet and the east end of it 514 feet south of the line called for in Hamblin's notes as the south line of section 13."

"Drew Hampton explains the fact that the east and west section lines of sections 9, 10 and 11, and section 14 have been run on lines lying approximately 660 feet south of the lines called for by Hamblin's notes as the corresponding lines of sections 12 and 13 in this way. The east line of township 43 south, range 29 east, as established by Apthorp has no excess length in it, but the west line of the township from the Caloosahatchee river to the corner originally established for its southwest corner is 660 feet longer than is called for by the field notes. The western line of section 9 (in which La Belle is situated) is less than two miles from the west line of the township, and four miles from the eastern line of the township, and because of this fact the lines of section 9 have been run from the west line of the township instead of from the east line. This has resulted in their being run 660 feet further south than lines run from the mile posts set by Apthorp on the east line of the township. He accounts in the same way for the location of the lines of sections 10, 11 and 14 south of the corresponding lines of section

160 Boundary Retracement

12. He further testifies that the chain used for making these surveys in 1872–73 was a chain 660 feet (ten chains) long, and that it was not uncommon for them to 'drop' a chain in keeping a count of the distance run. His testimony as to the excess length of the western line of the township is nowhere denied."

"It is also to be noted in this connection that Apthorp in his field notes for the survey of the south line of the township states that when he began at the southeast corner of the township and ran west on a trial line for the south line of the township be intersected the west line of the township six chains north of the corner theretofore established for its southwest corner. Under paragraph 5 of section 751, U.S.C.A., Title 43, Public Lands, heretofore quoted in footnote 1, if there was an excess in the length of the western line of the township, the surveyor making the subdivision thereof was required to so subdivide the township into sections so as to throw this excess into the west line of the northwest section of the township. If this was done, it necessarily resulted in the south line of the northern tier of sections being a broken line, unless there was a corresponding excess in the length of the eastern line, which there was not."

"We now come to the evidence upon which Gove based his opinion that Hamblin actually ran the south line of section 1711 feet farther south and its west line 410 feet farther east than his field notes call for."

"When he started his investigation, Mr. John H. Caldwell, Sr., took him to the point R on the diagram and showed him the bend in the road there as the reputed southwest corner of section 12. Assuming this to be the corner, he ran a line S 0 degree 54 E, and at 80.50 chains came to a 'lighter stake' in place. He continued on down the same line looking for the pond which Hamblin calls for on the west line of section 24. He found the pond, but it was to the west of the line on which he was running."

"He then came back to the point R and ran along the road N 0 degree 54' W 80 chains, the distance Hamblin calls for between the southwest and northwest corners of section 12, but found no evidence of an established corner. Putting in a temporary stake at this point he turned east and looked for marked trees. He found one, a 30 in. oak, which stood 2.15 chains east of the road. This tree had a blaze on both its eastern and western sides. (He later cut into this tree and found that it showed signs of the blazes for 'over fifty rings' from the other surface.)"

"After he had made certain examinations which need not be here noted, he went to the known corner post (N) on the south line of the township which Apthorp had set for the corner of sections 36 and 35, at which point Hamblin began his survey of the section lines of the eastern tier of sections in the township."

"From this point he ran N 0 degree 54' W which, making appropriate correction for the variation of the needle, is the line called for by Hamblin's field notes. At eighty chains from his starting point he found no evidence of a corner. He then sought for marked trees east and west of the line, and

Which Set of Footsteps Is Which? 161

running an east and west line from a point 78.77 chains from his starting point, he found blazed side line trees both east and west of his north and south line, indicating 'a line crossing' at that point. He cut into these trees and found evidences of the blazes at fifty-eight and fifty-five annual rings from their outer surfaces. On this evidence he accepted the point 78.77 (N 1) as Hamblin's corner to sections 36 and 25, though it was 1.23 chains short of the distance called for."

"Continuing the same course northward he found evidence that he was retracing the line run by Hamblin. At eighty chains north of N 1 he found none of the witness trees called for by Hamblin nor any evidence of a line crossing, but he did find an old corner stake lying on the ground there. He accepted this point (N. 2) as Hamblin's corner of sections 25 and 24, the southwest corner of 24."

"Continuing his north line he came to a pond, the center of which he located as 33.33 chains north of the southwest corner of section 24. Of this pond Gove says:

> "That is the only pond in that vicinity. It is about 200 feet in diameter and has a mud bottom. Most of these flats out there are of a hard sand bottom. The water during a period of rain will stand there a time and be probably shoe deep, but not a permanent pond, while this pond is a permanent pond. He calls for that pond at a distance of forty chains because he set a quarter corner in the pond; it was too deep to raise a mound and he simply set a stake. You can sink as deep in the mud as you do in the water." The plat he filed with his testimony shows a creek rising in and flowing south from this pond.
>
> "Continuing north in the same course across the pond, at 36.44 chains from the southwest corner of section 24, 'We crossed the border of grass land surrounding this pond into a pronounced line of pine and palmetto. He calls for forty-six chains.'"

Hamblin's field notes for the west line of section 24 read:

North between sections 23 and 24

32:00 to grass pond

40:00 set post in pond for 1/4 section corner. Could not raise mound.

46:00 cross pond to pine

80:00 set post for corner to sections 13, 14, 23, and 24.

"Upon these facts Gove makes this comment, which expresses the conclusion upon which his subsequent conclusions are largely based: 'I find that pond at 36:44 chains from the southwest corner of section 24 a difference there of 9:56 chains, showing he did not set his stake in the center of that pond because this is a distinct mark. I find then he has made an error just prior to reaching the quarter corner, he has dropped ten chains as evidenced by the fact he is 9:56 chains short of the call distance.' This is a conjecture,

162 *Boundary Retracement*

rather than a legitimate inference, and is based upon the assumptions that this pond was of the same size and the growth on the adjacent lands had the same characteristics in 1873 as in 1931; and that Hamblin meant by his call '46:00 cross pond to pine' what Gove found to be 'a pronounced line of pine and palmetto' at 36:44 chains from the southwest corner of section 24. According to Hamblin's notes the pond he found there was at least eight chains (528 feet) across, while at the time Gove made his survey he found that it was only 200 feet in diameter."

"He continued on his original line N 0 degree 54' W until he came to State highway No. 25, but after leaving the pond found no evidence of any corners, markings, or other evidence on the ground, to show that it had been run by Hamblin."

"At the point (M 3) 80:50 chains south of the point R to which he had run in his preliminary examination, he found a 'lighter stake' which someone had set at the southwest corner of section 13 (northwest corner of 24). 'Feeling,' he says, that this point was very close to being the southwest corner of section 13 as established by Hamblin, he came back to that point and found it to be 417 east of a point on his original north line and 74.40 feet north of the southwest corner of section 24.

"Running east from this point (N 3) on a line N 89 degree 06' E he found at 39:94 chains three trees, whose courses and distances corresponded with the trees called for by Hamblin as witness trees for the quarter section post on the south line of section 13, but none of them bore marks more than twenty years old. Continuing eastward he intersected the township line at a point where he found a stake someone had set 614 feet south of the corner post set by Apthorp for the corner to sections 13 and 24, the point we have designated as M 3."

"From point M 3 he ran a line west to his original north and south line, and found that it intersected that line at point N 4, which is 83:20 chains north of the southwest corner of section 24. Along this line he found no line markings of any kind. He projected this line eighty chains west of its intersection with his north line, but found no line markings on the projection. But along a line run 420 south of this projection he found these markings at these distances west of his north line. At 417 feet less than forty chains he found a 'hub' which had been 'recently' set for the quarter section post of sections 23 and 14; at 432 feet less than eighty chains a 'lighter post' apparently set for the western corner between sections 23 and 14; and at eighty chains plus 107 feet another 'lighter post' which someone had set also apparently as the western corner between sections 23 and 14.

"Coming back to the point on his original north line 83:20 chains north of the southwest corner of section 24, he measured eighty chains, which he found carried him to a point 669 feet north of the road shown on the diagram from S to R. Running N 89 degree 06' E from this point, at 418 feet, he crossed the road which is shown on the diagram as running from R to K; at 25:20 he passed fifty feet south of a ten inch pine tree which had thirty-five rings

Which Set of Footsteps Is Which? 163

and was blazed on its south side, which 'evidences some subsequent survey, whose I don't know.' He intersected the township line at the known point at which Apthorp had set the post for the corner to sections 13 and 12, where he found a 'scribed lighter post' and one pine tree and two pine stumps which fitted Apthorp's call for witness trees for this post."

"He then went back to the point R from which he had started his preliminary investigation. Assuming that the road running west from that point was on the south line of section 11, he proceeded westward along that road, and 'about 25:92' chains west of R he found a 12 foot bridge, which is over 'a recognized cross line of drainage.' Hamblin's field notes for the south line of section 11 call for the line to cross 'a slight lead' at a point 25:92 chains west of its southeast corner."

"Returning to the point R, he extended the line of the road eastward to the township line. At seventeen chains east of the road he found an eighteen inch pine on the line he was running. 'This tree showed very little, if any, evidence of interior markings, but a slight swell on each face, and on cutting into this tree I found the northwest and northeast faces both blazed, not directly opposite one another, but the blazes approximately the same angle on the northeast and northwest facing. * * * I cut two blazes from which I found fifty to fifty-five rings.' At 39:86 chains he found a fourteen-inch lighter post, and at 59:18 an eighteen-inch lighter post. Continuing east he intersected the township line at an unmarked point, 680 feet south of Apthorp's corner for sections 12 and 13."

"Running S 0 degree 54' E from the point R he found a 'lighter post' at 39:72 which someone had recently set (he says, he assumes in 1925 or 1926 as a quarter section post for sections 13 and 14), from which post a fence ran west."

"Running north from R along the road he found fences running to the road from the west at 19:86, forty and sixty chains."

"In addition to the facts above stated, Gove testified that he noted the natural features such as 'prairie,' 'pine,' 'pine and scattering cabbage,' 'wet prairie to pine land,' etc., called for in Hamblin's field notes along the lines of section 24, 13 and 12; and that the call distances for such features do not fit these features, as he observed them on the ground, as well when these lines are run in accordance with Hamblin's field notes, as they do when they are run upon the assumption that Hamblin's north and south line from the pond northward was run about 417 feet farther east, and the south and north lines of section 13 intersect the township line 614 and 680 feet, respectively, south of the points at which Apthorp set posts for the southeast and northeast corners of section 13. But it is to be remembered in this connection that fifty-eight years had elapsed between the time that Hamblin and Gove made their surveys, and such natural features as Hamblin mentioned may have undergone material changes between the time of the two surveys."

"From the above mentioned physical facts, and his conjecture that between the southwest corner of section 24 and the point at which he set his stake in

164   *Boundary Retracement*

the pond for the quarter section of section 24, Hamblin dropped ten chains (660 feet), Gove reaches the conclusion that when he left the pond Hamblin ran his north line 417 feet farther east than he was running before he reached the pond, and established his southeast corner of section 12 at point R, which is 669 feet farther south than his field notes call for."

"Acting upon this conclusion he measured north along the road the distance called for by Hamblin between the southwest and northwest corners of section 12, that is, eighty chains, at which distance he found no evidence of a corner having been established there by Hamblin. However, feeling that he had 'demonstrated that it' (the thirty-inch oak heretofore mentioned) 'should be from findings lower down on the north line of section 12,' and finding that the call distance (eighty chains) brought him to a point 2:15 chains west and twenty-one feet south of this oak, he moved up twenty-one feet north to a point in the road west of this oak. From this point X he ran a line N 87 degrees 06' E through the oak to a township line at point Y, which is 711 feet fourth of the post set by Apthorp as the corner to sections 1 and 12. At the point Y he found no evidence of a corner. In addition to this oak, he found on this line, at a distance of 2,412 feet (about 36.55 chains) east of Point X, a 12-inch pine with a twenty-year old blaze, which, he says, 'is not an original marking at all, but someone in their endeavor to locate the lines about that time has marked that tree as some subdivision marking.' But it is to be noted that a point 36:55 chains from a section corner is not the corner of any recognized subdivision of a section, or quarter section. The only 'natural feature' noted by him on this line was that the line at 7:80 chains east of X came to 'pine and palmetto.' The first call in Hamblin's field notes for the line between sections 1 and 12 is '7:00 to pine and scattering cabbage.'"

"Running north from the point X he crossed State highway No. 25 at 10:83 chains; and 25:38 chains north of the highway he came to a fence which he accepted as being on the north line of section 1. Hamblin's field notes call for the length of the western line of section 1 as 25:53 chains. Running north from point Y along the east line of the township he came at 711 feet to the point where Apthorp set a corner post for sections 1 and 12, and found that the distance from Apthorp's corner to State highway No. 25 at D is 24:20 chains. Hamblin's field notes call for the length of the eastern line of section 1 as 23:85 chains."

"The testimony of John H. Caldwell, Jr., corroborates the testimony of Gove as to the physical facts found by Gove on the ground; but otherwise adds nothing to the force or effect thereof."

"For the defendant to sustain his plea of set-off, the burden is upon him to prove that Hamblin in fact established the southern corners of section 1 and ran the south line of the section south of the corners and line called for by his field notes. Under the facts of this case the evidence introduced by the defendant for this purpose falls short of that clear and convincing evidence, which is, as a matter of law, required to overcome the presumption that Hamblin's field notes correctly described the corners and lines of section 1 as run by him, and that as he correctly located and identified the corner set by Apthorp

Which Set of Footsteps Is Which?

for the southeast corner of section 1. It is, therefore, insufficient to support a verdict finding that Hamblin established the corners of the section and ran the south line thereof on the ground substantial distances south of the corners and line called for by his field notes."

"For this reason we are of opinion that the court did not err in striking out the defendant's evidence, and that the judgment of the court should be affirmed."

When there is ambiguity in field notes, a procedure is outlined in the U.S. v. Champion Paper Inc. case, following.

Discrepancy between Field Notes and Plat

In the case of *Doe v. Hildreth,** the court reported, "If there was any variance between the plat and field-books the former must have controlled, for it represented the lines and corners as fixed by the surveyor general, and by which the land was sold; and the law declares that the corners and boundaries as returned *by,* not *to,* that officer, shall be the corners and boundaries."

In the PLSS, for the most part, field notes are filed with the appropriate agency. There are often two copies, one filed at the local level, the other at headquarters. A list of the repositories appear in the appropriate government publications.

Passing Calls

A passing call, one which notes a particular topographical feature along a surveyed line, is also known as an "incidental call" and not strictly a locative call. Locative calls are significant within court decisions since they position a feature on the surface of the earth.† They are sometimes called "topographical calls."

The case of *Davenport v. Bass, Com. App.*‡ contains a discussion of passing calls, stating, "where lines of surveys of porciones§ running north from

---

* Doe on the *Demise of the City of Madison v. Hildreth,* 2 Ind. 274, 1850.

† In harmonizing conflicting calls in a deed or survey or public lands, courts will ascertain which calls are locative and which are merely directory, and conform the lines to the locative calls; "directory calls" being those that merely direct the neighborhood where the different calls may be found, whereas "locative calls" are those that serve to fix boundaries. *Cates v. Reynolds,* 143 Tenn. 667, 228 S.W. 695, 1921.

‡ 137 Tex. 248, 153 S.W.2d 471, 1941.

§ A porcione in Spanish law, a part or a portion; a lot or parcel; an allotment of land. *Downing v. Diaz,* 80 Tex. 436, 16 S.W. 49, 1891.

Rio Grande river, which was a shifting stream, began on bank of river, fact that surveys called for a designated road on both boundary lines tended to show that road was regarded as of considerable locative value, and therefore consideration would be given to calls for the road, even though they were "passing" or "incidental calls" and not strictly "locative calls.""

Passing calls are frequently found in surveyor's field notes, especially within the PLSS surveys and in Texas surveys, although they may appear in records of surveys in any region. They may be useful to the retracement surveyor where they may be used to relocate lost or obliterated corners.[*] As noted in Wilson,[†] passing, or topographic, calls serve two functions:

1. They are a check on the work done by the original, or a previous, surveyor and relate to how reliable the previous measurements were done across the terrain.
2. They provide an index of any error to assist in corner search.

An additional value of analyzing passing calls is that they will relate some kind of story about how the previous surveyor worked, and what kind of care was exercised in making locations.

Since passing calls are generally not locative calls, they should not be afforded the dignity of the latter since their reliability tends to be less. In the absence of a locative call, a passing call may have probative force indicating the surveyor's footsteps (Figure 5.1).[‡]

Principles Concerning Field Notes

In the construction of the property description, "the trial court was not bound to give controlling effect to every call in the field notes, or even to strictly follow the ordinary priority of calls. Circumstances may vary their usual order of dignity." Reference in *Larsen v. Richardson*. (See also S.R.H. Corp. v. Rogers Trailer Park, Inc. 252 A.2d 713 (N.J., 1969).)

It is well settled that the beginning corner, as given in the field notes, is of no more dignity than any other corner; and the survey may be constructed from any corner found on the ground. *Cox v. Finks*, 41 S.W. 95 (Tex. Civ.App. 1897).

---

[*] Glossaries of BLM Surveying and Mapping Terms.
[†] Wilson, D. A. *Forensic Procedures for Boundary and Title Investigation*. Hoboken, NJ: John Wiley & Sons, Inc., 2008.
[‡] Robillard, W. G., D. A. Wilson, and C. M. Brown. *Evidence and Procedures for Boundary Location*. 6th Edition. Hoboken, NJ: John Wiley & Sons, Inc., 2011.

Which Set of Footsteps Is Which?

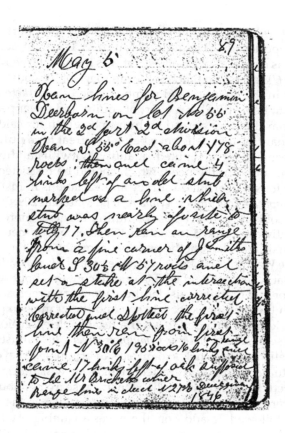

FIGURE 5.1
Field notes.

Misleading Evidence in Field Notes

There are a number of misleading calls in field notes, caused by a variety of factors. Its evidence comes in several forms:

Misidentified tree species

Blunder in a distance

Local attraction

Problems in documents with missing data, misspellings, strange words, handwriting that is difficult to read

Tree blazed high, such as on snowshoes

Marked tree fallen down face down

168   *Boundary Retracement*

In the case of *Quesnel* v. *Woodlief et al.*, decided by the Court of Appeals of *Virginia*, and reported in 2 Henning & Munford 173, the tract of land sold was estimated as containing 800 acres, but was found afterward to comprise but a little over 608 acres, "so that," in the language of the court, "both parties were mistaken in the quantity and number of acres contracted for, and the mistake ought to be rectified in a court of equity, and the appellant allowed a deduction from the price agreed by him to be given for said land for the deficiency in quantity, that deficiency being too great for a purchaser to lose under an agreement for a reputed quantity, notwithstanding the words 'more or less,' inserted in the deed, which should be restricted to a reasonable allowance for small errors in survey and variations in instruments."

Again, in *Campbell's Executors* v. *Wilmore*, 6 J.J. Marshall 209, where a tract of land had been sold and conveyed as containing 124 acres more or less, and it was subsequently ascertained to contain 132 acres, a judgment was ordered for the additional eight acres, at the price per acre at which the land had been sold. Mr. Justice Washington observed, in the case of *Thomas* v. *Perry*, 1 Peters' C. C. R. 49, "that when the land sold is said to contain about so many acres, both the grantor and grantee consider these words as a representation of the quantity which the grantor expects to sell and the grantee to purchase. The words more or less are intended to cover a reasonable exception or deficit."

The Texas case of *United States v. Champion Papers, Inc.** illustrates how the court system relied on field notes to retrace an original surveyor's footsteps. The case had to do with a line created in 1838 for which original evidence had disappeared. However, through extensive research of abutting parcels, along with intermediate surveys perpetuating evidence, the original footsteps could be followed. In addition, there was an ambiguity in the original surveyor's field notes. This issue was addressed in some detail in Chapter 2 dealing with ambiguities *per se* and again here to discuss the applicability and usefulness of field notes (Figures 5.2 and 5.3).

"The land which is the subject of the controversy is located in San Jacinto County, Texas. In 1938 the Government purchased most of the land comprising the Pleasant B. Riggs survey from Gibbs Brothers and Company. This land is a part of the Sam Houston National Forest located in Texas. In 1947 the defendant Champion Papers Company purchased the adjoining surveys of land. These surveys adjoin the southeast and southwest boundaries of the Riggs survey and are known as Washington County Railroad Company surveys No. 1, No. 2, and No. 5. The Riggs survey approximates a square, measuring about two miles in length on each side. It is laid out so that the corners are located at approximately true north, south, east and west. The property in all of these surveys has been utilized as forest land with its timber being cut at various times. Historically, fences have rarely, if ever, been built in the area of dispute."

---

* 361 F.Supp. 481, 1973.

Which Set of Footsteps Is Which?

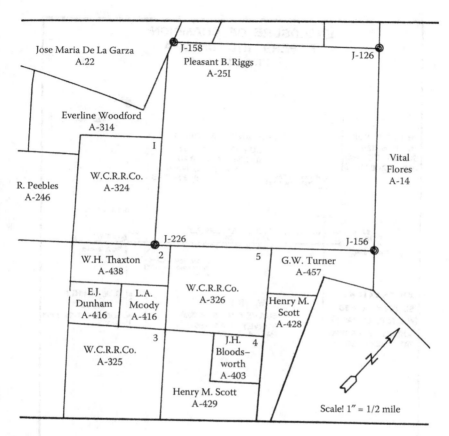

FIGURE 5.2
Overall depiction of the boundary question with surrounding parcels.

"However, defendant has recently constructed a fence which follows its claimed boundaries in the disputed area."

"The Riggs survey was originally surveyed for the Republic of Texas by Job S. Collard in February of 1838. The Washington County Railroad Company surveys were originally surveyed for the State of Texas by John Wade in 1861."

"While contested in some respects, the following sketch most accurately depicts for illustrative purposes the general location of the Riggs survey in relation to the adjoining surveys:

"The land which is in dispute constitutes approximately 32.85 acres. The Government as owner of the Riggs survey claims its southeast boundary runs from monument J-226 to monument J-156. Defendant Champion Papers, Inc., as owner of Washington County Railroad Company surveys No. 2 and No. 5, claims its northwest boundary runs from monument SJ-143 to monument SJ-145, and it has constructed a fence along such line. The land in contention is illustrated on the accompanying diagram:

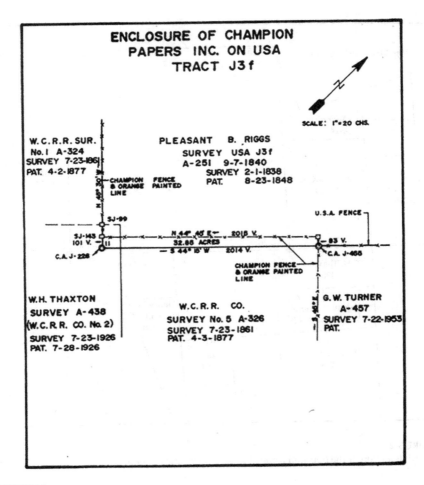

FIGURE 5.3
Enlargement of the boundary in question.

"The 135 year old original field notes of the surveyor Job S. Collard read as follows (Figure 5.4):

"As can be seen from the claims of the parties and an examination of the diagrams, the precise issue before the Court is the location of the southeast boundary of the Riggs survey which, in turn, is dependent upon the location of the south corner of the survey. The problem is essentially one of retracing the footsteps of the original surveyor as he conducted the Riggs survey in 1838."

"The Government contends that the south corner of the Riggs survey is at the point monumented by monument J-226, which is located approximately 100.98 varas southeast of defendant's monument SJ-143. Reliance is placed upon a 1929 resurvey conducted at the behest of Gibbs Brothers and Company. The proof adduced by the Government consisted

Which Set of Footsteps Is Which?

Survey for Pleasant B. Riggs of sixteen Labors 782131⁵⁄₁₀ square varas of land situated in Montgomery Co. west of the east forth of San Jacinto on Nebletts Cr. being a part of the quantity of land to which he is entitled by virtue of a certificate No. 98 issued by the board of land commissioners for the County of Jefferson—Beginning on the S. West boundary of Andrew Briscoe's survey 850 varas from the S.S.W. corner of said survey set post from which a white oak 30 inches in diameter bears N. 5° E. 8⁵⁄₁₀ varas dist and an ash 12 inches in diam bears S. 85° W. 18⁵⁄₁₀ varas dist.

varas	
	Thence S. 44° W.
560⁵⁄₁₀	To Neblett's Creek 4 varas wide runs S. 75° E.
4096⁵⁄₁₀	To 2d corner set post from which a pine 12 inches in diam bears S. 89° W. 4⁵⁄₁₀ varas dist and a pine 36 inches in diam bears N. 23° W. 6⁵⁄₁₀ varas dist.
	Thence N. 46° W.
4096⁵⁄₁₀	To 3d corner set post from which a sweet Gum 10 inches in diam bears N. 22° E. 9⁵⁄₁₀ varas dist and a sweet Gum 8 inches in diam bears S. 22° W. 9⁵⁄₁₀ varas dist.
	Thence N. 44° E.
470	To Nebletts Creek 3 varas wide runs S. 75° E.
4056	Intersected said Briscoe's survey set post from which a Spanish oak 10 inches in diam bears S. 62° W. 5⁵⁄₁₀ varas dist and a Spanish oak 8 inches in diam bears S. 38° E. 4³⁄₁₀ varas dist.
	Thence S. 46° E. with said survey
700	To Spring branch 2 varas wide runs N. 45°E.
4104⁵⁄₁₀	To the beginning.

FIGURE 5.4

One hundred and thirty-five-year-old field notes of Job S. Collard.

essentially of (1) evidence with reference to the remains of certain bearing trees recited by surveyor Collard which consist of stump holes; (2) evidence with reference to a passing call by Collard to Neblett's Creek; (3) evidence with reference to a corner of an adjoining senior survey at which surveyor Collard began his survey; (4) evidence with reference to an adjoining junior survey which consisted of a projection of a line, the Bybee line, from Little Caney Creek; and (5) evidence with reference to various resurveys of the Riggs survey."

"Defendant concedes that (1) the stump holes are in the locations asserted by the Government which are in the relative positions that surveyor Collard called for bearing trees in his field notes; (2) the distance from the eastern corner of the Riggs survey which is located by monument J-156 to Neblett's Creek in a southeastern direction more nearly corresponds to the distance called by surveyor Collard than the distance from the east corner of the Riggs survey as established by the 1912 resurvey of J. W. Oliphint; and (3) a projection of the Bybee line in a northeastern direction would be close to monument J-226."

"Defendant, however, contends that the Government cannot establish by a preponderance of the evidence that the Collard survey established the southeast boundary of the Riggs survey to correspond to the line between J-226 and J-156. Defendant contends that the testimony has failed to locate any corner of the Riggs survey by original patent evidence and that the Bybee line is in conflict with the original patent notes of the Washington County Railroad Company surveys. It is asserted that there is as much evidence to establish the south corner of the Riggs survey at SJ-99 as there is for establishing it at J-226. In this regard it is strenuously urged that there is no substantial confirmable original patent evidence to be located at either monument. It is also asserted that there is evidence which can support a finding of the south corner at SJ-143. As a result of the inability to find original patent evidence, defendant contends that the course and distance from the stipulated west corner of the Riggs survey should control the establishment of the south corner."

The Court makes the following Findings of Fact and Conclusions of Law:

Findings of Fact

I. The East Corner of the Riggs Survey

1. Surveyor J. S. Collard commenced the 1838 survey on the east corner of the Riggs survey at a point 832 1/2 varas from the southwest boundary of the Vital Flores or Briscoe survey. This point is to the northwest of Neblett's Creek and was witnessed by an ash tree bearing S 85~ W, 18 4/10 varas in distance and a white oak tree bearing N 5~ E 8 6/10 varas in distance.

2. The ash and the white oak trees have since disappeared.

3. A stump of an ash tree located in 1954 by surveyor Baldwin and in 1955 by the surveyor Morgan is the apparent sprout of the original ash tree marked by surveyor J. S. Collard.

4. As a result of the soil composition in the area, ash trees will not ordinarily grow any farther N 45~ W than 850 varas from the common corner of the Vital Flores survey and the Edward Russell survey.

Which Set of Footsteps Is Which?
173

5. A white oak stump hole which is 8 6/10 varas from a point called for by surveyor J. S. Collard is actually the stump hole of the original bearing tree called for by this surveyor. Surveyor Baldwin located both the white oak tree stump hole and the ash tree stump hole.

6. US Forest Service monument J-156 marks the true location of the east corner of the Riggs survey as established by surveyor J. S. Collard.

7. In 1838, Neblett's Creek crossed a line projected on a course S 44-> W from monument J-156 at a distance of 560 6/10 varas. In 1967 this creek crosses a line projected on a course S 44~ W from monument J-156 at a distance of 507 varas. As a result, since 1838 the Creek has moved in an easterly direction leaving several old runs from a distance of 560 6/10 varas to its present course.

8. The east corner of the Riggs survey cannot be located farther up the southwest boundary of the Vital Flores survey, since any line projected from such a corner on a course S 44~ W would cross Neblett's Creek at a distance that could not reasonably comply with the distance called for by surveyor J. S. Collard.

9. The defendant has presented no evidence locating the east corner of the Riggs survey.

10. The defendant did not survey or report any findings along a line projected N 44~ E from its monument SJ-99.

11. The defendant did not survey or report any findings along a line projected N 44~ E from its monument SJ-143 past the intersection of such line with a line projected N 45~ W from the boundary between Washington County Railroad Company survey Nos. 5 and 6.

II. The South Corner of the Riggs Survey

1. Surveyor J. S. Collard located the south corner of the Riggs survey by marking two trees "P.R." These trees included a 12 inch pine located 4 6/10 varas S 89~ W and a 36 inch pine located 6 4/10 varas N 23-> W from a post at the established corner.

2. Two stump holes with bearings and distance of S 89~ W 4 6/10 varas and N 23~ W 6 4/10 varas were located from the intersection of well-marked lines by surveyors J. V. Butler and J. W. Oliphint in 1929.

3. At the intersection of the well-marked lines, there was one line running from the northeast on a course S 44~ W and continuing through such point.

4. A second well-marked line was on a course running from the northwest S 46~ E. However, this line did not extend beyond the intersection.

174 *Boundary Retracement*

5. These stump holes are holes left by the trees marked "P.R." by surveyor J. S. Collard in 1838 and called for by surveyor Wade in his field notes for the east corner of the Washington County Railroad Company survey No. 1 in 1861.
6. US Forest Service monument J-226 now marks the corner of the Riggs survey as first established by surveyor J. S. Collard. These two stump holes are on a marked line, the course of which is S 44~ W from US Forest Service monument J-156.
7. There is no original patent field note evidence at defendant's monuments SJ-143 or SJ-99.
8. A projection of a line S 44~ W from US Forest Service monument J-226 intersects the Riggs survey at a point on the south bank of Little Caney Creek. This point on the south bank of Little Caney Creek was originally located by surveyor Wade in 1861 and witnessed by a 12 inch pine marked "X" bearing S 6~ E varas and a 12 inch pine marked "X" bearing N 3~ E 4 varas.
9. The point on the south bank of Little Caney Creek was located in 1910 by surveyor Clinton Bybee as reflected in the field notes of his survey of the W. H. Rigell homeplace.
10. The south corner of Washington County Railroad Company survey No. 1 was established by the District Court of San Jacinto County, Texas, on November 12, 1909, in Cause No. 1219, styled B. H. Bassett v. W. H. Rigell, as follows:
    At a stake for corner on the south bank of Little Caney Creek the same being the Southwest Corner of said Section No. One, Washington County Railroad Survey from which a large pine snag 38 in. dia mkd X (old mark) brs. N 42 E 30 ft. and a Sweetgum 20 in. dia mkd X brs S 42 E 11 ft in the north boundary line of R. A. Peebles Survey and also corner of Section No. 2.
11. The south corner of Washington County Railroad Company survey No. 1 was established by the District Court of Montgomery County, Texas, on May 20, 1932, in Cause No. 17,364, styled W. H. Rigell v. Ed Mays, as follows:
    At the most Southern Corner of Section No. One Washington County Railroad Survey in San Jacinto County, Texas, a stake from which a sweet gum mkd x brs. N 25 E 17 1/2 vrs.
12. The south corner of Washington County Railroad Company survey No. 1 was located by surveyors J. W. Oliphint, J. V. Butler, and W. T. Robinson on April 30, 1929, in their search for confirmative evidence of the south corner of the Riggs survey. Similarly, this corner was located by surveyor Morgan in 1967 as reflected in his field notes.
13. The south corner of the Washington County Railroad Company survey No. 1 is on an extension of a line from US Forest Service monument J-156 to J-226.

Which Set of Footsteps Is Which?

14. Surveyor Wade intended the east corner of the Washington County Railroad Company survey No. 1 to be common with the south corner of the Riggs survey.

15. The east corner of the Washington County Railroad Company survey No. 1 is common with the south corner of the Riggs survey and is presently marked or monumented by US Forest Service monument J-226.

III. The Southwest Boundary of the Riggs Survey

1. The southwest boundary of the Riggs survey runs from US Forest Service monument J-226 on a course N 46~ W to the west corner. The latter corner is at a point of intersection with the northwest boundary of the survey. This corner actually falls within the adjoining de la Garza survey which is senior to the Riggs survey.

2. A US Forest Service monument J-158 and a monument established by Foster Lumber Company mark the point of intersection of the southwest boundary of the Riggs survey with the de la Garza survey.

3. A line southwest of and parallel to the southwest boundary of the Riggs survey was surveyed on the land of the Washington County Railroad Company, the Eveline Woodford survey, and the de la Garza survey by surveyor J. W. Oliphint in 1912. However, the line surveyed by this surveyor does not extend to or intersect the southeast boundary of the Washington County Railroad Company survey No. 1.

4. Surveyor J. W. Oliphint was in error when he surveyed this line and established a corner for the Riggs survey some 101 varas northwest of the southeast boundary of Washington County Railroad Company survey No. 1. This corner is also some 109 varas southwest of the southwest boundary of the Riggs survey.

5. It appears that surveyor J. W. Oliphint corrected his original survey of 1912. This surveyor was a member of the survey crew of the 1929 Gibbs Brothers and Company survey when the stump holes fitting the description of the witness trees marked by surveyor J. S. Collard were found and a stake set for the corner. This surveyor was also a member of the survey crew of the 1933 Gibbs Brothers and Company survey. This survey set a stake for the south corner of the Riggs survey at the intersection of old marked lines adjacent to pine stump holes which fit the description of the patent field note trees. The latter survey crew then further perpetuated the corner by marking other trees.

176   *Boundary Retracement*

IV. The Northwest Boundary of the Riggs Survey

1. The northwest boundary is well established by the US Forest Service, and its location is not controverted by any evidence which was forthcoming at the trial of this cause.

2. This boundary runs on a course S 44~ W from US Forest Service monument J-126 to J-158.

V. The Northeast Boundary of the Riggs Survey

1. The northeast boundary is well established by the US Forest Service, and its location is not controverted by any evidence which was forthcoming at the trial of this cause.

2. This boundary runs on a course S 46~ E from US Forest Service monument J-126 to US Forest Service monument J-156.

VI. The Riggs Survey in General

1. It appears clear that with reference to the land within the Riggs survey, there have been no disputes between the grantors to the Government, Gibbs Brothers and Company, and the adjoining owners of the land within Washington County Railroad Company surveys Nos. 1 and 5.

2. It appears clear that there have been no disputes regarding the southeast boundary and the southwest boundary of the Riggs survey between the government and adjoining survey owners until the defendant purchased the Washington County Railroad Company surveys from Ed Mays in 1948.

3. The government went into possession of the land within the Riggs survey on August 12, 1939.

4. Former adjoining survey owner Ed Mays recognized the boundaries monumented by the US Forest Service and, in fact, instructed his employee, Ray Griffith, not to cut timber across the then US Forest Service marked line.

5. The primary use of the property in the Riggs survey has been to the boundaries marked by US Forest Service monuments J-156 and J-226.

6. The distance along the various boundaries of the Riggs survey are found to be as follows: Southeast boundary 4133.0 varas; southwest boundary 4203.3 varas; northwest boundary 4001.4 varas; and northeast boundary 4162.1 varas.

Conclusions of Law

1. This court has jurisdiction over the parties and the subject matter of this cause.

Which Set of Footsteps Is Which?

2. The purpose of the inquiry in a boundary dispute action is to locate and follow the footsteps of the original surveyor. Goodson v. Fitzgerald, 40 Tex.Civ.App. 619, 90 S.W. 898 (1905). Various rules of construction for purposes of ascertaining boundaries have been adopted by the courts to aid in following the surveyor's footsteps. In accordance with these rules, the priority of the calls which are found in the original surveyor's field notes is as follows: (1) natural objects; (2) artificial objects; (3) courses; (4) distances; and (5) quantities. Newsom v. Pryor's Lessee, 20 U.S. (7 Wheat.) 7, 5 L.Ed. 382 (1822); Stafford v. King, 30 Tex. 257 (1867); Thomas Jordon, Inc. v. Skelly Oil Co., 296 S.W.2d 279 (Tex.Civ.App., Texarkana 1956, writ ref'd n.r.e.).

"The purpose of the rules of priority of calls in an original survey is to aid the Court in finding the best evidence of what the original surveyor actually did on the ground. In the event the footsteps of the original surveyor can be more accurately traced or his intention more accurately ascertained by following a call of lower order, then the rules of priority are inapplicable. *See* Linney v. Wood, 66 Tex. 22, 17 S.W. 244 (1886)."

"These rules of construction are designed to carry out the intention of the parties. The intention of the parties is considered to be essentially the same as that of the surveyor. Strong v. Sunray DX Oil Co., 448 S.W.2d 728 (Tex.Civ.App., Corpus Christi 1969, writ ref'd n. r. e.). The surveyor's intention is to be ascertained by scrutinizing what he actually did in making the survey as reflected by his field notes and the attending totality of circumstances of the survey. Blake v. Pure Oil Co., 128 Tex. 536, 100 S.W.2d 1009 (Tex.Com.App.1937); Finberg v. Gilbert, 104 Tex. 539, 141 S.W. 82 (1911); Atwood v. Willacy County Navigation District, 271 S.W.2d 137 (Tex.Civ.App., San Antonio 1954, writ ref'd n. r. e.). It is well established that the various calls contained within the field notes must be harmonized and as few calls as possible disregarded so that the calls which result in the least conflicts in the total survey are given precedence. Wilson v. Giraud, 111 Tex. 253, 231 S.W. 1074 (1921). *See* Orn, Vanishing Footsteps of the Original Surveyor, 3 Baylor L.Rev. 273 (1952).

3. It is readily apparent that there is an ambiguity in the description contained within surveyor Collard's field notes, since the survey fails to close as it should and does not coincide with the sketch on the original field notes. The various subsequent resurveys do not coincide with either the original field note description or the field note sketch. It is therefore necessary to apply the various rules of construction and admit parol evidence to resolve the conflicts and to establish the lines which were actually run by the surveyor. *See* Bond v. Middleton, 137 Tex. 550, 155 S.W.2d 789 (1941).

178   *Boundary Retracement*

4. The burden of proof is upon the Government to locate and prove the boundary by a preponderance of the evidence. *See* Brown v. Eubank, 378 S.W.2d 707 (Tex.Civ.App., Tyler 1964, writ ref'd n. r. e.); Hancock v. Bennett, 230 S.W.2d 328 (Tex.Civ.App., Waco 1950, no writ hist.). In establishing the existence and location of original witness trees, the Government must introduce sufficient evidence "[to] carry the case beyond coincidence and into the realm of [a reasonable certainty]." Kirby Lumber Corp. v. Lindsey, 455 S.W.2d 733, 740 (Tex.1970). *See* Gates v. Asher, 154 Tex. 538, 280 S.W.2d 247 (1955).

5. There is a presumption that surveyor Collard actually surveyed all the lines of the Riggs survey, ran the course and distances and marked the boundaries as called for in the field notes. A survey must be so construed unless later surveyors following the original survey-or's footsteps demonstrate that the original surveyor's calls consti-tute a mistake or are otherwise incorrect. Gibson v. Universal Realty Co., 378 S.W.2d 115 (Tex.Civ.App., Houston 1964, writ ref'd n.r.e.); State v. Sullivan, 127 Tex. 525, 92 S.W.2d 228 (Tex.Com.App.1936).

6. A period of 135 years has elapsed since the Riggs survey was located. Time has destroyed the original witness trees marked by surveyor J. S. Collard. However, the calls contained within the field notes for two pine witness trees at the south corner of the survey can still be identi-fied with reasonable certainty. Two stump holes have been identified by credible evidence as being at the intersection of old and well-marked boundary lines. Construing the field notes for the patent of the Riggs survey in light of all the surrounding circumstances, these calls are the most certain and positive factors available in locating the footsteps of the original surveyor. *See* Kirby Lumber Corp. v. Lindsey, 455 S.W.2d 733 (Tex.1970); East Texas Pulp and Paper Co. v. Cox, 381 S.W.2d 78 (Tex.Civ.App., Beaumont 1964, writ ref'd n. r. e.); Muldoon v. Sternenberg, 139 Tex. 22, 161 S.W.2d 783 (Tex.Com.App.1942); Miller v. Southland Life Ins. Co., 68 S.W.2d 558 (Tex.Civ.App., El Paso 1934, no writ hist.); Smith v. Turner, 13 S.W.2d 152 (Tex.Civ.App., El Paso 1928), rev'd on other grounds, 122 Tex. 338, 61 S.W.2d 792 (1933); Plowman v. Miller, 27 S.W.2d 612 (Tex.Civ.App., El Paso 1930, writ dism'd); Petty v. Paggi Bros. Oil Co., 254 S.W. 565 (Tex.Com.App.1923); Antone v. Hoffman, 256 S.W. 656 (Tex.Civ.App., Texarkana 1923); Wm. Cameron & Co. v. Taylor, 288 S.W. 268 (Tex.Civ.App., Beaumont 1926); Weatherly v. Jackson, 98 S.W.2d 1037 (Tex.Civ.App., San Antonio 1936, writ dism'd); Findlay v. State, 238 S.W. 956 (Tex.Civ.App., Austin 1921), aff'd, 113 Tex. 30, 250 S.W. 651 (1923). *Compare* Kirby Lumber Co. v. Adams, 127 Tex. 376, 93 S.W.2d 382 (Tex.Com.App.1936); Runkle v. Smith, 63 Tex.Civ.App. 549, 133 S.W. 745 (1911).

7. The resurvey conducted by Gibbs Brothers and Company in 1929 and the resurvey of surveyor Morgan in 1967 tend to locate and

Which Set of Footsteps Is Which? 179

reestablish the corners and the boundary line of the original survey. *See* Barnes v. Wingate, 342 S.W.2d 352 (Tex.Civ.App., Beaumont 1960, writ dism'd); East Texas Pulp and Paper Co. v. Cox, 381 S.W.2d 78 (Tex.Civ.App., Beaumont 1964, writ ref'd n. r. e.).

8. Since the footsteps of surveyor J. S. Collard have been identified with reasonable certainty, it is apparent that the calls for distances in the original field notes are erroneous. The first call of 4096.6 varas actually should be 4133.0 varas, the second call of 4096.6 varas actually should be 4203.3 varas, the third call of 4056.0 varas actually should be 4001.4 varas, and the fourth call of 4,104.6 varas actually should be 4162.1 varas. *See* Sansing v. Bricka, 159 S.W.2d 142 (Tex.Civ.App., Amarillo 1941, writ ref'd w. o. m.).

9. A period of more than 100 years has elapsed since the adjoining Washington County Railroad Company survey No. 1 was located. The original witness trees marked by surveyor J. M. Wade have deteriorated and vanished. However, a stake on the south or southwest bank of Little Caney Creek has been perpetuated through the years by resurveys. One such resurvey was by Clinton Bybee in 1910. Both Gibbs Brothers and Company in 1929 and surveyor Morgan in 1967 cited and reported the Bybee survey witness trees. Surveyor Morgan also found a stump hole of one of the original witness trees marked by surveyor Wade. A line from US Forest Service monument J-156 on a course S 44~ W would pass through the south corner of the Riggs survey and intersect the south corner of Washington County Railroad Company survey No. 1. As a result, the south corner of the Riggs survey at US Forest Service monument J-226 is further established.

10. The southeast boundary of the Riggs survey, as marked by monuments J-156 and J-226, has been acquiesced in by the adjacent survey owners for a comparatively long period of time. *See* Kirby Lumber Corp. v. Lindsey, 455 S.W.2d 733 (Tex.1970); Bolton v. Lann, 16 Tex. 96 (1856).

11. The declarations of Ed Mays as prior owner of an adjoining survey which acknowledged the boundary line proffered by the Government are supportive evidence of the actual boundary. *See* Goodson v. Fitzgerald, 40 Tex.Civ.App. 619, 90 S.W. 898 (1905).

12. The defendant is bound by its actions in purchasing property in Washington County Railroad Company survey No. 1 by deeds from Ed Mays in 1948 and L & M Lumber Company in 1961 as well as a special warranty deed in 1968. These recorded deeds describe the property as having its south corner on Little Caney Creek. The defendant is also bound by its actions in leasing property in Washington County Railroad Company survey No. 1 to L. A. Carnes by leases dated July 22, 1958, March 5, 1963, and June 26, 1970. In these leases,

180   *Boundary Retracement*

the description of the south corner of the survey is by reference to the defendant's deeds. *See* Reynolds v. Bradford, 233 S.W.2d 464 (Tex. Civ.App., Fort Worth 1950, no writ hist.); Pierce v. Schram, 53 S.W. 716 (Tex.Civ.App.1899); Linney v. Wood, 66 Tex. 22, 17 S.W. 244 (1886).

13. The defendant presented no evidence establishing or tending to establish the footsteps of surveyor J. M. Wade in running the original surveys of the adjoining Washington County Railroad Company survey No. 1 through No. 6. Defendant's evidence reflects that no patent evidence was available, and as a result it failed to identify the sources and to connect the source of its evidence to the original patent witness calls.

14. The government has presented substantial credible evidence locating the east corner of the Riggs survey at a point now monumented by US Forest Service monument J-156. There is no evidence before the court other than that presented by the government which locates the east corner of the Riggs survey.

15. The government has presented substantial credible evidence locating the south corner of the Riggs survey and the east corner of the Washington County Railroad Company survey No. 1. The defendant has presented no credible evidence tending to locate the south corner of the Riggs survey or the east corner of the Washington County Railroad Company survey No. 1.

16. The greater weight of the credible evidence adduced at the trial of this cause establishes the southeast boundary of the Riggs survey to be at a line connecting US Forest Service monuments J-156 and J-226.

"The foregoing constitute the Court's Findings of Fact and Conclusions of Law. Counsel will prepare and submit an appropriate judgment within ten (10) days incorporating these Findings of Fact and Conclusions of Law. The Clerk will notify counsel."

Ambiguity in the Field Notes

Since there was found to be ambiguity in Collard's field notes, because "the survey failed to close as it should and did not coincide with the sketch on the original field notes," the court provided a directive as to how to reconcile the problem. "The various subsequent resurveys do not coincide with either the original field note description or the field note sketch. It is therefore necessary to apply the various rules of construction and admit parol evidence to resolve the conflicts and to establish the lines which were actually run by the surveyor. *See* Bond v. Middleton, 137 Tex. 550, 155 S.W.2d 789 (1941)."

6

Figures, Numbers, and Symbols

Figures don't lie, but liars will figure.

unknown

In analyzing evidence to determine the appropriate set of footsteps to follow, or once decided on a set to follow, invariably there will be interpretations to make, conflicts and adjustments to resolve, and corrections to consider as well as other reconciliations. This chapter addresses the mathematics of original surveys: directions, distances, and adjustments as well as interpretations of various forms of evidence.

One of the first decisions to make in following any set of footsteps is determining how the survey was made, which should indicate the type of equipment that was used. Either using the same type of equipment as the original surveyor, or considering the differences between then and now, an attempt should be made to follow field notes or other descriptive information, or, in their absence, to traverse between two known points from the original work. A comparison can then be made between the two systems of survey, and an evaluation of the previous work can be done.

The types of survey that could be expected, somewhat dependent on the previous surveyor's training and experience, are as follows. Each method dictates the precision to be expected, and provides insight as to the errors inherent in the survey.

Types of Survey

For older surveys, most of the work was accomplished using some combination of the following instrumentation. Each had its own procedures and resulting attributes as well as deficiencies.

Compass-chain. Done with either a handheld or similar compass, or a staff compass mounted on a Jacob's staff or a tripod (the latter tending to be more accurate) and some type of chain.

Compass-tape. Done with a compass similar to the forgoing and tape, a cloth tape (not usual), or a continuous steel ribbon, of some length.

181

Plane table-chain. Done with a plane table, and some form of measuring device, usually a chain of some length.

Plane table-stadia. Done with a plane table and an alidade for taking readings on a graduated rod to determine distances.

Plane table-intersections. Done with a plane table intersecting angles to points from at least two stations.

Transit-tape. Use of some type of transit and a steel tape of some type, and length.

Transit-stadia. Angles measured with some type of transit, and distances determined by reading a graduated rod. Many traverses were done with by using transit and tape while additional detail and locations were determined through use of stadia.

Each of the foregoing carries with it its own characteristic precisions and types of error. When it becomes necessary to reconcile errors in an original survey, it is often necessary to determine what method was used to conduct the survey. If it is not possible to determine the type of equipment used from the surveying record, speculation may be field-tested and compared with reported results.

Mathematics provides the means for following a previous surveyor from point to point. To follow footsteps implies following procedures to get to and identify the same points the previous, or original, surveyor used, or established.

Mathematics becomes even more important when setting a point from courses and distances in the absence of a corner. Raw data, or even adjusted data, may not be a reliable roadmap depending on how the data were produced.

Or when, as stated in *Lugon,* or in *Mercer,* putting the marker or running the line where the previous surveyor *would have put it.* In order to do that, it is necessary to know how that particular surveyor did his work, or how he would have done it, and what kind of precision might be expected.

The goal is to get to where the marker is, where it was, or where it should have been placed if it had been set.

Marks on the ground constitute the survey; courses and distances are only evidence of the survey. Myrick v. Peet, 56 Mont. 13, 180 P. 574 (1919), citing 9 Corpus Juris § 210; Hunt v. Barker, 27 Cal. App. 776, 151 Pac. 165; Woods v. Johnson, 264 Mo. 289, 174 S.W. 375. As frequently as numbers are used, and as often as they appear, they are secondary, contain error, and are otherwise unreliable other than as an aid to finding corners. While taxing authorities, deed writers, and landowners place a great deal of importance (and faith) on distances and acreages, as demonstrated numerous times throughout the system, courts have traditionally found them the most unreliable pieces of information.

Later treated in a separate section is the point that there is no such thing as a perfect measurement, and that all measurements have error, some small and insignificant, others large and troublesome. A few are compensating, in whole or in part, such that they are unidentifiable, for example, an acceptable

Figures, Numbers, and Symbols

183

error of closure is found, but there are two or more gross errors within the traverse, mathematically compensating for each other or, canceling each other out for the most part so that a false sense of reliability is encountered.

One of the earliest and most comprehensive discussions of retracement is found in the Texas decision of *Stafford v. King*,* wherein the court summarized the duty of the retracement surveyor, and reasons for the priority of rules.

> It is the duty of the surveyor to run round the land located and intended to be embraced by the survey and patent, and see that such objects are designated on it as will clearly point out and identify the locality and boundaries of the tract, and to extend a correct description of these objects (natural and artificial, with courses and distances) into the field-notes of the survey, in order that they may be inserted in the patent, which will afford the owner, as well as other persons, the means of identifying the land that was in fact located and surveyed for the owner, and until the reverse is proved, it will be presumed that the land was thus surveyed and boundaries plainly marked and defined. And if any object of a perishable nature, called for in the patent, be not found, the presumption will be indulged that it is destroyed or defaced; but if it be established, by undoubted evidence, that the land was not in fact surveyed, yet, as the omission was the fault of the government officer and not the owners, it would seem extremely unjust to deprive him of the land, by holding the patent to be void, if the land can, by any reasonable evidence, be identified. And if course and distance alone, from a defined beginning-point, will, with reasonable certainty, locate and identify the land, that will be held sufficient.
>
> Of all the *indicia* of the locality of the true line, as run by the surveyor, course and distance are regarded as the most unreliable, and generally distance more than course, for the reason that chain-carriers may miscount and report distances inaccurately, by mistake or design. At any rate, they are more liable to err than the compass. The surveyor may fall into an error in making out the field-notes, both as to course and distance, (the former no more than the latter,) and the commissioner of the general land office may fall into a like error by omitting lines and calls, and mistaking and inserting south for north, east for west. And this is the work of the officers themselves, over whom the locator has no control. But when the surveyor points out to the owners rivers, lakes, creeks, marked trees, and lines on the land, for the lines and corners of his land, he has the right to rely upon them as the best evidence of his true boundaries, for they are not liable to change and the fluctuations of time, to accident or mistake, like calls for course and distance; and hence the rule, that when course and distance, or either of them, conflict with natural or artificial objects called for, they must yield to such objects, as being more certain and reliable.
>
> There is an intrinsic justice and propriety in this rule, for the reason, that the applicant for land, however unlearned he may be, needs no

---

* 30 Tex. 257, 94 Am.Dec. 304, 1867.

184

Boundary Retracement

scientific education to identify and settle upon his land, when the surveyor, who is the agent of the government, authoritatively announces to him that certain well-known rivers, lakes, creeks, springs, marked corners, and line constitute the boundaries of his land. But it would require some scientific knowledge and skill to know that the courses and distances called for are true and correct, and with the aid of the best scientific skill mistakes and errors are often committed in respect to the calls for course and distance in the patent. The unskilled are unable to detect them, and the learned surveyor often much confused.

Although course and distance, under certain circumstances, may become more important than even natural objects—as when, from the face of the patent, the natural calls are inserted by mistake or may be referred to by conjecture and without regard to precision, as in the case of descriptive calls—still they are looked upon and generally regarded as mere pointers or guides, that will lead to the true lines and corners of the tract, as, in fact, surveyed at first. "The identification of the actual survey, as made by the surveyor, is the *desideratum* of all these rules." The footsteps of the surveyor must be followed, and the above rules are found to afford the best and most unerring guides to enable one to do so.

There is another rule to be observed in estimating these natural and artificial calls. They are divided into two classes: descriptive or directory, and special locative calls. The former, though consisting of rivers, lakes, and creeks, must yield to the special locative calls, for the reason that the latter, consisting of the particular objects upon the lines or corners of the land, are intended to indicate the precise boundary of the land, about which the locator and surveyor should be, and are presumed to be, very particular; while the former are called for without any care for exactness, and merely intended to point out or lead a person into the region or neighborhood of the tract surveyed, (9 Yer., 55,) and hence not considered as entitled to much credit in locating the particular boundaries of the land when they come in conflict with special locative calls, and must give way to them.

A call for course and distance is the most unsatisfactory and unreliable of all calls, and is considered the least important. This is true because of the liability of chain carriers to error. Judge Gaines, in *Wyatt v. Foster & Rafferty*,* said "but of all the calls in a survey that for distance is of the least dignity and ordinarily must yield to the calls for the corners and lines of the surrounding surveys."

Problems with Directions

Unless the compass is beyond the influence of disturbing causes, and the surveyor is very careful in adjusting it properly, and in noting

---

* 79 Tex. 413, 15 S.W. 679, 1891.

Figures, Numbers, and Symbols 185

minutely the variation at which the line is run—and we know the date of the survey—so that the increase or decrease of the variation since, can be added or deducted, no surveyor can ever feel confident that he is running even very near to the line traversed by his predecessor, and by whose minutes he is working.

M'Clintock v. Rogers, 11 Ill. 279 (Ill. 1849)

The biggest problem with directions is in dealing with the basis of magnetic north, which is constantly changing in position. Most original surveys, with the exception of the original large grants, were made with a magnetic instrument. Many of the original grant lines were located by astronomical observation. One very obvious example of this is the famed Mason–Dixon line, separating Maryland from Pennsylvania, or the Calvert grant from the Penn grant. It took the two astronomers 5 years to complete their survey.

In 1874, Elihu Quimby (1826–1890), Professor of Mathematics and Civil Engineering at Dartmouth College, published a paper entitled *A Paper on Terrestrial Magnetism Designed for the Use of Surveyors*. In his introduction, Professor Quimby stated:

an examination of the records of surveys made within the last fifty years will show that there is need either of more general knowledge on this subject, or of a better use of what is known. It is quite unusual to find in any of these records the slightest reference to magnetic declination; and there is reason to believe that surveyors sometimes rely too implicitly upon the needle in retracing old lines by their former magnetic bearings. It will appear by the behavior of the needle that, while it is a valuable aid, it can never be depended on for such purposes, and should, in all cases, be used with caution, and only when extreme accuracy is not required.*

Quimby wrote,

it is true that, when only a comparison of directions is required, as in the survey of a field to determine its figure and area, it is of no consequence what the declination is, provided it remains the same during the progress of the survey and for all points where he needle is used; but even then, to make the survey useful in retracing the same lines at a future time, the declination should be known and recorded.

By observations upon the needle of a well constructed magnetometer, the following facts relating to the declination will appear, some of which will be indicated even by the ordinary compass needle.

The declination is not the same in all places.

1. For a given place, it is subject to a secular change of unknown period, but requiring at least several hundred years for its completion.

---

* I would state that very little, if anything, has changed in the past 140 years.-DAW.

186

Boundary Retracement

2. It has a diurnal change, with a maximum and minimum for each day.

3. It has also an annual maximum and minimum, changing with the seasons of the year.

4. It is subject to irregular disturbances, called magnetic storms, being more or less affected by every meteorological change.

He finished his treatise with the following paragraph:

> The most difficult problem ever presented to the surveyor is that which asks him to retrace a lost line, with but one point known, and the bearing from some old deed. To add to his perplexity, the parties in interest are usually too much excited by the apprehension of being robbed of a square rod of rocky pasture, or of swamp rich in mud and brakes, to be able to give correctly such facts as might be serviceable in the solution of the problem. In such case, if the parties cannot be induced to agree upon a second bound and thus determine the line, there is no way but to *"run by the needle,"* after making due allowance for change in declination since the previous survey. Running in this way may lead to the discovery of some old landmark, nearly obliterated, and thus settle the dispute; but if not, though the error in the bearing is likely to be 15′ to 30′, it is *better than a lawsuit*; and if, in such case, the parties in their ignorance believe that to be "true as the needle to the pole" is to be true enough it is certainly an occasion where "tis folly to be wise."

By the Needle

The question often arises as to how to retrace the direction taken from an ancient deed description. The following two cases lend some insight.

In some states, the courses in a deed are invariably presumed to have been run according to the magnetic meridian unless there is something in the instrument to indicate a different method. It was held in *Wells v. Jackson Iron Manufacturing Co.** that the use of such words as "due north" will not justify the inference that a different method was intended.

Obviously, surveys in the PLSS states should have been run in accordance with the instructions in use at the time. The question that then arises, where bearings were to be run by the true meridian, was true north calculated, or otherwise determined, correctly.

In *Brooks v. Tyler*[†], Prentiss, J. wrote, the land was described "by courses and distances only, without any reference to the lines of the lot of which the land is alleged to be a part, or to any certain or natural monuments; and whether or not the defendant was in possession of any part of the land described, could be determined only by actual survey. When there is nothing stated

---

* 47 N.H. 235.

† 2 Vt. 348, Vt., 1829.

Figures, Numbers, and Symbols

to control the courses and distances, the lines must be run by the needle.—(*McIver v. Walker*, 13 U.S. 173, 3 L.Ed. 694, 9 Cranch 173.)—It is true, that the magnetic course is subject to variation and uncertainty; and in weighing evidence of recent surveys of ancient lines, regard may be had to the variation of the needle. But the declaration in the case contains no reference to the lines of the lot, or to any ancient survey, and we cannot understand the courses stated in the declaration, to mean courses as designated by the needle forty or fifty years ago, instead of courses designated by it now."

While the difference between magnetic north (constantly changing) and true north (stable) can be calculated, and therefore determined at any point in time, many retracement surveyors do not undertake the necessary exercise, even though it is a requirement of the law, and therefore fail to find the corner. Subsequent action by those individuals relies on courses and distances (which is allowable, *but only in the absence of the corner*), to determine a point, easily leading to unnecessary "pincushion corners."

Definition of Course

A "course," as used with reference to boundaries, is the direction of a line run with a compass or transit and with reference to a meridian.*

Which Meridian

While the most common north meridians relied on for directions are true north and magnetic north, there are others. Occasionally, especially with azimuths (which may be converted to bearings), the south meridian has been used, both true south and magnetic. It has also been common practice for surveyors to relate their surveys to existing lines defined by someone else's survey. Useful for orientation have been government surveys, railroad surveys, highways, and others. The basis of directions and choice of meridian were not always checked, but more likely taken for granted. If directions are based on the true meridian, how was it determined? And if based on the magnetic meridian, how reliable were the observations, and when were they made?

Type of Compass Used

There have been a variety of compasses used over the years, by people who were novices, to highly trained professionals. It is often thought that large Rittenhouse-type compasses were used either on a staff or a tripod; however, that was not always the case. Likely, the more precise and therefore more costly instruments were used by surveyors, and in the hands of a capable person could produce excellent, reliable results. Handheld compasses, and

---

* *M'Iver v. Walker*, 9 Cranch (US) 173, 3 L.Ed. 694.

compasses of other types were also used on a need basis, and often their accuracy left a lot to be desired.

Declination

The court system has been no stranger to the changes in declination overtime.

In *M'Iver's Lessee v. Walker*, 1815, 9 Cranch 173, 13 U.S. 173, 3 L.Ed 694, Mr. Chief Justice Marshall said:

> It is, undoubtedly, the practice of surveyors, and the practice was proved in this cause [the surveyors in the instant case confirmed the practice], to express in their plats and certificates of survey, the courses which are designated by the needle, and if nothing exists to control the call for course and distance, the land must be bounded by the courses and distances of the patent, according to the magnetic meridian.

We need not presently decide whether in the construction of deeds the use of magnetic north instead of "true" north is to be presumed as the New Hampshire Court ruled in *Wells v. Jackson Iron Manufacturing Company*, 1866, 47 N.H. 235. But, in weighing evidence of recent surveys of ancient lines, consideration must be given to the variation of the needle in determining a magnetic course. *Brooks v. Tyler*, 1829, 2 Vt. 348.

It is somewhat surprising how many people, surveyors included, who do not attempt to correct for the change in magnetic declination overtime. A change of 1° will displace a mile-long line (the length of a section line in the PLSS) 92 feet (see Figure 6.1).

Considering an error of 1° per mile of line, translate that into a longer line, such as are found in the larger grants of townships and of the larger private land grants such as are found in Kentucky and Tennessee. See chapter 13 for treatment of overlapping grants.

Impact on Successful Corner Location

Also, in swamps or otherwise difficult conditions such as heavy brush, etc., more than a few feet of difference between where one is searching and

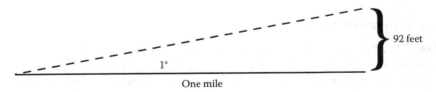

FIGURE 6.1
A horizontal error (such as in difference in declination) of just 1° will result in a horizontal error of 92 feet at the end of a mile (the length of a section line). For shorter lines, that translates to an error of about 10 feet in a 500-foot line.

Figures, Numbers, and Symbols   189

where the corner is actually located, deteriorated, or otherwise hidden and obscured evidence may not be seen, or found, resulting in failure to locate the corner. That difference could be as little as a foot or two, or as much as a hundred feet or more.

In addition to the inherent natural change in the location of the magnetic north pole as a basic point of reference, there are variations between compasses, so ideally the retracement surveyor should attempt to determine what was used as a device for observing the original directions in any given case.

The first isogonic chart was constructed by Edmond Halley (1701), the astronomer who discovered the famous comet. The British Royal Navy assigned him a vessel, *The Paramour* with which he traversed the Atlantic Ocean from Latitude N 51° to Latitude S 51° in 1699 and 1700. For some years following, the lines of equal declination were known as Hallean lines, whereas today they are commonly known as Isogonic lines. The line of equal declination is termed the Agonic line.

Local Attraction

Another concern for the retracement surveyor is the attraction caused by local conditions. Consider those conditions where and under which the surveyor is working. If a surveyor was using a magnetic compass throughout the survey, there would be one result; on the other hand, if the surveyor started at a known base line then turned angles (and perhaps later calculated directions) throughout, the result might be quite different.

One value of running a line that a previous surveyor has run is that any local attraction due to natural magnetic occurrences or disturbances may be identified. However, if the surveyor had metal on his person, or parked an axe or a rifle near the compass when making his observations, the following surveyor would not know that. But the difference between the current bearing and the previous bearing should present itself.

Something that should always be done is comparing the previous surveyor's bearing with the current bearing on a common, known, line. As soon as two points are identified that are the same for both surveys, a line can be used for comparison. They do not have to be consecutive points so long as they are common to both surveys, the previous (or original) and the current retracement.

Taking into account local attraction may require some thought. There are three possible situations:

Local attraction existed when the original survey was done, but not now.

Local attraction did not exist when the original survey was done, but does now.

Local attraction existed then and still does, however, it may not be of the same magnitude as it was then, or it may otherwise be different—not even the same source of the attraction.

Conducting a survey and encountering local attraction is identifiable, provided both backsights and foresights are taken on all legs. Encountering it, the following procedure is standard in correcting the bearings to compensate for the attraction.

When taking backsights and foresights, if the two observations do not agree on the same line, particularly by a significant amount, local attraction may be suspected. Continue to observe directions both ways until encountering a line where both observations agree. The local attraction will be affecting the observations where the two do not agree, and will be isolated between the two legs where they do agree. Then, computing the angles between the observations, apply the angles to the correct bearing following through all the legs until the correct line is reached further on. At any one point, the attraction will affect both foresight and backsight equally, thereby not disturbing the angles between them.

Should local attraction affect all the lines of the measured figure, run line from the figure to one side until a point is reached that both foresight and backsight agree. Then apply the angular correction described above to the affected lines.

Courts understand this concept and its inherent problems, and therefore have not only discussed it numerous times, sometimes to great length, but also set a standard of what must be done. In The North Carolina case of *McCourry et al. v. McCourry*, the court stated "in running a line established in 1885, allowance must be made for the variation of the needle."* Its analysis was as follows:

> The question in the case had to do with the dividing line between the plaintiff and the defendant in the partition proceedings of their father's land in 1885 which was defined as being "from the poplar (an admitted point) west 190 poles to the top of Griffith ridge." The defendant contended that the line should be so run without any allowance for the variation of the needle. The plaintiffs contended that, owing to the variation of the magnetic needle, this would not be the line actually laid off in 1885, and that the true line can now be laid off only by allowing for such variation 1 3/4 degrees; i.e., the line will now read "N. 88 1/4° west," 190 poles.
>
> John M. Houck, the surveyor appointed by the court in this case, testified that he made the survey, that he had been a surveyor between 50 and 60 years, and that it is the custom of all surveyors since 1805 to allow one degree variation of the needle for 20 years, and that a line which was

---

* 180 N.C. 508, 105 S.E. 166, 1920.

Figures, Numbers, and Symbols

laid off in 1885 to run due west would now run N. 88 1/4° west, owing to such variation in the needle.

Mr. Young, witness for the defendant, testified that he had been county surveyor for 16 years. He stated that "it is the usual custom to allow a variation of one degree for every 20 years," and on cross-examination he said: "The proper variation for a line run in 1885 is 1 3/4 degrees." Though in his opinion the line could have been more properly ascertained by making the survey in a different manner, which he indicated, on this, which was the pertinent point in issue, he agreed with Mr. Houck; and the judge properly told the jury that "if they believed the evidence" to find the issue accordingly.

It is common knowledge that there is a regular variation in the compass, which is different at different places on the globe, and the testimony of both surveyors is that in this locality the magnetic north is moving westward at present one degree for every 20 years. Therefore a line which ran due west in 1885 would now run N. 88 1/4° west. In the course of time the variation will begin to swing back. Authoritative tables are from time to time printed by the governments of the world, showing the variation at different places; but in this case the evidence of the two surveyors was uncontradicted that for this locality the above is the customary and proper allowance which was doubtless based upon scientific data.

Until the discovery of the magnetic needle, which became known in Europe just before the discovery of America by Columbus, ships dared not put boldly to sea, but coasted along from headland to headland, rarely out of sight of land. Something over 100 years ago it was discovered that there was a variation in the needle from the true north year by year, and varying in different localities.

Besides the complete revolution of the earth on its own axis every 24 hours and its annual sweep around the sun, the earth has nine other regular movements—eleven in all—one of the latter being a change in the position of the poles of the earth moving in an ellipse by which the "north star" which should be in exact line with a line through the two poles of the earth gradually shifting its position, relative to our north pole which gradually moves to the west, and then in an ellipse returning to the east, and thus back to its original position. This compels a regular variation in the magnetic needle which can be calculated years in advance, and all well-informed surveyors act upon and allow for such variations. There are some other variations due to local causes, such as iron in the ships, or in the ground, and the direction of valleys and streams, which need not be considered here. If the uncontradicted testimony of the experts on this case, the surveyors, is to be believed, the verdict and judgment have correctly located the line as it was actually laid off in 1885.

Light moves at the speed of 186,300 miles a second, a speed which would carry it around the earth more than six times in the tick of a second by the clock. Light comes from our sun, which is over 93,000,000 miles away, in a little over 8 minutes. The nearest fixed star (and all the fixed stars are suns) is 275,000 times further from us than our sun, and it therefore takes light from it 4 1/2 years to reach us, but the North star is

192 *Boundary Retracement*

more than eight times further off, and the light from it takes over 36 1/2 years to reach us. How and why that body, at such an incredible distance, should so control the magnetic currents on this tiny planet upon which we live, and why the magnetic needle in all compasses vary with the slow wobbling of our poles relative to the North star, is not yet known; but all ships at sea at their peril must take notice of it for safe voyaging and all surveys on land to be accurate must conform.

It may be that the Pole star has not this influence, but it always marks the true north. It never sets or rises like other stars, and the only change in its relative position to the earth is caused by the elliptic revolution of our north pole.

Owing to scientific facts, the line in dispute which was properly laid out as "due west 190 poles" in 1885 can be identified now only by setting the compass "N. 88 1/4° west." If this were not done, the defendant would have gained, and the plaintiff would have lost, a strip of land covered by the variation and the timber cut thereon.

As early as 1809, the Kentucky court had an understanding of errors in surveys, and laid down a series of rules for guidance. In addition to the hierarchy of rules and procedures, detailed guidance was presented regarding courses and distances. Concerning courses, the court stated that "allowances to be made for variation of the needle." And, "a mistake in one course, not to be presumed to have affected any other course." Also, "court bound to take notice that there is magnetic variation from the true meridian," and "surveyors generally took their courses from the magnetic meridian."[*]

As recently as 2006, the case of *Rapides Parish Police Jury v. Grant Parish Police Jury*[†] contained instructional discussion of the inherent perils in surveys by the compass. The court wrote, "We have detailed the inaccuracies in the GLO Plat previously outlined by Mr. Willis in the section on legislative intent." Mr. Willis further testified that the GLO surveyor did the work in the field using a survey compass on a tripod, and made field notes as he took measurements. Then a draftsman at some later time would then draw the map based upon the field notes of the surveyor. Mr. Willis explained the adjustments that were required in using such a compass:

And little known by most people is this magnetic north is not true north, it varies from year to year, from day to day. It even varies during the day, we call that diurnal variation.... Right now north is about 4° off to the east of where true north is. So there is a little adjustment knob. The surveyor was supposed to, sometimes they apparently didn't, but they're supposed to look at Polaris, which is the north star, and get this thing oriented and get to where when he's reading magnetic north it gives him an adjusted reading which converts it to true north. But Polaris is moving in a circle up in the sky counterclockwise and if you don't hit it at

---

[*] *Bryan v. Beckley*, 16 Ky 91, 1809.

[†] 2005-268 La.App. 3 Cir. 2/22/06, 924 So.2d 357, La. App. 3 Cir., 2006.

Figures, Numbers, and Symbols

the right time, you're not looking north then either. ...and you get errors like that and those errors ... reflect the type of natural or instrumental or personal errors that can go on during a survey even by the most competent of GLO surveyors.... When the General Land Office surveyor says that he is running north 99.99% of the time, he's running at what he thinks is about the best he can do to get north and that deviation can be substantial.

I've seen it—the most he could get here would be 6 inches in 300 feet, so that's the precision ... [a]nd if it changes during the day or if he gets off a little bit or if this screw gets loose, there's no telling what's going to happen and that is why it is so important to understand the premise of doing a retracement survey by following the footsteps of the old surveyor ... as I mentioned ... [t]he survey is what's on the ground... [t]here's only so much you can do with this thing.

Mr. Willis further stated that a hiccup or a loose screw could cause serious survey errors when using the antique equipment of that era. Accordingly, the reliability of the GLO Plat itself is questionable, particularly under the circumstances of this case. The court, in *Texas International Petroleum v. Delacroix Corp.*, 94-1426 (La.App. 4 Cir. 1/31/95), 650 So.2d 815, *writ denied*, 95-0467 (La. 4/21/95), 653 So.2d 567, determined that the reliability of a GLO plat which is unsupported by any effort to retrace the lines on the ground is questionable at best.

The New Hampshire decision of *Wells v. Jackson Iron Mfg. Co.** referenced earlier had to do with the retracement of a large tract of land, which amounted to an entire granted township, which contained Mount Washington, New Hampshire's tallest mountain and a significant landmark. The case bears a study by any retracement surveyor for its scientific value in both making and interpreting surveys run by the compass needle, as well as being reported as such.

Noah Barker† was a witness called by defendant, and the court was satisfied that he was an expert in surveying and therefore entitled to give his opinion as such. He gave the following testimony, that part in brackets being subject to plaintiff's exception:

In 1853 and 1854, employed by David Pingree, surveyed Sargent's Purchase and other adjoining tracts. Surveyed the Dorcas Merrill (same as Dorcas Eastman) lot according to the grant and according to the actual location of it—surveying it according to the grant, I started from a point shown me by John Emery and by Hardison L. Emery, (who is dead, and who lived on the next lot down the branch)—that point is on Rocky

---

* 47 N.H. 235, N.H., 1866.
† Noah Barker (c. 1808–) was a Land Agent for Maine (late 1850s and early 1860s) and a highly respected land surveyor.
   In 1864, he published a treatise titled *An Essay on the Cardinal Points: Being a Collection of Authorities in Explanation of the Terms: "Due North", "Due South," "Due East," and "Due West" as Applied to Land Surveying.*

194   *Boundary Retracement*

Branch from 60 to 80 rods above her buildings—there was no bound or mark there—there was a pine stump a few feet from it cut many years ago and scored into—I marked the stump as a witness—from that point I ran south 81 degrees east by needle 100 rods, thence north 9 degrees east 160 rods, thence north 81 degrees west 100 rods, thence south 8 degrees west to place of beginning—taking that as the Dorcas Merrill lot according to the description in the grant, and starting from the northeast corner of it, I surveyed Sargent's Purchase according to the description in the grant, using a meridian of longitude as a due north and south line, and marked trees and bounds all round Sargent's Purchase—the south line was 3 1-4 miles and the west line 11 miles 76 chains and 4 links—the north line to Green's Grant was 4 miles 15 chains and 87 links—these bounds made 25,000 acres, excluding 150 acres for Eastman tract—the ground was so rough I had to triangulate some.

[Triangulating is as accurate as measuring by chain—commenced about middle of June, 1854, was at work July 4th—had five or six men—lived on the ground except one day—made plans afterwards.]

The Tip-Top house, on the summit of Mt. Washington, is about 9 3–4 miles north of the south line thus run—and the north line about 2 miles north of said house—made monuments all around the tract thus surveyed. In 1861, I found what was called the north line of the Dorcas Merrill lot as located and occupied—the northwest corner was a rock maple tree about four rods east of Rocky Branch—in 1863, I chained this north line and found trees with ancient spots—cut out chips—the line was marked before 1832—traced that line to within few rods of where we made a corner—there was nothing where the corner should be—there had been fires and new growth—they had cleared near to the maple tree—from that tree I found a spotted line 1 mile and 53 chains, extending nearly to Bald Ledge—this line was north 81 degrees west in 1853—in 1863 I found the needle there 1–4 degree (and perhaps more), further west than in 1853—I made out this line to be due west by the polar meridian, calculating from the north star—the maple and ash, of which Meserve testifies, are the northwesterly corner of the Dorcas Merrill lot as located—I did not find the original northeast corner as located—found a line up to within few rods of it—this northeast corner is 100 rods from the maple tree.

On cross-examination Barker testified as follows, the parts in brackets being subject to defendant's exception:

After running on the Willey line in 1853 I abandoned it—could not start from the maple and make out Dorcas Merrill's lot—[went with Meserve to Willey, and Willey described the maple tree to me as the corner of the lot]—Meserve was there as agent of defendant, and said he had been there before and had known where it was—Willey said in making the Sargent grant he ran the marked line with a compass—that he commenced at the maple by Rocky Branch and ran west to Bald Ledge, where he could see Mt. Washington, and found 3 1-4 miles would

Figures, Numbers, and Symbols

extend beyond Mt. Washington—that he had trouble with his needle, and his son was sick, and he never returned to finish the survey—we run all lines by compass—I understand by "north," north by the needle, but by "due north," north by the polar meridian, and Gibson, Flint, and other authors on surveying, treat it so. In 1853, near the corner of Dorcas Merrill lot, the meridians varied 9 degrees—and on the west line of Sargent's Purchase, in places, they varied more—in some places could not depend on the needle—if the grant had been "north" and "west" and not "due north" and "due west," I should have surveyed by the magnetic meridian of 1832—the meridian of longitude does not vary—the magnetic meridian does vary.

I made my surveys of Sargent's Purchase in 1854, independent of the line run by Willey. The variation of the meridians at the Glen house in 1863, was 11 1–2 degrees—in one place in Maine, in 1847, I found the variation to be 16 degrees 38 minutes—and between Canada and Maine in July, 1850, 15 degrees 16 minutes—those variations are much higher than the average—taking the average variation of all Maine and New Hampshire, from extreme northeast to extreme southwest, perhaps the average might be from 11 degrees to 12 1–2 degrees.

Loomis' book shows the following variations in Maine: Farmington, 11 degrees 20 minutes; Dixfield, 12 degrees; Waterville, 12 degrees 8 minutes; Raymond, 9 degrees 45 minutes; Umbagog Lake, 13 degrees; Rumford, 11 degrees; Belfast, 13 degrees; West Thomaston, 12 degrees; and the following in other places: Hanover, N. H., 9 degrees 15 minutes; Burlington, Vt., 7 degrees 36 minutes; Williamstown, Mass., 6 degrees 15 minutes; Dorchester, Mass., 9 degrees 6 minutes; one place in Georgia, 5 degrees east; New Haven, Ct., 6 degrees 10 minutes; Champlain, N. Y., 9 degrees 30 minutes; Albany, N. Y., in 1826, 6 degrees 14 minutes; Albany, N. Y., in 1831, 6 degrees 32 minutes; Detroit, Mich., 3 degrees east. All but the last two are west—there is a line through North Carolina, Pennsylvania and Ohio, where the meridians are one and the same—west of that line the variation is east—east of that line the variation is west—the general rule is that the farther you go north and east the greater is the variation, and the further south and west the less it is. In going north to make the west line of Sargent's Purchase, I allowed 1 minute to a mile for increased variation, which is the rule and practice. To find the polar meridian, observation of the north star must be taken. (The witness described the manner of taking such observations.)

The maple was marked "S. 1832," with spots on the east and west sides—starting from the maple and running by the magnetic meridian, it is my impression, but I am not certain, that the top of Mt. Washington would not be in Sargent's Purchase.

I ran the Dorcas Merrill lot by compass because "due" was not in her grant—that lot had been located a few years before 1832.

Alexander Wadsworth testified, the part in brackets being subject to plaintiff's exception:

"Live in Boston—been a surveyor 38 years—[in running a line "due north" I should run on a line of longitude, and for "due west" should run at right angles on a parallel of latitude—should ascertain the true north by the north star—never ran a line by the meridian of longitude, nor calculated the difference between the meridians.]"

"Defendant then introduced evidence tending to show possession of Mt. Washington in defendant since 1853."

"It was agreed that the Dorcas Merrill lot had been located and occupied before 1832 by bounds essentially varying from those described in the grant."

"Defendant claimed that the first bound mentioned in the grant of Sargent's Purchase was the northeast corner of the Dorcas Merrill lot, as located at the date of the Sargent grant, and that the meridian of longitude should be used in surveying Sargent's Purchase."

"Plaintiff claimed that said first bound was the northwest corner of said Merrill lot as located at the maple tree, because Willey had fixed and located it as said first bound and had run a line westerly from it as a part of the south line of Sargent's Purchase—that the line so run by Willey was the south line of Sargent's Purchase as far as it went—that the remainder of the south line should be run west at right angles with the magnetic meridian—and that the west and north lines of Sargent's Purchase should be run by the magnetic meridian."

"On these points the court ruled in favor of plaintiff and the defendant excepted."

All the depositions, papers and plans used at the trial may be referred to in argument.

Amendment to Case

Plaintiff also offered to show, that, at the date of said grant to Sargent and others, said Willey, in writing said deed, used the word "due" with reference to the magnetic meridian, and that the only method of surveying land, then known or practiced in this State, was by the magnetic meridian, and that the word "due" as then used in describing land boundaries applied uniformly in this State to the magnetic meridian. The court rejected this evidence and plaintiff excepted.

"Above amendment allowed.
Binghams, Benton & Ray, for plaintiff.
Burns & Fletcher and *Heywood*, for defendant.
BARTLETT, J.

The copy of the deed from Willey was properly received as part of the plaintiff's chain of title. *Harvey v. Mitchell*, 31 N.H. 582. The first objection to the caption of the depositions of Charles Faulkner and others is well taken, as it does not state whether on the 12th of April the defendant did or did not object; *Rand v. Dodge*, 17 N.H. 355; but we think the other two objections are unfounded as the caption explicitly states that the defendant was present on the 13th and did not object, and it shows with sufficient certainty that each of the deponents took the proper oath.

Figures, Numbers, and Symbols

The objection that the deposition of George A. Whitney was not properly taken on interrogatories seems without foundation in fact; and the exceptions to the second and fourth interrogatories must be overruled, for neither of them, upon any fair construction, is leading, and certainly it was competent for the plaintiff to prove the genuineness of the signatures, and the answer to the third interrogatory shows the witness qualified to give his opinion.

We are unable to appreciate the force of the objection that "the deeds from Charles Bellows to Cady, and from Cady to John Bellows, are a source of title different from what had been before introduced," for we see nothing in the fact that the plaintiff has set up a claim under conveyances from Thompson and Meserve to prevent him from showing a tax title also, if he has acquired such, or from relying on mere possession. The deed Cady to Bellows is said to have been a quitclaim, but that furnishes no legal ground of objection to its admissibility, and it would be color of title, even if the deed from Charles Bellows conveyed no interest to Cady; *Minot v. Brooks*, 16 N.H. 374; *Rand v. Dodge*, 17 N.H. 343; and an entry by a grantee under such a deed would give him possession of the whole tract described in it. *Tappan v. Tappan*, 31 N.H. 53; *Gage v. Gage*, 30 N.H. 425. As these deeds were offered as part of a chain of title, and appeared by official certificates upon them to have been regularly recorded, it was unnecessary to prove their execution. *Bellows v. Copp*, 20 N.H. 502; *Knox v. Silloway*, 1 Fairf. 202; 1 Green. Ev. Sec. 571, n.

A practical location is but an actual designation by the parties upon the ground of the monuments and bounds called for by the deed. *Clough v. Sanborn*, 40 N.H. 316; *Colby v. Collins*, 41 N.H. 304; *Peaslee v. Gee*, 19 N.H. 274; 4 C. & H.'s Phil. Ev. 549; *Jenks v. Morgan*, 6 Gray 448; *Cleaveland v. Flagg*, 4 Cush. 76; *Kellogg v. Smith*, 7 Cush. 382; *Knapp v. Marlborough*, 29 Vt. 282. The testimony of Thompson did not tend to show a practical location of the land conveyed by Willey's deed. The transaction he states was not a designation of the monuments, & c., called for by that deed, for the deed was not then in existence; *Sanborn v. Clough, Peaslee v. Gee*; and the prior negotiations must be taken, so far as the construction of the deed is concerned, to have been merged in that instrument, "the conclusive presumption being that the whole engagement of the parties, and the extent and manner of it, were reduced to writing." *Nutting v. Herbert*, 35 N.H. 121; *Cook v. Combs*, 39 N.H. 597; *Galpin v. Atwater*, 29 Conn. 97; *Parkhurst v. Van Cortland*, 1 Johns. Ch. 282; *Clark v. Northy*, 19 Wend. 323; 4 C. & H.'s Phil. Ev. 519. The deed contained no reference to any monument established by Thompson and Willey, or to any survey by them, (Sanborn *v. Clough*, 40 N.H. 239,) and the effect of the evidence at most could be merely to show that Willey and Thompson intended a different tract of land from that afterward conveyed by the deed, if the lines of their exploration are found to differ from the calls of the deed; and its reception to control the deed would be in violation of a principle quite elementary. *Bell v. Morse*, 6 N.H. 208; *Furbush v. Goodwin*, 25 N.H. 426; *Dean v. Erskine*, 18 N.H. 83; *Clough v. Bowman*, 15 N.H. 514; *Cook v. Babcock*, 7 Cush. 526; *Curtis v. Francis*, 9 Cush. 421; *Knapp v. Marlborough*, 29 Vt. 282; *Linscott v. Fernald*, 5 Greenl. 426; *Flagg v. Thurston*, 13 Pick. 150;

198   *Boundary Retracement*

Allen v. Kingsbury, 16 Pick. 235; *Dawes v. Prentice,* 16 Pick. 435; *Pride v. Lunt,* 1 Appl. 115.

Besides, Meserve who was one of the grantees in the deed was not a party to this transaction by Willey and Thompson, and there is no evidence that he ever authorized or ratified it. *Prescott v. Hawkins,* 12 N.H. 27. This evidence was therefore incompetent to affect the construction of the deed; and it does not tend to show that the summit of Mt. Washington is within the tract conveyed by it, as there is nothing in the testimony of Thompson tending to show that the westerly line, over which they passed, was on the easterly line of Chandler's Grant; and although the subsequent entry by Thompson under the deed gave possession of all the tract conveyed by it, yet there is no evidence that Mt. Washington is part of that tract.

But the motion for a nonsuit was properly denied, for the case finds that the evidence of Spalding and Davis tended "to show John Bellows' possession of Mt. Washington at various times between 1851 and 1859," and this is evidence of his muniment as against the defendant, for at the time of the motion no evidence of title in the defendant appeared; *Rand v. Dodge,* 17 N.H. 343; *Wendell v. Blanchard,* 2 N.H. 456; *Woods v. Banks,* 14 N.H. 113; *Jones v. Merrimack Co.,* 31 N.H. 384; *Parker v. Brown,* 15 N.H. 185; *Lund v. Parker,* 3 N.H. 50; *Graves v. Amoskeag Co.,* 44 N.H. 464; *Straw v. Jones,* 9 N.H. 402; *Sparhawk v. Ballard,* 1 Met. 95; and the deed from John Bellows to the plaintiff would give the latter such seizen as would enable him to maintain this action against one who showed no evidence of title. *Edmunds v. Griffin,* 41 N.H. 532; *Tappan v. Tappan,* 36 N.H. 120; *Carter v. Beals,* 44 N.H. 413; *Ward v. Fuller,* 15 Pick. 185.

If it was necessary under the statute to prove the handwriting of both of the subscribing witnesses to the signature of Coues, in order to show that his title passed, (see *Cram v. Ingalls,* 18 N.H. 616, *Melcher v. Flanders,* 40 N.H. 156,) still no objection is suggested to the proof of the execution by Pingree, and the deed was admissible to show the conveyance of his interest to the defendant.

"Parol proof of the appointment and commission of Selden would seem incompetent, but it was quite sufficient to show him an acting commissioner or notary." *Bellows v. Copp,* 20 N.H. 503; *Prescott v. Hayes,* 42 N.H. 56; *Forsaith v. Clark,* 21 N.H. 422. The second interrogatory in Selden's deposition is not leading, and the objection that no deed was enclosed seems not sustained in fact. Had objection been taken to the regularity of that mode of taking depositions, it is unnecessary now to say whether it ought to have been sustained. Obviously, the practice of founding an interrogatory or answer in a deposition upon a deed merely "enclosed" is very loose and not to be encouraged, as, aside from its inconvenience, it might open a wide door for fraud or mistake; see *Brown v. Clark,* 41 N.H. 245; but no question upon this point has been reserved.

There is nothing tending to show that the deed was fraudulently or intentionally withheld when the depositions were filed; and as an examination of the depositions would have given the plaintiff notice as to the

Figures, Numbers, and Symbols

deed, so that he might have procured an order for placing it on file, if it ought to have been filed, and need not have suffered by the omission, we see nothing in the mere circumstance of the neglect to file the deed, that should exclude the deposition. Besides the depositions seem to have been filed under the twenty-sixth rule of court for the purpose of limiting the adverse party in the time and manner of objecting to the caption; and in such case the only effect of a failure by the defendant fully to comply with that rule would seem to have been a failure to obtain the restriction of the plaintiff under the rule.

Numerous questions have been raised in reference to the evidence of a tax title introduced by the defendants, but several of them have not been argued by counsel, and may possibly not arise or may be obviated upon a trial of the cause; and upon the merits of some of these we have not deemed it advisable to pass at this time.

Sargent's Purchase, though uninhabited, might properly be taxed, *Wells v. Burbank*, 17 N.H. 393, *Russell v. Dyer*, 40 N.H. 173, Laws 1831, p. 26, Laws 1805, p. 448, and was made liable to a tax by the legislature. Laws November, 1840, p. 173. In *Wells v. Burbank*, 17 N.H. 394, it was decided that "it is not necessary to post an advertisement of a sale for taxes in an unincorporated place, which is uninhabited"; "and we do not understand the authority of this case upon that point to have shaken by any subsequent decision." *Russell v. Dyer*, 42 N.H. 399. We do not now see any sufficient reason for overruling the case in that particular. It cannot, as in *Russell v. Dyer*, 40 N.H. 184, be presumed that the legislature did not intend to subject such an uninhabited place to the statutes relative to taxation, since this and other similar places have for many years been specially named in our statutes as objects of taxation, and the taxation without the power of collection by sale would seem futile; so that upon a careful comparison of the objects and provisions of the statutes in question here, with those considered in *Russell v. Dyer*, as well as in reasons of public policy, we find sufficient grounds for a distinction between the latter case and *Wells v. Burbank*. Under these circumstances, as the doctrine of *Wells v. Burbank* does not seem likely to work any real practical injustice, and as it is probable that a very considerable number of titles to real estate acquired during the twenty years since the decision in that case was made, and while it has been unquestioned by the court and undisturbed by legislation, may depend upon the rule there laid down, we should deem it our duty under the law not now to question its correctness unless for more cogent reasons than appear to exist, in cases under the acts in question; Broome Leg. Max. sec. 109, *et seq.*; Fearne on Rem. Sec. 134; and we must, therefore, regard the authority of that case as decisive here. If, therefore, Sargent's Purchase was uninhabited, it was unnecessary to post any advertisement of the sale within its limits, and it would be immaterial where upon the Purchase, or when the notices were put up or taken down, or whether they were ever returned to the clerk's office. The original warrant was returned to the State treasurer, and as nothing further appeared, its contents could not properly be proved by parol. If the loss of the record in the clerk's office was shown, its contents could be proved by parol, certainly so far "as the case does not from its nature disclose the existence of other and better evidence." 1 Greenl. Ev. Sec. 509; *Scammon v. Scammon,* 33 N.H. 59; *Forsaith v. Clark,* 31 N.H. 418.

As the sale was in January, 1843, and the Revised Statutes did not take effect till the following March, (Rev. Stat. P. 474, sec. 1,) we are to look to the statutes in force prior to the Revised Statutes for the provisions to govern the proceedings. By the statute then in force the sheriff was required to deposit with the clerk the lists and other papers containing evidence of his proceedings in the sale of lands for taxes, and it was made the duty of the clerk to receive and preserve them, and to make and certify copies thereof as of other papers on file in the office. Act Dec. 16, 1824; Laws 1830, p. 572. As the statute requiring the deputy secretary of the State to retain in his office a certified copy of the list returned to the collector was not in force at the time of these transactions, (Laws 1847, ch. 495,) we need not inquire whether in case of the loss of the original, resort should be had to that before introducing parol evidence. See 1 Greenl. Ev. Sec. 84, & n; 4 C. & H.'s Phil. Ev. 285; *Melvin v. Marshall,* 22 N.H. 382.

As already stated, upon proof of the loss of the originals in the office of the clerk, their contents may be proved by any secondary evidence, where the case from its nature does not disclose the existence of other and better evidence, and we do not find that any exception in the case of records like these is made by the common law or by our statutes; and therefore the plaintiff's objection that "the record should have been made up anew under the direction of the court," & c., and that the record so made up would be the only competent evidence, cannot be sustained. He does not cite any authority for his position, or point out any law or show any usage requiring such a renewal of the record; and it is to be observed that the records so deposited in the clerk's office are not proper records of the court itself, for they are not made by its officers as such, and do not contain its transactions; they are merely deposited in the office of the clerk; so that we see nothing in the nature of the case that should require what neither the common law nor our statutes have prescribed, nor any well settled usage established.

Where a single sum is assessed upon an unincorporated place, the treasurer's warrant is a list within the meaning of the statute, *Wells v. Burbank,* 17 N.H. 407; *Homer v. Cilley,* 14 N.H. 100; and if a copy of it was duly returned by the collector, this was a sufficient compliance with the second section of the act of July 4, 1829; Laws 1830, p. 564; and the certificate of the deputy secretary upon it was competent evidence of the time of filing and return; *Wells v. Burbank,* 17 N.H. 409; *Smith v. Messer,* 17 N.H. 430; and this having been destroyed, its proof would fall within the rule already stated.

The evidence tends to show that the sheriff, within ten days after the sale, delivered to the clerk a copy of his sale, but there is no evidence that it was accompanied by his charges according to the provision of section 4 of the act of 1829, Laws 1830, p. 565; that section, however, does not provide that the account of the sale shall be under oath, nor have we found any such requirement prior to the Revised Statutes; Rev. Stat. Ch. 4609; but under the act of 1829 the copy of the sale was to be attested.

Figures, Numbers, and Symbols

201

As Meserve testified that on the day after the sale he filed with the clerk a copy of the record of sale, "with the Patriot and Democrat, and all other papers," the jury might have found that within ten days after the sale he so filed the copy of his list, if that were essential; but it may not be altogether clear that this is required by the statute. Section 4 of the act of 1829 only provides for the filing of "an attested copy of the sale," with charges of sale, within ten days after the sale, and section 7 makes it the duty of the collector to lodge with the town clerk, within ten days after the sale, the newspapers containing the advertisement of such sale, and the advertisement which may have been posted up in such town with a certificate accompanying the same, under oath that it was posted up according to law, which advertisement and certificate shall be recorded by the town clerk, and a certified copy of such record shall be deemed sufficient evidence of those facts in any court of law; and the said newspapers shall be kept on file by the clerk. The act of July 1, 1831, Laws, p. 26, gave the sheriff in a case like this, "the same power and authority with respect to the taxes committed to him to collect, which collectors of towns have or may from time to time by law have with respect to the taxes of non-residents"; and provides that "he shall observe the same directions as collectors of towns are or may from time to time be bound by law to observe in collecting the taxes of non-residents," & c.; with a proviso requiring an advertisement in the shire town of the county, as well as in the place where the lands lie. The first section of the statute of Dec. 16, 1824, enacted "that the lists returned by the receiver of non-resident taxes and other papers containing evidence of the proceedings of any former or future sheriff of any county in this State, relating to sales of land by him as sheriff, for State and county taxes, be deposited in the office of the clerk of the Superior Court," & c.; and that it "be the duty of the said clerk to receive and preserve the same, and to make and certify copies thereof as of other papers on file in said office"; "and the second section provided that such copies might be used as evidence in courts of law in all cases in which the originals might be used," and with the "same force and effect." Laws 1830, p. 572. And in *Wells v. Burbank*, 17 N.H. 410, it is decided that the act of July, 1831, did not repeal this act of 1824.

The eighth section of the act of 1829 relates merely to lands redeemed. These would seem to be all the provisions of the statute upon the subject then in force. The act of 1831 only provides what the sheriff shall file, and when he shall file it, by reference to other laws. The act of 1824 seems to be the only statute at that time requiring the list to be filed with the clerk of the court, and it contains no express provision as to the time; while the act of 1829, which is the only statute fixing the time for filing the papers by the collector, does not appear to include the list; and as the land might be redeemed within one year from the sale by a tender to the collector, (Laws 1830, p. 565, sec. 4,) it may admit of a doubt whether the copy of the list was required to be filed with the clerk within ten days after the sale; but it is unnecessary now to decide this question.

Besides, it is not entirely certain that the neglect of the sheriff after the sale seasonably to file a copy of his sale, or his list, or the list of lands redeemed, should defeat the title of a bona fide purchaser acquired under a previous sale legally made to him, where such purchaser has himself been in no fault; see *Smith v. Messer*, 17 N.H. 428; *Smith v. Bradley*, 20 N.H. 120; *Scammon v. Scammon*, 28 N.H. 432; *Pierce v. Richardson*, 37 N.H. 310, 312; *Hayes v. Hanson*, 12 N.H. 290; *Cardigan v. Page*, 6 N.H. 193; *Tucker v. Aiken*, 7 N.H. 113; *Pinkham v. Murray*, 40 Me. 587; *Lane v. James*, 25 Vt. 481; *Taylor v. French*, 19 Vt. 49; *Sumner v. Sherman*, 13 Vt. 609; but we do not propose to pass upon this question at the present time.

The clerk at the auction was not an officer making or in any way controlling the sale, and we see no legal objection to his becoming the purchaser.

It is said that a part of Sargent's Purchase was annexed to Jackson in 1837. If chapter 336 of the Laws of that year is referred to in this statement, that fact does not appear on the face of the act which merely establishes the location of certain lines of the town. If, however, the effect of that statute was as stated, then the part so annexed would thereafter, for purposes of taxation, cease to be part of Sargent's Purchase, and become part of Jackson, and liable to taxation as part of that town; and the residue of the original purchase would for such purposes remain Sargent's Purchase, precisely as in the case of any town in the State, after a farm has been severed from it and annexed to an adjoining town; and the apportionment of Dec. 22, 1840, Laws, p. 499, which fixes the proportion of Sargent's Purchase at two cents for each thousand dollars to be raised by the State, must be taken to mean Sargent's Purchase as it existed for purposes of taxation, in the same way as it denoted the towns and other places as they legally existed, or might exist, for such purposes. Whether, in the absence of any statutory provisions, the objection that the deed included less than was sold, could avail between these parties, we have not inquired; for by the sixth section of the act of 1829, it is provided that when two or more persons are interested in any tract of land so sold, every individual may redeem his own part thereof by paying or tendering his proportion of the taxes and cost for which the said land was so sold, and this proportion shall be according to the number of acres in the tract of land sold. Section fourteen provides that when any estate of non-residents shall be sold by virtue of this act, and the money necessary for the redemption thereof shall not have been paid or tendered within one year from the sale thereof, the collector shall then execute a good and sufficient deed of such estate to the purchasers of the same, & c.; and it prescribes the form of the deed with covenants that the collector has conformed to the requirements of the law in making the sale, and that as collector he has "good right, so far as that right may depend on the regularity of his own proceedings, to sell and convey the same in manner aforesaid." Laws 1830, p. 567. The collector, then, is to execute a deed of what has been sold and not seasonably redeemed, and if he excepted from the conveyance those lots redeemed within the year, he seems to have followed the statute, for although they had been sold, yet as they had been seasonably redeemed, they did not come within the description of the fourteenth section, while the residue of the land sold did. If any one of those interested in the tract had paid his proportion before the sale, his share was properly omitted in the sale.

Figures, Numbers, and Symbols

203

Laws 1830, p. 565, sec. 3. The objection that they sold the whole of the original Sargent's Purchase, including the parts on which the taxes had been paid, and the part annexed to Jackson, if there were such, does not seem very clearly supported by the case, for such a fact is by no means necessarily to be inferred from the evidence stated, which is at least quite as susceptible of a different interpretation.

Whether upon Meserve's testimony that he sold "the whole tract except what had been paid on," naming and excepting the lots upon which payments had been made, and stating "the tract and the amount of taxes and costs," in connection with his deed, after proof of the loss of the records, the jury could have found that the taxes had been paid for the lots excepted from the sale, and that the residue had been sold for the remainder of the taxes and costs; or whether it was necessary to show these facts more specifically, so that the amounts paid for each portion, the number of acres in each of such portions and in the residue or the relative interests of the owners in the tract, and the amount for which as taxes and costs such residue was offered and sold, should appear, and whether similar facts should have been shown as to the lots described in the deed as redeemed, (see *Cardigan v. Page,* 6 N.H. 193; *Pierce v. Richardson,* 37 N.H. 315; *Smith v. Bodfish,* 27 Me. 289,) we do not deem it advisable now to inquire, as possibly such questions may not arise upon a trial of the case.

If the objection that Sargent's Purchase was never allotted rests merely upon the position that the lots were not actually marked upon the ground, it is not well taken, for an allotment by plan might indicate with certainty the location of each and every lot, although no lot lines had in fact been run out; and if the lots could be made certain, that would be sufficient; *Darling v. Crowell,* 6 N.H. 424; *Smith v. Messer,* 17 N.H. 428; *Wells v. Burbank,* 17 N.H. 412; *Corbett v. Norcross,* 35 N.H. 119; but whether it was ever properly so allotted, we cannot decide upon the case before us. If it was material for the defendant to prove this fact either for the purpose of showing that the locus was not excepted or for any other reason, (see *Smith v. Bodfish,* 27 Me. 289,) it would seem that there is better evidence of the allotment than that offered, and no reason is shown why it should not be produced.

The constitution of this State provides in article 44, that "every bill which shall have passed both houses of the general court, shall, before it becomes a law, be presented to the Governor; if he approve, he shall sign it," & c.; and in article 45, that "every resolve shall be presented to the Governor, and before the same shall take effect, shall be approved by him," & c. The court will take judicial notice of the fact that David L. Morrill was Governor of the State in 1826, and of the genuineness of his signature. 1 Greenl. Ev. Sec. 6. Pinkham's Grant was a monument described in the titles set up by each party, and the resolution by which it was granted would seem admissible in determining its location, but there is nothing in the case to show that the grants to Rogers and others, to Green and to Martin, were in any way material.

The defendant claims that Sargent's Purchase is to be run out by the sidereal or astronomical meridian, or, as it is sometimes called, by the "true" meridian, and not by the magnetic meridian. Unquestionably, in this State, the courses in a deed are to be run according to the magnetic meridian, unless something appears to show that a different mode is intended in the instrument. 4 Kent 466, and n.; 4 C. & H.'s Phil. Ev. 550; *M'Iver v. Walker*, 9 Cranch 177; S. C. 4 Wheat. 444; *Brooks v. Tyler*, 2 Vt. 348; *Owen v. Foster*, 13 Vt. 267; *Riley v. Griffin*, 16 Geo. 147; *Young v. Leiper*, 4 Bibb 503; 1 U. S. Dig. 476, n. 51, and see *Jackson v. Stoats*, 2 Johns. Cas. 352; *Wilson v. Inloes*, 6 Gill 163; *Clark v. Northy*, 19 Wend. 324; *Loring v. Norton*, 8 Greenl. 69, and *Pernam v. Weed*, 6 Mass. 133. We do not understand the defendant seriously to question this, but he rests his position upon the use of the word "due" in connection with the words descriptive of the courses, claiming that "due north" means north by a sidereal meridian. It is observable, that, in the description of Sargent's Purchase, the last course from the southwesterly corner of Jackson is merely "south," and that the defendants' view if logically followed out, would seem to require the southern, western and northern boundary lines of the Purchase to be run out by a sidereal meridian, while a portion of the eastern boundary is to be laid down according to the magnetic meridian; a result not particularly desirable in point of convenience, or very likely to have been in fact intended by the parties. The word "due" in this connection means merely "exactly," and in fact adds nothing to the description of the point of compass, for "due north" is exactly north, and so is simple "north." As the designation of the points of compass is conventional, the word "due" applies with equal propriety to those points as referred to either meridian. We find no evidence that either the law or usage in this State has appropriated the term specially to the sidereal meridian. In various cases in our reports, and in many more before our courts, deeds and pleadings have shown a use of the term like that in the deed in question; see *Corbett v. Norcross*, 35 N.H. 100; *Bowman v. Farmer*, 8 N.H. 402; yet this is the first time, so far as we are aware, that such an effect has been claimed for it; and we find the word used elsewhere in legal language by courts entitled to the highest respect, without regard to any such distinction as the defendant claims. *Jackson v. Reeves*, 3 Caines 293; *Brandt v. Ogden*, 1 Johns. 156. We find nothing in the strict meaning of the term, in its popular acceptation, or in its legal or scientific use, or in our own usages or history, to sustain the defendants' claim, and he has directed us to no authority for his position, which seems to us untenable.

It is impossible for us to ignore the fact as matter of history and of common knowledge, that in this State private and even town boundaries have almost, if not quite, uniformly been run out according to the magnetic meridian, and we must hold it part of the common law of this State that the courses in deeds of private lands are to be run according to the magnetic meridian, when no other is specially designated, and this seems impliedly admitted by the defendant when he places his claim solely upon a supposed effect of the word "due," which, as we have seen, does not in our view belong to it.

Any resort to proof of the actual intention of the parties or the surveyor would be attended with all the mischiefs which have heretofore,

Figures, Numbers, and Symbols

by the familiar general rule, excluded parol evidence of intention in the construction of deeds, and in our judgment is not permissible. We do not understand that proof was offered of any local custom, and the court, in construing a deed, can hardly need the aid of opinions from experts in surveying, either as to the meaning of the word "due," or as to the custom or the common law of this State in relation to the general mode of surveying. The present method of surveying the public lands of the United States can have no bearing upon the question here, as it was specially adopted, at a comparatively recent date and long after the system of surveying private boundaries in this State had been established by ancient and long continued usage, if not originally fixed by the common law. The manner in which extensive public boundaries, like those between States and nations, have been surveyed, can have but little weight in the determination of the present question, for in such surveys regard is had to accuracy, permanency and certainty of verification, rather than to the expense or difficulty of the method or its practical convenience or adaptation for common use in our ordinary surveying. As the point in the present case is to be decided according to the law as already established in this State, for the determination of private boundaries, the relative advantages of the two systems are not in question before us, but perhaps upon examination it will be found that in practice neither mode gives perfect theoretical accuracy, and where the question of the adoption of a system is open, it would seem a question of relative accuracy and general convenience rather than of entire exactness. We are of opinion that in the grant to Sargent the courses are to be construed as referring to the magnetic meridian notwithstanding the addition of the word "due" to their description.

The starting point for ascertaining the bounds of Sargent's Purchase is the northeast corner of Dorcas Eastman's grant, and the authorities already cited are decisive that in this action at law it cannot be shown that the word "northeast" was inserted in the deed by mistake for "northwest." If nothing more appears, that corner is to be ascertained by looking to the terms of her grant; but if it appears that at the date of the grant to Sargent it had been practically located upon the ground in a manner to bind the parties to it, then the northeast corner as thus located is to be taken as the northeast corner intended in the deed to Sargent. *Hall v. Davis*, 36 N.H. 569; *Breck v. Young*, 11 N.H. 489; *Kellogg v. Smith*, 7 Cush. 376. Even if Willey had run a line for the south line of Sargent's Purchase, as it is not in any way referred to in the deed and as the act of Willey alone could not amount to a practical location of the grant, even if after the execution of the deed, and certainly not if prior to that, this could not control the description in the deed. Whether the northeast corner of Dorcas Eastman's lot is necessarily the southeast corner of Sargent's Purchase, or merely a point of beginning to run the due west line of three and a quarter miles, and whether that corner of Sargent's Purchase is to be found at the intersection of a line due south from the southwest corner of Jackson, with a due east and west line drawn through the northeast corner of Dorcas Eastman's grant, it is unnecessary now to inquire. The case is to be discharged.

In New Hampshire the courses in a deed are to be run according to the magnetic meridian, unless there be something in the instrument showing that a different mode is intended.

A different mode will not be inferred from the fact that in giving the courses of the several lines, there is prefixed to the course of part of them, the word "due," as thence "due west," or "due north"; the word *"due"* in this connection meaning merely *exactly* north, or *exactly* west, and it applies with equal propriety to these points, whether the magnetic or the sidereal meridian be referred to. It must be considered as part of the common law of the State that the courses in deeds of private lands are to be run according to the magnetic meridian when no other is specially designated.

Directions Given in Reverse Order

Frequently, when a direction was close to the East-West meridian, a bearing would be expressed with that letter first, followed by the number of degrees of arc, then either the North or South meridian (see Figure 6.2).

For example, in the case of *Jakeway v. Barrett**:

> The defendant introduced as evidence a copy of a deed, duly certified, from Harvey Church to Aaron Gibbs, dated December 10th, 1828, conveying land bounded by the following description, viz: "Bounded on the east by a line running from a birch tree standing in the south line of Aaron Gibbs' land one hundred rods west ten degrees north of a stake and stones standing in the south-east corner of Aaron Gibbs' land; thence south ten degrees west sixty-three rods to a stake and stones;

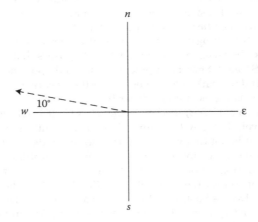

FIGURE 6.2
Direction stated as west 10 degrees north, which translates to north 80 degrees west.

---

* 38 Vt. 316, Vt., 1865.

Figures, Numbers, and Symbols

thence W. 30° 41' N. about 60 rods

FIGURE 6.3
Excerpt from a deed description citing bearing as west 30° 41' north.

thence west ten degrees north thirty-six rods to a stake in the marsh; thence south ten degrees west seventy-four rods to the west bank of the marsh adjoining Lake Champlain; bounded on the south by the lake or land of Joseph Smith, deceased; on the west by Lake Champlain, and north by land owned by Aaron Gibbs, containing fifty acres of land, be the same more or less."

One must constantly be on guard when searching records as occasionally some person unaware of this practice may have assumed a mistake had been made and unilaterally made an inappropriate conversion. This can introduce considerable error into the mathematics of the survey, or surveyed definition, of the land description resulting in a failure to find an existing corner. Depending on the nature of the conversion, the parcel may still close mathematically, therefore not revealing the introduced error(s). The only way to avoid getting caught in such a mistake is to trace the records back, or otherwise make a comparison, of a given description with the original description, which should be based on the original survey. Even then there is no guarantee that any numbers had been transferred correctly from the survey to the first written description of that survey. Therein lies yet another example of there being no substitute for having the original survey and making reference thereto (see Figure 6.3).

Problems with Distances

Distances (lengths of lines) offer a multitude of problems. As was explained in the early Texas case of *Stafford v. King*,* "Of all the indicia of the locality of the true line, as run by the surveyor course and distance are regarded as the most unreliable, and generally distance more than course, for the reason that chain-carriers may miscount and report distances inaccurately, by mistake or by design. At any rate, they are more liable to err than the compass."

There is an extensive variety of length measurement units, and many of them have more than one definition. There are two standard lengths for the foot, several for the English rod, and several for the chain and therefore also their composing links, a number of definitions for both the vara and the

---

* 30 Tex. 257, 94 Am. Dec. 304, 1867.

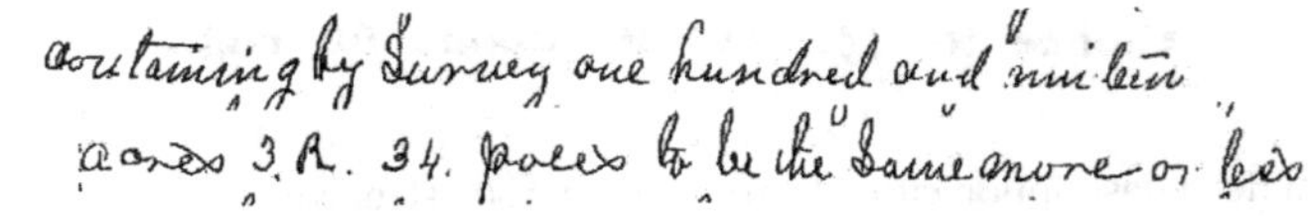
containing by Survey one hundred and nineteen
acres 3 R. 34. poles to be the same more or less

FIGURE 6.4
Excerpt from deed stating the area as 119 acres, 3 roods, 34 poles.

arpent, in addition to several other units including the mile, many of which are used locally.

One must also be especially wary concerning conversions. Sometimes encountered is a change in units from one deed to the next, due to someone attempting to modernize the description, or to put it in more easily understood units. A common occurrence is the conversion from rods and links to feet, tenths and hundredths. Sometimes the conversion is not quite as refined, due to rounding of the numbers. First, most people assume that the rod is equivalent to 16.5 feet, which is not always the case. Second, early measurements may have included an allowance for good measure, and therefore it is not a true, direct, conversion. Third, which most people do not consider is the tolerance of a measurement in the mathematical sense. Plus or minus a rod (assuming a 16.5 foot rod) is plus or minus 8.25 feet; plus or minus a foot is equivalent to 0.5 feet, or 6 inches. A conversion such as this can be very misleading, when used to determine a precise point from the information of an original survey. Such a change has inserted unnecessary error into the work.

AUTHOR'S NOTE: In addition to the use of the term rod, there is a unit of area measurement known as the rood. Occasionally, it will be found where someone assumed a misspelling, and changed roods to rods, thereby again inserting an unnecessary error. This more often affects the area figure, since there are also (square) rods generally merely stated as "rods" but used as a statement of area, apart from its use as a measurement of length. Both may be found in the same description, but signifying two different measurements. A rood is equivalent to 40 square rods, or ¼ acre (see Figure 6.4).

Measurement Allowances

It was standard practice to make allowances in measuring both distances and area. Frequently, the allowance was given in the instructions to the original surveyor, but otherwise there were accepted standards within the surveying community.

Because of inherent chaining errors, especially those of sag and alignment, early surveying instructions, or customary survey practice, would specify an allowance to be made. If a return of survey was made, sometimes the allowance used would be reported (see Figure 6.5).

The Vermont court in the case of *Neill v. Ward** addressed this situation:

* 103 Vt. 117, 153 A. 219, 1930.

Figures, Numbers, and Symbols 209

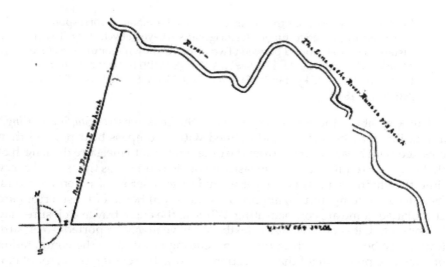

FIGURE 6.5
Return of survey.

Percy G. Smith, a civil engineer, and a witness for the plaintiff, was employed by the plaintiff to survey his farm and locate the dividing line between lots 59 and 60. A plan which he made of the plaintiff's farm from his survey, and which shows the dividing line between these lots as claimed by the plaintiff.

He testified in direct examination that he surveyed on the lines of lots 53–59, inclusive, and of other lots, and that he followed the range line between the first and third divisions, which is the northerly boundary of lots 57, 58, 59, and 60, from the Berlin town line northwesterly.

It appeared that in the papers of the Surveyor General there was item of the year 1784; "That in perambulating or running town lines throughout the State, one thirtieth part be allowed for swag of chain"; and Smith testified that the lots he measured figured that way.

He testified in cross-examination that he measured the range line from the Berlin town line across seven lots to the northwest corner of lot 59, as claimed by the plaintiff, with a steel tape, and the distance was 895 rods; that the actual distance from the Berlin town line to the northwest corner of lot 59 according to the town plan and field book, and with the allowance of one-thirtieth part added, was a little over 839 rods; that his actual measurements overran the distance given in the field book by about 56 rods.

The evidence showed that the northwest line of lot 59, according to the town plan and field book, is substantially where the defendant claims it is.

In the PLSS, the instruction concerning retracing an original survey with regard to the distances stated is as follows.

210   *Boundary Retracement*

In all cases of resurvey the chain used must be made to correspond to that used by the government (or original surveyor), by testing it with distances on the ground between two or more known monuments. (See *Mason v. Braught* (33 S.D. 559, 146 N.W. 687 (S.D. 1914), *Henrie v. Hyer* (92 Utah 530, 70 P.2d 154 (Utah 1937) and *Hess v. Meyer*, 88 Mich. 339, 50 N.W. 290 (Mich. 1891).

Thinking about the reasoning stated in the Texas case of *Stafford v. King*,* a long line may be sighted and defined with a compass bearing, and then checked with a back bearing from the other end. But to measure the length of that long line requires using a measuring device numerous times. And every time the chain, or tape, is used it is subject to a series of potential errors. (Sag, tension, temperature, alignment, and so on) Some of these errors are cumulative, some are compensating. While at the end a traverse closure may be computed, it is only what the textbooks have always reported—a closure (difference between starting point and ending point) then the computation of a relative precision (of that collection of work) by relating the error of closure to the entire length of traverse. That does not account for compensating errors, so there may be gross error in the entire work which is unidentifiable. Another good reason for finding the corners. As one court said; when the footsteps are found, the case is solved.

Errors in Chaining

Early chains were heavy and therefore susceptible to considerable sag (sometimes known as "swag"), and to compensate, surveyors would apply the aforementioned allowance which was usually standardized, but sometimes arbitrary on the part of the individual surveyor. Individual links in the early chain were susceptible to wear, and sometimes got bent. Occasionally, one is found with an entire link missing. And, regardless of the measuring device being used, a problem measuring through think vegetation and heavy brush can introduce alignment errors, which are not constant, depending on the amount of interference. However, it is often found with original surveys, particularly if done by careful surveyors, that vegetation was cleared along the line of sight prior to the chaining.

Errors in Taping

Even though also known as a chain, primarily because of its units of length, the modern chain is manufactured as a continuous steel ribbon, and more often known as a tape. It is much lighter so not as susceptible to sag, but is subject to many of the same problems as other devices. Depending on conditions, temperature, alignment, pull (stretch) can affect the measurement

---

* 30 Tex. 257, 94 Am.Dec. 304, 1867.

Figures, Numbers, and Symbols 211

of distance. Knowing the conditions under which the work was done can be helpful, which, again, is why field notes and survey returns can be very useful. A single line measured twice, once in below freezing conditions and again during the heat of summer, can produce results that are widely separated. Obviously, the longer the line, the greater the difference between the two measurements.

Presumption of a Straight Line

As presented in Chapter 3, a line described as running from one point to another is presumed to be a straight line, unless a different line is described in the instrument, or marked on the ground. By ascertaining the points at the angles of a parcel of land, boundary lines can at once be determined.* The rule of surveying, as well as of law, is to reach the point of determination by the line having the shortest distance.[†]

In 1881, the US Supreme Court delivered an opinion concerning original township grants in the New Hampshire case, stating some very important principles. The case was brought to recover possession of a certain tract of land in Grafton County, described as follows (see Figure 6.6):

> Beginning at the northwest corner of the town of Albany, and thence running north about 3 degrees east, 3 miles and 65 rods, to a spruce tree marked; and from thence north about 6 degrees east, 4 miles and 95 rods, to a fir tree marked; and from thence south about 87 1/2 degrees east, to the westerly line of Hart's Location, and to the easterly line of Grafton County, as established by the act approved July 3, 1875, entitled "An Act establishing the east line of Grafton County;" and from thence along the east line of Grafton County to the bound begun at, and containing 8,000 acres of land, more or less.
>
> The defendant filed a plea, defending his right in, and denying disseisin of, all the land described in the plaintiff's writ which is included in the following-described tract:
>
> "Beginning at the northwest corner of the town of Albany, formerly called Burton, and thence running north about three degrees east, three miles and sixty-five rods, to a spruce tree marked; and from thence north about six degrees east, four miles and ninety-five rods, to a fir tree marked; and from thence south about eighty-seven and one-half degrees east, to the westerly line of Hart's Location; thence southerly by the westerly line of Hart's Location to the point in said westerly line nearest to the northwest corner of said Albany; thence in a straight line to the northwest corner of said Albany."
>
> "The demandant, on the trial, produced and deraigned title under a quitclaim deed from James Willey, land commissioner of the State of New Hampshire, to Alpheus Bean and others, dated Nov. 26, 1831, made

---

* *Halstead v. Aliff*, 78 W.Va. 480, 89 S.E. 721, 1916.
[†] *Bartlett Land, etc. Co. v. Saunders*, N.H. 103 U.S. 316, 26 L.Ed. 546, 1881.

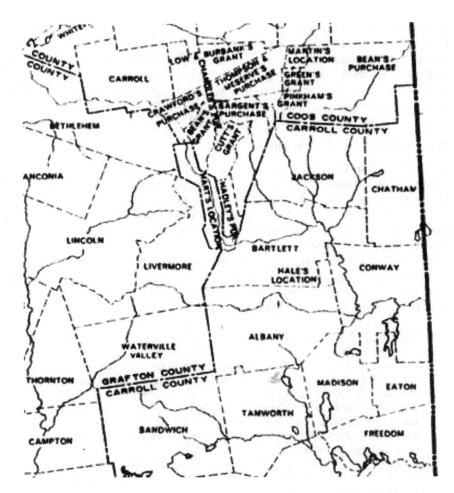

FIGURE 6.6
New Hampshire townships.

by authority of a resolve of the legislature, which included the lands claimed in the writ."

"He also produced a prior deed, under which the defendant claimed the land described in his plea, being a deed from Abner R. Kelly, treasurer of the State of New Hampshire, to Jasper Elkins and others, dated Aug. 31, 1830, and made by authority of a resolve of the legislature, which deed purported to convey the following-described tract in the county of Grafton, New Hampshire, to wit:----

"Beginning at the northeast corner of the town of Lincoln, and running east seven miles and one hundred and seventeen rods to Hart's Location; thence southerly by the westerly boundary of said location to a point so far south that a line drawn thence due south shall strike the northwest corner of the town of Burton; thence south to said northwest

Figures, Numbers, and Symbols

corner of Burton; thence westerly along the northern line of Waterville to the eastern boundary of Hatch and Cheever's grant; thence northerly and westerly by said grant to the east line of Thornton; thence by said line of Thornton northerly to the line of Lincoln, and along this line to the point first mentioned."

"The principal question in the cause was whether the premises thus granted to Elkins and others by the last-named deed embraced the land described in the defendant's plea; if they did, as was held by the judge at the trial, the defendant's was the elder title to the land in dispute, and the title of the demandant failed, and there is no error in the instructions as to the documentary title."

"The beginning corner of the premises granted to Elkins and others was conceded to be a well-known point, and the general position of the first line of survey, which is described as "running east 7 miles and 117 rods to Hart's Location," was not disputed; nor was the position of the northwest corner of the town of Burton (now Albany) disputed, it being a common point to which both parties referred; nor were the lines of the Elkins survey from the northwest corner of Burton, "westerly along the northerly line of Waterville, &c., to the point first mentioned," brought in question. The only point in dispute was the eastern boundary of the Elkins tract; the defendant contending that, by virtue of the deed of 1830, it extended eastwardly to Hart's Location, covering the disputed territory; and the demandant contending that it did not extend further to the eastward than the northwest corner of Burton (or Albany), and a line drawn north from that point.

"The language of the grant is, "east 7 miles and 117 rods to Hart's Location; then southerly by the westerly boundary of said location to a point so far south that a line drawn thence due south shall strike the northwest corner of the town of Burton; thence," &c. Now, if, when the grant was made, there was a track known as Hart's Location lying easterly and in the vicinity of the land granted, and if it had a westerly boundary to which the granted tract could by any reasonable possibility extend, no more apt language for this purpose could have been adopted. It would be a monument which would control courses and distances. If more or less distant from the point of beginning than seven miles and one hundred and seventeen rods, still it would control the survey. If a line drawn due south from any point of its western boundary would not strike the northwest corner of Burton, then they must be connected by a line not running due south. The line of shortest distance between said boundary and said northwest corner would be the proper one, and this is the one that was adopted. Hart's Location is called for, and to that location we are bound to go."

"The evidence was overwhelming and uncontradicted to show the existence and notoriety of Hart's Location. It is a large tract of land lying on both sides of the Saco River, directly to the eastward of the Elkins tract. On the 27th of April, 1772, this tract was granted by Governor Wentworth, in the name of the king, to one Thomas Chadbourne. The plaintiff produced in evidence a copy of that grant, having a plat or survey of the tract annexed to it. The premises granted are described as follows:----

214   *Boundary Retracement*

Beginning at a birch tree being the southwesterly corner bounds of a tract of land granted to Mr. Vere Royse; from thence running north four hundred and seventy rods, from thence extending westerly the same breadth of four hundred and seventy rods, the distance of two hundred and eighty-five rods, from thence running northwesterly six hundred rods, from thence running nearly a north course thirteen hundred rods until it meets the notch or narrowest passage leading through the White Mountains lying upon Saco River.

The plat, or survey, annexed to the grant shows the Saco River running through it. It follows the river on both sides from the beginning of the survey up to the mountains. It is conceded that the beginning corner is well known; and the general location of the tract is undisputed. By the name of Hart's Location it has been well known for nearly a century past. Its census has been published in the laws like that of a regular township, and it seems to have been treated in some sort as a quasi township. In the State census published with the laws of 1815, and again in 1820, the population of Hart's Location is put down as thirty-five for the year 1810, and at sixty-five for 1820. In the acts for the apportionment of the State tax among the several townships of the State, the pro rata share of Hart's Location was fixed at eight cents on a thousand dollars in 1816; at twelve cents in 1820; at ten cents in 1824; and at eight cents in 1829. By an act approved Dec. 24, 1828, it was resolved, "That Hart's Location, in the county of Coos, be annexed and classed with the towns of Bartlett and Adams, in said county, for the purpose of electing a representative to the general court, until the legislature shall otherwise order." The demandant's principal witness stated that it had been a political organization at one time, and sent a representative to the general court.

But it was claimed by the demandant, and proof was offered to show, that the western boundary of Hart's Location, being in a wild and mountainous region, had never been located on the ground in 1830, and could not be located from the description contained in the grant, because it was too vague and uncertain to admit of a fixed and definite survey. But the plat annexed to the grant, and referred to by the grant for greater certainty, did show a boundary line, laid down to a scale. If there was no other evidence on the subject, this would be sufficient to show that Hart's Location had a boundary, and a definite one, whether it was ever actually run out on the ground or not. In or about 1803, on occasion of a general perambulation of the townships of the State, made in pursuance of an act of the legislature, a survey of Hart's Location was made by one Merrill, by public authority, and deposited in the office of the secretary of state. This was also produced in evidence on the trial, and showed a well-defined map of the location, laid down to a scale,--differing somewhat from the plat annexed to the original grant, but not more than might be naturally expected if the original was not used.

There can be no doubt, therefore, that when Hart's Location was referred to in public acts and resolves, whether for the purpose of taking the census, taxation, or political jurisdiction, it was referred to as a defined tract or portion of territory, within the bounds of which the State claimed no proprietary interest. In 1830, when the legislature, by a

Figures, Numbers, and Symbols

resolve, authorized, and by its treasurer made, to Elkins and his associates, a grant of land to extend from the town of Lincoln on the west to Hart's Location on the east, the exterior line extending along "by the westerly boundary of said location," it is difficult to find any ground for uncertainty or ambiguity in the grant, or to imagine how, after that, the State, or any persons claiming under the State, could, with any show of reason, claim that there was no such thing in being as a Hart's Location having a western boundary; or that the Elkins grant did not extend to and bound upon it. All rights of the State up to and adjoining said location were as clearly disposed of as if the two grants, that of Hart's Location and that to Elkins and others, had been made in the same instrument,—granting to one party, first, Hart's Location as described in Chadbourne's patent, and then granting to Elkins and his associates all the residue of the lands westward to the town of Lincoln between designated side lines on the north and south.

The truth is, that Hart's Location itself was the monument indicated, whatever might be the location of its western boundary. The existence of the location as a territorial subdivision of New Hampshire was as notorious and certain as the existence of any township in the State. It must of necessity have had a boundary, whether that boundary had ever been actually surveyed on the ground or not. The State owned all the land lying westerly of it,—between it and the township of Lincoln,—and this land had never been granted to any person. It was wild, mountainous land of little value. The whole area, equal to the extent of a large township, and containing probably seventy or eighty square miles, was in 1830 valued at only $800. All this tract thus lying to the west of Hart's Location was granted to Elkins and his associates. They may have been under an erroneous impression as to the true location of the western boundary of Hart's Location, but, whatever it was, and whenever found, that was to be the boundary of the grant.

It may be true, as stated by the Supreme Court of Massachusetts in Morse v. Rogers (118 Mass. 573), that where a boundary is inadvertently inserted or cannot be found, or an adherence to it would defeat the evident intent of the parties, "the boundary may be rejected, and the extent of the grant be determined by measurement, or other portions of the grant." But that is not the case here. The evident intent of the parties was to go to Hart's Location as a territory or known body of land, without particular regard to a marked, designated, and visible line. It was their intent to leave no land belonging to the State between that territory and the tract granted. This was clearly the principal object in view; and as Hart's Location must necessarily have a western boundary somewhere, and as its limits and bounds were shown, whether correctly or incorrectly, by public maps in the archives of the State, it could not be said that this boundary was incapable of ascertainment. To hold this, and abandon the call of the deed for Hart's Location, and to confine the grantees to courses and distances, would defeat instead of furthering the intention of the parties. If the western boundary of Hart's Location had never been surveyed on the ground, it could be surveyed; or it could be located by agreement between the owners of it and the owners of the Elkins grant. They were the only parties who after that grant had any interest in the matter.

It may well be asked, if the call for Hart's Location and its western boundary can have no significance in the Elkins grant in 1830, how does it suddenly acquire significance in 1831, in the grant under which the demandant claims? The language used is almost exactly the same: "thence easterly to Hart's Location; thence southeasterly by said Hart's Location," &c.

With the accumulated evidence on the subject which was presented in the demandant's case, most of it of such a character as not to admit of contradiction, we think that the judge was perfectly right in assuming that Hart's Location was a monument sufficiently definite to control the courses and distances given in the grant. Indeed, we do not see how he could have done otherwise. The fact that the town of Burton, which lay to the south of Hart's Location, extended so far westerly that its northwest corner would not be met by a line drawn due south from any part of Hart's Location, cannot prevent the Elkins grant from extending to Hart's Location as its eastern boundary, as called for in the deed. As before stated, the connection between this location and the northwest corner of Burton, if it cannot be made by a line drawn due south as called for, must necessarily be made by the line of shortest distance between them. This is the surveyors' rule and the rule of law. Campbell v. Branch, 4 Jones (N. C.) L. 313. It is constantly applied when trees or monuments on or near the margin of a river are called for in a deed where the river is a boundary.

We think that the judge did not err in relation to the construction and effect of the Elkins deed.

But the demandant raised another point at the trial, namely, that the owners of the Elkins grant had estopped themselves from claiming under it any land eastwardly of a line running north from the northwest corner of the town of Burton, or Albany. The evidence offered on this point tended to show that about or soon after the date of the Elkins grant the grantees or some of them employed surveyors to ascertain the extent and boundaries of the grant, and that a line was run directly (or nearly) north from the northwest corner of Burton to the north line of the grant, as the supposed eastern boundary adjoining Hart's Location; but that this was done without any communication or agreement with the proprietors of Hart's Location or any other parties having an interest in the adjoining lands, and in ignorance of the true western boundary of that location on the land. The evidence consisted of the testimony as to the declarations of some or one of the grantees, as to the running of such line, made over forty years before, and of a recent examination of marked trees which indicated a date corresponding with the period referred to.

We think that the judge was right in holding that this evidence was totally insufficient, under the law of New Hampshire, or any other law, to show such a settlement of the line as to estop the owners of the grant from claiming to the extent of the description contained in the deed. Conceding that everything was proved which the evidence

Figures, Numbers, and Symbols

217

tended to prove, it would only show that the grantees made a tentative effort to find the limits of their property in a mountainous and almost inaccessible wilderness, without consultation or communication with any other parties, and without doing any act or thing that could in the least commit them in relation to such parties. The only line shown to have been the subject of any agreement was that located by Wilkins in 1850, parallel to, and two hundred and thirty-five chains from, the Saco, which was concurred in by Walker, the agent of the owners of the Elkins grant, and one Davis, who professed to own one-half of Hart's Location.

It is alleged by the counsel of the demandant that the law of New Hampshire on the subject of estoppel as to boundary lines is peculiar; that an agreement settling such lines, though made by parol, is binding upon the parties and all those claiming under them. Conceding this to be true, not the slightest evidence was offered to show any agreement whatever, or even any communication, between the adjoining owners prior to 1850, and the line then agreed upon coincides substantially with that which is now claimed by the defendant.

It is contended, however, that the running of the hypothetical line northerly from the Burton corner was an estoppel as regards the State; that the State, upon the faith of this line being run and marked by the Elkins grantees, entered upon the land eastward of it, and granted the same to Bean and others. That is, the State, by legislative resolve and solemn grant, having in 1830 granted to Elkins and others all the land west of Hart's Location, had the right to re-enter upon some eight thousand acres of the same land in 1831 and grant it out to third parties, because the Elkins grantees, in making an ex parte survey, had mistaken the position of the west boundary of Hart's Location. There is no pretence, certainly no proof, that this survey was made by any concurrence of the parties, or that there was even any communication between the agents of the State and the Elkin grantees. The agents of the State simply lay by and watched the operations of Elkins and company, and finding, or supposing, that they had made a mistake, and had left a vacant tract of land between the line they ran and Hart's Location, stepped in and made another grant to other parties of nearly a sixth part of the tract granted to the Elkins party. Not a particle of evidence was produced to show any acquiescence on the part of Elkins and his associates in this proceeding, or that they had any notice or knowledge of it. So far as appears, they have never acknowledged the right of these new grantees, nor have they ever admitted that any one had any right to interfere with the extension of their land eastwardly to Hart's Location. We think no case can be found that would make out an estoppel under such circumstances as these.

We have been referred with much confidence to the case of The Proprietors of Enfield v. Day, 11 N.H. 520. We have carefully examined this case, and do not find in it anything to support the proposition contended for. There the State interposed, after due notice to the parties and an inquiry by the legislature in reference to the true and right ownership of a certain gore between two adjoining townships, which by an alleged mistake of a figure had not been included in the grant (of Enfield), in which it was intended to be. The south line was south 68° east in the deed, when it should have been south 58° east. The grant of Grantham was made a few years afterwards, binding on Enfield, but having the right course (south 58°⋙ east) for its north line. On the application of the proprietors of Enfield and adjoining townships, the legislature was applied to correct this error, and commissioners were appointed to run the true line, and the disputed gore was granted to Enfield. The parties acquiesced for twenty years, and the question was whether Enfield had sufficient seisin and color of title to claim the benefit of the Statute of Limitations; and the court held that it had. But the court expressed itself with great caution as follows: "In this case we are clearly of opinion the seisin would not pass by the mere effect of the second grant; but was there not such a previous re-entry and assertion of right on the part of the government as to constitute, together with the grant, a conveyance with livery of seisin? An entry upon the land by the government agents, and the running anew and re-marking of lines, with the express design of a reconveyance to rectify a former mistake, would seem to be evidence sufficient to show an actual possession in the government of any given tract." Was anything of this kind done in the present case? Were the Elkins grantees notified of any error or mistake? Were they informed of the intention to regrant a portion of the tract granted to them? Did they acquiesce in such proceedings? Nothing of the kind. But the court adds: "The proceedings of the legislature were had on public notice and actual service on the proprietors of Grantham. They also had full knowledge of the subsequent proceedings of the proprietors of Enfield, in their entry upon and frequent sales of portions of this gore of land, claiming the whole under the grant from the State, and must be regarded as acquiescing in such adverse possession and claim. It is now too late for the proprietors of Grantham to assert their title." It is obvious that the cases are totally distinct; and it is unnecessary to discuss the subject further.

The judge, on this part of the case, instructed the jury that there was no evidence before them to stop or bar those claiming under the Elkins grant from maintaining their line by the westerly side of Hart's Location; and in this we think he was right.

Slope versus Horizontal Measurement

In some parts of the country, it is not uncommon to find that measurements were made along the ground, or along the slope, instead of horizontally. This became the custom in these areas, and occasionally is found elsewhere, sometimes where least expected. The decisions dealing with slope as opposed to horizontal measurement are primarily from Virginia, West Virginia, North Carolina, Tennessee, and Kentucky.

In the North Carolina case of *Duncan v. Hall*,* the court wrote "It is a fact of which the courts must take and have taken notice that the measurements

---

* 117 N.C. 443, 23 S.E. 362, 1895.

Figures, Numbers, and Symbols

of boundary lines in making the original surveys for deeds and grants are often, if not always, inaccurate. Those discrepancies between the distance called for and the actual measurement occur much more frequently, too, in an undulating or mountainous section, because, as is a matter of general knowledge, it often happens that, in the original surveys of grants, only two or three lines of a square or parallelogram were actually run, and that the earlier surveyors, at least, universally adopted surface measurement. In running long lines from the top of one high and precipitous mountain to that of another, the area or acreage sold by the state to its citizens would have appeared much less than it actually was if the level measurement had been adopted in laying off large grants. It is therefore a well-known fact that, owing to inaccuracies in measurement, different results will follow from adopting one or the other of the two methods of surveying where many of the old monuments have perished or been removed. In determining which is correct, the courts proceed upon the idea that the object of legal investigation and inquiry is to find the lines, corners, and monuments which were agreed upon by the parties to the original conveyance, and that, in order to attain that object, the lines should be run in the direction and order adopted by them. *Harry v. Graham*, 1 *Dev. & B.* 78, 79; *Norwood v. Crawford*, 114 N.C. 519, 19 S.E. 349. There are some exceptional instances, in which it is manifest that reversing a line is a more certain means of ascertaining the location of a prior line than the description of such prior line given in the deed, but such cases are the rare exceptions to a well-established general rule. *Harry v. Graham*, supra, and Norwood v. Crawford, 114 N. C., at page 521, 19 S.E. 349; Safret v. Hartman, 7 *Jones [N. C.]* 203. The general rule is an established law of evidence, adopted as best calculated to ascertain what was intended to be conveyed, and it is incumbent on a party asking the courts to depart from it to show facts which bring the particular case within the exception to the rule."

Actual Survey versus Paper Survey

This same court, in the case of *Cody v. England*,* stated: "while there is authority in this state, *Duncan v. Hall*, 117 N.C. 443, 23 S.E. 362, and *Stack v. Pepper*, 119 N.C. 434, 25 S.E. 961, to the effect that there is a presumption, founded on custom sanctioned by judicial opinion, that surveyors used surface measure in the early surveys of entries on which grants were issued, particularly in the mountain sections, there is no factual basis for such presumption where it appears, as here, that no survey was made." The custom of a surveyor to use surface measure in surveying entries has no probative value in cases where no actual survey was made. In such event the rule of correct measurement must be

---

* 221 N.C. 40, 19 S.E.2d 10, 1942.

220   *Boundary Retracement*

applied. The authorities agree that horizontal measure is the correct and accurate method of measurement in the survey of land. 8 Am.Jur. 794, Boundaries, Sec. 67; *Gilmer v. Young,* 122 N.C. 806, 29 S.E. 830; *McEwen v. Den,* 24 How. 242, 16 L.Ed. 672. Manifestly, a line platted on the plane of paper is horizontal.

> The U.S. Supreme Court stated the principles in the case of *McEwen v. Den** concerning a Tennessee grant, and compared the rule of custom with the rule of law for measurement of distance: "It was proved at the trial, and is admitted here, that no line was originally run and marked but the first one; and that at H there is a marked poplar corner tree, which is a line mark of the grant". It being admitted that the first line is established, and that it is regarded as a north and south line, and that the other lines of the tract were not run or marked, it follows they must be ascertained by course and measurement. How they are to run is matter of law; and on this assumption, the Circuit Court instructed the jury as follows: "To identify the land appropriated, the jury must look to the calls, locative and directory, the foot of the mountain, the creek, the coal bank, the marked trees, courses and distance, number of acres demanded and paid for, &c.; and they will look to the survey, full or partial; that assuming the correct mode of survey to have been by horizontal measurement, and that the surveyor based his identification of the land entered on surface measure, in accordance with his custom and the custom of the mountain range of country in which he resided, this would not of itself defeat the location of the land, and the boundaries of the grant as indicated by the survey, calls, and other evidence, to all of which they would look in adjusting the boundaries of the plaintiff's grant." To this charge, exception was taken. We think the instructions given were too vague and general to afford the jury any material aid in ascertaining the true boundaries of the land granted. The first line calls for two corners admitted to exist; this line must govern the three others. 1 Meigs's Digest, 154. It falls short of the distance called for, being only about 800 poles long. Its course being found, the next line running west must be run at right angles to the first one. In ascertaining the southwest corner of the tract at 894 poles from the poplar corner, the mode of measuring will be to level the chain, as is usual with chain-carriers when measuring up and down mountain sides, or over other steep acclivities or depressions, so as to approximate, to a reasonable extent, horizontal measurement, this being the general practice of surveying wild lands in Tennessee. The reasonable certainty of distance, and approximation to a horizontal line, is matter of fact for the jury to determine.
>
> The 3d line running north, from the ascertained western termination of the second, must run parallel with the first line, and be continued to the distance of 894 poles; the chain being leveled as above stated. The 4th line will be run from the northern terminus of the 3d line to the beginning near Bowling's Mill.

---

* 24 How (U.S.) 242, 16 L.Ed 672, 1860.

Figures, Numbers, and Symbols

The surveyor who made the survey on which grant No. 22,261 is founded, deposed at the trial, "that no actual survey was made in 1838 of said land, except the first line from A to H. That the other three lines of the grant were not run, but merely platted. That the proper mode of making surveys was by horizontal measurement, but that he had not been in the habit of making them in that way; that in making the line from A to H, in this survey, he had measured the surface; that the custom of the country was to adopt surface measure; and that he had made the survey in accordance with such custom."

The grantee was bound to abide by the marked line from A to H; but the other lines must be governed by a legal rule, which a local custom cannot change. Should this custom be recognized as law, governing surveys, it must prevail in private surveys, in cases of sales of land, when the purchaser who bought a certain number of acres might, by surface measure across a mountain, lose a large portion of the land he had paid for. And such would be the case with this grantee, were he restricted to surface measure; whereas, by the terms of his patent, the Government granted to the extent of lines approximating to horizontal measurement. How far the act of limitations will affect the plaintiff's title, will depend on the fact whether Evans's coal bank falls within the boundary of the patent sued on, as it is not claimed that the other possession at a different place on grant No. 22,261, and for which trespass the recovery was had, was seven years old when the suit was brought.

As recently as 1960, the Kentucky court in *Justice v. McCoy** dealt with a situation of slope measurement. To support its decision, it relied on precedence from several earlier decisions from the same court.

The question in this case is whether, in measuring the depth of a residential town lot which lies on a sharply sloping hillside, under a deed which calls for a depth of 60 feet "up the hill," a horizontal measurement or a surface measurement should be used. The circuit court, in an action between the owner of the lot and the owners of two adjoining lots lying above on the hillside, held that the surface measurement should be employed.

The land in question lies between High Street and Kentucky Avenue in the City of Pikeville. The two streets are about 100 feet apart, by surface measurement, but the level of High Street is around 40 feet above that of Kentucky Avenue.

The lot of the appellants faces Kentucky Avenue. The lots of the appellees face High Street. Each of the two lots facing High Street has a house on it, which houses were built at a time when all three lots were owned by one family. The lower lot was sold by this family to the appellants' predecessors in title in 1939 and the upper two lots were sold to the appellees in 1941 and 1944, respectively.

By reason of the sharp slope of the land, a horizontal measurement of 60 feet would produce a surface depth of around 70 feet for the appellants' lot. This would mean that the back several feet of the appellees' houses would be on the appellants' land.

---

* 332 S.W.2d 846, 1960.

In order to reach the conclusion that the parties to the deed intended a horizontal measurement to be used, we would be required to believe that the grantors intended to convey, and the grantees to buy, not only a vacant piece of hillside land but also several feet off the rear of the grantors' houses. We would also be required to ignore the provision of the deed for measurement "up the hill."

The appellant maintains that it is not a question of intent, but that there is an absolute rule calling for the use of horizontal measurement. We find authority, however, for the proposition that a surface measurement is proper where it is a custom of the locality or where it is dictated by the circumstances of the case. 11 C.J.S. Boundaries Sec. 9, p. 551. Also, it has been held that straight line measurement will not be employed where language of the deed (such as "along the road") or other circumstances, or local customs, indicate a different intent. *Hite v. Graham*, 5 Ky. 141; *McKee v. Bodley*, 5 Ky. 481; *Whitaker v. Hall*, 4 Ky. 72.

The appellants' surveyor testified that people not ordinarily versed in surveying would generally measure a slope such as this by surface distance. All of the circumstances here point to the use of surface measurement—the fact that the parties obviously did not intend a conveyance of part of the grantors' houses; the fact that the deed calls for a distance "up the hill"; and the fact that for some 13 years after the deed was executed the parties treated the boundary line as being from a cherry tree behind one of the appellees' houses, approximately 60 feet by surface measurement from Kentucky Avenue. Furthermore, one of the grantees in the original 1939 deed (a predecessor in title to the appellants), testified that it was the grantees' understanding that the distance called for in the deed was by surface measurement, and that the line was to run to a tree behind one of the grantors' houses.

It is our opinion that the findings and judgment of the circuit court are correct.

Distance Deficiencies

All lines are supposed to be actually surveyed, and the intention of the grant is to convey the land according to the actual survey; consequently, distances must be lengthened or shortened and courses varied so as to conform to the natural objects called for. (*M'Iver's Lessee v. Walker*, 17 U.S. 444, 4 Wheat. (U. S.) 444, 4 L.Ed. 611.)

In *Johnson v. M'millan**, the court addressed the issue of distances not fitting perfectly by stating, "The great principle which runs through all the rules of location, is, that where you cannot give effect to every part of the description, that which is more fixed and certain shall prevail over that which is less so." The court added, "Distances may be increased, and sometimes courses departed from, in order to preserve the boundary; but the rule authorizes no

---

* 1 Strob. Law, 143, S.C., 1846.

Figures, Numbers, and Symbols 223

other departure from the former, than such as is necessary to preserve the latter."

In 1809, the Kentucky court in the case of *Bryan v. Beckley,** laid down a series of rules, which included allowing for the unevenness of the ground, and also stated that a "mistake in distance originally committed in one line, could have affected only the opposite." Refer to the previous chapter for this court's discussion of directions.

Compare Measurements with Original: The Rule

In all cases of resurvey the chain used must be made to correspond to that used by the government (or original surveyor), by testing it with distances on the ground between two or more known monuments.

See *Mason v. Braught,* 146 N.W. 687 (S.D., 1914), *Henrie v. Hyer,* 92 Utah 530 (1937), and *Hess v. Meyer,* 73 Mich 259, 11 N.W. 422.

Problems with Area Recitations

In those situations where no survey was performed for the conveyance, recitations of area have been found to be estimated, while some are based on distances derived by pacing. Where more sophisticated determinations have been made, area figures have been derived by dividing a plotted perimeter into squares and rectangles, calculating their areas individually, then adding them together. Others have been determined by plotting on a grid and knowing the area of a square on the grid, adding up the number of squares. Some were calculated without closing the figure of the perimeter, thereby providing a false number. Some were "force closed" on paper; others had unreliable results depending on whether distance and direction errors (multiplied to determine the enclosed area) were compensating or cumulative overall.

Some areas were determined with allowance causing them to frequently overrun. This was a deliberate act, in order that grantees did not get less than the stated area, but rather more. It is not unusual to see early parcels larger in area than the recited figure. Others were deliberately understated to avoid paying full tax amounts. Modern practice is to mathematically close the survey figure to compute area, but in doing so the adjustment routine alters the value of the component directions and distances, the more error in the basic data the greater the change. With modern software and computational routines based on measurements with precise equipment, resulting errors are minimal, therefore changes to raw measurements are likely to be small. However, this does not account for compensating errors. Not only can we not make perfect measurements, the resulting data shown on plats and in descriptions is not only imperfect, but also generally somewhat

---

* 16 Ky 91, 1809.

224 *Boundary Retracement*

manipulated. This is another very good reason for being especially careful and persistent in finding corners so that secondary information, known to be imperfect and unreliable, does not have to be relied upon.

Often stated as area in formal presentations, the term in usual parlance is often called acreage where acres are used to express the quantity of land. In legal texts, *quantity* is frequently the favored term.

Problems with Mathematics

Adjustments and Getting Rid of Closure Errors

One of the first mathematical analyses to be performed is to determine whether any adjustment(s) were applied to raw measurements in order to provide a perfect (mathematical) closure, and if so, what method may have been used. Any type of adjustment will alter the data obtained in a survey, by how much will depend on the magnitude of the total error in the work. Errors, particularly gross errors at critical locations can distort the final figure obtained, sometimes by a significant amount. Using the reported information as opposed to the actual measurements taken in the field can mean a difference between success and failure of finding the location of a corner. Where stones were used to mark corners in areas containing an abundance of stones, where wooden stakes were set or trees marked in heavily wooded areas, not to mention swamps and other wetlands, or where the landscape has been altered, the retracement surveyor can be within a very short distance, even inches, of an existing corner, and be misled into believing there is nothing there.

Adjustments or Not: Shifting Positions

Where adjustments have been applied, the reported data are not the true data that were measured, they have been manipulated to provide for a closed final figure. How much manipulation will depend on the type and magnitude of the errors, and the size of the figure that was measured. Ten feet of error is a lot more difficult to find, or locate, in a township of 36 square miles opposed to a section of 640 acres.

Working with older surveys, especially in the form of plans, which can expose one to a major hazard, is without some background knowledge of a clue whether distances were made along the slope or horizontal, and whether any form of adjustment or even a forced closure was made. In more recent surveys, especially those done by trained engineers, there is some likelihood that an adjustment routine was applied, but unless there is some statement on the plan, which is seldom included, the reader cannot know without further investigation. Generally, there were four possibilities to choose from—the

Figures, Numbers, and Symbols   225

transit rule, popular with engineering training; compass rule (which, incidentally, produces results very close to the results derived by the method of Least Squares); Crandall method, popular when distances were determined through the use of stadia, less precise than angle measurements; and arbitrary whereby the surveyor or draftsman would apply more weight to angles in cases of poor set-up or more weight to distances when taken over difficult terrain. The rule of Least Squares adjustment, which with some computer programs allows the assignment of weights to points and lines, is a more modern procedure.

Usually, the closure error, which resulted from an accumulation of random errors, compensating errors, and cumulative errors, would be discovered when plotting the survey. That is, the resulting figure would not graphically close on paper. In such cases, one of two things would take place: the draftsman would "force close" the figure by manually adjusting some or all of the lines, or else ignore the closing line altogether, resulting in a final figure with one missing line, the last, or closing, line. Particularly, with today's computational aids, these errors can be identified, both in magnitude and likely location (unless there are two or more gross errors) especially if they are gross errors. However, compensating errors are unidentifiable, and do not present themselves when computing a mathematical closure. For example, two 10-foot errors, one each in two lines parallel with each other or nearly so, would mathematically cancel each other out, resulting in an acceptable overall closure. But the reality is that there would be 20 feet of unidentifiable error in the figure. Remeasurement and cross checking are possible methods for checking for such an occurrence.

As previously described in Chapter 5, although colonial surveyors did not usually have a detailed set of survey instructions as is familiar in the PLSS, many of them did have certain instructions for laying out lots, which depended on the jurisdiction and the time. A review of the governing statutes, as well as a review of proprietors' instructions to their surveyors, is essential to understand how the original surveyors were instructed to do their work. A retracement surveyor would be under a severe handicap without that information, and would be likely to fail to find original evidence. As also discussed in Chapter 5, following a surveyor and analyzing known evidence can impart valuable clues as to an individual's procedures.

The intent of parties to a conveyance must be gathered from the words of the instrument not only in light of the surrounding circumstances, but also in light of the law in existence at the time, including applicable statutes.* In addition, a deed, or any instrument of conveyance for that matter, should be construed in light of the law existing when the deed was executed.†

---

* *Morton v. State*, 104 N.H. 134, 1962.
† *Stuart v. Fox*, 129 Me. 407, 152 A. 413, 1931.

7

Resolving/Reconciling Errors

"It is a well-known fact that surveyors are apt to differ from each other, and surveyors employed by the United States government are not immune from the frailties of their profession."

United States v. State Investment Co., 264 U.S. 206
cited in Hagerman et al. v. Thompson et al., 235 P.2d 750 (Wyoming, 1951)

Land Surveying as an Art

Many, if not most, land parcels contain numerous problems regarding incorrect or improper boundary calls or miscalculations from earlier surveys, title examinations, and related interests of such easements. In addition, many parcels have been created from a series of divisions of larger tracts over time. With each maneuver the risk of problems increases. The result often manifests itself through abutting parcels not physically touching as intended, leaving gaps and overlaps. More often than not the retracement surveyor must overcome puzzles and mazes with considerable handicaps.

In such cases, solutions are based on hours of research, data collection, and interpretation of records. Solutions in retracement work only result from perseverance and continual study, making interpretations and conversions where necessary, to bring about a modern view of yesterday's processes.

With any retracement survey, there are, at minimum, two sets of errors, the errors of the current retracement survey and those of the original survey. If there intermediate retracements, or other activity, there will be additional sets of errors. All of these must first be recognized, then resolved.

The retracement surveyor has some control over minimizing errors, and being able to reconcile those that do occur, but has no control over the accumulation of errors in the original survey, nor without a detailed analysis, has any insight as to what and where the errors are, or their consequences. However, when pursuing the location of a corner, or as the court stated in the *Lugon* case, placing the corner "where the surveyor would have put it," some analysis may need to be undertaken to learn what the original surveyor was subject to. While PLSS surveyors at least had a set

228 *Boundary Retracement*

of instructions, colonial surveyors were not so fortunate. Even those where there sometimes were instructions, they were not very comprehensive, and often there were none at all concerning how survey work should be done. Comparing present day surveys and measurements with yesterday's is like trying to compare apples with oranges. Only by independent study, are we able to learn who the original surveyor was, and how he did, or may have done, his work.

If there are other attempts at retracement which are either inadequate or incorrect, the retracement surveyor's problems are exacerbated. There are additional footsteps to follow, many of which result in dead ends. Others may affect the title, and through their misleading property owners cause improper occupation or other actions by subsequent landowners, or otherwise cloud a clear title, adding to the burden or causing costly litigation to correct the problem(s).

The key to resolving an error is to first, recognize the error, and second, to understand how the error occurred, then finally determining how to deal with it. As doctors practice, first is to identify the ailment, then prescribe the cure. It is a simple concept, but not necessarily one easily carried out: identify the problem, then seek a way to fix it.

Problems are rampant, and many of them go undetected mostly because those who encounter them are often not equipped with either the education or experience to recognize or to identify them. Consequently, it goes without saying that these same people do not understand the impact of the problem(s). To emphasize, heed may be taken from an early Mississippi case:* "Practically, no deed or conveyance of land was ever made, however minute and specific the description, that did not require extrinsic evidence to ascertain its location; and this is so, whether the description be by metes and bounds, reference to other deeds, to adjoining owners, water courses, or other description of whatever character. Looking through the authorities, it will be seen that, under every conceivable state of facts, and in every imaginable circumstance, the cases may be counted by hundreds, if not thousands, where contracts, wills, and deeds are made effective by the identification, by extrinsic evidence, of the person or subject intended, yet in no wise violating the rule, that such evidence cannot be admitted to contradict, add to, subtract from, or vary the terms of a written instrument."

This leads to the reasoning behind why monuments control. Simply, there is no error in an original monument. And several courts have stated that a monument does not lose its significance merely because its marker has disappeared.[†]

---

* *Peacher v. Strauss*, 47 Miss. 353, 1872.
† *Lloyd v. Benson*, 2006 ME 129, 910 A.2d 1048, Me., 2006. (See discussion on missing monuments in Chapter 3).

Resolving/Reconciling Errors

Errors in Measurements

Brinker and Wolf* four inherent problems concerning errors:

No measurement is exact.

Every measurement contains errors.

The true value of a measurement is never known.

The exact error present is always unknown.

Errors occur in a variety of forms, and may be categorized as follows:[†]

Errors

Inaccuracy in any one measurement due to the type of equipment used or in the way the equipment was used. The difference between the measurement and the true value is known as the **resultant error**. The resultant error in a measurement consists of the individual errors from a variety of sources, some causing the measurement to be too large, others making it too small. Some are positive, while some are negative, giving rise to both cumulative and compensating errors, none of which are identifiable but can be fixed within probable limits.

Error cannot be entirely omitted, but conscientious and careful surveyors have minimized and will attempt to minimize error.

Discrepancies

The difference between two measurements of a given quantity, or the difference between the measured value and the known value. A small discrepancy between two measured values indicates that likely there are no mistakes, and accidental errors are small. It does not indicate the magnitude of systematic errors.

It may indicate the precision with which the measurements were made. For instance, the value the previous surveyor put on direction(s), distance(s) or area(s), compared to your results, so long as you are both measuring between the same points.

---

* *Elementary Surveying,* 7th ed., 1984.
[†] Adapted from Wilson, *Forensic Procedures for Boundary and Title Investigation,* John Wiley & Sons, 2008.

Errors of Closure

Also known as **closing error**, it is the misclosure of a survey or closed figure in a mathematical sense. The error of closure divided by the total distance measured is equal to the **relative precision** of the work. Standards are set to allow for an acceptable amount of error, based on the relative precision of the type of work performed. Unfortunately, some work is accepted as being in accordance with the standard, whereas due to compensating (accidental) errors the error of closure is misleading and not a true indicator of the amount of error in the work.

Random Errors

Errors that obey no mathematical or natural law other than chance. Sometimes called **accidental error** or **uncertainty**. The magnitudes and algebraic signs of random errors are matters of chance. There is no absolute way to compute or eliminate them. Random errors are also known as **compensating errors**.

Index Errors

Also known as **instrumental errors**, they arise from imperfections or faulty adjustment of the instrumentation used in making measurements. For example, a tape may be too long, or more likely, too short, affecting each measurement taken. Places where a tape is kinked, repaired, or bent may introduce significant error in longer lines. Any instrument out of adjustment from being in perfect working order will introduce error.

Personal Errors

These occur through the observer's inability to manipulate or read the instruments exactly. An observer's eyesight may affect readings each time they are taken. At an unusual setup, a person's stance such that an observation is at an angle can introduce error.

Resolving/Reconciling Errors

Natural Errors

Natural errors occur from variations in the phenomena of nature such as temperature, humidity, wind, gravity, refraction, and magnetic declination. Changes in temperature can significantly affect steel tapes, and in some cases, electronic equipment. Changes in magnetic declination and other magnetic disturbances affect compass readings, upon which the majority of early surveys and titles in many regions are based.

Systematic Errors

Errors that occur in the same direction, thereby tending to accumulate so that the total error increases proportionally as the number of measurements increases. The magnitude of the angle does not affect the size of the error, the length of distance may, such that a line had to be measured in segments, thereby repeating the error each time the measuring device was used, and thereby causing errors to be cumulative.

Systematic errors are mechanical errors, and result from imperfections in the equipment or the manner in which it is used. Under the same conditions, they always have the same size and algebraic sign. Although they often introduce serious errors in results, since they generally have no tendency to cancel, they are often difficult to detect.

Accidental Errors

Errors that occur randomly in either direction, thereby tending to cancel one another so that, although the total error does increase as the number of measurements increases, the total error becomes proportionally less, and the accuracy becomes greater as the number of measurements increases.

Accidental errors are the small unavoidable errors in observation that an observer cannot detect with the equipment and methods being used. Greater skill, more precise equipment, and better methods will reduce the size and overall effects of accidental errors. However, consistent use of the same equipment in the same way will not result in any improvement. A surveyor using the same equipment and techniques throughout

232 *Boundary Retracement*

his practice is likely to achieve the same type of results, containing the same or similar errors.

Following enough of this particular surveyor's work and getting to know it intimately will impart valuable insight into what to expect in a given situation.

Computational Errors

Errors that occur in the computation process, which include gross mistakes (blunders), transposition of figures, simple math mistakes, rounding errors, and so on. A small error in one part of the computational process may result in a large error affecting the end result.

Mistakes

Inaccuracies in any one measurement because some part of the operation is performed improperly. Mistakes are caused by a misunderstanding of the problem, carelessness, or poor judgment. Large mistakes are often known as **blunders**. Blunders are the result of human error and cause a wrong value to be recorded. When a blunder is found, there may be an opportunity for correction. When a legitimate error is found, it must be dealt with in proper fashion. Transposition of numbers is a very common blunder, as is making a mistake of a whole foot, ten feet, or a whole chain length. Such mistakes are usually the result of miscounting. Blunders can usually be discovered, however, occasionally two blunders will cancel each other out and therefore go unnoticed.

Mistakes, or blunders, can only be corrected if they are discovered. Today's surveyors find themselves with an unexpected problem when they are the ones discovering the mistakes made by one or more previous surveyors. Some of these earlier problems have escalated into serious boundary and title concerns when they have gone on for a long time, or especially when other unwary surveyors have relied on the mistaken work believing it to be all right.

Blunders that may not be discovered are such things as staking out the wrong lot in a block or even on the wrong street, misreading a number on a plan, or counting the wrong number of tape lengths in very nearly parallel sides of a traverse.

Certain kinds of mistakes are not blunders. These include mistakes due to lack of judgment or lack of knowledge.

Resolving/Reconciling Errors 233

Total Error

Total error is the sum of the inaccuracies in a finished operation. Since inaccuracies are either positive or negative, the total error is the algebraic sum of all the errors.

Reliability

Measurements taken under the same conditions are equally reliable. A survey where measurements are made at different times warrants a different consideration since the reliability of the various measurements may not be directly comparable and therefore unable to be treated the same. For example, taped distances from two different days where the temperature difference is large may result in two different sets of errors, each of which needs to be considered separately. It was the practice in the past for some surveyors not to bother to correct for temperature when taping, especially where a relatively high degree of precision was not required. This was especially true for woodland surveys.

So-called "mountain surveys" in some of the southern states are in another class, where original surveyors had their own formulae for dealing with steep terrain and like conditions.

Lower Precision Surveys

It is human nature, and survey standards reflect this, that high-value property requires more care in making measurements than low-value property. However, this may become a significant problem at some time in the future when the use of property changes, and low value all of a sudden becomes high. Retracing lines of a woodlot is difficult in itself, but when it is soon destined to be a multi-million dollar shopping center, it becomes increasingly difficult, and locating correct points becomes extremely critical.

In the past, the tendency was to survey woodlots, wetlands, and other low value land with low precision equipment and to take only what care was necessary, sacrificing precision to save time and produce a survey at low cost. Retracing these today is difficult, particularly in comparison to urban and suburban parcels, which generally were surveyed to a higher precision. Woodlots include sighting problems resulting in short traverse lines, alignment problems because of trees and brush, and a host of other inherent

234   *Boundary Retracement*

difficulties. They were often surveyed with a compass where angles were measured to 1° or perhaps to 15 min of arc, and each observation was subject to local attraction and other influences which were not often identified. In contrast, urban surveys were measured carefully using equipment of higher precision resulting in better closures, and survey costs were justifiably higher. The resulting errors from the various types of survey and measurements must be considered in their appropriate context.

Different surveyors operate under different standards, if standards are considered at all. One might select an acceptable closure of 1 part in 2500, while another would insist on nothing greater than 1 part in 5000. The purpose to which the property may have been considered may have changed over the years whereby subsequent surveys demanded a greater precision, or smaller error of closure. One must be careful not to compare apples with oranges in comparing two different surveys with each other. One must also remember that even though a survey closes mathematically within the desired or required limits, there is no guarantee against compensating errors.

Measure of Precision

There is no measure for accuracy. Either the property is located or it is not. There is, however, a measure for the precision of a closed survey figure. Mathematically, a figure which is intended to be a closed figure, should end at the same point begun at. Although physically, in the field, this occurs, mathematically there is error so that the beginning point and the ending point, while very close to one another, are not exactly the same. This is because of the error, which can be calculated. Standards dictate how much error is acceptable and, eventually, adjusted out by some method. When the **closing error**, or **error of closure**, is more than acceptable limits, either the work must be done over, or sometimes, if the error is mostly in either one angle or in one line, it can be isolated, and that part corrected to bring the entire work within acceptable limits. Compensating errors, however, cannot be detected this way since there is no gross error of closure.

Errors in the Collection of Data

These errors may be found in two forms, surveying errors as described above, and transcription errors where numbers are transcribed incorrectly, such as into field notes or onto sketches.

Errors in Reporting Data

Again, there are two types of errors, plat errors where numbers were placed on a resulting plat, sketch, or other type of drawing incorrectly, and description errors where numbers were put into a land description incorrectly. Generally, both of these problems are attributable to persons other than the original surveyor. A glaring example appears back in time where anyone could write or dictate a land description into a title document. Many of these so-called "scriveners" had little or no training, and were without knowledge of the later legal and surveying consequences of their inability to relate the correct story.

One of the biggest problems in present day retracement is misidentifying the error, and applying a correction, or the wrong correction. Comparing a modern survey with an ancient survey is like trying to compare apples with oranges. Common situations can be found:

Making incorrect assumptions, probably due to lack of experience: ("This is what the surveyor meant, this is a blunder, I can fix it.")

"I can measure better than he could, therefore my survey is the correct one."

"Equipment is better today, we can do a better job of measuring."

"This is what they intended, or meant to do, but failed to do it."

The courses and distances in a deed always give way to the boundaries found upon the ground, or supplied by the proof of their former existence when the marks or monuments are gone. So the return of a survey, even though official, must give way to the location on the ground, while the patent, the final grant of the state, may be corrected by the return of survey, and, if it also differs, both may be rectified by the work on the ground. One of the strongest illustrations of this rule is to be found in the instance of the surveys of the donation lands, set apart for the soldiers of the Pennsylvania line in the Revolutionary War. The law required the tract to be identified by marking the number of it upon a tree within and nearest to the northwestern corner. It was held that this number controlled all the remainder of the description in the patent, so as to wrest it entirely from its position and adjoiners, as described in the patent and general draft. *Smith v. Moore, 5 Rawle (Pa.)* 348; *Dunn v. Ralyea, 6 Watts & S. (Pa.)* 475. Chief Justice Gibson, in the former case, stated the general principle thus: "It is a familiar principle of our system, and one in reason applicable to this species of title, as well as any other, that it is the work on the ground, and not on the diagram returned, which constitutes the survey, the latter being but evidence (and by no means conclusive) of the former." * * * Staub v. Hampton (Tenn.)

236 Boundary Retracement

AUTHOR'S NOTE: The phrase [monuments found] "or supplied by the proof of their former existence" when the marks or monuments are gone is significant. Such proof comes in the form of land descriptions, maps, field notes, parol testimony, and evidences of long-term recognition and occupation, among other information.

Courses may be run in a reverse direction where by so doing a difficulty can be overcome and the known calls harmonized; but such a practice should be resorted to ordinarily only when the termini of the call cannot be ascertained by running forward.

As discussed, the weight of authority is to the effect that the beginning corner of a survey does not control more than any other corner actually well ascertained, and where a disputed or lost line or corner can thereby be established more nearly in conformity with the terms of the instruments and with the intent of the parties as gathered therefrom, it is competent to ascertain such line or corner by first ascertaining the position of some other bound and tracing the line back from that by reversing the course and distance.*

If an insurmountable difficulty is met with in running the lines in one direction, and is entirely obviated by running them in the reverse direction, and all the known calls of the survey are harmonized by the latter course, it is only a dictate of common sense to follow it. *Ayers v. Watson*, 11 S.Ct. 201, 137 U.S. 584, 34 L.Ed. 803.

The Texas court addressed this several times. In the case of *Gulf Production Co. v. Spear*,[†] the court stated: "The reversal of calls is but another rule of construction. It may be used to correct the defect, but it does not destroy or eliminate it. It is of greater dignity or controlling effect than the rule that a general description must be called to aid a doubtful or defective particular description. Both rules serve the same purpose, to discover the real intention."

And, in *Howell v. Ellis*, the court stated, "The only reason for reversing calls in field notes is to follow better the surveyor's footsteps, and mere running of lines according to course and distance does not locate the surveyor's footsteps in absence of any marked lines or established corners.[‡]

The Kentucky case of *Combs v. Jones*[§] lends some insight to a solution of this type of problem.

"Appellee Jones' claim of superior paper title must fail for two reasons: First, he utterly failed to prove any paper title; and, second, it is clear that the John A. Duff patent does not include any of the land in dispute. This patent was issued in 1852 and called for 75 acres in Perry county on Lost creek, a

---

* In locating boundaries, beginning at a definite corner or monument running reverse course is permissible, if necessary, to harmonize the calls in the description. In locating boundaries, it is permissible to begin at any definite corner or monument and to run a reverse course, if necessary to harmonize all the calls in a description. *P'pool v. State*, 93 Fla. 378, 112 So. 59, Fla., 1927.
† 125 Tex. 530, 84 S.W.2d 452, 1935.
‡ *Howell v. Ellis*, 201 S.W. 1022, Tex. Civ. App., 1918.
§ 244 Ky. 512, 51 S.W.2d 672, Ky. App. 1932.

Resolving/Reconciling Errors

branch of Troublesome. The only surveyor introduced as a witness testified that he went upon the ground and, starting at the beginning point, ran the three lines of the Duff patent which lay nearest to the land in dispute, and that the nearest point in the Duff patent to the land claimed by the appellants was more than 400 feet distant. Appellee introduced a number of witnesses who testified that they had seen surveyors run the lines of the Duff patent on several occasions, and that the line of the sixth call crossed the line of the first call several poles from the beginning point, and, as thus surveyed, included a portion of the land in dispute. They all agreed, however, that, when the lines were so run, the survey failed to close by several hundred feet. It is clear that one of the calls in the survey is erroneous. The surveyor who testified in the case filed with his deposition a plat which is a copy of the plat which appears with the original survey which was made on October 2, 1843. The Duff patent as described in that plat does not conflict with the Nicholas Combs patent of 1856 of which the land in dispute is a part. A mistake in the calls of a patent may be corrected by reference to the original plat, and, in ascertaining boundaries, attention should be given to the figure of the survey in the absence of any other controlling influence. *Howes v. Wells*, 110 S.W. 245, 33 Ky. Law Rep. 212; *Brashears v. Joseph*, 108 S.W. 307, 32 Ky. Law Rep. 1139; *Combs v. Virginia Iron, Coal & Coke Co.*, 106 S.W. 815, 32 Ky. Law Rep. 601; *Hogg v. Lusk*, 120 Ky. 419, 86 S.W. 1128, 27 Ky. Law Rep. 840.

> "In the instant case, if the calls are reversed, starting at the beginning point, it is also clear that the description in the Duff patent does not conflict with the description in the Nicholas Combs patent. Where there is a known error, the calls may be reversed in order to correct the error and close the survey. *Simpkins' Adm'r v. Wells*, 26 S.W. 587, 16 Ky. Law Rep. 113; *Combs v. Valentine*, 144 Ky. 184, 137 S.W. 1080."

Note that the court did not address the problem of the description failing to close, other than to note it. The issue was whether the description included other land, or encroached on other land, and the court ultimately found that clearly was not the case.

The case of *Cornett v. Kentucky River Coal Co.** provides a well-explained solution to such problems, lending guidance for dealing with similar issues.

In this case, the appellant contended "that the proper location of the land covered by the 500-acre patent issued to Robert Cornett April 21, 1849, under which he holds title, includes 918 acres, while under the construction contended for by appellee and adopted by the court it is held to include only 590 acres, of which, however, under either construction, 247 acres in Leslie county is to be excluded as not covered by the contract, leaving to be conveyed, under the contract, if appellant's construction of the patent is correct, 671 acres, but only 343 acres under the construction advanced by appellee

---

* 175 Ky. 718, 195 S.W. 149, Ky. App. 1917.

238   *Boundary Retracement*

and adopted by the court. This difference in the location of the boundary results from the fact that the last six calls in the patent, do not call for any established object, and that, if run by the courses and distances, the patent will not close. There are thirteen lines called for in the patent, the first seven of which are located by marked trees and about the location of which there is no dispute affecting this controversy. However, in running the call between the sixth and seventh corners in the patent, it was found necessary to extend the line approximately 80 poles beyond the distance called for in order to reach the timber which it is agreed marks the seventh corner, and that this error in the recorded distance of this line is responsible for the failure of the patent to close is the contention of appellant, who also claims that the effect of this error is compensated for and adjusted by extending the distances called for on the courses given in the last two lines of the boundary until they will intersect. This theory is approved by W. A. Ward, a practical surveyor of 12 years' experience, who first surveyed the land for appellee and was introduced by it as its witness. The theory contended for by appellee and adopted by the court for closing the patent is advanced and supported by the testimony of R. L. Blakeman, a thoroughly qualified surveyor who made the last survey of the land in question for appellee, and was also introduced by it as a witness. His theory is to reverse the last call from the beginning corner and run the course and distance called for in the patent, and the result he accepts as the proper location of the thirteenth corner. He then begins at the eighth corner, which is the last marked corner, and run the courses and distances of the eighth, ninth, tenth, and eleventh lines, the result of which he accepts as establishing the twelfth corner. He then connects the twelfth and thirteenth corners by a straight line, which varies from the course of that line called for in the patent by 7 1/2°, but which approximates the distances called for for that line, being lengthened almost exactly the same distance as the error in the sixth line. In other words, Blakeman's theory sacrifices the course called for in one line and attributes to that particular line whatever error that may have been responsible for the failure of the patent to close. Appellants' theory makes the patent, which called for 500 acres, include 918 acres, and sacrifices the distances called for in the twelfth and thirteenth lines, but retains their courses and extends them until they intersect, a distance out of all proportion to the error in the sixth line. He extends the twelfth line from 806 poles, called for to 1,058 poles, and the thirteenth line from 90 poles, called for, to 320 poles.

"It has long been the rule in attempting to locate lost lines to give preference to the courses rather than to the distances, and to close the survey, if possible, by lengthening or shortening the distances rather than changing the courses, but in so doing the error in distances must be reasonable and should be apportioned to all of the lost lines rather than arbitrarily placing it in only some of them. This rule of sacrificing distances to courses is only a general rule, and is subject to many exceptions where, from the evidence of a particular case, some other more satisfactory method of adjusting the error is

Resolving/Reconciling Errors

239

disclosed. This rule was first announced in *Beckley v. Bryan, Ky. Dec.* 93, and has been followed and recognized in many cases, but its limitations have been called to attention and it has been departed from about as frequently as it has been followed. A frequent reason given for departing from this general rule is where, by following it, a figure is produced that does not correspond with the original survey and plat upon which the patent is issued, and which always may be looked to in determining the proper mode to be adopted in closing the survey. *Steele v. Taylor, 3 A. K. Marsh.* 226, 13 Am.Dec. 151; *Bruce v. Taylor, 2 J. J. Marsh.* 160; *Harris v. Lavin, 6 Ky. Law Rep.*304; *Hagins v. Whitaker,* 42 S.W. 751, 19 Ky. Law Rep. 1050; *Goff v. Lowe,* 80 S.W. 219, 25 Ky. Law Rep. 2176. Neither party in this case offered to introduce in evidence the original patent or survey and plat, although they might have been exceedingly helpful in a correct solution of the problem. Both Mr. Blakeman and Mr. Ward testified that their respective theories for closing the patent were in accord with an unauthenticated copy of the original survey and plat with which they compared their surveys, but neither was able to say whether or not the copy was correct. This evidence was therefore of no value.

"Another reason for departing from the above general rule is where a known error in established lines may be offset by a corresponding change in the lost line opposite, when the course or distance called for in the lost line opposite the known error, or both, may be changed to close the survey. This exception to the general rule was announced and applied in the case of *Preston Heirs v. Bowmar, 2 Bibb,* 493, decided soon after the Beckley v. Bryan opinion was delivered, and has frequently been used since in closing surveys. In that case the survey was closed by changing the course rather than the distance called for, and we think the facts here are sufficiently analogous to call for the application of the same method for closing this survey.

> The proven error here is in the sixth line, which forms a part of the irregular northern boundary line of the survey, and the twelfth line is the lost line opposite forming the southern boundary of the patent. To change this twelfth line by the same distance as the error in the sixth line and by 7 1/2 degrees closes the survey so as to include approximately the number of acres called for in the patent, and while the number of acres is not controlling evidence, it has evidential value, as has the fact that the known error is in the opposite line, both of which support appellee's theory, while appellant's theory is supported by no evidence whatever, but depends entirely upon the application of a rule of construction manifestly not applicable here because it requires an extension of the opposite line out of all proportion to the known error in the sixth line, and extends the thirteenth line, only 90 poles in length as called for, to nearly four times its prescribed length without any known error in its opposite line. Appellants argue there was a proven error in the length of the second line of 201 poles which, added to the error in the sixth line of 80 poles, makes the total error in the length of the northern boundary lines approximately the same as the extension they propose in the

240   *Boundary Retracement*

twelfth line, but there is no competent, if any, evidence of any such error in the second line. The original patent is not in evidence, and this line, as described in the contract between the parties, is 225 poles, and as run is 223 poles, in length. So there is no error proven in its length, and we do not know upon what authority the statement of error in this line is predicated.

There is no hard and fast rule for closing a survey, but such rules as are employed are but rules of construction in aid of an effort to relocate lost lines as they were located in the original survey, which is always the problem for solution, and rules of construction must give way to competent evidence disproving their applicability to a given case.

Two rules of construction often recognized and applicable here are that reversing calls is as lawful and persuasive as following their order (*Pearson v. Baker, 4 Dana*, 323), and that when a party is claiming under a survey where the course or distance must yield without data to determine whether the mistake was in the one or the other, the mode of closing the survey must be adopted which operates most unfavorably to the party claiming under it (*Preston Heirs v. Bowmar*, supra, and *Pearson v. Baker*, supra)."

Words from the Courts

The courts have long been cognizant of inherent errors in early surveys, and understand that there is no such thing as any kind of measurement without error. At times, judges have been quite vocal about making statements to that effect and the following are selections from various decisions containing choice statements.

Summary of the View by the Courts on the Subject of Surveys and Their Errors

That the subject of disputed boundaries has been a fruitful source of litigation since property rights were first recognized finds proof in the prodigious mass of literature to be found in the books upon the subject. The difficulty is not to find authority, but to select cases which best express the rule to be applied to the facts in issue. Innumerable cases involving boundary lines can be traced to loose description, faulty surveys, and excessive areas created in marking off governmental subdivisions—the bane of all tribunals called upon to reconcile discrepancies in the surveys of the public lands. *Myrick v. Peet*, 180 P. 574 (Mont., 1919).

Resolving/Reconciling Errors

It is a well-known fact that surveyors are apt to differ from each other, and surveyors employed by the United States government are not immune from the frailties of their profession. *Hagerman v. Thompson, et al.*, 235 P.2d 750 (Wyo., 1951).

In the North Carolina case of *Duncan v. Hall*,* the court wrote "It is a fact of which the courts must take and have taken notice that the measurements of boundary lines in making the original surveys for deeds and grants are often, if not always, inaccurate.

The United States Supreme Court recognized the existence of surveying errors as early as 1865, stating, "ordinarily, surveys are so loosely made, and so liable to be inaccurate, especially when made in rough or uneven land or forests, that the courses and distances given in the instrument are regarded as more or less uncertain, and always give place, in questions of doubt or discrepancy, to known monuments and boundaries referred to as identifying the land."†

In *Rowell v. Weinemann*,‡ the Iowa court said, "It is well known that the original surveys were faulty in many respects, and that they will not stand the test of careful and accurate retracing. It is not the purpose of such actions as this, or of any other, for that matter, to straighten out lines, or to remove unsightly crooks, however desirable such a result might be. *Rollins v. Davidson*, 84 Iowa 237, 50 N.W. 1061. Hence everything yields to known monuments and boundaries established by the government surveyors."

In the case of *Fay v. Crozer*,§ after detailing the history of disposing of lands and settlers relocating to the west, the court stated, "The result of this loose, cheap, and unguarded system of disposing of public lands was that in less than 20 years after the adoption of the system nearly all of them were granted, the most part to mere adventurers, in large tracts or bodies, containing not only thousands, but in many cases hundreds of thousands, of acres in one tract. Often the grantees were nonresidents, and few of them ever saw their lands or expected to improve or use them for purposes other than speculation. The entries and surveys under warrants so cheaply and easily obtained were often made without reference to prior grants, thus creating interlocks, thereby covering land previously granted, so that in many instances the same land was granted to two or more different persons. Sometimes upon one survey actually located others were laid down by protraction—constructed on paper by the surveyors, without ever going upon the lands, thus creating on paper blocks of surveys containing thousands of acres, none of which were ever surveyed or identified by any marks or natural monuments. Thus, while the state was rapidly disposing of her large domain at 2 cents per acre, it was not attaining the main objects in view—the settlement of

---

* 117 N.C. 443, 23 S.E. 362, 1895.
† *Higuera's Heirs v. United States*, 72 U.S. 827, 5 Wall. (U. S.) 827, 18 L.Ed. 469.
‡ 119 Iowa 256, 93 N.W. 279, 1903.
§ 156 F. 486, S.D.W. Va., 1907.

the country and revenue from the owners of lands. It was found that the grantees not only failed to settle upon and improve their lands, but in most instances they wholly neglected to pay the taxes due thereon, whereby revenue failed and the improvement of the territory was embarrassed and retarded. Nonoccupation of the lands and nonresidence of the owners made a resort to the lands themselves the only fund for delinquent taxes."

No doubt every retracement surveyor has read the words of Justice Thomas Cooley from the case of *Diehl v. Zanger:** "Nothing is better understood than that few of our early plats will stand the test of a careful and accurate survey without disclosing errors. This is as true of the government surveys as of any others, and if all the lines were now subject to correction on new surveys, the confusion of lines and titles that would follow would cause consternation in many communities. Indeed the mischiefs that must follow would be simply incalculable, and the visitation of the surveyor might well be set down as a great public calamity."

The Early Surveys in Hawaii

According to Alexander (1882), the old surveys made under the direction of the Land Commission, commonly known as "kuleana" surveys, had the same defects as the first surveys in most new countries. There was a lack of proper supervision, no Bureau of Surveying, and the President of the Land Commission was overwhelmed with work such as there was no time for supervision. In addition, there was little money to pay for the work and little time to accomplish it.

No uniform rules or instructions were given to the surveyors, who were "practically irresponsible." Few of them were regarded as thoroughly competent surveyors, while some were not only incompetent but "careless and unscrupulous."† The result was that almost every possible method of measurement was adopted. Some used 50-foot chains, while others used four pole (66-foot) chains divided into links; some attempted to survey by the true meridian, others by the average magnetic meridian, while most made no allowance for local variations of the needle. Some recorded surveys have been found to have been made with a ship's compass and even a pocket compass. Few made the effort to mark corners or to note topographic features.

---

* 39 Mich. 601, 1878.

† Contrast this work with that undertaken in the original 13 colonies, where today's retracements produce results that are not only finding the original work highly reliable, but sometimes shockingly accurate, leading to high praise for accomplishments under some of the most trying conditions such as interference from the natives and discomfort from severe weather conditions.

Resolving/Reconciling Errors

In addition, rarely was one section or district assigned to one surveyor. It has been reported that over a dozen were employed in the surveying of Waikiki, for instance, not one of whom knew what the other surveyors had done, or made any attempt to make his surveys agree with theirs where they adjoined one another. As one might expect, overlaps and gaps became the rule rather than the exception, so that it became generally impossible to put these old surveys together correctly on paper without determining their true relative locations through actual measurements on the ground.

Course and Distance in Perspective

"Almost all locations, where there are calls as well as course and distance, are locations with a double aspect, because the course and distance seldom, if ever, agree with the calls."[*]

As early as 1851, the Massachusetts court, in the case of *Kellogg v. Smith*,[†] recognized several of the inherent, everyday problems of the surveyor:

> "Variation of the compass, imperfection of the instrument, unskillfulness in the use of it, roughness of surface, and other causes, inevitably produce, in every instance, more or less uncertainty of result."

As explained by the court in *Riley*, 16 Ga. at 148, "any natural object, when called for distinctly, and satisfactorily proved—and the more prominent and permanent the object, the more controlling as a locator—becomes a landmark not to be rejected, because the certainty which it affords, excludes the probability of mistake," whereas "course and distance, depending, for their correctness, on a great variety of circumstances, are constantly liable to be incorrect. Difference in the instrument used, and in the care of surveyors and their assistants, lead to different results." *See also McCullough v. Absecon Beach Co.*, 48 N.J. Eq. 170, 21 A. 481 (1891):

> "I think either of the modes just indicated are not improbable, in view of the well known practice of that kind which prevailed in West Jersey for so many years in making proprietory surveys. It is a matter of history that many of those surveys, when run by the monumental calls, contained several times as many acres as they called for and as their strict courses and distances would include. The proprietors were, by this means, defrauded of vast quantities of land, it being uniformly held by the courts that the monuments must govern. They finally refused to act upon any survey which called for more than one monument, and that

---

[*] *Carroll v. Norwood's Heirs*, 5 Har. & J., 155, Md., 1820.
[†] 61 Mass. 375, Mass., 1851.

must be the beginning point. Instances of these surveys are found in *Lippincott v. Souder*, 8 N.J.L. 161; *Curtis v. Aaronson*, 49 N.J.L. 68, 7 A. 886; and Justice Ford refers to them in *Corlies v. Little*, 14 N.J.L. 373, 374.

"The Verree survey, by which the title to the lands in question passed from the proprietors, states that it contains fifty-nine acres, but, in point of fact, it contains three or four hundred acres. And so late as 1869, Samuel S. Downs, a deputy surveyor, made a proprietary survey, for Charles H. Tatum, of the southwest end of Absecon beach, including both lots Nos. 31 and 32 and a part of lot No. 30, and, as clearly appears by his map, intended to include the whole of the point, yet when his survey is put upon the ground it will not reach it by a quarter of a mile. In fact, the practice of using monuments to compress into given courses and distances two or three times the quantity of land they will contain is a familiar one in West Jersey, and is, in my judgment, a sufficient explanation of the peculiarities of the description which have given rise to this controversy. It at the same time illustrates the great values which our courts have always set upon monumental calls as against mere courses and distances."

And earlier, in *Cherry v. Slade*, etc.,* "Where there are no natural boundaries called for, no marked trees or corners to be found, nor the places where they once stood ascertained and identified by evidence; or where no lines or courses of an adjacent tract are called for; in all such cases, we are of necessity confined to the courses and distances described in the patent or deed; for however fallacious such guides may be, there are none other left for the location."

In summary, "The original plats, maps, and surveys of western cities and villages, in respect to figures of measurement, and courses and distances marked thereon, in a large majority of cases have been found notably imperfect, incorrect, and unreliable. At almost any time in the course of municipal history, to rely upon the figures, courses, and distances of the original plat and survey, or upon a resurvey upon the *data* thereof, would be utterly subversive of the rights of real property, and of public and private interests."[†]

---

* 3 Murph. 82, 1819.
† *City of Racine v. J.I. Case Plow Co.*, 56 Wis. 539, 14 N.W. 599, Wis., 1883.

8

Recognizing What Was Left Behind

As the law requires every official surveyor to see the survey plainly marked by trees or natural boundaries, the presumption is, that every survey has been thus marked, or bounded, when made, though the abuttals may not now be found.

Beckley v. Bryan and Ransdale 1 Ky 91 (1801)

Recognizing evidence of past activity requires study, training, and experience. Each preceding surveyor or scrivener had their own particular habits, procedures, type of equipment, or other tools and, in the absence of instructions as to how surveys would be done, results in a wide variety of possibilities. Where some uniformity exists, such as where surveyors were generally instructed as to how to perform a survey, or in those instances where a standard treatise was used for study or for guidance, there is some presumption that those instructions were followed, but there is no guarantee. The basic presumption is that surveyors did follow their instructions and took their jobs seriously.[*]

This was noted in *Mercer, etc. v. Bate, etc.*[†] "Court will presume surveyor's report to be accurate, and will accredit his official acts."

Again referring to *Stafford v. King*, Smith, J. of the Texas court in a lead off discussion of how original surveys were performed stated, "It is the duty of the surveyor to run round the land located and intended to be embraced by the survey and patent, and see that such objects are designated on it as will clearly point out and identify the locality and boundaries of the tract, and to extend a correct description of these objects (natural and artificial, with courses and distances) into the field-notes of the survey, in order that they may be inserted in the patent, which will afford the owner, as well as other persons, the means of identifying the land that was in fact located and surveyed for the owner, and until the reverse is proved, it will be presumed that the land was thus surveyed and boundaries plainly marked and defined."

---

[*] Until it is otherwise shown, it is presumed the surveyor did his duty in making a survey. *Phillips v. Ayres*, 45 Tex. 601.It cannot be assumed that a surveyor failed to do his duty in actually running a line. *Castleman v. Pouton*, 51 Tex. 84.

[†] 4 J.J. Marsh. 334, Ky, 1830.

Evidence

Today's retracement surveyor is faced with finding and locating evidence, evidence of past events, of a variety of sorts. Among the many definitions of evidence, depending on how it is used, one seems appropriate: in short, that which tends to prove or disprove something; or grounds for belief.

Evidence may be found in a variety of forms, and evidence may be found at, or merely nearby, a corner, or a boundary, depending on who placed it there, and their intentions.

Tree Names

Not all surveyors were trained in tree identification, many learned to tell one tree from another in the course of their career. The resulting accuracy of that educational process depended on who was doing the instruction, and whether they had it correct. The one big exception were those surveyors trained in forestry, or who gained their experience from timber-related activities. Early surveyors were instructed to assess the quality of the land as they went, and to do so required a knowledge of indicator plant and tree species as well as the merchantability and the value of the timber on the land being evaluated.

The goal is to identify the particular tree that was called out either as a corner or as a witness. Since there are a number of colloquial names for every tree species, which vary, sometimes considerably, from locale to locale, some insight must be gained as to who was calling what species by what name. Often people in the lumber business had their own set of names, while scientists had a technical name, leaving farmers, surveyors, and others somewhere in between. Common examples are the groupings of hard pines and soft pines and hard maples and soft maples, each of which contain a number of actual tree species. Two very common examples are Eastern Hemlock and Eastern Larch, known colloquially and by field people as Spruce Pine and Juniper, respectively.

Today, there is an online site whereby a colloquial name may be entered and the appropriate correct name given, thereby narrowing the search. Other factors such as relationships to other corners and similar evidence will help to verify the correct tree, no matter whether identified correctly or not. If the tree was marked as the corner or as a witness, this serves to aid in its identification as well. The bottom line is to select the correct tree, whether named, or called out, correctly or not.

Blazed Tree

Trees may be used as corners (see Figure 8.1), or may be marked to indicate a line. Not too many years after their selection, the challenge becomes in finding them, or identifying them. Paint fades, falls away, axe marks such as blazes (chaps) become grown over leaving barely discernable scars, and trees are cut down, blown over, and otherwise removed.

A line will be run to a marked tree called for, though it departs from the course call for, and materially varies the shape of the original survey. *Wash v. Holmes*, 1 Hill, Law, 12 (S.C., 1833).

Witness Trees

A witness tree is not an established corner, but merely an object by means of which, in connection with the field notes, if correct, the corner may be found. The course and distance from a witness tree given in the field notes are just as liable to be erroneous as any others. In fact, it is the common experience of surveyors that the course, or, as the witnesses call it, the "angle" from the

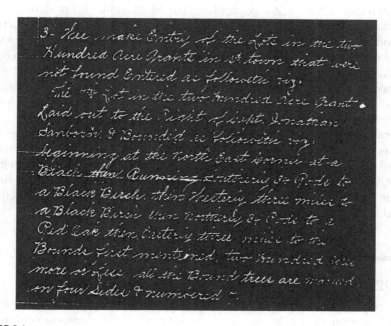

FIGURE 8.1
Excerpt from original 1720 grant by Kingston, New Hampshire calling for trees, Beach (sic), Black Birch, and Red Oak, at the corners.

248 *Boundary Retracement*

witness tree to the corner designated in the field notes is very often erroneous. *Stadin v. Helin*, 76 Minn. 496, 79 N.W. 537 (Minn. 1899).

The Colorado decision in the case of *Davies v. Craig** details the value of tree identification. The court wrote in its decision, "The supervisor of Arapahoe Forest testified that in 1917, at the southwest corner of section 6, he found bearing trees with blazes, and that upon cutting into the trees he found 36 rings of annual growth since the blazes were made, from which he determined that the cuttings had been made 36 years prior to his investigation; that is, in 1881, the date of the original survey. He found trees likewise blazed at the south quarter corner, and some near the CC stone on the west side of the lake, all showing 36 years of growth since the blazing. He testified further that there was a plainly marked line from the south quarter corner eastward to said CC stone; that he found near that stone a dead tree blazed, showing 32 rings, and in a live tree blazes showing 36 rings. He further testified that he had had experience in relocating survey lines in the forest."

Scribe Marks

Trees, stumps, and wood posts, when scribed, may tell a story about the corner itself, or a corner it is a witness (accessory) to. Many symbols are standard, and some are unique. It was common in nonrectangular states for the surveyor who marked the object to include his own personal symbol. This can be a valuable tool to lead a retracement person to the surveyor's notes, or other records.

Many surveyors, especially when doing woodland surveys, carried a special tool for making scribe marks. A sophisticated tool was capable of making straight lines and circles, so that any combination could produce letters, numbers, and special characters (see Figure 8.2).

Eventually these scribe marks, like tree blazes, become grown over with bark as the tree continues to grow and put on girth as well as height. The trained eye may discern scarring, or there may be other indications that evidence exists inside the tree, and therefore needs to be examined. A slab of outside wood may be removed, revealing the convincing evidence well preserved within the tree (see Figure 8.3).

Stump Holes

After trees have disappeared due to cutting, age and decay, fire, insect attack, and other anomalies, often the remains or traces of the stump are still recognizable, generally in the form of a depression in the landscape. If remains are

---

* 70 Colo. 296, 201 P. 56, Colo., 1921.

Recognizing What Was Left Behind

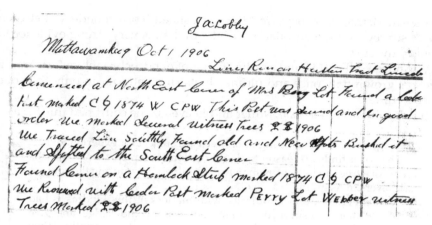

FIGURE 8.2
Excerpt from a set of field notes, which were taken during a retracement survey done in 1906. The scribe marks led to uncovering evidence of this surveyor and the previous surveyor being retraced.

found in the correct configuration as called for as witnesses in the original field notes, they may serve to locate the position of the corner they are witness to.

Tree Rings

The age of a tree can reveal whether the tree was in existence at the time when the description first called for it, helping to identify whether it is the correct tree. Marks on a tree may be dated, and therefore related to survey activity when in agreement, whether they result from blazing the tree, scribing the interior wood, or otherwise wounding the tissue such as when a fence is attached.

Ring patterns on stumps, logs, and fence posts can be matched to living trees to determine where in time they once fit as a living thing, thereby indicating the point in time when they were cut.

> "The annulations of the line trees are entitled to great weight, as evidence of the time when a survey was made. The court stated the reason was that this evidence 'arises from the progress of nature itself, without danger of corruption.'"
>
> *Bulor's Heirs v. James M'Cawley, 10 Ky (3 A.K. Marsh. 573 (1821).*

FIGURE 8.3
Excerpt from deed description noting scribe mark on a tree.

250   *Boundary Retracement*

Also, in *Mercer, etc. v. Bate, etc.*,* the court stated "correspondence between age of survey, and the annulations of the timber, is very strong evidence to identify the survey."

As Andrew Ellicott Douglass, the "father of dendrochronology", stated, "a tree has grown only once and ultimately its ring pattern can only fit at one place in time."

In *Baker & Sons v. Sherman*, Miller and Leavens,[†] the court had this to say:

> The age of marks or spotting upon trees is a subject for expert evidence.
>
> The court finds no reason to doubt the long-accepted theory that the years of a tree can be approximately determined by counting the concentric layers in its grain.
>
> Such experience and familiarity with matters of this kind as is had by woodsmen and surveyors, constitute peculiar knowledge, and give special skill in determining the age of trees, and such, especially if they have been accustomed to examine with the glass, may testify as experts.

Wooden Posts

Many times wooden posts were set, especially in wooded areas. In addition, in those areas with extreme snow cover, a wooden post would be visible above the level of the snow. Wood posts frequently are identified with scribe marks.

Post is an "artificial monument." *U.S. v. Gallas*, 269 F.Supp. 141 (D.C. Md., 1967).

In fixing limits, where the original boundary posts can no longer be seen, new posts must be fixed, but placed where the former limits stood, without regard to the title papers. *Zeringue v. Harang*, 17 La. 349 (La., 1841).

Where "a post" is called for in the survey of boundary lines in an improved country between individual holders, it is such a monument as will control course and distance. *Lessee of Alshire v. J.R. Hulse*, 5 Ohio (5 Ham) 534 (1832).

Wooden Stakes and Their Remains

Wooden stakes placed by a surveyor on the land to mark corners of lots or the intersection of boundaries and measuring lines constitute "monuments." *Sellman v. Schaaf*, 269 F.Supp. 160, 26 Ohio App.2d 35 (1971).

---

* 4 J.J. Marsh. 334, Ky, 1830.

† 71 Vt. 439, 46 A. 57, 1899.

Recognizing What Was Left Behind 251

After the disappearance of a wood post or stake due to natural deterioration or fire, sometimes its former location may be identified from bits of remaining wood or wood fibers, or else from a discolored area in the soil. An important key to the identification is the shape of the stain, where posts are usually round or nearly so, their points have often been "squared off" with an axe. The same may be said about stakes, which are more often square than round. Pieces of preserved wood are sometimes found below the ground surface in which cuts made by an axe may still be visible.

Stake and Stones

Most corner posts and stakes were set within a ring of stones. However, one must be on guard for the call "stake and stones" (plural) as opposed to the call "stake and stone" (singular). The latter would likely have been a post with a single stone placed alongside it to continue to mark the position after the post had disappeared. With the former, usually the stake, or post, is gone but the ring or pile of stones remains. Frequently, there is a discernable gap in the middle of the pile, and underground at that location is sometimes found the remains of the stake or post.

Depending on site conditions, it often requires a careful eye to recognize stone piles after they have sunk into the earth and have become otherwise obscured with brush and vegetation. Sometimes, in the northern areas, it is possible to search subsequent to a light snow after the sun has starting its melting process. The snow will melt from the stones the soonest, before melting from bare ground, making them readily visible in contrast to their surroundings.

Pits and Mounds

It is in evidence and not disputed, that in surveying and subdividing townships the government surveyor is required to, and usually does, establish a section corner by a mound, four pits, and a stake; and a quarter section corner by a mound, two pits, and a stake or post; the stakes or posts to be properly marked, to indicate the corner represented. A court will presume, therefore, that such government surveyor, in the absence of evidence to the contrary, performed his duty in establishing such corners. That presumption is strengthened in this case by the evidence of several witnesses, who, at a very early day, and before the township was settled, saw, in different parts of the township, mounds, and, in some instances, pits and stakes, some one or more of the stakes having marks

252   *Boundary Retracement*

upon them. As early as 1868-69, Generius Thompson and Ole Thompson testify that, in passing over the township, they saw several such mounds, with one or more pits, and, in some cases, four pits and stakes, and were able to follow them, in making their way over the township. Several of the first settlers found mounds with one or more pits and one or more stakes in 1872, when they went into the township to locate land. It is true that not many of the mounds seemed, at that time, to have four pits, or even two pits, and but a very few stakes were to be seen. But the mounds appeared to be old mounds, "grown over with grass." Some eight or ten witnesses on the part of the plaintiffs testify to seeing such mounds. A number of the early settlers, witnesses on the part of the defendants, testify that they saw mounds, but that they did not have the two or four pits and stakes usually found at government corners. That most of the stakes should have disappeared, and that the mounds and pits should have become indistinct and difficult to find, 8 or 10 years after the original government survey, and after exposure during that time to the elements, in a prairie country, would be quite natural. But these monuments, having been placed at the original corners to mark the boundaries, though somewhat indistinct and imperfect, must control in all subsequent surveys, if they can, in any possible manner, be identified.

Section 694, Comp. Laws, as we have seen, distinctly requires that, 'in retracing lines or making any survey the surveyor should take care to observe and follow the boundaries and monuments as run and marked.' That a number of these old monuments were in existence at the time of the Van Antwerp survey, we think, is established by a preponderance of the evidence. A number of such old corner mounds were testified to by the witnesses for the plaintiffs. It is true that the witnesses for the defendants, while admitting that they saw mounds in various parts of the township, seemed to take the view that they were not government mounds, for the reason that they did not find connected with them the required number of pits and stakes properly marked, and for the further reason that other surveyors, in locating parties in that township subsequently to 1872, had in some cases erected new mounds to mark the corners of their survey.*

Other Witnesses

A witness to a corner can be almost anything, a nearby tree, a rock, the corner of a building, other corners on the subject, or nearby parcels.

The California court recognized a nearby house as a monument in *Wise v. Burton.*[†] "There is one controlling monument in the field-notes of the Jesus Maria survey. This is the Santa Lucia house. The existence of this house is

---

* *Randall v. Burk Tp. of Minnehaha County*, 4 S.D. 337, 57 N.W. 4, S.D., 1893.
† 73 Cal. 166, 14 P. 678, 1887.

Recognizing What Was Left Behind

testified to by the witness Harris, called by plaintiffs, and by Jesse, Lewis, and Cooper, called by defendant. The field-notes of the Jesus Maria show that the survey was made in 1859; and Lewis, who assisted in making it, says that it was then occupied by Burton. It is stated in the field-notes of the Jesus Maria that this house is west of the line of that ranch, and distant from it about twenty chains. This house is a monument, as a witness-tree is a monument. The map and the objects on it are to be regarded. (See *Vance v. Fore,* 24 Cal. 435; *Serrano v. Rawson,* 47 Cal. 55; *Black v. Sprague,* 54 Cal. 266; *McIver's Lessee v. Walker,* 9 Cranch, 173; *Chapman v. Polack,* 70 Cal. 487.) It is a *witness-house* called for in the field-notes, and delineated on the map and field-notes, which are in the patent, constituting a part of it."

Corner Accessories

From Manual of Surveying Instructions, 2009

4-83. The purpose of an accessory is to evidence the position of the corner monument. A connection is made from the corner monument to fixed natural or artificial objects in its immediate vicinity, whereby the corner may be relocated from the accessory. Thus, if the monument is destroyed or removed, its position may be identified by any remaining evidence of the accessories. One or more kinds of accessory are employed at each corner established in the public land surveys (except for corners of minor subdivisions and where specifically not required by the manual, or omitted by the special instructions).

Accessories consist of (1) bearing trees or other natural objects such as notable cliffs and boulders, permanent improvements, reference monuments; (2) mounds of stone; or (3) pits and memorials. Aside from availability, selection is based on their order of permanence.

4-84. The surveyor cannot perform any more important service than that of establishing permanent and accurate evidence of the location of the corners in his survey. Where the accessories cannot be employed, other means should be adopted that will best serve the purpose.

Bearing Trees and Bearing Objects

4-85. Bearing trees are selected for marking when available, ordinarily within a distance of three chains of the corner; a greater distance if important. One tree is marked in each section unless a tree in one or more positions may not be available. A full description of each

254 *Boundary Retracement*

bearing tree is given in the field notes. This includes the species of each tree, its diameter at breast height, the exact direction from the monument, the horizontal distance counting to the center of the tree at its root crown; and, the exact marks scribed for the identification of the corner.

Almost any nearby natural object that can be readily identified should be recorded by description, course, and distance. Such objects may not be of a character that can be marked, exception in the case of a rock cliff or boulder. These are supplemental to the marking or bearing trees, or to fill out a quota where trees are not available in some sections. The description of the cliff or boulder should provide ready identification, including the marking of a cross (X) plainly and deeply chiseled at the exact point to which the direction and distance are recorded.

Another desirable accessory, especially where the usual types are not available, nor suitable on account of the side conditions, such as at a corner that falls in cultivated land, is to record accurate bearings to two or more prominent landmarks.

4-86. The marks upon a bearing tree are made upon the side facing the monument, scribed in the manner already outlined for marking tree corner monuments. The marks embrace the information suggested in the schedule hereinafter given, with such letters and figures as may be appropriate for a particular corner, and will include the letters "BT." A tree will always be marked to agree with the section in which it stands, and will be marked in a vertical line reading downward, ending in the letters "BT" at the lower end of the blaze approximately 6 inches above the root crown.

4-87. There is a great difference in the longevity of trees, and in their rate of decay; trees should therefore be selected, if possible, with a view to the length of their probable life, their soundness, favorable site conditions, and size. Sound trees, not matured, of the most hardy species, favorably located, are preferred for marking. Trees 5 inches or less in diameter should not be selected for marking if larger trees are available, and it is generally better to avoid marking fully matured trees, especially those showing signs of decay. Trees 4 inches in diameter, or less, if no better trees are available, are marked with the letter "BT" only at the base, and an "X" at breast height, facing the monument. The species, size, and exact position of the bearing trees are of vital importance, as this data will generally serve to identify a bearing tree without uncovering the marks, or even to identify two or more stumps after all evidence of the marks has disappeared.

4-88. Generally, only one tree is marked in each section at a particular corner, but in certain instances, two trees are required in a section. In

Recognizing What Was Left Behind

255

such cases, it is better to select trees of different species or of widely different size, direction, or distance. If the trees are of the same species, in order that confusion may be avoided in the future identification of a remaining tree where the companion tree has disappeared, one is marked with an "X" only (and "BT" at the base).

4-89. A cross (X) and the letters "BO" are chiseled into a bearing object, if it is a rock cliff or boulder; the record should enable another surveyor to determine just where the marks will be found. The rock bearing object is the most permanent of all accessories; it is used wherever practicable, and within a distance of five chains.

4-90. A connection to any permanent artificial object or improvement may be included in this general class of corner accessories. The field notes should be explicit in describing such objects, and should indicate the exact point to which a connection is made, as "southwest corner of foundation of Smith's house," "center of Smith's well," "pipe of Smith's windmill," etc. No marks will be made upon private property without the consent of the owner.

Memorials

4-91. Where there is no tree or other bearing object, as above described, and where a mound of stone or pits are impracticable, a suitable memorial is deposited alongside the monument. A memorial may consist of any durable article which will serve to identify the location in case the monument is destroyed. Such articles as glassware, stoneware, a marked (X) stone, a charred stake, a quart of charcoal, or pieces of metal constitute a suitable memorial. A full description of such articles is embodied in the field notes wherever they are employed as a memorial. When replacing an old monument with a new one, such as substituting an iron post for an old marked stone, the old marker is preserved as a memorial.

Mounds of Stone

4-92. Where native stone is available and the surface of the ground is favorable, a mound of stone is employed as an accessory to a corner monument, or to surround it, even though a full quote of trees or other bearing objects can be utilized. A mound of stone erected as a

256   *Boundary Retracement*

corner accessory should be built as stable as possible, should consist of not fewer than five stones, and should be not less than 2 feet base and 1½ feet high. Where the ground is suitable, the stone mound is improved by first digging a circular trench, 4–6 inches deep, for an outer ring, then placing the base of the larger stones in the trench. In stony ground, the size of the mound is sufficiently increased to make it conspicuous. The position of the accessory mound is shown in the schedule following. The nearest point on its base should be about 6 inches distant from the monument. The field notes show the size and position of the mound.

4-93. Where it is necessary to support a monument in a stone mound, and if bearing trees or other objects are not available, a marked (X) stone or other memorial is deposited alongside the monument.

A stone mound accessory, in addition to the mound surrounding a monument, is built wherever this will aid materially in making the location conspicuous.

Pits

4-94. Where the full quota of trees or other bearing objects are unavailable for marking, the position of the monument is, under favorable conditions, evidenced by pits. No pits should be dug in a roadway, or where the ground is overflowed for any considerable period, or upon steep slopes, or where the earth will wash, or in loose or light soil, or where there is no native sod, or where suitable stone for a mound is at hand.

A firm soil covered with a healthy native sod is most favorable for a permanent pit. Under such conditions, the pits will gradually fill with a material slightly different from the original soil, and a new species of vegetation will generally take the place of the native grass; these characteristics, under favorable conditions, make it possible to identify the original location of the pits after the lapse of many years.

4-95. All pits should be dug 18 inches square and 12 inches deep, with the nearest side 3 feet distant from the corner monument, oriented with a square side (and not a corner) toward the monument, arranged as shown in the schedule following. The earth removed is scattered in such a way that it will not again fill the pits. A description of the pits is embodied in the field notes, and should include a statement of their size and position.

Recognizing What Was Left Behind 257

Accessories to Special-Purpose Monuments

4-96. Accessories to special-purpose monuments are selected and marked as follows:

Witness corners. Formerly, the accessories for witness corners were the same as though the monument had been established at its true point, but the marks upon the bearing trees or other objects were preceded by the letter "WC" and the selection number was made to agree with the section in which the tree or object actually stood. The rule now is that bearing objects, if available, are treated as for a regular corner. Bearing trees, with direction and distance from the monument, are marked with an "X" at breast height, on the side facing the monument and the letters "BT" at the base. Mounds of stone are treated as though the monument were located at the true corner.

Reference monuments. All bearing objects and bearing trees, including marks, refer to the position of the regular corner, as that location will be occupied as an instrument station.

Witness points. No requirements are set up as to the accessories for a witness point other than to mark a bearing tree or a bearing object, if available, at important locations or to record bearings to more distant natural objects or improvements.

Depending on conditions, subsurface and above ground, various species of wood in the form of posts and stakes can last a long period of time. A case in point is the Popham Colony in Maine, established in 1606 and abandoned in 1607. While the residents were at the site they erected homes and other structures, the remains of which are being uncovered today. For example, wooden posts supporting the church at the site have been found and verified during the past few years. Even after near total deterioration, or after total deterioration, evidence of their remains are still identifiable.

Stones. Where stones are readily available, they are often used to mark corners. Generally, they are sizeable and set well into the ground. They often mislead people searching for a corner, especially where the monument is not called out in the description, since they look just like any other rock of local origin, which they once were. If they are marked in some fashion with initials, hieroglyphics, or otherwise, they may be easily identified. If not, they make require a trained retracement surveyor to locate them and relate them to other, related corner evidence, either on site or off-site, such as on an abutting parcel.

Metal objects. Metal objects, for the most part, are quite durable, and may be almost anything the marker of the corner had available. Calls for

258 Boundary Retracement

corners include, buggy axles, gun barrels, iron pins, pipes and posts, welding rods, pieces of television antennae, plugs, discs, railroad spikes, nails, and the like. They may be called out in the description or noted on the plat, or not. Metal detectors are indispensable in finding these items, especially when they are obscured by vegetation, or are now underground due to sinking, being pushed down, or having fill deposited over them.

Metal detectors are basically of two types: those that detect ferrous metals due to the magnetic field established through oxidation, and those that detect other metals, such as brass or aluminum.

Memorials

A memorial is a durable article deposited in the ground at the position of a corner to perpetuate the position in case the monument is removed or otherwise destroyed. It is often placed at the base of the monument and may consist of anything durable, such as glass or stoneware, a marked stone, charred stake, or a quantity of charcoal.

It was sometimes the practice of surveyors, when setting a wood post, to break a glass bottle, or a piece of stoneware in the hole before putting in the post. Other durable materials include pottery chards and broken pieces of brick. Charcoal was a common addition, as well as charring the end of the post being set in the ground, for durability and later recognition. These types of items are evidentiary clues at the location of corners and boundaries.

An interesting example of an accessory is found in the decision of *Thomsen v. Keil*,* quoting Clark on Surveying and Boundaries, at section 373, says:

> 'Where a monument is obliterated, the accessories furnish the highest evidence of the location of the original monument, and therefore such accessories are of prime importance in relocating such obliterated monument. The term accessories, includes all witness trees, line trees, mounds, pits, streams, bodies of water, ledges, rocks, or other natural features to which the distance from the corner or monument are known. These natural features furnish unmistakable evidence of the location of the monument, the nearer to the required point, the stronger the evidence.'
>
> In the instant case the road called for in the field notes was in a ravine—the only place, according to the evidence, it could be—and the place where it is shown to have been. This certainly must be considered as an accessory—a call from which the point of the original location of the quarter corner could be established. Particularly is this true when it is remembered that the point at which the marked stone was found and that designated in the field notes as the location of this quarter corner is only about 70 feet from this road and in clear view thereof.

---

* 48 Nev. 1, 226 P. 309, Nev. 1924.

Recognizing What Was Left Behind 259

No doubt, too, the court attached considerable significance to the rock to which we have alluded. It is true that the rock was not found in a mound or in such a position as to indicate that it had been placed at the point found, but, in view of all of the testimony, particularly that as to cloudbursts in that locality, the court was justified in concluding that it had been carried a short distance from the point at which it was originally placed. The effort on the part of the plaintiff to conceal this stone is a circumstance justifying the court in looking with suspicion upon his case.

We think, in view of all of the evidence, the court was justified in establishing the quarter corner with reference to the call for the road in the field notes where it did.

Fences

It has been said that people build fences for one of three reasons, namely, to keep things in, to keep things out, and to mark a property boundary. An additional reason would be for convenience. Several courts have raised this issue when it came to whether a fence was the appropriate location of a boundary, or it was merely near the boundary, depending on who built it, and why.

Fences may be found made of a variety of materials, the most common being wood and metal, or a combination of the two (e.g., wire strung on wood posts, or stapled to trees, or both). However, where such materials were lacking, fences were sometimes erected by piling cut sods to a desired level, planting trees as a barrier, and creating lines of tree stumps with their roots as fields were cleared. The history of fences, including the various types, has been written about extensively.

Even after fences have seemed to have disappeared, it is often possible to recover their remains. Remnants of fence posts, pieces of wire (in trees and underground), and slight differences in the landscape may serve to demonstrate the location of a former fence line. Fruit trees, such as wild cherry, growing in a relatively straight line may indicate where birds sat and digested a recent meal. Single stones, either in a straight or a staggered line, may serve to indicate "heel stones" placed to keep wooden rails off the ground for longer life.

Fences can be valuable evidence to the retracement surveyor seeking to identify boundary lines. However, too often people quickly assume that a fence, or the remains of one, marks a boundary and no further inquiry is necessary. Not all fences have been placed exactly on boundaries, many are deliberately placed elsewhere and many others erected for other reasons, with no thought of having any relation to a boundary.

The following decisions demonstrate the thinking of the court with regard to fences. For the thinking regarding why a fence should not be accepted as boundary evidence, the reader is referred to Chapter 9 concerning boundary agreements.

260   *Boundary Retracement*

However, in considering the value of a fence for boundary purposes, one would do well to remember the words of Justice Thomas Cooley in *Stewart v. Carleton*[*]: "The city surveyor should have directed his attention to the ascertainment of the actual location of the original landmarks set by [the original surveyor], and when those were discovered they must govern. If they are no longer discoverable, the question is where they were located; and upon that question the best possible evidence is usually found in the practical location of the lines, made at a time when the original monuments were presumably in existence and probably well known. As between old boundary fences, and any survey made after the monuments have disappeared, the fences are by far the better evidence of what the lines of a lot actually are."

In determining whether a corner was lost or obliterated, the court in *U.S. v. Doyle*[†] stated that before a corner could be considered lost, all means for ascertaining its location must first be exhausted. Means to be used to locate lost monuments or corners include collateral evidence such as boundary fences that have been maintained, which should not be disregarded by the surveyor.

A building or a fence constructed according to stakes set by a surveyor at a time when these were still in their original locations may become a monument after such stakes have been removed or disappeared, and next to the stakes, may be the best evidence of the true line.[‡] In this case, based on a new survey of a section of the city, a landowner was told that his long-standing fence, in front of lots 11 and 12, was within the right of way limits of one of the city's streets. This section of the city was subdivided and platted in 1842. The claim was that the fence was within the east side of Wisconsin Street, 2 feet at one corner of said lots, and 2 feet 8 inches at the other corner. The defendant has been present with this fence standing substantially where it is since 1848, so that, if the fence is an obstruction to Wisconsin Street as claimed, it is a very ancient one and the defendant has been guilty of maintaining it over 40 years.

What precipitated this action was a new survey by the city in 1881, which replatted that part of the city. By this plan, the lines of lots were materially changed, and, according to the city and their surveyors, the location of fences and buildings had to be materially changed to suit the new lines. Monuments were set at various points by this new plan and resurvey in 1881, and in 1890, the city engineer, by order of the city, ran the lines of Wisconsin Street according to the said monuments and the newly found distances, and found that the fence of the defendant was within that street, as above stated.

> As early as 1844 the lots in this part of the city were occupied by lessees or purchasers, and fences were built along Wisconsin street according to stakes set to indicate the lines according to the old plat, and such fences,

---

[*] 31 Mich. 270, 1875.
[†] 468 F.2d 633, Colorado, 1972.
[‡] *Racine v. Emerson*, 85 Wis. 80, 35 N.W. 177, 1893.

Recognizing What Was Left Behind

or many of them, still stand in the same places; and shade trees were set out and buildings erected on or according to such lines. The question then is, Where is the east line Wisconsin street in front of the lot in question, according to the original plat of 1842. The question is not, Where is such line according to any subsequent survey or plat.

"All resurveys or subsequent surveys are of no effect except to determine that question. A resurvey that changes lines and distances and purports to correct inaccuracies or mistakes in the old plat is not competent evidence in the case. There are only two questions: (1) Where is the true line fixed by the original plat? (2) Is the fence in question on that line? A resurvey that changes or corrects the old survey and plat can never determine the first question. A resurvey must agree with the old survey and plat to be of any use in determining it. The survey made on the arbitrary plan established by the common council in 1881 does not agree with the old plat in courses or distances, in the dimensions of blocks and lots, or in the lines of the streets. It seems to have been made to correct the old plat, to straighten the streets, and make a better plat than the old one. Resurveys for the lawful purpose of determining the lines of an old survey and plat are generally very unreliable as evidence of the true lines. The fact, generally known and quite apparent in the records of courts, is that two consecutive surveys by different surveyors seldom, if ever, agree; and the greater number of surveys, the greater number of differences and disagreements will occur. When two surveys disagree, the correct one cannot be determined by still another survey. It follows that resurveys are of very little use in such a case as this, except to confuse it. In *Miner v. Brader*, 65 Wis. 537, 27 N.W. 313, there were two surveys, and they disagreed; and the court had to resort to the evidence of a practical location of the lines by monuments. Monuments set by the original survey in the ground, and named or referred to in the plat, are the highest and best evidence. If there are none such, then stakes set by the surveyor to indicate corners of lots or blocks or the lines of streets, at the time or soon thereafter, are the next best evidence. The building of a fence or building according to such stakes, while they were present, become monuments after such stakes have been removed or disappeared, and the next best evidence of the true line."

The public and private owners have acquiesced in the lines established by the first and original survey and plat, and by practical location and undisturbed possession for a great many years, and there does not seem to have been any necessity to disturb them at this late day.

This same court stated in an earlier case,* "original plats, maps, and surveys of western cities and villages, in respect to figures of measurement, and courses and distances marked thereon, in a large majority of cases have been found notably imperfect, incorrect, and unreliable."

"The early settlers, who first buy and build upon the lots, do not attempt to ascertain their lines by a computation of measurements of all the other

---

* *City of Racine v. J.I. Case Plow Co.*, 56 Wis. 539, 14 N.W. 599, Wis., 1883.

lots and blocks by the figures on the plat, or stated in the certificate of survey, or the courses and distances marked thereon, or by a resurvey from the starting point of the first one. But they consult the stakes, and other monuments and land-marks, either natural or artificial, fixed and placed at the time of the original survey, if any, and such is generally the case, and such is the method adopted by those who buy and build afterwards, if such land-marks still exist; and afterwards, and after such monuments or land-marks have been destroyed or removed, such lines are ascertained by constructions of a permanent character which were built according to such original monuments, and finally, as time goes on, long usage, prescription, antiquity, and reputation may be the only means of determining the true lines and boundaries, and these methods in this order are to be preferred to courses and distances and figures marked on the original plat and survey, as the higher degrees of evidence. At almost any time in the course of municipal history, to rely upon the figures, courses, and distances of the original plat and survey, or upon a resurvey upon the *data* thereof, would be utterly subversive of the rights of real property, and of public and private interests."

9

Boundary Agreements

Boundary agreements, frequently known as boundary line agreements, although often inappropriate, perhaps even in violation of legal requirements, offer another set of footsteps. If appropriate, they may provide evidence of a line, if not inappropriate, then they are obviously nothing more than a misleading set of footsteps that need to be dealt with and possibly done away with. Unfortunately, some of them become a legal matter and must be dealt with through the court system, taking time, and unnecessarily costing money to third, or otherwise subsequent, often unsuspecting, parties.

Where the real expensive and time-consuming problem arises is when parties have decided to shortcut the process, executed a boundary line agreement, then later unwary and unsuspecting parties find the location of the true boundary, either as a result of a subsequent survey, or a like investigation. Because of the underlying requirements for boundary agreements so that they do not violate what is known as the statute of frauds, these agreements may actually be void. And worse, they may create a cloud on the titles affected by the boundary line agreement.

The problem created is that they are found on the record, and presumed to have been done correctly. Many would find fault with a deed that appeared to be deficient, yet accept a recorded boundary line agreement without hesitation or question.

Somehow, through some means, a message came through that a boundary agreement was a "quick fix" and could circumvent otherwise time-consuming and costly endeavors such as surveys and research projects, even court actions, that would be necessary and in accordance with appropriate rules and practice. These latter were deemed unnecessary in favor of an agreement.

The attempt has been to use a boundary line agreement to "fix the problem," primarily because people either do not wish to spend the time or the money to determine the true boundary. While this may appease the current landowners, it does nothing to cure the title problems. It merely adds another set of footsteps and inserts one more legal entanglement, causing multiple problems for future surveyors, lawyers, and ultimately the court system at one level or another. The basic problem is a title issue, not a boundary issue, and land title cannot be passed with any form of agreement. Title can, however, pass by way of adverse possession if the requirements for the same have been met. With adverse possession, possession may ripen into title in the name of the possessor, which is a brand new title, and "perfect in every way."* But agreements of any kind only serve to "fix" the location of a boundary line, and each has its own set of requirements. And each requires that the location of the true boundary is "in dispute, unknown or unascertainable." Far too often the boundary is said to be unknown or unascertainable, when in fact it is not, thereby making the situation a title issue rather than a boundary issue, and results in affecting *a parcel of land*. This is exactly what the statute of frauds is in effect to prevent (Figures 9.1 through 9.4).†

Many either do not understand, or else choose not to believe, that (1) a boundary line agreement is not a method of relocating a line for the sake of convenience, nor is it a method for sidestepping the requirements for investigating the true location of a boundary, and (2) an agreement may only be used when satisfying a rigid set of requirements dating back to at least the formation of the statute of frauds, or perhaps even earlier. To do otherwise,

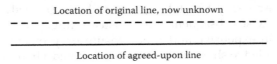

FIGURE 9.1
Possible outcome of boundary line agreements. Placing the line some place offset from the original location of the true boundary.

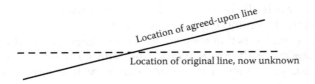

FIGURE 9.2
Possible outcome of boundary line agreements. Placing the line such that it physically crosses the location of the true boundary.

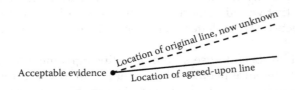

FIGURE 9.3
Possible outcome of boundary line agreements. Placing the line by accepting evidence at one end.

---

* Title acquired by adverse possession is a new and independent title by operation of law and is not in privity in any way with any former title. Generally it is as effective as a formal conveyance by deed or patent from the government or by deed from the original owner. In fact, it is a good, actual, absolute, complete, and perfect title in fee simple, carrying all of the remedies attached thereto. 3 Am Jur 2d, Adverse Possession, § 298.
† The original title of the statute was *An Act for Prevention of Frauds and Perjuries*, which was passed in England in 1677.

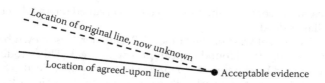

FIGURE 9.4
Possible outcome of boundary line agreements. Placing the line by accepting evidence at the other end.

according to many courts, may not only render the agreement void, but also create problems for subsequent title holders relying on it.

An attempt to convey title to a tract or parcel of land with a form of agreement is contrary to the statute of frauds in most instances. A perfect illustration of a problem discovered after the fact is the California case of *Williams v. Barnett*[*]:

> In January, 1953, Stephen W. Bradford, a licensed surveyor, made a resurvey (actually a retracement) of Lots 12, 13, and 14 of The Anaheim Extension. "At the trial of this action Mr. Bradford was called as a witness on behalf of the appellant. Mr. Bradford established the datum points for his survey by reference to the Hamel map and other maps in the offices of the City Engineer of Anaheim and the County Recorder of Orange County. He found two iron bolts marking the intersections of the center lines of Santa Ana Street and East South Street with the center line of Placentia Avenue, thus definitely establishing the northeast and southeast corners of the area embraced within Lots 12, 13 and 14 and thereby established the east boundary line of said area. He also established the northwest and southwest corners of the same area by reference to the same and other maps and data.
>
> Mr. Bradford prepared a plat of his survey. It is in evidence as plaintiff's Exhibit "4." In making his survey, Bradford found a 1 1/2-inch iron pipe in the ground 24.75 feet south of the center line of Santa Ana Street and 25.15 feet west of the west line of the area indicated on his plat as an area of 10 acres plus 10 feet. He also found a 4" × 4" redwood stake 24.75 feet north of the center line of East South Street at approximately the same distance west of the west line of the 10 acres plus 10 feet. Neither the iron post nor the redwood post bore any surveyor's identification marks. They appeared to be surveyor's markers. Mr. Bradford drew on his plat a broken line from north to south between these two markers and designated this line as "Line of Occupation," indicating that line to be 1,477.85 feet in length. These markers probably had been in place for a period of from 30 to 40 years. Mr. Bradford found no other posts or markers along the "Line of Occupation." The evidence does not establish who installed these markers.
>
> "Mr. Bradford found that the area which he marked on his plat as the property of appellant, comprised the 10 acres plus 10 feet called for by her deed and that the additional width of 25.15 feet described by the complaint and computed by him to comprise an additional. 85 acre was

---

[*] 287 P.2d 789, 135 Cal.App.2d 607, 1955.

not necessary to comprise the 10 acre plus 10 feet called for and described in appellant's deed.

About 5 or 6 feet west of the "Line of Occupation," as shown by the Bradford plat, Bradford found a eucalyptus windbreak. This windbreak was not claimed by either party to mark a boundary line. He also found that the western part of appellant's garage was on the strip of land in dispute, and that appellant's driveway to her garage ran from north to south within said strip for about 150 feet. He found no fence on or near the "Line of Occupation." He found cultivated orange trees on both sides of the line of occupation and of the eucalyptus windbreak.

We believe that from the facts recited hereinbefore the trial court was justified in drawing the following conclusions: That the 1 1/2-inch iron pipe found at the north end and the 4" × 4" redwood stake found at the south end of Bradford's "Line of Occupation" were not set by Hamel, but were set 45–50 years after the 1868 Hamel survey by a person or persons unknown; that the line drawn by Bradford as "Line of Occupation" did not purport to mark the true west line of appellant's property; that neither the cultivation by appellant of a part of the disputed strip nor the location of a part of appellant's garage and appellant's driveway thereon established exclusive possession by her of said strip; that the said "Line of Occupation" was not definitely marked by fence or otherwise; that the true west boundary line of appellant's property was definitely established by Bradford as 25.15 feet east of Bradford's "Line of Occupation," as shown by his plat; that no real uncertainty existed as to said true boundary line; that if any uncertainty had existed as to said true boundary line, such uncertainty was readily resolvable, and in fact was resolved by Bradford's survey; that there was, in fact no agreement, express or implied between the parties fixing a boundary line other than the true boundary line.

The doctrine of our law that coterminous owners may by agreement, implied from acquiescence, establish and fix their mutual boundary line, is applicable only where the true line is otherwise unknown or uncertain. There is no occasion for asserting that a boundary has been established by agreement, unless the description in the conveyance in reality designates a different boundary. (*Mello v. Weaver*, 36 Cal.2d 456, 462.) "Of course, agreements of this type cannot be used where the true boundary line is known." (*Vowinckel v. N. Clark & Sons*, 217 Cal. 258, 261.)

The owner of property must be presumed to have been familiar with the terms of the instrument which constituted his muniments of title. If he did not actually know the extent of his property and had the means of knowledge within reach, he would not be heard to say that a fence was located upon an accepted division line. (*Janke v. McMahon*, 21 Cal.App. 781, 788.)

Where there is an acquiescence in a wrong boundary, when the true boundary may be ascertained by the deed, it is treated both in law and equity as a mistake, and neither party is estopped from claiming to the true line. The boundary is considered definite and certain when by survey it can be made certain from the deed. (*Janke v. McMahon*, supra, 21 Cal.App. 781, 788.)*

---

* This is a sound principle of law explained earlier in *City of North Mankato v. Carlstrom.*

Boundary Agreements

A requisite to make applicable the doctrine of agreed boundary line is that the line acquiesced in by adjoining property owners must be specified, definite and certain. The agreement is not controlling if that line is left indefinite, uncertain or speculative. (*Garrett v. Cook*, 89 Cal.App.2d 98, 103).

Appellant and respondents are in agreement upon the principle that resurveys in no way affect titles taken under a prior survey.

The judgment is affirmed.

It is noteworthy that the court in this case emphasized four principles: (1) the doctrine of our law that coterminous owners may by agreement, implied from acquiescence, establish and fix their mutual boundary line is applicable only where the true line is otherwise unknown or uncertain; agreements of this type cannot be used where the true boundary line is known.* (2) Where there is an acquiescence in a wrong boundary, when the true boundary may be ascertained by the deed, it is treated both in law and equity as a mistake, and neither party is estopped from claiming to the true line. (3) The boundary is considered definite and certain when by survey it can be made certain from the deed.† In other words, the court said, the boundary may not actually be known, but "the means are within reach." (4) A requisite to make applicable the doctrine of agreed boundary line is that the line acquiesced in by adjoining property owners must be specified, definite, and certain. The agreement is not controlling if that line is left indefinite, uncertain, or speculative.

Whereby the court determined that since the requirements for a boundary agreement were not met, specifically in this case leaving an unmonumented agreed-upon boundary, the agreement was void.

See the recent New Hampshire case of *DRED v. Dow Sand & Gravel*‡ below.

The referenced case, *Mellow v. Weaver*, discussed background details, and stated the following:

> The requirements of proof necessary to establish a boundary by agreement are well settled by the decisions in this state. (See *Hannah v. Pogue*, 23 Cal.2d 849, and *Martin v. Lopes*, 28 Cal.2d 618, where the numerous cases are collected). Mere agreement to locate a boundary known to be different from that called for by the deeds is insufficient, since such an agreement would be tantamount to a conveyance by parol, an unrecognized method of transfer of real property. But as early as *Sneed v. Osborn*, 25 Cal. 619, where the position of the initial point in the description was uncertain, a division of land between coterminous owners as in accordance with the deed description, and mutual acquiescence in their practical location of the common boundary over a long period of time, was held to constitute the location of the true boundary as called for by the

---

* Known means capable of being determined. That is certain which can be made certain. *City of Mankato v. Carlstrom*. A boundary is considered definite and certain when by survey it can be made certain from deed. *Williams v. Barnett*, 287 P.2d 789, Cal. App., 1955.

† This principle was also adhered to for different reason in the case of *City of Mankato v. Carlstrom*.

‡ *NEW HAMPSHIRE DEPARTMENT OF RESOURCES AND ECONOMIC DEVELOPMENT v. E. Milton DOW* and E. Milton Dow d/b/a Dow Sand and Gravel, 148 N.H. 60, 803 A.2d 581, N.H., 2002.

deed. The doctrine grew out of the need for stability and repose in the matter of titles to real property.

The agreement need not be express, but may be implied from long acquiescence. But since it is valid only for the purpose of settling an uncertainty in a common boundary, the implied agreement must have been based on a doubtful boundary line. A dispute or controversy is not essential, but it may be evidence of the existence of a doubt or uncertainty. Nor is it a requirement that the uncertainty should appear from the deed or from an attempt to make an accurate survey from the calls in the deed. The fact that an accurate survey is possible is not conclusive of the question whether a doubt existed as to the location of a common boundary. Thus the doubt may arise from a believed uncertainty which may be proved by direct evidence or inferred from the circumstances surrounding the parties at the time when the agreement is deemed to have been made; and if in good faith the parties resolve their doubt by the practical location of the common boundary it will be considered the boundary called for by the deed.

Obviously, the court has left open the reality of the situation, which is when the true boundary line is, or can be found, and differs from the agreed upon line, then one party gains and the other one loses, or a combination of both, depending on the course of the agreed upon line and its relation to the true line. This is exactly what the courts want to avoid, since the parcel between the two lines is identifiable, and in fact is the subject of an attempt of transferring title to it in contravention of the statute of frauds. Many courts have stated that such action causes the boundary line agreement to be void.

Too often parties will agree on a line for the sake of convenience, since they believe that pursuing the correct boundary would be time consuming, expensive, and, depending on who is doing the investigation, may ultimately fail to achieve the answer. There seems to have been an abundance of this type of activity in the recent past. Agreements have been suggested for the sake of convenience and haste to reach a conclusion. Like everything regarding titles and boundary location, boundary agreements are possible, but carry strict requirements for their application.

A requisite to make applicable the doctrine of agreed boundary line is that the line acquiesced in by adjoining property owners must be specified, definite, and certain. The agreement is not controlling if that line is left indefinite, uncertain, or speculative. (*Garrett v. Cook*, 89 Cal.App.2d 9.)

It is also not controlling if it is not acquiesced in by both (all) parties on both sides of the agreed-upon line. In one of the more important decisions, *Osteen v. Wynn*,* the court said "a line is not fixed or located by verbal agreement unless actual possession is had up to the line, or something be done to execute the agreement in the direction of physical identification, as the erection of monuments, fences, marking of trees, or the line." In the *Farr v.*

---

* 131 Ga. 209, 62 S.E. 37, Ga., 1908.

Woolfolk,* the Georgia court stated "It is indispensable that a parol agreement as to a dividing line be accompanied with possession, or that something else should be done by the parties to execute the agreement" (citing several cases from several jurisdictions).

The Oregon court in the case of *Powers Ranch Co., Inc. v, Plum Creek Marketing Inc.*[†] summarizes nicely the requirements that must satisfied in order to have a valid agreement:

> First, there must be an initial uncertainty or dispute as to the "true" location of the boundary. The stated purpose of this requirement is to prevent the agreement from falling within the Statute of Frauds or violating other real property conveyancing requirements, for it establishes that the parties are resolving a dispute by mutually fixing an unknown boundary rather than by making a conveyance of land.
>
> Second, the uncertainty must be resolved by an agreement, express or implied, to recognize a particular line as the boundary. The boundary recognized must be mutually intended as permanent, not as a tentative or temporary boundary or as a mere barrier. The parties must intend to resolve the uncertainty; an attempt to locate the 'true' line cannot change the boundary described in the deed.
>
> Finally, the parties must evidence their agreement by subsequent activities. If the agreement is memorialized in writing, it may be recorded in the chain of title to establish the recognized dividing line. If there is an express oral agreement, courts have required occupation to the boundary line in question.

These requirements vary from jurisdiction to jurisdiction, and somewhat according to different time frames. Any wishing to rely on an agreement of any sort would be well-advised to be well-versed with the applicable legal requirements.

Statute Law in New Hampshire

New Hampshire is one of the few states having a statute regarding boundary line agreements. Most other states have an abundance of court decisions (case law) which is relatively consistent, although a few courts deviate slightly for their own reasons. One would have to review the decisions on a state-by-state basis to get an appreciation for how a particular jurisdiction leans.

The New Hampshire statute[‡] reads as follows:

> Whenever the **boundary line** between the land or estates of adjoining owners is **in dispute**, and the location of the same as described

---

* 118 Ga. 277, 45 S.E. 230, Ga., 1903.
† A142396 (ORCA), 2011.
‡ Chapter 472, 1992.

in the deeds of said owners or of their predecessors in title **cannot be determined** by the monuments and boundaries named in **any** of said deeds, the parties may establish said **line** by **agreement** in the following manner, and not otherwise.

RSA 472:1. It further provides that the **agreement** shall be in writing, reciting that ... the division **line** between their lands is in dispute, that the **line** described in their respective deeds or in the deeds of any of their predecessors in title cannot be located on the ground by reason of the loss or obliteration of the monuments and boundaries therein named and described, and containing a full and complete description of the **line** thus agreed upon and established....

RSA 472:4. In addition, the statute provides that "[t]he **line** agreed upon shall be surveyed and established by courses and distances, and suitable and permanent monuments shall be placed at each end and at each angle of the **boundary** so agreed upon." **RSA 472:3.**

While there are several decisions in this jurisdiction dealing with boundary line agreements, many of which are consistent with most other jurisdictions, the following decision is the most recent: *NEW HAMPSHIRE DEPARTMENT OF RESOURCES AND ECONOMIC DEVELOPMENT v. E. Milton DOW and E. Milton Dow d/b/a Dow Sand and Gravel.** The court in this case stated, "Having considered RSA chapter 472 as a whole, we interpret it as conditioning the validity of a boundary line agreement upon satisfying all of the statutory formalities. To otherwise uphold a boundary line agreement as valid would eviscerate the statute, rendering its mandatory language meaningless. We will not construe the statute in such a manner."

The implications of that analysis may be far-reaching. This court flatly stated that all of the statutory formalities had to be satisfied. In most jurisdictions, the courts have stated that the first requirement to be satisfied in a consideration of a boundary line agreement is that the boundary line be in dispute, or otherwise unascertainable.

What happened in this case was that the agreed upon line was not monumented as required by the statute. Leaving an agreed-upon line unmarked leaves the line, as the California court noted, indefinite, uncertain, or speculative.†

The Meaning of Uncertainty

Nothwithstanding "that is certain that can be made certain," Dean and McEntyre‡ stated that "there is not a complete consensus among the various jurisdictions concerning the requirements that the boundary line in question

---

* 148 N.H. 60 (N.H. 2002); 803 A.2d 581.
† *Garrett v. Cook*, 89 Cal. App. 2d 98, 103.
‡ Establishment of Boundaries by Unwritten Methods and the Land Surveyor.

Boundary Agreements

be in dispute, uncertain, or unascertainable. Most courts agree that a boundary line with any of the following elements: (1) unmarked and unknown, (2) two different positions contended for by adjoiners, (3) one position contended for one adjoiner, but disapproved by the other, or (4) an ambiguous and irreconcilable description of the boundary line will be sufficient to indicate that the boundary is in dispute, uncertain, or unascertainable."

> Some courts require these terms to meet the objective test of their meaning. They require that the dispute must arise from an uncertainty or unascertainable line, that is, the description must be ambiguous or not capable of being located from the description by a surveyor. Other courts take a less stringent stand and require the dispute, uncertainty, or unascertainability to be in the minds of the adjoiners. This latter court position meets the subjective test of the meaning of the requirement.

Clearly, the following subdivision would not meet any set of requirements, and demands an agreement of the parties (Figure 9.5).*

Parcel A: Bounded on the south by the highway, bounded on the west by land of Smith, bounded on the north by land of Jones, and bounded on the east by land I sold this day to B. Parcel B: Bounded on the south by the highway, bounded on the east by land of Brown, bounded on the north by land of Jones, and bounded on the west by land I sold this day to A.

That is the sum total of the descriptions—no dimensions given, no acreage recited, and no evidence whatever on the ground of use or occupancy.

If surveyors and others had properly done their job, located the line and the corners, or as several courts have appropriately noted, the places where they stood, most boundary line agreements existing today would be unnecessary. As it is, they invoke a title problem, and a survey issue. As with apportionment, people sometimes are way too quick to resort to a rule that is not only

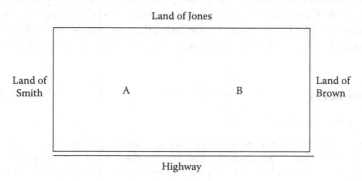

FIGURE 9.5
Subdivision of a tract of land with vague descriptions.

---

* The original title of the statute was *An Act for Prevention of Frauds and Perjuries*, which was passed in England in 1677.

inappropriate, but also in some cases unlawful (in violation of the statute of frauds) and causing, or at least contributing to, a title problem, perhaps even contributing to, or being, the underlying cause of the title problem. This often results in a cloud on the title of one of the parcels, in some cases, both ownerships.

In the area of unwritten rights, five doctrines come to the forefront: adverse possession, parol agreement, acquiescence, practical location, and estoppel. While to the uninitiated and the casual viewer, the results of each of them look the same, they are actually quite different from one another.

To begin with, adverse possession is an adverse relationship between two, or more, parties. Depending on the particular state, or jurisdiction, adverse possession is a creature of statute, which clearly bars the bringing of an action for recovery after certain specified requirements are met. It results in possession ripening into title through the passage of time, and creates a brand new title in the adverse possessor, which is not in privity in any way with any former title. The title is a new and independent title by operation of law and is not in privity in any way with any former title. Generally, it is as effective as a formal conveyance by deed or patent from the government or by deed from the original owner. In fact, it is a good, actual, absolute, complete, and perfect title in fee simple, carrying all of the remedies attached thereto.

> The title acquired will pass by deed. After the running of the statute, the adverse possessor has an indefeasible title which can only be divested by conveyance of the land to another, or by a subsequent ouster for the statutory limitation period.
>
> *3 Am. Jur. 2d., Adverse Possession, § 298*

Second, the other four doctrines, with a few exceptions, generally have little or nothing to do with title, or the passage thereof. They are separate doctrines which serve to fix the location of a boundary line which is in dispute, unknown, or unascertainable. Each results in the definition of the boundary location, and each carries its own, separate, set of requirements. Briefly, they are as follows.

Adverse Possession

> The actual, open, and notorious possession and enjoyment of real property, or of any estate lying in grant, continued for a certain length of time, held adversely and in denial and opposition to the title of another claimant, or under circumstances which indicate an assertion or color of right or title on the part of the person maintaining it, as against another person who is out of possession.
>
> *Black's Law Dictionary*

Boundary Agreements 273

Acquiescence

A silent appearance of consent; failure to make any objections.

Black's Law Dictionary

Establishment of Boundary by Acquiescence

Where adjoining owners occupy their land up to a certain line, which they mutually recognize and acquiesce in as the boundary line, for a long period of time, they and their successors in interest are precluded from claiming that the boundary line thus recognized and acquiesced in is not the true one, though such line may not be in fact the true line according to the calls of their deeds. Generally, the acquiescence must be continued for a period of time at least equal to that prescribed by the statute of limitations in cases of adverse possession.

Am. Jur. Proof of Facts, Boundaries

Estoppel

A preclusion, in law, which prevents a person from alleging or denying a fact, in consequence of his own previous act, allegation, or denial of the contrary.

Black's Law Dictionary

Estoppels are of three kinds:

1. By deed
2. By matter of record
3. By matter *in pais*

Estoppel *in pais* (or estoppel by conduct) is an equitable estoppel, and arises when one represents by word of mouth, conduct, or silent acquiescence that a certain state of facts exists, thus inducing another to act in reliance upon the supposed existence of such facts, so that, if the party making the representation were not estopped to deny its truth, the party relying thereon would be subjected to loss or injury.

It is the species of estoppel which equity puts upon a person who has made a false representation or a concealment of material facts, with knowledge of the facts, to a party ignorant of the truth of the matter, with the intention that the other party should act upon it, and with the result that such party is actually induced to act upon it, to his damage.

Black's Law Dictionary

Generally, the following must occur for there to be an equitable estoppel:

- An admission, statement, or act inconsistent with the claim afterward asserted
- Action by the other party on the faith of such admission, statement, or act
- Injury to such other party for allowing the first party to contradict or repudiate such admission, statement, or act

Parol Agreement

Oral, or verbal, agreement of the location of boundary line.

Effect of Statute of Frauds

The fixing of a boundary by parol agreement is generally not considered within the statute of frauds, and, provided the elements of uncertainty or dispute, agreement on a definite line, consideration, execution, occupancy, possession, or improvement, and acquiescence are present, the agreement will be binding on the parties. This is because no estate is created by the agreement, the parties holding title by virtue of their title deeds and not by virtue of parol transfer. The effect of such an agreement is not to pass title, but merely to ascertain the line to which the parties' respective lands extend.

Necessity for Uncertainty or Dispute as to Location of Boundary

As a requisite to the validity of a line established by parol agreement, it must be shown that the location of the true line is or has been uncertain or in dispute. This does not mean that there need be a dispute in the sense of a quarrel or ill feeling between the parties. A boundary line is in dispute where the

Boundary Agreements

respective owners are not in agreement as to its location; it is not necessary that they be in actual controversy in or out of court. The showing of a dispute or uncertainty as to the location of the true line is important for two reasons. First, if the line is known and undisputed, the agreement, rather than being a mere location of a boundary, becomes a parol transfer of property within the statute of frauds. Second, the agreement must be supported by a consideration, and this is generally found in the mutual surrender of any claims the respective parties might have to land beyond the agreed line.

Agreement upon a Definite Line

To be effective, the parol agreement must be upon a definite line, and must be unconditional. The actual location is the thing to which the parties must agree.

Am. Jur. Proof of Facts, Boundaries

Practical Location of Boundary

By practical location is meant an attempt by adjoining landowners to designate and mark on the ground the true dividing line between their properties. It differs from a location by parol agreement or acquiescence in that it is not the marking of an arbitrary line to settle an uncertain or disputed boundary, but rather a marking of that line which the parties understand to be the true one. When a practical location has been made it will be binding on the parties and those claiming under them even though the line so established may subsequently be found to vary from that called for in their title deeds, unless the making of the location was induced by fraud or mistake.

Am. Jur. Proof of Facts, Boundaries

It is often found in practice, that boundary line agreements have been executed, and subsequently relied on, inappropriately. Obviously, the surveyor involved did not fulfill his or her "sole duty, function, and power," but chose to get creative, or attempt to fix something they had no authority to get involved with. Surveyors should be concerned about failure to fulfill their duty, since it may result in negligence. Usually, the underlying causes of such an action result from:

Not understanding one's duty
Carelessness

Being in a hurry

Being controlled by a budget

Yielding to time constraints

If executed by nonsurveyors, or nonlawyers, likely the feeling was against spending any money, and merely coming to an agreement. As can readily be seen, the law does not support such careless activity.

Many of the above are because the field investigator(s) fail to dig sufficiently either at the suspected location of the corner, or at some other location where the corner actually is due to anomalies in the previous surveyor's procedures. A conscious surveyor understands that it is easy to inadvertently fall into someone else's trap.

With an agreed line, of whatever nature and however created, there may be two sets of "footsteps," the true line and the agreed-upon line. This leads to a consideration of at least two titles, one or more on each side of the line, and whether there is a violation of the requirements set forth in the statute of frauds.

In theory, there would be one line, the agreed upon line fixing that which is unknown, or unascertainable, that would be located where the line actually exists. However, in reality this is seldom, if ever, the case, and too often the surveyor discovers the original marks, and therefore locates the correct boundary, finding it in variance with the agreed-upon line. Long-term occupation and intent may serve to change the line of overall ownership, depending on the jurisdiction and the circumstances. The reported decisions are replete with examples from every jurisdiction, not all of them in agreement. Careful study is required when encountering such a situation, as to the rule(s) in the particular jurisdiction, and long-term consequences involved depending on what the law governs.

There are numerous examples whereby subsequent surveying of a parcel has uncovered the original (true) boundary, thereby placing the agreed boundary in violation of the legal requirements for such, and rendering the agreement void. Obviously, the more time that passes after such an agreement, and more conveyances based on the erroneous agreement there are, the more of a problem it becomes. Many of these lead to costly litigation. Many do not get addressed in an attempt to avoid such expense, and therefore become worse as land values increase and improvements on the lands come into existence, since such problems do not "go away" by themselves.

The following example is a trap for the unwary. Such a scenario is difficult, if not impossible, to predict and will result in extra responsibility on the part of the surveyor discovering the problem. It may also serve to add considerable expense to the project, and could easily necessitate a court proceeding to resolve the problem (Figure 9.6).*

---

* The original title of the statute was *An Act for Prevention of Frauds and Perjuries*, which was passed in England in 1677.

Boundary Agreements

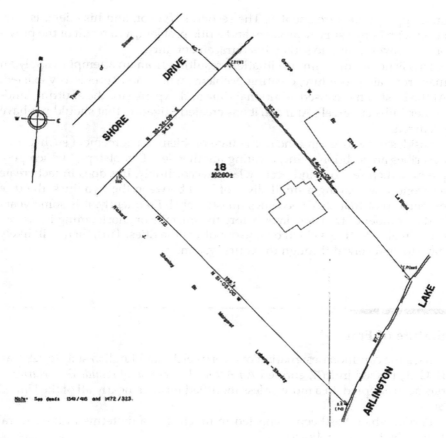

FIGURE 9.6
A parcel of land depicted on a recorded plat.

A surveyor, attempting to locate (survey) an abutting parcel, followed the appropriate state surveying standards and researched the abutting titles. In the process, he uncovered the above recorded plat. In his attempt to locate this parcel, he quickly uncovered a discrepancy, in that not only did he find the markers noted on this plat, but he also found and located the original markers set when the parcel was originally created. The two sets of markers did not agree with one another, and the period of time between the two events did not satisfy the requirement for the statute of limitations (in this case, 20 years). The surveyor and his clients executed (inappropriate) boundary agreements, which were also on record.

Because of the inadequacy of the previous retracement survey of this tract, an overlap resulted onto the parcel now being surveyed. Titles have been approved, mortgages transacted and recorded, and insurance policies written, all based on incorrect information resulting from a faulty survey based on inappropriate agreement(s). The second surveyor, and his client, is now faced with both a survey problem and a title problem, as a result of the previous surveyor failing to satisfy his retracement duty.

This is not an uncommon situation, resulting from an attempt to apply an inappropriate procedure, creating problems where none previously existed. At the least, it has caused additional time and expense for the abutting landowner (subject parcel). At most, it has created litigation that should not have occurred.

Besides not addressing the immediate problem, an agreement such as this describes an additional line, creating another set of "footsteps," which people may then rely on, and occupy lands accordingly. If it does in fact create a second line, leaving a small sliver of land between the two lines, the true ownership of that sliver becomes questionable. Frequently, it is some years and a number of transfers later before the mistake or shortcoming is discovered, likely creating a cloud on one or both of the titles. This, then, will likely have to be resolved through the court system.

Statute of Frauds*

This is the common designation of a very celebrated English statute, (29 Car. II. C. 3,) passed in 1677, entitled *An Act for Prevention of Frauds & Perjuries*. It has been adopted, in a more or less modified form, in nearly all of the United States.

Part 4, which has been reenacted in much the same terms in the several United States, is as follows:

> "No action shall be brought (1) whereby the charge any executor or administrator upon any special promise to answer damages out of his own estate; or (2) whereby to charge the defendant upon any special promise to answer for the debt, default, or miscarriages of another person; or (3) to charge any person upon any agreement made upon consideration of marriage; or (4) upon any contract or sale of land, tenements, or hereditaments, or any interest in or concerning them; or (5) upon any agreement that is not to be performed within the space of one year from the making thereof; unless the agreement upon which such action shall be brought, or some memorandum or note thereof, shall be in writing and signed by the party to be charged therewith, or by same person thereunto by him lawfully authorized."

Number 4 is of particular interest here as it relates to *"any* contract or sale of *land, tenements, or hereditaments,* or *any interest in* or *concerning them."* This

---

* *Black's Law Dictionary.*

Boundary Agreements

279

is why unwritten rights may be honored without a writing, since (1) adverse possession (ripening of possession into title through passage of time) is provided for with its own statute, one of limitations governing actions for recovery; and (2) forms of agreement (acquiescence, estoppel, parol agreement, and practical location) do not serve to transfer title to anything, but are *actions by the appropriate parties to fix the location of a common boundary*, under the right circumstances. Otherwise, such activity would result in an attempt to transfer title without a writing, would be contrary to the statute of frauds, and likely void. If the appropriate requisites are not in place, any transaction or reliance, it would seem, would likely be deemed void.

However, a few agreements may be enforceable though not in writing. For example, easements by implication: easements, which are agreements that permit the use of real estate by someone who has no property interest in the land, may be created by operation of law rather than by written instrument. This may happen where, for example, a piece of land is partitioned between owners and pre-existing utilities routes or access paths that would otherwise be trespassory over one of the plots is reasonably necessary for enjoyment of the other plot. In such case, the pre-existing use must be apparent and continuous at the time of the partition for an easement to be created by implication. The implied easement constitutes an interest in land that does not require a writing to be enforceable.

US law has adopted the 1677 English law known as the statute of frauds. Every state has some type of statute of frauds; the law's purpose is to prevent the possibility of a nonexistent agreement between two parties being "proved" by perjury or fraud. The statute of frauds in various states come in two types:

1. Those that follow the English statute and provide that "no action shall be brought" on the contract or the contract "shall not be enforced"
2. Those that are declared "void"

The Montana case of *Myrick v. Peet*,* referenced earlier, had this to say:

> The burden of proof is always upon the party attempting to show the existence of an agreement fixing the location of a boundary line, and that the boundary so fixed had been accepted and acquiesced in. 4 R. C. L., title, Boundaries, § 66; *Jones v. Pashby*, 67 Mich. 459, 35 N.W. 152, 11 Am. St. Rep. 589. The boundary must be one between contiguous lots, and must be doubtful and uncertain. 4 R. C. L. § 67, supra; *Randleman v. Taylor*, 94 Ark. 511, 127 S.W. 723, 140 Am. St. Rep. 141, note; *Turner v. Baker*, 64 Mo. 218, 27 Am. Rep. 226; *Galbraith v. Lunsford*, 87 Tenn. 89, 9 S.W. 365, 1 L. R. A. 522. In the absence of a real dispute, an agreement purporting

---

* 56 Mont. 13, 180 P. 574, 1919.

280 *Boundary Retracement*

to establish the boundary between the lands of adjacent proprietors, at a line known by both to be incorrect, and the result of which, if it be given effect must be to transfer to the one lands which both know do not belong to him, is without consideration and within the statute of frauds, and consequently void. 4 R. C. L., title, Boundaries, § 67; *Lewis v. Ogram*, 149 Cal. 505, 87 P. 60, 10 L. R. A. (N. S.) 610 note, 117 Am. St. Rep. 151.

It is also well settled that where two adjoining proprietors are divided by a fence which they suppose to be the true line, they are not bound by the supposed line, but must conform to the true line when ascertained. *Jacobs v. Moseley*, 91 Mo. 462, 4 S.W. 135; *Schraeder Min. Co. v. Packer*, 129 U.S. 688, 9 S.Ct. 385, 32 L.Ed. 760; 4 Am. & Eng. Ency. Of Law, 866, and cases cited; *Kimms v. Libby*, 87 Neb. 113, 126 N.W. 869; *Foard v. McAnnelly*, 215 Mo. 371, 114 S.W. 990; *Voigt v. Hunt (Tex. Civ. App.)* 167 S.W. 745; *Jahnke v. McMahon*, 21 Cal.App. 781, 133 P. 21; *Lind v. Hustad*, 147 Wis. 56, 132 N.W. 753.

A similar situation was addressed in *Pilgrim v. Kuipers*, 209 Mont. 177, 679 P.2d 787 (1984), where the claim was that "the old 'fox farm' fence" was a monument that established the boundary. The court disagreed, explaining:

There is a critical distinction between a fence which establishes a boundary line, and a fence that merely separates one side of the fence from the other. The former is a monument as well as a fence, while the latter is merely a fence. Unlike the highway right-of-way and the Beaverhead River, there are no calls in the legal description to the "fox farm" fence. There is no evidence that the fence line was surveyed or that the fence was built to conform to a surveyed line. One witness testified that the fence was built zig-zag apparently around trees and without any pattern at all. Another said it "jogged" by as much as 20 feet. In contrast, the legal description calls for a straight line. There simply is no evidence to support the fence as a monument. Nor does a fence establish a boundary line when it does not conform to the true line, even though the property owners thought it was the boundary. Where two adjoining properties are divided by a fence, which both owners suppose to be on the line, such fence is a division fence, as between them, until the true line is ascertained, when they must conform to the true line. *Pilgrim*, 209 Mont. at 181-82, 679 P.2d at 790 (internal quotation marks omitted).

Not Meeting the Basic Requirement

In *Watrous v. Morrison*, 33 Fla. 261, 14 So. 805 (Fla. 1894): where the owners are not uncertain as to the true boundary, the statute of frauds applies, and the doctrine announced above is inapplicable. *Nichol v. Lytle's Lessee*, 4 Yerg. 456; *Jackson v. Douglas*, 8 Johns. 286; *Vosburgh v. Teator*, supra; *Terry v. Chandler*, 16 N.Y. 354. Still it seems, though we do not say it is applicable here, that

Boundary Agreements

the acquiescence in an actual location of a line may be of such a nature and of such continuation as to be evidence of an express agreement. *Rockwell v. Adams*, 7 Cow. 761; *Kip v. Norton*, 12 Wend. 127; *Jordan v. Deaton*, supra; *Jackson v. McConnell*, 19 Wend. 175.

Another principle coming within the discussion of this case is that, in cases of mistake as to the true line between adjoining lands, the real test as to whether or not a title will be acquired by a holding for the period of 7 years (Florida's statute of limitations) is the intention of the person holding beyond the true line. If such occupation is by mere mistake, and with no intention upon the part of the occupant to claim as his own land which does not really belong to him, but he intends to claim only to the true line, wherever it may be, the holding is not adverse. If, however, the occupant takes possession, believing the land to be his own up to the mistaken line, and claiming title to it, and so holds, the holding is adverse. The intent to claim title up to the line is an indispensable element of adverse holding. The claim of right must be as broad as the possession. Simple acquiescence, or lying by, without objection, for the statutory period, in case of such adverse holding, will bind the party so lying by to the line, though not the true line. *Liddon v. Hodnett*, 22 Fla. 442.

In *Glenn v. Whitney*, 116 Utah 267, 209 P.2d 257 (Utah 1949):

> The cases and text writers in stating the general rule announce the principle that the question as to whether an established fence line has become the true boundary line separating two adjoining tracts of land is one of fact and the court must evaluate the facts in each case. Before doing so, we find it necessary to define the meaning of certain terms in view of the fact that there seems to be some confusion in the minds of the litigants as to what elements are necessary to establish a boundary line in a suit of this character. If it was not clear before the case of *Tripp v. Bagley*, 74 Utah 57, 276 P. 912, 69 A. L. R. 1417, it was expressly recognized there and in all Utah cases in point handed down subsequent to it, see *Home Owners' Loan Corp. v. Dudley*, 105 Utah 208, 141 P. 2d 160; and *Smith v. Nelson*, 114 Utah 51, 197 P. 2d 132, that there must be some uncertainty or a dispute between adjoining owners as to the location of the true boundary line before a fence which they subsequently erect to resolve their differences and in which they acquiesce for a long period of time, may be taken as the agreed boundary line. Using the terms "uncertainty" and "dispute" loosely, we might say that the parties here were uncertain as to the location of the boundary line inasmuch as neither of them had attempted to locate it prior to the survey made by the plaintiff. This, however, is not "uncertainty" as this term was meant to be used in this connection for as is said in Thompson on Real Property, Section 3309:
>
> > If an owner ignorant of his true boundaries by mistake acquiesces in a line as a boundary, he and his grantees are not thereby precluded from afterwards claiming to the true line, and it has been held that one who has no knowledge that the adjoining owner has encroached upon his land cannot be held to have lost his rights by acquiescence in such occupancy no matter how long continued, for one cannot waive or acquiesce

282 *Boundary Retracement*

in a wrong while ignorant that it has been committed, especially where each party has equal means of ascertaining the correct line.

Thus, lack of knowledge as to the location of the true boundary is not synonymous with uncertainty. This being true, it cannot be said that the parties here were uncertain as to the location of the true boundary line, for there is nothing in the record before us to indicate that either of them had any idea as to the true location of the boundary line apart from an assumption that some existing fences separating the lands of other owners in the area might mark the section lines.

Furthermore, the fence was not erected to settle any uncertainty or dispute between the litigants or their predecessors in interest for according to the undisputed testimony of Mr. Bishop, he erected the fence merely to prevent the escape of his livestock to the east, and he did not attempt to erect a boundary line between the properties now involved or to settle any doubt or uncertainty as to the location of the true boundary line. According to defendant and his father, from whom defendant deraigns his title, they had merely assumed that the fence which existed at the time defendant's father purchased the property, was on the boundary line. The theory under which a boundary line is established by long acquiescence along an existing fence line is founded on the doctrine that the parties erect the fence to settle some doubt or uncertainty which they may have as to the location of the true boundary, and they compromise their differences by agreeing to accept the fence line as the limiting line of their respective lands. The mere fact that a fence happens to be put up and neither party does anything about it for a long period of time will not establish it as the true boundary. *Peterson v. Johnson*, 84 Utah 89, 34 P. 2d 697; *Tripp v. Bagley*, supra. We conclude that the defendants failed to establish title to the strip of land in question on the theory that the fence line was the true boundary line by erection of the fence and long acquiescence of the parties in its location.

Consentable Lines

In Pennsylvania, boundary lines by agreement are termed "consentable lines" and much has been written about the term, in both professional articles and in treatises.[*]

In the case of *Miles v. The Pennsylvania Coal Co.*,[†] the court stated, "it must be conceded on all sides, indeed it is conceded, that if a consentable line marked on the ground was recognized and acquiesced in by the adjoining landowners since 1868, a permanent boundary was established, and with that boundary line thus fixed there is no basis for the contention made by appellants

---

[*] See Elsesser, George M., Jr., *Consentable Lines in Pennsylvania*, 54 *Dickinson Law Review* 96, 1949, and Kline, Kristopher, *Unmistakable Marks*, POB, Sept., 2012.

[†] 245 Pa. 94, 91 A. 211, Pa., 1914.

Boundary Agreements

here. In the light of our own cases, no one can seriously question this settled rule of law. As far back as *Brown v. McKinney*, 9 Watts 565–567, this court said: 'It cannot be disputed that an occupation up to a fence on each side by a party or two parties for more than twenty-one years, each party claiming the land on his side as his own, gives to each an incontestable right up to the fence, and equally whether the fence is precisely on the right line or not.'"

> Our courts have always favored the settlement of disputes of this character by recognizing consentable lines established by the parties themselves, and this without regard to whether the line agreed upon conforms to the exact courses, distances and bounds of the original surveys. This rule applies with convincing force to the facts of the case at bar. In the original surveys there was a large amount of surplus land, most of which was included within the boundaries of the tract held by appellants, when the consentable line was marked on the ground. They are not in position to say that they were overreached in the division of the surplus land when the line between the two tracts were marked on the ground in 1868. So far as the division of the surplus land between the two adjoining tracts is concerned, the equities are with the appellee, but aside from this, and without reference to the division of the surplus land, the fact is that a line was marked on the ground at that time, and the evidence shows that it has been maintained, to some extent at least, and acquiesced in by the interested parties from that time to the present. Appellants undertook to discredit the line established in 1868 and to show that it had not been marked and recognized in such a way as to make it binding on the parties. The convincing weight of the evidence is against the contention of appellants, but the most favorable view that could be taken of the facts as disclosed by the record is that the marking of the line in 1868, the maintenance of fences, and the acquiescence of the adjoining landowners in the line thus established, were questions for the jury and they were so submitted. In our opinion the evidence is absolutely convincing that this consentable line was established in 1868 and that it has been recognized and acquiesced in by all interested parties until within a very recent period, when appellants undertook to disregard it.
>
> Similarly, in Georgia, the case of *Osteen v. Wynn** discusses the consentable line in this manner. "Where there is room for controversy as to the location of a dividing line, the coterminous proprietors, independently of the cited Code section, may orally agree upon the line, and if the agreement is accompanied by possession to the agreed line, or is otherwise duly executed, such agreement will be valid and binding, and the line thus defined will thereafter control their deeds. However, it is not necessary that possession under the agreed line should be had for 20 years, to give validity to the agreement, though the agreement derives additional weight from long acquiescence. A parol agreement between adjoining landowners to fix a boundary line between their respective

---

* 131 Ga. 209, 62 S.E. 37, Ga., 1908.

tracts theretofore unascertained, uncertain, or disputed, is not within the operation of the statute of frauds, for the reason that no estate is created. When a boundary line is established by consent, the coterminous proprietors hold up to it by virtue of their title deeds, and not by virtue of a parol transfer of title. *Hagey v. Detweiler*, 35 Pa. 409. A line is not fixed or located by verbal agreement unless actual possession is had up to the line, or something be done to execute the agreement in the direction of physical identification, as the erection of monuments, fences, marking of trees, or the like. But when adjacent landowners make a consentible line between their respective tracts, it is not necessary that possession be had to the line for 20 years in order to establish the agreed line as the divisional line. In some jurisdictions, a parol agreement between adjoining proprietors as to a dividing line is considered within the statute of frauds, and will not be enforced unless acted upon up to such an extent as to make it inequitable for either party to set up the true boundary. *Meyers v. Johnson*, 15 Ind. 261. But in this state, on the authority of the cases cited in *Farr v. Woolfolk*, supra, and upon the strength of their reasoning, we have accepted the doctrine that, in the case of disputed or uncertain boundary between coterminous landowners, an executed parol agreement is not within the statute of frauds, and will suffice to establish a dividing line.

Conditional Lines

Lines termed "conditional lines" are commonly found in Kentucky, but may be found elsewhere. The following are a few examples.

The Kentucky decision of *Hoskins Heirs v. Boggs** discusses the courts' views on this topic. In this dispute, the patent that included the disputed tract "contained thirteen calls to, and from, the beginning point, 'two dogwoods, [a] black oak and hickory at the head of the right hand fork of [Lewis Creek].' Of these thirteen calls, *only one*, the first call from the beginning point 'to two black oaks and [a] maple' was actually surveyed on the ground. The other twelve calls were projected to 'stakes.' This type of patenting technique used in Kentucky's earlier years is referred to in common parlance as a 'stake patent.'"

> "Stake patents', although valid in Kentucky, have nevertheless engendered a multitude of litigation, both as to the location of their boundaries, as well as, overlaps with other adjacent patents.
>
> It will be observed that the two first corners of the tract are corners of Smith's survey, and that all the other corners of the patent are located at stakes. The fact that no timber is called for [, only stakes,] would seem to

---

* 242 S.W.3d 320, 2007.

Boundary Agreements

indicate that this patent was, perhaps, laid out by protraction, and that the surveyor did not in fact run the lines.

"*Creech v. Johnson*, 116 Ky. 441, 76 S.W. 185, 187 (1903). Yet, as was noted in *Uhl v. Reynolds*, 23 Ky.L.Rptr. 759, 64 S.W. 498 (1901):

> [A] patent could not be questioned collaterally by anything dehors the patent, in the absence of a statutory provision authorizing it. It was therefore incompetent to show by parol testimony in a collateral proceeding like this that no survey of the land included in the patent was actually made by the surveyor. Therefore the only question which remains to be considered is whether the exterior boundary of the patent relied on by appellant... can be definitely located and determined.
>
> 'Stake patents', being as imprecise as they were, led to the use of "conditional line agreements" between adjoining neighbors, so as to avoid the inconvenience and expense of boundary line litigation. "A conditional line in eastern Kentucky is a line made by agreement of [the] parties, generally without the aid of a surveyor." *Martin v. Hall* 30 Ky.L.Rptr. 1110, 100 S.W. 343, (1907). Problematically, "conditional line agreements" were often unrecorded, yet marked and known on the ground by their creators—and *just sometimes*, their heirs.
>
> Indeed, one might say it was tough creating land titles in the mountainous ranges of Eastern Kentucky at the time; yet, it was a beginning. It was also a time before copy machines, thus, deeds recorded in the various county court clerks' offices were handwritten into the clerk's deed books, copied from the tendered original by the clerk or deputy clerk, when time was available. We note this point in explanation of why the wording of the old documents, as well as the signatures, acknowledgments, and certificates of the time appear to be in the same handwriting. Generally, they are.

In the case of *Brown v. Lyons*,* the Kentucky court was involved not only with a conditional line, but also a conditional corner: "In 1870, John Tuggle was the owner of a tract of land situated on Beaver creek in Wayne county, Ky. containing about 400 acres. He sold the land to his son, Washington Tuggle, and his son-in-law, Franklin Hammon, who proceeded to divide the land between themselves by establishing a *conditional or division line* which they designated by certain objects, courses, and distances, etc. Pursuant to that division, John Tuggle executed separate deeds to them according to the division and lines agreed on."

"In 1933, a dispute arose between appellant and appellee respecting one call or line in the deed, beginning on 'Privitt Rock' and running a southerly direction and intersecting with the exterior boundary line of the Tuggle land, designated in the deed as the 'upper line.' The location of this upper line is not in dispute between the parties. It is the southwestern boundary

---

* 257 Ky. 1, 77 S.W.2d 438, Ky. App., 1934.

286   *Boundary Retracement*

line of the John Tuggle land, the eastern end of which line is designated by three pines on the top of the ridge and, running a northwesterly direction to a cliff, a distance of 160 poles. According to the division of the land agreed on by Washington Tuggle and Franklin Hammon, and, as deeded to them by John Tuggle, each tract was bound on the south or southwest by the upper line. The Washington Tuggle deed under which appellant claims runs from the Privitt Rock a southwesterly direction to a cliff, the western end of the upper line; thence S. 60 E. 50 poles (on the upper line) to a stone, a *conditional corner* between the Hammon and Tuggle line and with Hammon's line N. 5 W. 100 poles back to the Privitt Rock; but the corrected survey, as shown by the map, shows that the line is on a degree of N. 10 W. and 143 poles, instead of N. 5 W. 100 poles, as called for in the deed."

The following description was the subject of the case of *Wells v. New York Mining And Manufacturing Co.*[*] "By the title bond Jesse Baker sold to D. M. Vance 'a tract of land lying and being in the county of Wise, on the waters of the Stone Gap fork of Powell's river on both sides of a small branch that runs through said Baker's farm. Beginning with a conditional line made and agreed on between the said Baker and John B. Cooper, running with a ridge to a chestnut and chestnut oak, thence northwardly to a large chestnut near the top of the *main ridge* (evidently Rogers' ridge); thence running westwardly, *crossing a branch to a gap in the ridge to a chestnut at or near a line of A. & D. Vance's; thence with the top of a ridge southwardly to a conditional line marked and agreed on*; thence eastwardly with the said conditional line to the beginning.'"

Recognition and Acquiescence

Discussion that summarizes the courts' thinking is presented in the Wyoming case of *Carstensen v. Brown*.[†] The doctrine of recognition and acquiescence of a boundary line is upheld by many authorities. 9 C. J. 244; Tiffany, Real Property (2nd Ed.) sec. 295; Thompson on Real Property, sec. 3112. It is sometimes referred to as acquiescence in, or as a practical location of, or as an implied agreement as to, a boundary. Considering all of the various jurisdictions in the United States, the doctrine is still in a chaotic condition, and no one has yet undertaken to point out definitely the circumstances under which it is applicable. Some of the authorities consider long acquiescence only as evidence of a boundary, which may be contradicted. Tiffany, supra, sec. 295. Other authorities say that an agreement may be inferred or presumed from such acquiescence.

---

[*] 137 Va. 460, 119 S.E. 127, Va., 1923.
[†] 32 Wyo. 491, 236 P. 517, Wyo., 1925.

10

Proper Procedure

In order to locate a tract of land, it is not necessary that the surveyor should begin his survey at the beginning corner. He may begin at any point which can be satisfactorily established; and, when one point has been settled upon, he may fix the other if he can.

Dugger v. McKesson,
100 N.C. 1, 6 S.E. 746 (N.C. 1888).

Procedure

What Makes an Effective and Successful Retracement Surveyor

First and foremost, a successful retracement surveyor is a person who understands the title as well as possible. Many people look for, or concentrate on, a deed. A title consists of a series of transfers between entities from the creating source to present day. Any document in the chain of ownership may have an impact on the property in question. See Chapter 2 for the various elements and sources of title with details on interpretation.

Second, the successful retracement artist considers any retracement an investigation, adhering to the rules and protocols of investigate procedure. Evidence is collected, documented, and preserved, the same as for a chain of custody in a criminal investigation.

Third, the investigator must be organized, keeping a careful record of procedures and findings, collecting appropriate copies of documents, taking statements, and photographing evidence.

The Original Creation and the Original Surveyor

One of the most effective ways to be successful in the recovery of original evidence and original corners is to understand as nearly as possibly the procedures of the original surveyor, and follow them as closely as possible. Doing so requires an understanding of not only the individual, but also the times in which he worked. If there were specific rules and directions he was

288 *Boundary Retracement*

to follow, consider the same, whether in the PLSS, or in the colonial states, contemporaneous with, or preceding, the PLSS.

Consider the equipment with which the original surveyor worked and duplicate his procedures. If difficulty arises with the current survey and the property it is associated with, retrace the same surveyor at a different location. Read his journals, any available field notes, anything else he may have written, or been involved with. Consider his education, training, background, and experience. And attempt to determine the textbook he used for guidance. Sometimes a review of his probate records will produce an inventory listing his equipment, maps, and books, all providing insight as to how he did his work.

Generally, the older the original work, the more difficult it may be to follow. Landscapes have changed, evidence has been altered or has deteriorated, documents fade or are difficult to decipher, and memories not only fade, but disappear. However, as the court stated in *Turnbow v. Bland,** "titles to land are not to fail merely because old markers may have disappeared, or because it may be difficult to trace the footsteps of the surveyor." Again, it is about the sanctity of titles (*Cragin v. Powell†*) and the stability of boundary lines. (*Froscher v. Fuchs*; 12 Am. Jur. 2d., Boundaries, § 61, Resurveys).

In *Cragin,* the court emphasized not interfering with property rights:

> ... great confusion and litigation would ensue if the judicial tribunals, state and federal, were permitted to interfere and overthrow the public surveys on no other ground than an opinion that they could have the working the field better done and divisions more equitably made than the department of public lands could do.
>
> *Justice Catron in Haydel v. Dufresne,*
> 17 How. 30, cited in Cragin v. Powell.

In cases deciding the boundary between two parcels of land, the law is settled that it is the duty of the surveyors to follow the original survey lines under which the property and neighboring properties are held notwithstanding inaccuracies or mistakes in the original survey. The purpose of this rule of law is that stability of boundary lines is more important than minor inaccuracies or mistakes. *Froscher v. Fuchs*, 130 S.2d 300 (Fla., 1961).

Boundaries once fixed (established) remain as originally located, forever. While property ownerships and resulting boundaries change, individual parcel boundaries do not change, and remain fixed. If it were not that way, boundaries would be in a constant state of flux, subject to ill-equipped locators and meddlesome troublemakers. As Justice Cooley noted in his two famous cases, which has been quoted many times since, if the law sanctioned such a course of action, there would be continual "consternation in the neighborhood."

---

* 149 S.W.2d 604, Tex.Civ.App., 1941.
† 128 U.S. 691, La., 1888.

Proper Procedure 289

Some of today's people, including some surveyors, believe that with modern sophisticated equipment, a better job can be produced. Worse yet, they can overcome legitimate inherent errors of the past. Without the original points, any measurements subsequent to the original are meaningless. As written in *Boundaries and Landmarks*,* "it is far better to have a somewhat faulty measurement where the line truly exists, than to have an extremely precise measurement where the line does not exist at all."

In a modern view, the court stated in the case of *Ivalis v. Curtis v. Harding*,[†] previously referenced, "The line determined by [the current surveyor] represents a sort of technical arrogance: shot in a matter of hours by sophisticated EDM laser transits accurate to a fraction of a second of arc, it is undoubtedly accurate but stands in stark contrast to the arduous work performed by the original government surveyors, and the early county surveyors after them who hacked their way through a section confined to a line by compass and dragging a survey chain behind them."

Since the original survey, or the "original deed writers" created the original title, it is the responsibility of subsequent surveyors and researchers to honor that creation. Again, several courts have pointed that out, in that the underlying factors are the preservation of the sanctity of titles and the stability of boundary lines.

In General

Proper procedure generally may be taken from the case of *Newfound Management v. Sewer*.[‡] "After a surveyor has completed a comprehensive review of all available records, deeds and prior surveys, the surveyor begins the field survey. Once in the field, the surveyor has a duty to make a diligent search for all monuments referenced directly or indirectly in the deed or property description that either occur naturally or were put in place by prior surveyors or other persons."

While seemingly, at first glance, this decision details excessive requirements, its guidance is helpful discussion for the retracement surveyor. First, a "comprehensive review of all available records, deeds and prior surveys" is the first order of business. Without such a study, one cannot know a number of things which are likely indispensable in a retracement survey. As previously discussed, the beginning definition of the title along with its establishment of boundaries is the subject of what the retracement surveyor is supposed to be following, and honoring above all. Changes that have taken place over time such as acquisitions, outsales, encumbrances, and the like may serve to change the boundaries of the overall ownership, but may also insert errors or otherwise be misleading. Subsequent surveys may or may

---

* A.C. Mulford, Van Nostrand, 1912.
† 173 Wis.2d 751, 496 N.W.2d 690, Wis.App., 1993.
‡ 885 F.Supp. 727, U.S. Dist., 1995.

not have been done correctly and may lead to confusion, improper locations, conflicts, and title concerns, depending on how accurately and thoroughly the work was accomplished.

Next, the retracement surveyor "has a duty to make a diligent search for all monuments referenced directly or indirectly." Without a comprehensive search of the property history, this directive cannot be met. Since monuments are items of the highest dignity, and control other elements, it is necessary to insure they have been identified and given proper consideration.

The word "duty" should not be taken lightly. In a question of negligence on the part of any professional, the first item to be determined is whether there is a duty owed, and second, was that duty breached. This court, stating that the surveyor has a *duty* to undertake certain obligations, should be considered carefully when conducting a retracement. There is usually an abundance of evidence, documentary, and physical, sometimes oral information that may need to be considered. Knowing a standard at the outset can aid in assisting the surveyor in properly fulfilling one's duty.

Again, the court at the federal level has stated concisely what the duty of the surveyor is to be, the other decision being *Rivers v. Lozeau,* wherein the court stated the retracement surveyor's "*sole duty, function and power* is to locate on the ground the boundaries corners and boundary line or lines established by the original survey; he cannot establish a new corner or new line terminal point, nor may he correct errors of the original surveyor. He must only track the footsteps of the original surveyor. The following surveyor, rather than being the creator of the boundary line, is only its discoverer and is only that when he correctly locates it" (citing Clark on Surveying and Boundaries, Tracking a Survey).

Surveyors may be on notice of certain things: events and occurrences, documents, previous owners, physical evidence, previous surveys, even claims, and the like. The three types of notice, actual notice, constructive notice, and inquiry notice have been discussed in detail in Chapter 2.

The recent Montana case of *Larsen v. Richardson* has also been discussed, but in a different context, that of what are the footsteps when there is no original survey? Another part of the decision in this case is a statement by the court as to proper procedure by the retracement surveyor:

> Both [surveyors] began their respective analyses by, first, obtaining the deeds in the chain of title and the chains of title of the adjacent landowners. This is standard practice in conducting a retracement survey.

Known and fixed monuments will control though they conflict with courses and distances, even where there are two fixed artificial monuments in conflict with each other and both entitled to the same credit and value as a monument. In such case of ambiguity, the one should be taken which corresponds with the descriptions and distances. *Ziebold v. Foster,* 118 Mo. 355. The primary object to be attained in the establishment of lost corners is to

Proper Procedure 291

put them, if practicable, in the exact spot they were put by the government, and this object can best be accomplished when all traces of the corners are gone and there is no fixed monument called for, by observing the courses and distances called for in the field notes of the original government survey. *Major v. Watson*, 73 Mo. 66. So long as the monuments placed upon the earth's surface by the United States surveyors can be identified, there are no lost corners, and they control, no matter what more recent surveys, by courses and distances disclose. *Jacobs v. Moesley*, 91 Mo. 464; *Knight v. Elliott*, 57 Mo. 317.

Original corners as established by the government surveyors, if they can be found, or the places where they were originally established, if they can be definitely determined, are conclusive, without regard to whether they were located correctly or not. 5 Cyc. 873. The general duty of a surveyor in instances like this is plain enough; he or she is not to assume that a monument is lost until after he has thoroughly sifted the evidence and found himself unable to trace it, and exhausted every recourse to which surveyors have access, as the original government surveys must govern. *Stewart v. Carlton* 31 Mich. 270; *Diehl v. Zanger*, 39 Mich. 601; *Dupont v. Sparring*, 42 Mich. 492.

In the case of *Williams v. Tschantz*, 88 Iowa 126, 55 N.W. 202 (Iowa, 1893), the question to be determined was not, where the government corner ought to have been located, but where was it in fact located? "Once found, or the place of its location identified, it must control, regardless of the fact that the actual location of the corner may result in deflecting the section line from a straight course between government corners located east and west of said supposed lost corner. This proceeding is not instituted for the purpose of straightening lines, thereby removing unsightly crooks in roads, no matter how desirable such a result might be, but it is to ascertain the location in fact of the government corner."

In *Davis v. Curtis*, 68 Iowa 66, 25 N.W. 932, the proceeding may be had as to any lost, destroyed, or disputed corner and boundary, whether established by government or other legal survey. It often is a matter of importance to adjoining owners to know where the true corners and boundaries of their lands are, as their rights depend on them.

The Kentucky decision of *Fordson Coal Co. v. Napier** nicely summarized the role of the surveyor. "It is a well-recognized rule that where natural objects are called for in a survey and they cannot be reached when the lines are run according to the courses and distances called for, the latter must give way to the natural objects." *Fore v. Gilliam*, 256 Ky. 591, 76 S.W.2d 893; *Asher v. Fordson Coal Co.*, 249 Ky. 496, 61 S.W.2d 20; *Fordson Coal Co. v. Osborn*, 245 Ky. 539, 53 S.W.2d 937; *Scott v. Thacker Coal Mining Co.*, 191 Ky. 782, 231 S.W. 498; *Williams v. Brush Creek Coal Co.*, 149 Ky. 188, 148 S.W. 372.

> It is a rule equally well established that mistakes in the calls of a patent may be corrected by referring to the original plat, and, in extending

---

* 261 Ky. 776, 88 S.W.2d 985, Ky.App., 1935.

boundaries, attention should be given to the figure of the survey in the absence of any other controlling influence." *Combs v. Jones,* 244 Ky. 512, 51 S.W.2d 672. And, where it appears, as in the present case, that the surveyor did not actually run the survey out on the ground, but located the lines by protraction, the plat made by the surveyor will afford strong evidence of the shape and size of the survey. *Swift Coal & Timber Co. v. Sturgill,* 188 Ky. 694, 223 S.W. 1090; *Bryant v. Strunk,* 151 Ky. 97, 151 S.W. 381, 383; *Strunk v. Geary,* 217 Ky. 113, 288 S.W. 1053. In *Bryant v. Strunk,* it was said: It is true the rule is that the calls of a patent for course and distance must give way to known or established objects found on the ground. But, after all, the rules that have been laid down on this subject are for the purpose of establishing the actual location of the lines and corners of the original survey, and they have little application where the lines were not run out in the original survey, but were simply laid down by the surveyor by protraction as was evidently the case in the patent before us. When the lines were not in fact run, we have little to guide us except the calls of the patent and the plot of the land accompanying the original survey. The plot accompanying the original survey is potent evidence in the determination of the general shape of the tract of land intended to be patented.

In the case of *J.R. Buckwalter Lumber Co. v. Wright,** the court summarized in its decision the role of the surveyor and the proper procedure. "As [this] case must go back for a new trial, we direct attention to the insufficient proof of a correct survey by the county surveyor. It may be that without this survey there was enough testimony to go to the jury, but when a surveyor is called upon to establish a land line between coterminous owners, the surveyor should begin at an established government corner called for by the field notes and find the evidences or some of them sufficient to show that it was a correct corner, and he should then reestablish the line and should find that at the other end of the line is a recognized government corner, or one re-established by another surveyor, or one that is so well established and recognized by all persons in the community that there can be no doubt of its correctness. The original survey, whether correctly made or not, is the true boundary between sections, and the surveyor must locate the original lines as run by the original surveyor. In case he misses the recognized corner, he should not divide the difference between the owners, because this does not reestablish the true boundary line. It may have the effect of giving each landowners the number of acres to which he is entitled, but it does not necessarily do so and it does not establish the true line. He should readjust his instruments, take his bearing from the true corner, and continue his survey until he correctly traces the boundary between the two landowners."

The Pennsylvania case of *Knupp v. Barnard*† is an appeal from the lower court wherein the court discussed a procedure for what, in Pennsylvania, is

---

* 159 Miss. 470, 132 So. 443, 1931.
† 206 Pa. 280, 55 A. 981, Pa., 1903.

Proper Procedure 293

known as a "block survey." In the first trial, the judge instructed the jury as follows:

> To bring this definition more clearly before you I desire to refer to the definition of blocks as defined by the late Judge WILLIAMS of the Supreme Court, in the case of Ferguson against Bloom, 144 Pa. 549. Judge WILLIAMS was known to be an able land lawyer, and speaking of the manner in which warrants were granted by the commonwealth and the manner in which the surveys were made, he states: "When more land was desired than could be included in one tract, the person wishing to buy made application in the names of the members of his family, his servants and employees, as well as his own. When this happened, the deputy surveyor would sometimes locate the entire batch of warrants in a body, as one tract, and such a body of surveys, made at one time, for one owner, was called a block." If the surveyor discharged his duty, and marked the lines of each tract, the word "block" was sometimes used to describe the body of lands held by one owner; but in such cases, the location of the tracts comprising the block was made on what may be called the individual system. If only exterior lines of the block were marked, the location of the separate tracts was practicable only upon what we have called the "block system." Further defining these two systems, he says: "If no tract corners are marked on the block lines, they must be run in accordance with the returns of survey. If corners are to be found on the block lines, these will control the courses and distances given in the returns, and the interior lines must be protracted across the block in accordance with them." Further he states: "In *Mock v. Astley*, 13 S. & R. 382, it was said a foundation for its application in any given case must be laid by showing two facts: First, the existence of a block must be established 'by the production of documents showing title to the whole body. This shows a grant by the commonwealth of her title to the whole body composing the block. Next, the evidence must show that the tracts composing the alleged block were located in a body without interior lines.'" After reciting other cases, he says: "Upon the authority of these cases, and upon the principle, it is clear that if there are any interior lines on the ground they are of equal value with external ones, as they are equally the 'footprints of the surveyor,' made at the location of the tracts, and for interior members of the block. The general rule is that when there is work on the ground, made for and peculiar to a given tract, such work will control the location of the tract and fix the places of its lines." Again, "We have thus two systems for ascertaining the proper location of a survey; the natural and general one, which rests on such of the original marks made for the tract as can be found, and on the legal presumption arising from the return of survey as to such as cannot be found; and the block system, which is applicable where no internal lines were run, and where, therefore, no marks exist to guide in the location of internal tracts or lines, except such as may be found on the exterior of the block."

> Now, gentlemen of the jury, from the definitions which I have read to you from the decision of the Supreme Court in *Ferguson v. Bloom*, which is almost the last case in which the systems are clearly defined

294   *Boundary Retracement*

by the Supreme Court, you will see that a legal block, in the legal sense I mean now, is where the exterior lines only are run and marked upon the ground. However, in the popular sense in which a block is used, it may be used where not only the exterior lines of the block or batch of surveys are run and marked upon the exterior lines, but the warrants themselves, composing the batch of surveys, the lines or corners are marked, and sometimes the lines are run upon the ground.

The appeals court, after its review, stated the following:

Taking the general lines and return of the block as the evidence and the only evidence of the relative location of those two tracts the weight of it would probably sustain plaintiff's claim; but the marks of a block consist of the marks, if such are found, of every tract of the block, and the marks if originally intended as corners for a particular tract become marks for locating the whole block. The simple question is, did the early surveyor in locating the block make the marks of these two surveys at that date as members of the block? He may establish his leading warrant by undoubted monuments maintained to this day; in running his long block lines from these monuments he may mark corners on these lines for the lines of interior tracts; in running these block lines now, it may be found that these corners are not just where designated in the block lines as returned; the distance from the leading warrant may be longer or shorter than that in the block line; it may not be in the exact course; but this variance of itself does not destroy its significance as a monument; it only demonstrates what is well known, the looseness and want of accuracy in the early surveyors, and the negligence exhibited by them in plotting and returning their surveys. In that day, land in Warren county was cheap; now the production of millions of dollars worth of oil from beneath it has made it very high priced, but nevertheless, we are compelled in ascertaining titles to take into consideration the loose methods of the early surveyors, otherwise we may make disastrous mistakes. Here, we do not undertake to decide that beyond doubt the line from the white oak to the river was the true dividing line between the two tracts; to our minds the evidence is not clear; but there was competent evidence adduced by defendants, that it was; the jury believed it as they had the right to do and we cannot disturb their verdict.

We do not concur in the distinction made by the learned judge of the court below between a popular block of surveys and a legal block. The location of a block of surveys may be established from a single undoubted monument of the block on the ground if there be no others, by the courses and distances in the return; the interior tracts must then be located relatively wholly from the return of the block; but the return may show marks for corners of the interior tracts; if these be found upon the ground they establish the lines of these interior tracts although this may to some extent disturb the lines of the block; such a location of an interior tract of a block although it may somewhat change the course of the exterior line as plotted in the return or shorten that line running from the leading warrant yet giving effect to that fact, does not disregard

Proper Procedure

295

the established rule, that a member of a block cannot be wrested from its position and be located outside of it; it is not thereby wrested from the block but its position is relatively the same as in the return, although one of the exterior lines of the block has been for a short distance deflected from its course to accord with the established monuments on the ground of the interior tract common to it and the block of which it is a member; but there can be no block of surveys in either a popular or legal sense where the tracts are not contiguous. It was not intended by us in *Ferguson v. Bloom*, 144 Pa. 549, nor in any other case to change or modify the law as long settled governing the location of a block of surveys and the different members of a block. While we do not concur with what the learned judge of the court below says on this subject, or rather his understanding of the language of the cases cited, we can only say that his instruction did plaintiff no harm, for he distinctly told the jury, that they must be controlled in their verdict by the evidence of the marks made in 1795 on the ground.

Appellants used this language in their brief, referring to a survey made by Jones: "In 1929, before visions of oil had changed these old land lines."

The Texas court, having the benefit of deciding numerous retracement cases, have stated two very choice principles: Titles to land are not to fail merely because old markers may have disappeared, or because it may be difficult to trace the footsteps of the surveyor,* and search must be made for the footsteps of the surveyor, and that, when found, the case is solved.†

The lengthiest discussion of the rules of evidence applicable in cases like this one is found in *Taylor v. Higgins Oil & Fuel Co., Tex.Civ.App.,* 2 S.W.2d 288, 300, writ dismissed. In this case, the rule is stated to be:

Every rule of evidence laid down for guidance in boundary questions is for the purpose of ascertaining the true location of the line in dispute, by which is meant the place at which the original surveyor ran the line. After 90 years have elapsed and time has destroyed in large measure the evidence left by the original locater, it is then permissible, not only permissible, but of necessity is required, that we resort to any evidence tending to establish the place of the original footsteps which meets the requirement that it is the best evidence of which the case is susceptible."

In the case on appeal, all of the surveyors on both sides, in their efforts to find where the original surveyors actually went, searched for tree roots, buried stumps, stumps that were not buried, trees lying on the ground, stump holes, old fence lines and corners, creeks and old abandoned roads, and markings on trees made a hundred years ago; and all of them studied the field notes of surrounding surveys as well as corners

---

* *Turnbow v. Bland,* 149 S.W.2d 604, Tex.Civ.App., 1941.
† *Stafford v. King,* 30 Tex. 257, 94 Am. Dec. 304, 1867.

296   *Boundary Retracement*

and lines of surrounding surveys on the ground. And, in turn, they all testified concerning these matters. We find nothing from the testimony of appellants' surveyors, or from any other circumstance in the case, to warrant the criticism so made of Garrett's testimony. In short, the surveyors on both sides offered their testimony. The trial court saw and heard them, and found as he did. All we can see in the case is a conflict of testimony, the weighing of which was left to the trial court.

In all of the cases, from *Stafford v. King*, 30 Tex. 257, 94 Am.Dec. 304 (1867) on down, the law has been that search must be made for the footsteps of the surveyor, and that, when found, the case is solved.

Government Procedure, Manual of Surveying Instructions

The printed Manual of Surveying Instructions, dated January, 1902,* referred to the Florida Statute 2399, contains the following and also provides for special instructions to meet particular cases:

246. When new surveys are to be initiated or closed upon the lines of old surveys, which although reported to have been executed correctly, are found to be actually defective in alignment, measurement, or position, it is manifest that the employment of the regular methods prescribed for surveying normal township exteriors and subdivisions would result in extending the imperfections of the old surveys into the new, thereby producing irregular townships bounded by exterior lines not in conformity with true meridians or parallels of latitude, and containing trapezium-shaped sections which may or may not contain 640 acres each, as required by law.

247. Therefore, in order to extend such new surveys without incorporating therein the defects of prior erroneous work, special methods, in harmony as far as practicable, with the following requirements, should be employed, viz.:

The establishment of township boundaries conformable to true meridian and latitude lines.

The establishment of section boundaries by running two sets of parallel lines governed respectively by true meridians and parallels of latitude, and intersecting each other approximately at right angles at such intervals as to produce tracts of square form containing 640 acres each.

"The reduction to a minimum of the number of fractional sections in a township, and consequently of the amount of field and office work."

In sections 283 et seq. of the manual, provisions are made for surveying "Hiatuses and Overlaps."

---

* Occasionally, the instructions in use at the time are the guidance. This Manual has since been revised, in 1947, 1973, and 2009.

Proper Procedure 297

The Ideal Line

The Kentucky decision of *Rowe v. Kidd*, 249 F. 882 (1916) was discussed in some detail in Chapter 4, but in part repeated here for its beneficial insight as to proper procedure in retracement of this nature.

As early as 1916, the Federal District Court discussed a situation in Kentucky that had several noteworthy issues. It spoke of the ideal line along with following the original surveyor's footsteps and protracted lines. This court stated that in the description being followed, contained what was called "ideal lines," and the surveyor "should have run those lines according to the courses and distances called for, and not have altered them in order to reach those lines of those surveys."

> The contention is that where there is a call for a line to run a certain course and distance to an ideal line of another survey, and a line run according to the course and distance called for will not strike such line, the call for such line is to be disregarded, and that to the course and distance followed. By ideal line is meant not necessarily one that was not run, but one that is not actual. It is an open line as contrasted with one not marked. Of course a line that was not run is always an ideal line. It is open and not marked.

In this case, the fifth or closing line of the survey relied on was an ideal line. It was open and not marked, and the reasonable inference is that it was not run. The first four corners called for were timber corners, which had been identified. The fifth corner was a stake, which indicated that the surveyor did not run beyond the fourth corner, but merely protracted the fourth and fifth lines. Likewise, the 11th line of the second survey relied on was then also an ideal line. It was open and not marked, and the reasonable inference was that it was not run. The first nine corners of that survey were timber corners, a number of which were identified, and the five other corners were stakes, indicating that the surveyor did not run beyond the ninth corner, but merely protracted the next six, including the 11th lines.

As an argument, the plaintiff made the argument that four decisions by this court supported their argument in that the call for another survey was disregarded in favor of following the course and distance stated. In one of the cases, *Mercer v. Bate*,* the judge said:

> But there is, in this respect, a palpable and essential difference betwixt actual and an ideal line or a marked and open line. And as, in the one case, Madison might be bounded by the marked line wheresoever it might be (if he made no mistake), so, in the other, he must be restricted to the line *as it appeared to be*, and *as he believed it was when he called to adjoin it*. In the first

---

* 4 J.J. Marsh 334, Ky, 1830.

298

Boundary Retracement

case, he would have a right to the marked line, because, being visible, he knew where it was, and therefore intended that, as marked, it should be his boundary. In the last case, for the very same reason, wherever he supposed the invisible line to run, he must be bounded, because he intended when he made his survey to be, and therefore was bounded by it.

A related decision frequently referenced is the Florida case of *Savannah, Florida and Western Railway v. Geiger.** "In the eye of the law every man's land was inclosed and set apart from his neighbor's, either by a visible and material fence as one field is divided from another by a hedge, or by an ideal invisible boundary existing only in contemplation of the law as when one man's land adjoins another's in the same field. Every such entry or breach of a man's close carried necessarily along with it some damage or other, for if no other special loss could be assigned, still the words of the writ itself specified one general damage, the treading down and bruising his herbage. A man was moreover answerable not only for his own trespass but that of his cattle also, for if by his negligent keeping they strayed upon the land of another (and much more if he permitted it or drove them on) and they there trod down his neighbor's herbage or spoiled his corn or trees, this was a trespass for which the owner must answer in damages; and the law gave the owner of the land in such cases a double remedy by permitting him to distrain the cattle thus damage feasant till their owner should make him satisfaction, or else by leaving him to the common remedy of trespass quare vi et armis clausum."

The Corner Never Set or the Line Never Run

As noted several times, a corner never set should be located where the surveyor would have placed it (had he done so). *Lugon v. Closier.* And, where the purpose "is not to ascertain the position of lines and corners once actually run and established, but to construct a survey by making two lines never run, these lines should be fixed where the surveyor would have made them if he had run them out." *Mercer v. Bate, 4 J. J. Marsh.* 334 (1830).

About Corners and Witness Trees

Francis Hodgman, author of *A Manual of Land Surveying*[†] and a county surveyor for many years, presented a paper to the Michigan Association of Surveyors and Civil Engineers in 1881. The paper was titled "About Corners"

---

* 21 Fla. 669, 1886.
† Climax, Michigan: The F. Hodgman Co., 1913.

Proper Procedure 299

and outlined in some detail the proper procedure for the recovery of original corners.

Hodgman's paper* first discussed the witness or bearing tree. He wrote, "These are marked, and their directions and distances noted in order to assist in finding the corner posts set on the survey. These bearing trees are marked with a blaze and a notch near the ground on the side facing the corner. The measures were taken from this notch. At this time most of the living witness trees have grown to such an extent that only a scar remains in sight, to indicate the point where the notch was cut. In order to get at the notch the superincumbent wood, which is in some cases a foot in thickness, will have to be cut away. It will not often be necessary to do this, as we can come sufficiently near the correct point to find the stake without it. But if the stake has been destroyed, or there are several stakes near, we shall need to be exact and measure from the notch. If the tree has been cut down and a sound stump remains the marks will be easily exposed. Sometimes the mark is gone, but a part of the stump is left. At others the stump is gone, but a dish like cavity remains in the earth to show where the tree once stood. We can almost always find under and around those cavities, places where the large roots have penetrated the subsoil, and thus be able to locate within a foot or so the position of the bole of the tree when standing. In looking for a corner post we may frequently assume for the time being, that a certain stump or a cavity where the tree had stood was the stump of, or the place occupied by a bearing tree. If we then measure the required direction and distance, and find a stake, we may reasonably conclude that our assumption was correct. Such assumptions are frequently of great assistance in finding corners. There may be, and I know there are cases, where the original corner stakes have been destroyed, and can be more nearly restored to their original position by measurements from old stump bottoms or holes in the ground than any other way. But bearing trees, however good their condition, are by no means infallible witnesses as to the location of a corner. Mistakes in laying down their directions or distances, or both, are not rare." (See *McClintock v. Rogers*, 11 Ill. 279.) A direction may be given as north instead of south, east instead of west or vice versa. The limb may have been wrongly read 64° for 56°. The figures denoting the bearing may have been transposed in setting down, as 53 for 35. So, too, the chain may have been wrongly read as 48 for 52, the links having been counted from the wrong end. Or, they have been counted from the wrong tag as 48 for 38. Mistakes of the nature of those mentioned are common, so that in working from a bearing tree to find a corner, and not finding the stake at the place indicated in the notes, it will be well to test all those sources of error before giving up the search, for as I have said before, *the post planted at the time of the original survey is the best evidence* of the corner it was intended to indicate.

---

* *About Corners.* Paper delivered to the *Second Annual Meeting of the Michigan Association of Surveyors and Civil Engineers,* Lansing, January 11–13, 1881.

300

Boundary Retracement

I next consider fences in their relation to corners. *Potts v. Everhart*, 26 Penn., St. Rpt. 493. Taken as a whole very little reliance is to be placed on them, although whether any particular fence may be depended on to indicate the true line will depend on the particular circumstances attending that case. In a general and rough way a fence will indicate to the surveyor where to begin looking for his corner. But the practice has been, and still is common, for the first settlers on a section to clear and fence beyond the line in order to have a clear place on which to set their permanent fence when they get ready to build it. Afterward they forget where the line is and set the new fence where the old one stood. Many fences, too, were set without any survey or any accurate knowledge of where the line was, and left there to await a convenient time to have the line established. So, too, where the land has been long settled and occupied, it is a common custom for adjoining land owners by consent to set the fence on one side of the true line, there to remain until they were ready to rebuild, the one party to have the use of the land for that time in consideration of clearing out and subduing the old fence row. The original parties frequently sell out or die, and the new owners have no knowledge of the agreement and suppose the fence to be on the true line. For those reasons fences should be looked on with suspicion, unless corroborated by other evidence, and the surveyor should enquire pretty closely into the history of a fence before placing any great reliance on it to determine the position of a corner. It may be the best of evidence, and it may be utterly worthless.

It not unfrequently happens that there are no trustworthy marks near a corner to direct the surveyor to his search for the post or from which to replace it if it be destroyed. In those cases he must visit the nearest corners he can find in each direction (varying with the circumstances), go through the process of identification with each of them, and then make his point so that it will bear the same relation to those corners as did the original corner post. Many very intelligent gentlemen suppose that if the surveyor can but find one of the corners of the original U.S. survey he can readily determine the position of all the rest from it. They were never more mistaken in their lives. The continual change in the direction of the magnetic needle, the uncertainty as to what its direction was when any particular line was run, the difference in the length of chains, and the difference in the men who use them, introduce so many elements of uncertainty into the operation as to render it on of little value, and not to be resorted to except in the absence of trustworthy evidence nearer at hand.

Lastly I shall consider the evidence of living persons. *Weaver v. Robinett*, 17 Mo. 459; *Chapman v. Twitchell*, 37 Maine 59; *Dagget v. Wiley*, 6 Florida 482; *Lewen v. Smith*, 7 Port. (Ala.) 428; *McCoy v. Galloway*, 3 Han. (Ohio), 283 and *Stover v. Freeman*, 6 Mass. 441. Conceding all men to be equally honest in their evidence, there is a vast deal of difference among them with regard to their habits of observation and their ability to determine localities. Some have an exceedingly acute sense of locality, if we may so call it, and can determine very accurately the position of any object which they have been accustomed to use; while others seem to have little or no capacity of that sort. In twelve years' almost constant practice in making

Proper Procedure

his class of investigations, I have found many men who would describe accurately the sort of monument used to perpetuate a corner, and who would tell you that they could put their feet on the very spot to look for it; but when the trial came I have found but few of them who could locate the pint within several feet, unless they had some object near at hand to assist the recovery, and even then they would frequently fail.

It may happen where a corner post has been destroyed that its location can be more nearly determined by the testimony of persons who were familiar with it when standing, and can testify to its relations to other objects in its vicinity, than in any other way. But the surveyor in the habits of accurate observation, and the memory of localities possessed by the person testifying, in order to know how much weight to give his testimony.

Reversing Course

Courses may be run in reverse direction whereby so doing a difficulty can be overcome and the known calls harmonized; but such a practice should be resorted to ordinarily only when the termini of the call cannot be ascertained by running forward.* Since the beginning corner of a survey does not control more than any other corner ascertained, and where a disputed or lost line or corner can thereby be established more nearly in conformity with the terms of the instrument and with the intent of the parties as gathered there from, it is a well-accepted procedure to ascertain such line or corner by first ascertaining the position of some other bound and tracing the line back from that by reversing course and distance.

The North Carolina case of *Jarvis v. Swain*[†] is a brief, simple analysis of the principle, in which the court stated, "where the beginning corner in a description cannot be located, but the second corner can, the beginning corner may be established from the second corner by reversing the first call." The analysis in this case is instructive:

> The rule is, in running the calls of a deed, to begin at the beginning corner if it is known or established, and to follow the calls in their regular order, and it is said in *Harry v. Graham*, 18 N.C. 76, 27 Am. Dec. 226, and approved in *Gunter v. Mfg. Co.*, 166 N.C. 166, 81 S.E. 1070, that there is no case in our reports where the court has given its sanction to the correctness of a survey made by reversing the lines from a known beginning corner; but it is equally well established that, if the beginning corner is uncertain and the second corner is known or established, the first line may be reversed in order to find the beginning, and the same rule prevails as to the other corners and lines. *Dobson v. Finley*, 53 N.C. 495;

---

* 11 C.J.S. Boundaries, § 9(3).
† 173 N.C. 9, 91 S.E. 358, N.C., 1917.

Norwood v. Crawford, 114 N.C. 513, 19 S.E. 349; *Clark v. Moore,* 126 N.C. 1, 35 S.E. 125; *Hanstein v. Ferrall,* 149 N.C. 240, 62 S.E. 1070.

In *Dobson v. Finley,* which has been frequently cited and approved, the beginning was at two pines on the south side of a hill, and the second corner was a pine, Thomas Young's corner. The two pines at the beginning had disappeared, and the beginning corner could not be found, but the pine at Young's corner was found and established, and the judge of the superior court permitted the jury to reverse the first line to find the beginning corner. This rule was approved by the Supreme Court, the court saying:

Supposing the pine to be established as the second corner, could the first, a beginning corner, be located by reversing the course and measuring the distance called for, from the pine back; that is, on the reversed course? His honor ruled that the beginning corner could be fixed in this way. We agree with him. If the second corner is fixed, it is clear, to mathematical certainty, that by reversing the course and measuring the distance you reach the first corner; so there is no question about overruling either course or distance by measuring the line, and the object is to find the corner by observing both course and distance.

This authority is directly in point, except that the facts in this record are more favorable to the contention of the plaintiff than in the Finley Case, because here the beginning corner is at a stake, an imaginary point, while in the Finley Case it was at two pines.

In *Simpkins v. Wells,** the lands in question were uninclosed or "wild" lands, and the description in question reads, "Beginning at the white oak near the gap of a ridge," which seemed to be satisfactorily established as the beginning corner, the calls of the patent and those of the certificate of survey on which it was issued agreed for nine consecutive calls. The tenth call in the patent was "N. 75 E., 610 poles;" in the survey it was "N. 75 E., 60 poles." Thereafter, to the beginning, the calls agreed. No natural object was called for after the sixth call, although water courses and well-defined ridges were necessarily crossed. Continuing the plat by the calls of the patent, the last, or eighteenth, call was reached, which was "N. 65 W., 20 poles," which finds the reader 327 poles from the beginning, and the course was S. 84 W. Moreover, the patent would contain 784 1/2 acres, instead of 100 acres. When so run out, the patent embraces the land in dispute. Going back to the tenth call and running by the survey, reaching the last call—"N. 65 W. 20 poles"—it brings us 220 poles from the beginning instead of 20, and the course is N. 61° 31′ E. instead of N. 65 W, and the quantity is 366 1/2 acres instead of 100. Moreover, the lines cross each other when run out by the survey, and the land is located on the west of the calls down the Brushy Fork of Daniels creek, instead of on the east, where the testimony shows it to be. This running does not embrace the land in dispute. The court determined that it was impossible to locate the land intended to be embraced in the patent by either of the above plans of running it out.

"Therefore, it was suggested that, after running to the tenth call, to return to the beginning, and, reversing the calls, run to the variant or

---

* 26 S.W. 587, Ky. App., 1894.

Proper Procedure 303

disputed call, and, if so, it will force the survey and patent to a close by a
call only slightly differing from the patent and survey call, and the patent would then embrace something over 100 acres in its boundary. The
junior patent would not then interfere with the senior, and the resultant
figure be quite similar to the tracing from the surveyor's book of the survey made for the patent. This plan was adopted in correcting the patent
under consideration in *Alexander v. Lively,* 5 T. B. *Mon.* 159.

There it was said: If we commence at the beginning and run to the third
corner, and then return to the beginning and reverse the last and two next
preceding lines, we will form an uninclosed survey, or one which may be
closed by one additional line, and, when so closed, it will include the quantity desired," etc. Thus, a new line was added, and the patent sustained.
Here, the court stated, "we add no new line, though the precise course and
distance supplied are unknown to any call, either of the patent or survey.
Yielding to necessity, and as affording the only solution of the difficulty, we
suppose this plan may be adopted, and the validity of the patent upheld.
There is then no interference with it by the William Wells patent save to the
extent of a few acres, where no timber was cut, and the quantity corresponds
substantially with that demanded by the patent and certificate of survey.
Another corroborative circumstance is that the long parallelogram found
at the southern border would extend nearly to the dividing ridge between
the counties—then Floyd and Lawrence, now Johnson and Martin—without projecting into Lawrence or Martin county, as would be required under
the contention of the appellees. Whether the patent would then include the
improvements made by the ancestor of the appellees at the mouth of the
Brushy Fork we can only tell by the process of protraction, but these they hold
unquestionably by long adversary possession. They do not, however, hold
adversely up Daniel's creek beyond the west line of the junior patent. The
proof is abundant that to that line, and to it only, in that direction, the ancestor of the appellees claimed, and did so for many years. The line was found
to be well marked, the knotty poplar especially being well known as one of
the corners of the William G. Wells' land. This line—the western boundary of
the junior patent—may be regarded as the established line between the conflicting claimants, save as to the few acres of lap mentioned heretofore. By this
adjustment of the lines of the patent and survey, and the recognition of the
adverse holding of the appellees to line mentioned, we reach the conclusion
that the lands on which the timber was cut belong to the appellants."

Government Surveys

If a surveyor, by applying the rules of surveying, can locate the land as
described in a deed, the description is sufficient; and a deed will be sustained

304 Boundary Retracement

if it is possible to ascertain and identify the land intended to be conveyed. *Ansley v. Graham*, 73 Fla. 388, 74 So. 505, and cases cited. The government rules do not require the location of boundaries to be ascertained by tracing the line from the starting point merely. It is permissible to begin at any definite corner or monument, and run a reverse course if necessary to harmonize all the calls of the description. *Ayers v. Watson*, 137 U.S. 584, 11 S.Ct. 201, 34 L.Ed. 803; *Simmons Creek Coal Co. v. Doran*, 142 U.S. 417, 12 S.Ct. 239, 35 L.Ed. 1063.

Locating Unrecorded Maps, Plans, and Sketches

> A plat is not a mark on the land, but a representation of the land on paper, appealing to the eye by means of lines and memoranda, rather than by words alone.
>
> *Justice Lumpkin, Thompson v. Hill,*
> 137 Ga. 308, 73 S.E. 640 (1912).

Whenever a survey has been done, there is a strong likelihood that it resulted in a map, or at least a sketch. The earlier the survey, the less chance there is of easily finding the map, since in the early times there either were no recording facilities available, or within a convenient distance. If copies were made, they would have been done by hand drawing (copying). Consequently, if recordings were made, they may have been made somewhat later than the survey itself. However, since many of the maps were one-of-a-kind, and there were no convenient reproduction facilities until more recent times, if such a map still exists, it may be in an obscure location. Hopefully, people realized the value of such a map, and took steps to preserve it.

In recent times, steps have been taken to preserve and catalog surveyors' collections, and occasionally one is found, sometimes in an unlikely place.

The Unrecorded Plat

The Florida case of *Froscher v. Fuchs** began its discussion by re-iterating the standard for a disputed boundary:

> In case deciding the boundary between two parcels of land, the law is settled that it is the duty of the surveyors to follow the original survey lines under which the property and neighboring properties are held notwithstanding inaccuracies or mistakes in the original survey. The purpose of this rule of law is that stability of boundary lines is more

---

* 130 So.2d 300, 1961.

Proper Procedure 305

important than minor inaccuracies or mistakes. This rule was firmly established in Florida by *Akin v. Godwin*, Fla.1950, 40 So.2d 604.* See also *Wildeboer v. Hack*, Fla.App.1957, 97 So.2d 29[†] and *Bishop v. Johnson*, Fla. App.1958, 100 So.2d 817.

The testimony of plaintiff's witnesses established the existence of an unrecorded plat, prepared in 1913. The testimony "establishes without contradiction that this plat has been used by substantially of the surveys made" [in this area]. Four recorded plats were introduced, each following the lines established by the unrecorded plat. None of the four plats covered the area in question in this case, they did locate neighboring lands.

The court emphasized, "the long established use of the unrecorded survey, together with the fact that property rights have been acquired pursuant to recorded plats following this unrecorded survey, is sufficient to establish that the unrecorded survey was recognized by the surveyors of the area as the survey which established the procedure to be following in locating boundary lines within [the immediate area]."

This decision demonstrates the value of an earlier survey, even though unrecorded. To be valid, and of use, until relatively recently plats were not required to be recorded. Even now, for the most part, private surveys do not have to be recorded unless they fall under the subdivision act, or some other special regulation (boundary line agreement, court order, etc.). Even though not on record, they have evidentiary value, and may demonstrate the establishment of title lines, and offer the original survey of those lines.

In the *Froescher* case, the parties knew about the plat, some had relied on it. An existing plat, though not on record, may still be of use, may even be necessary. If referenced in a deed, it is part of the deed, and a deed must be honored in its entirety. It does not lose its evidentiary value even though it may be difficult to find.

Other courts have made similar statements, and it is a well-known and accepted principle that, for example, "where a description refers to a map, plat or survey it is made apart of the deed as much as attached thereto." (*Chapman v. Pollack*, 11 P. 764; *Smita v. Young*, 43 N.E. 486, 489; *Black v. Sprague*, 54 Cal. 266; *Nixolin v. Schnerderline*, 33 N.W. 33; Cyc., vol. 13, page 633, note 7.)

Probate Records

Documents from the probate court, or other tribunals where family relationships are involved, can often be helpful. Early procedures included partitions and divisions among joint owners in order to allow everyone their fair

---

* See Chapter 1 for several brief discussions of this case cited as authority in other decisions.
† See Chapter 1, where this case is also discussed.

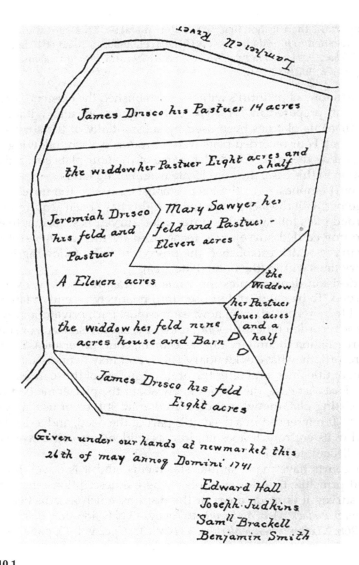

FIGURE 10.1
Plat of probate partition. The return of the survey described all of the lots with metes and bounds descriptions, providing the original survey data for each of the titles.

share. Land was divided by committee appointed by the court among heirs and joint owners according to both quantity and quality. Dower interest, the widow's interest, was usually a life estate with a reversionary interest or a remainderman. Both of these categories frequently involved an original survey, with proper instructions as to the procedure to be followed by the committee, or the appointed surveyor. Often the resulting parcels were monumented and mapped, although the plans may or may not be found with the probate file, but could still exist elsewhere if not within the probate court records. Since dower parcels were life estates, their monumentation was meaningful at one time, but sometimes ceased to be significant once the title holder had passed. Since no one went back to remove those monuments, they may still be found today, and are sometimes found to be misleading (Figure 10.1).

Historical Collections

Private collections of survey records are sometimes found in a variety of locations: colleges and universities, local and state historical societies, local libraries, state archives, museums, and similar repositories (Figures 10.2 and 10.3).

Where survey businesses and other collections have been acquired, records may have been preserved by private surveying companies. Fortunately, for the most part, the value of early records is appreciated, and efforts have been made to preserve and protect these one-of-a-kind resources.

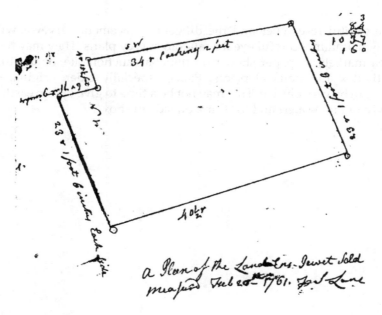

FIGURE 10.2
Ancient map located in a private collection.

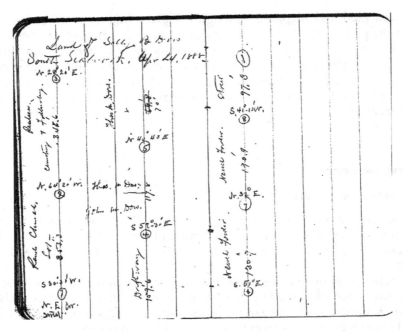

FIGURE 10.3
Copy of field notes from a private collection. From the data, a plat can be constructed.

The careful researcher and the diligent retracement surveyor will constantly maintain a careful eye for random survey plans. They may be found in flea markets, at paper shows, on the eBay auction site, or in otherwise unorthodox and unusual places. Plans, especially older, original survey plans, can be very elusive. This may not be a time to give up too easily or too quickly, but to be alert and to "think outside the box."

11

The Lost Corner

The most difficult work surveyors are required to do is to restore lost corners.

Clark, Third Ed., § 173

What If the Corner Is Truly Gone?

The primary object to be attained in the establishment of lost corners is to put them, if practicable, in the exact spot they were put by the Government, and this object can best be accomplished when all traces of the corners are gone and there is no fixed monument called for, by observing the courses and distances called for in the field notes of the original Government survey. *Major v. Watson*, 73 Mo. 66.

So long as the monuments placed upon the earth's surface by the United States surveyors can be identified, there are no lost corners, and they control, no matter what more recent surveys, by courses and distances disclose.

Original corners as established by the government surveyors, if they can be found, or the places where they were originally established, if they can be definitely determined, are conclusive, without regard to whether they were located correctly or not. 5 Cyc. 873. The general duty of a surveyor in instances like this are plain enough; he is not to assume that a monument is lost until after he has thoroughly sifted the evidence and found himself unable to trace it, and exhausted every recourse to which surveyors have access, as the original Government surveys must govern.*

Restoring Lost Corners

First and foremost, it is essential that the basic requirement is met: the corner is, in fact, lost and not just that those who have been searching have been able

---

* *Woods v. Johnson*, 264 Mo. 289, 174 S.W. 375, Mo., 1915.

309

310 *Boundary Retracement*

to find it. If the corner is indeed not lost, then the rules governing lost corners have no application, and any efforts have been for naught.

To stress again, a corner and a marker may not be equivalent. A marker may be in the position of a corner, or it may be nearby, for a variety of reasons. Again, according to the Tennessee court, a corner is "the intersection of two converging lines or surfaces; an angle, whether internal or external; as the "corner" of a building, the four "corners" of a square, the "corner" of two streets. A mere variation in a line does not constitute a "corner."*

It is not a matter than *you* can't find the corner, it is a matter of the corner being actually and truly lost, which means *it* can't be found, ever, no matter what. But, to the wary, too many so-called "lost" corners are magically "found" at a later date.

Some reasons for not finding an existing corner and incorrectly terming it a lost corner are the following. These reasons why you (or the surveyor(s) before you) did not find the corner include:

Searching in the wrong location (the obvious answer)

Did not adjust the distances to account for measurement errors

Used the incorrect tape or conversion

Did not correct for declination (a common fault)

Failed to take into account local attraction (then and now)

Original directions were incorrect or misleading

Didn't dig (confined search to the ground surface)

Corner disturbed or destroyed

Corner never set[†]

Little or no experience in deciphering the original or even a prior survey

Where the Corner Formerly Stood

Numerous decisions as well as formal instructions consider not only the physical location of the "corner", but also, when missing, the place where it formerly stood (see Chapter 3 for a discussion of missing monuments).

Hodgman[‡] devoted a chapter in his book to Re-location of Lost Corners. His words are gleaned from a variety of Supreme Court decisions. He begins with saying "a corner is not lost so long as its position can be determined by evidence of any kind without resorting to surveys from distant corners of the same or other surveys. Often after making a survey from a distant corner, the surveyor will come upon some traces or evidence which will enable him

---

* *Christian v. Gernt. et al.*, Tenn. Ch., 64 S.W. 399, 1900.
† According to *Lugon v. Crozier*, 78 Colo. 141, 240 P. 462, Colo., 1925: a corner never set cannot be either a lost or an obliterated corner.
‡ Hodgman, F. *A Manual of Land Surveying*. Climax: The F. Hodgman Co., 1913.

The Lost Corner

311

to determine the true position of the corner he is seeking. It is an uncertain way at the best to locate corners by running lines and measuring from distant corners, and should only be resorted to in absence of better proof of the original location of the corner sought."

> It will sometimes happen that the exact spot where a lost corner stood cannot be found or shown by evidence, but it can be proved that it stood within certain limits. In these cases, which are not rare, there is no question but that the corner should be placed at that point within the known limits which best agrees with all the evidence in the case.
>
> Failing of better evidence by which to determine the location of a lost corner, we may next resort to [specified] methods.
>
> The general rule is to "retrace the known lines of the description and find how the lengths and directions of these lines by your survey agree with those of the same lines as laid down in the original description. Then run the unknown lines and place the lost corners so that they will bear the same relation to the known lines and corners as they are required to do by the description of the original survey.

This paragraph is in keeping with several noted decisions which included when a corner has not been set, it can be neither lost nor obliterated, and should be placed where the original surveyor would have set it.[*]

Finally, Hodgman advises, do not give up a corner as lost while any means of finding its exact location are left untried.

Lost *v.* Obliterated Corners

There is a difference between missing a corner and a lost corner just like there is a difference between a lost and an obliterated corner, even though in all cases the "corner" is not visible, or at least is not obvious. In theory, a true corner is not visible because it is a mere point at the intersection of two lines, although its position may be determined. And if were to fall in the same category as the center of section, it may also be classed a monument. The case of *United States v. Champion Papers* discussed in Chapter 5 involved "missing" original corners, but surveyors over time perpetuated their positions by the nature of their work. Remember, a corner is only truly lost if such methods (there are others) totally fail to yield its position.

The Maine case of *Bean v. Batchelder*[†] discussed instructions to the jury, "that the lines run by Herrick upon the surface of the earth, as and for the boundaries of lot 5, would still be the boundaries of that lot, if their locality could be found; that the question for them to decide was the locality upon the surface of the earth of the lines actually run by Herrick in making the survey of that lot. The instruction was correct. *Esmond v. Tarbox*, 7 Me. 61, is express

---

[*] *Lugon v. Crosier*, 78 Colo. 141, 240 P. 462, Colo., 1925.

[†] 78 Me. 184, 3 A. 279, Me., 1886.

312 Boundary Retracement

authority for it. See, also, *Pike v. Dyke*, 2 Me. 213; *Williams v. Spaulding*, 29 Me. 112. The plan was merely a picture. The survey was the substance. The plan was not made to show where the lots were to be hereafter located, or how they were to be hereafter bounded. It was made as evidence of where they had before been located and bounded. The lot actually surveyed, bounded by the lines actually run, was the lot intended to be conveyed. The plan was named in the deed, rather as a picture, indicating the location and lines of the lot. Still the actual boundaries, rather than the pictured boundaries, were to be sought for. The picture might not be wholly accurate."

Even if the marker, even the one set by the original surveyor, is missing, the actual corner is still where it was originally created. Remember, a corner is not considered to be the marker, it is the intersection of two lines. The marker is to relate to the location of the corner, and when missing presents a difficulty in locating the corner, but the corner itself, being an imaginary point albeit a legal entity and of utmost significance, is still at the exact location where it was originally established.

How Lost Is "Lost"?

The recent decision of *United States v. Doyle** brings together several sources and succinctly answers the question. It states, "The authorities recognize that for corners to be lost '[t]hey must be so completely lost that they cannot be replaced by reference to any existing data or other sources of information.' *Mason v. Braught*, supra, 146 N.W. at 689, 690. Before courses and distances can determine the boundary, all means for ascertaining the location of the lost monuments must first be exhausted." *Buckley v. Laird*. 493 P.2d 1070, 1075 (Mont.); Clark, Surveying and Boundaries § 335, at 365 (Grimes ed. 1959); see advisory comments of the supplemental manual, supra at 10.

> The means to be used include collateral evidence such as boundary fences that have been maintained, and they should not be disregarded by the surveyor. *Wilson v. Stork*, 171 Wis. 561, 177 N.W. 878, 880. Artificial monuments such as roads, poles, fences and improvements may not be ignored. *Buckley v. Laird*, supra, 493 P.2d at 1073; *Dittrich v. Ubl*, 216 Minn. 396, 13 N.W.2d 384, 390. And the surveyor should consider information from owners and former residents of property in the area. See *Buckley v. Laird*, supra, 493 P.2d at 1073-1076. "It is so much more satisfactory to so locate the corner than regard it as "lost" and locate by "proportionate" measurement." Clark, supra § 335 at 365.
>
> "So completely lost that they cannot be replaced by reference to *any existing data* or other sources of information." Before courses and distances can determine the boundary, "*all means* for ascertaining the location of the lost monuments must first be exhausted."

---

* 468 F.2d 633, 1972.

The Lost Corner

313

"All means" and "any existing data" leave an investigation wide open, and invite specialists in forensic procedures to review the situation and suggest means of examining additional existing and sources of information, no matter how remote the chance for success may be, and no matter how far-reaching the investigation may become. It may amaze people how often the "impossible" is made possible, and how frequently "what cannot be found" is found.

In the referenced case of *Mason v. Braught*,* the court stated, "it is only where the section lines and corners were either never marked at all or where such lines and markings have become so completely lost that they cannot be retraced or replaced by the aid of any known or recognized natural object or permanent monuments. And to be lost, when applied to section or township corners, means more than that they have been merely obliterated, tampered with, or changed. They must be so completely lost that they cannot be replaced by reference to any existing data or other sources of information."

This has been the directive from the courts at least as far back as 1915, with the case of *Woods v. Johnson*:†

The primary object to be attained in the establishment of lost corners is to put them, if practicable, in the exact spot they were put by the Government, and this object can best be accomplished when all traces of the corners are gone and there is no fixed monument called for, by observing the courses and distances called for in the field notes of the original Government survey. (*Major v. Watson*, 73 Mo. 66).

So long as the monuments placed upon the earth's surface by the United States surveyors can be identified, there are no lost corners, and they control, no matter what more recent surveys, by courses and distances disclose. (*Jacobs v. Moesley*, 91 Mo. 464; *Knight v. Elliott*, 57 Mo. 317).

Procedure

In *People of the State of California v. Thompson & Gray, Inc.*,‡ "Proportionate measurement method of finding corners may be used only as last resort when original corner is 'lost' and cannot be relocated on the ground."

Corners of section were not "existent" and were "lost" where monuments could not be found in several surveys made over period of years, original plat and field notes did not reveal location or form basis for finding location, and property owners in area did not know location of

---

* 33 S.D. 559, 146 N.W. 687, S.D., 1914.
† 264 Mo. 289, 174 S.W. 375, Mo, 1915.
‡ 99 Cal.594; 22 Cal.App.3d 368, 1971.

314 Boundary Retracement

objects mentioned in field notes and therefore use of double proportionate measurement method was proper.*

The court, in the case of *County of Yolo v. Nolan*,† stated,

The rule as to restoring lost corners by putting them at an equal distance between two known corners has no application, if the line can be
retraced as it was established in the field.

As previously emphasized,

Original corners as established by the government surveyors, if they can
be found, or the places where they were originally established, if they
can be definitely determined, are conclusive, without regard to whether
they were located correctly or not. 5 Cyc. 873. The general duty of a surveyor in instances like this are plain enough; he is not to assume that a
monument is lost until after he has thoroughly sifted the evidence and
found himself unable to trace it, and exhausted every recourse to which
surveyors have access, as the original Government surveys must govern.
Stewart v. Carlton, 31 Mich. 270; *Diehl v. Zanger*, 39 Mich. 601; *Dupont v.
Sparring*, 42 Mich. 492.

This instruction summarizes and states the directive most clearly: the
original corners, or the places where they were originally established are
to govern; and cannot be regarded as lost until the surveyor has "exhausted
every recourse to which surveyors have access."

Again, "So long as the monuments placed upon the earth's surface by the
United States surveyors can be identified, there are no lost corners, and they
control, no matter what more recent surveys, by courses and distances disclose. *Jacobs v. Moesley*, 91 Mo. 464; *Knight v. Elliott*, 57 Mo. 317. Measurements
must yield to the true location controlled by the Government's monuments,
as indicated in the records. *Brown v. Carthage*, 128 Mo. 10. A Government corner must prevail over distances in the ascertainment of the true line. *Mining
& Smelting Co. v. Davis*, 156 Mo. 422; *Campbell v. Clark*, 8 Mo. 553. Government
corners must prevail. *Major v. Watson*, 73 Mo. 661; *Carter v. Hornback*, 139 Mo.
238; *Climer v. Wallace*, 28 Mo. 556; *Mayor of Liberty v. Burns*, 114 Mo. 426. Field
notes of the Government surveyors will control in ascertaining location of
corners and boundary lines, even when the established corner by the surveyor cannot be found. *Bradshaw v. Edelin*, 194 Mo. 640. Government surveys are conclusive. *Frederitzie v. Boeker*, 193 Mo. 229. Conflict between plat
and actual survey is controlled by the survey. *McKinney v. Doane*, 155 Mo.
289. Original corners as established by the government surveyors, if they
can be found, or the places where they were originally established, if they

---

* *Reid v. Dunn*, 20 Cal. Rptr. 273, 1962.
† 144 Cal. 445, 77 P. 1006, 1904.

The Lost Corner 315

can be definitely determined, are conclusive, without regard to whether they were located correctly or not. 5 Cyc. 873. The general duty of a surveyor in instances like this are plain enough; he is not to assume that a monument is lost until after he has thoroughly sifted the evidence and found himself unable to trace it, and exhausted every recourse to which surveyors have access, as the original Government surveys must govern." *Stewart v. Carlton,* 31 Mich. 270; *Diehl v. Zanger,* 39 Mich. 601; *Dupont v. Sparring,* 42 Mich. 492."

Circular 1452 United States Department of the Interior General Land Office, "Restoration of Lost or Obliterated Corners and Subdivision of Sections." *Cordell v. Sanders,* 331 Mo. 84, 52 S.W.2d 834."

The circular 1452, supra, states general rules about which there is no dispute, which we quote:

> "First. That the boundaries of the public lands, when approved and accepted, are unchangeable."
>
> "Second. That the original township, section, and quarter-section corners must stand as the true corners which they were intended to represent, whether in the place shown by the field notes or not."
>
> "Fourth. That the center lines of a section are to be straight, running from the quarter-section corner on one boundary to the corresponding corner on the opposite boundary."
>
> "Sixth. That lost or obliterated corners are to be restored to their original locations whenever it is possible to do so."

This is obviously the "catch-all" philosophy of retracement, or for following original footsteps: Locate a missing corner at the [exact] position where it was originally set. If not set, then it is to be located where the original surveyor would have set it (*Lugon v. Crosier*). While proportionate measurement is offered as a guideline, and has become a widely accepted method for positioning a corner, as one witness, a Mr. Pratt testified in the Missouri case of *Hale v. Warren,** that double proportionate measurement "is the method of last resort when all other methods of surveying have completely been exhausted ..." and it was employed if the original corner has been lost. He also related that a corner is considered lost if it is "lost beyond a reasonable doubt" and cannot "be monumented by acceptable evidence or testimony." He further testified that relocating corners using double proportionate measurement almost never "results in the corner being placed exactly where the old land surveyor put it in the original records," but, nevertheless, is the statutorily required method for re-establishing lost corners. As stated, the method does not result in the correct solution. And as Griffin aptly stated, apportionment (of any sort for any reason) "may yield the most probable location of the corner." Unless there is absolutely no hope, as several courts have stated, of locating the position of the original corner, apportionment should be considered inappropriate, since it defeats the whole purpose of

---

* 236 S.W.3d 687, Mo. App. S.D., 2007.

316   *Boundary Retracement*

following the original footsteps, and results in something less than preserving the sanctity of titles and the stability of boundaries. Griffin also noted that "a boundary once established, should remain fixed through any series of mesne conveyances." In fact, he stressed that "a boundary once established must remain fixed in its original position *ad infinitum.*"

This is obviously the easier solution (supposedly) as it just involves some verification of original corners, then a simple mathematical exercise. However, courts don't always go along with it depending on the circumstances. And, in every case, the underlying requirement, first and foremost, must be met—the corner is <u>lost</u>. The retracement surveyor should ask oneself: "what if it is not truly lost, but I merely think that it is." There are many examples within the court system and even more outside of the reported cases that fall into this latter category. When followed through, such practice only serves to create more of a problem, or one or more additional problems.

As stated in Clark, "The most difficult work surveyors are required to do is to restore lost corners."[*] This is equally true in both the PLSS and the metes and bounds states, as well as with metes and bounds descriptions of any sort, whether based on survey or protracted, that are superimposed on a rectangular system of any nature.

In the Arizona case of *Galbraith v. Parker*[†]:

> Plaintiff asserts that this is not an original monument. That must be conceded. It is a replacement monument, but it is located in the exact place mentioned on the plat. Extrinsic aids to show actual location of original monuments may be used. It is competent to prove by parol the location thereof and, if lost or destroyed, the places where they were set." (Turnbull v. Schroeder, 29 Minn. 49, 11 N.W. 147; Borer v. Lange, 44 Minn. 281, 46 N.W. 358; City of North Mankato v. Carlstrom, 212 Minn. 32, 2 N.W.2d 130.)
>
> Two buildings, the post office and the Methodist Church, were properly identified by plaintiff's witness Mr. Minium as having been located with this monument as a starting point. Since the replacement granite marker at Center street is now under the pavement and cannot be used as the starting point, that point would have to be determined by the corners where said buildings are now located. Mr. Minium states that this can definitely and accurately be done. Such a monument becomes conclusive in determining the starting point of a survey. City of Racine v. Emerson, 85 Wis. 80, 55 N.W. 177, 39 Am.St.Rep. 819; Arms v. City of Owatonna, 117 Minn. 20, 134 N.W. 298; 8 Am.Jur., Boundaries, p. 750, § 8.

From 11 C.J.S., Boundaries, § 13b, p. 555, we quote:

> Where the location of the original monument is lost, that is, where it cannot be discerned or established by evidence, the corner may be located or

---

[*] Clark, Third Ed., § 173.
[†] 17 Ariz. 369, 153 P. 283, Ariz., 1915.

The Lost Corner

established by a new survey made from points which can be determined and in accordance with the field notes of the original survey. In general, it is to be located by running lines from the nearest identified original standard corners found to the south and east to the nearest original standard corners found in the north and west and taking the point of intersection of such lines as the corner. Where a survey becomes necessary it must be made from the east, and not from the west, boundary line of the township, and in all cases of resurvey the chain used must be made to correspond to that used by the government, by testing it with distances on the ground between two or more known monuments.

In Wacker v. Price,* the court quoted,

as was said in the Oregon case, *Trotter v. Stayton*, 41 Or. 117, 68 P. 3; *Diehl v. Zanger*, 39 Mich. 601, that it is a matter of common knowledge that the great majority of original surveys are more or less inaccurate and since it has always been the rule that courts must resort and be bound by the best evidence available, it follows that the boundaries fixed by the property owners themselves in the absence of the inability of surveyors to definitely fix the monuments from which the original survey was made must control and that the city surveyor nor any other surveyor has any authority to establish new boundaries which must of necessity affect the property rights of all property owners concerned where they cannot establish title by adverse possession.

Bearing in mind that there is no evidence that an actual survey was ever made of Grand Avenue Subdivision or that any actual measurements were ever made from the Government monument at the intersection of 15th Avenue and McDowell Road and that our sole concern here is to determine if we can from the best evidence available what actually was the location of the lots in question as fixed by the plat of the Grand Avenue Subdivision, we are forced to the conclusion that their location is to be determined if at all from well-established long-standing monuments existing within the subdivision itself.

What is the best evidence in this case? Certainly the Government monument at McDowell Road cannot be treated as the best evidence of a starting point from which the actual location of the lots here involved may be accurately determined for the reason that there is no evidence that such monument was ever used as a starting point, in platting Grand Avenue Subdivision. On the other hand so far as the evidence discloses Cedar Street as it exists today is located exactly where the plat of the Grand Avenue Subdivision places it. It certainly is located exactly where the property owners in Blocks 31 and 32 by common consent have placed it since it was opened for use, some time after 1911. As stated above, if the side lines of Cedar Street were projected west they would be superimposed upon the side lines of Magnolia Street and vice versa. That is where the original plat places it. The extended side lines of a street

---

* 70 Ariz. 99, 216 P.2d 707, Ariz., 1950.

318   *Boundary Retracement*

constitute in law a permanent monument. *Carey v. Clark,* 40 Nev. 151, 161
P. 713. Therefore the side lines of Cedar Street are in legal effect a permanent monument in Grand Avenue Subdivision from which the location
of lots in Block 31 may be definitely determined. In fact, under the best
evidence rule it is the only monument from which the location of said
lots according to the Grand Avenue Subdivision plat can be made. The
accuracy of this monument has been confirmed by acquiescence by all
of the property owners in Block 31 except Lots 8, 6, 4 and 2 thereof, and
all of the property owners in Block 30 to the south of Cedar Street. This
acquiescence is clearly evidenced by location of homes on such lots and
the building of fences and growing of trees along the side lines thereof
for a long period of years far in excess of the statutory period required
to acquire title by adverse possession. This is true as to lots on both the
east and west side of Block 31 and especially directly west of Lots 8
and 6. These things were clearly visible to both parties to this litigation
when they purchased the lots from Shaw. The fences along the side lines
of the lots themselves may be treated as monuments. *Perich v. Maurer
et al.,* 29 Cal.App. 293, 155 P. 471. It was said in *Ralston v. Dwiggins,* 115
Kan. 842, 225 P. 343, that where a survey appears to conform with the
recognized boundary lines of city lots on which buildings have been
erected and expensive improvements made the boundaries so generally
accepted and recognized for many years lend some support to the survey approved by the court; citing *Tarpenning v. Cannon,* 28 Kan. 665. The
surveyor in that case testified that if the appellant's theory was adopted
"it would move every existing improvement in town 6 feet." If appellee's
survey is accepted it will move every lot line in Blocks 30 and 31 Grand
Avenue Subdivision south approximately 25 to 30 feet. Such a result
would be disastrous to all of said property owners were it not for the
fact that they have all acquired title to the property actually occupied by
them by adverse possession. The court in the Ralston case, supra, further
said: "The primary rules for locating city plats upon the ground are, in
order of precedence in application, as follow: (1) Find the lines actually
run and the corners and monuments actually established by the original
survey. (2) Run lines from known, established or acknowledged corners
and monuments of the original survey. (3) Run lines according to courses
and distances marked on the plat."

Citing *In re Richardson,* 74 Kan. 557, 87 P. 678. The court further said in that
opinion:

It is urged that the section line was the proper base line which the surveyor should have ascertained and from which his measurements and
calculations should have been made. That leaves out of consideration the
original survey as actually located upon the ground. A certain hedge
fence is spoken of as having been used as a base line in early days, but
time has erased that mark. The monuments and marks found by the
surveyor furnished reasonably good evidence in locating the original
survey. * * * We have conceded that Grand Avenue Subdivision was not

The Lost Corner

accurately platted but we are bound by that plat as we find it and not by what it should be if accurately platted. We are bound by the best evidence rule which must be held to be the monuments established by the plat itself, acquiesced in and confirmed by the property owners in Block 31 as evidenced by long-established property lines.

It follows that the location of Lots 8 and 6 must be determined in accordance with the best evidence rule, that is, by measurements from monuments irrefutably established by the original plat itself as confirmed by the property owners of Block 31.

The findings of the trial court are not supported by substantial evidence. The judgment of the court should therefore be reversed and the cause remanded to that court for the purpose of determining the amount of the damage suffered by appellant, in the manner specified in Justice STANFORD'S opinion.

"DISSENT BY: UDALL

UDALL, Justice, (dissenting).

"We consider that the decision of the majority not only does a gross injustice to the appellees but the principles of law applied to this situation, if followed, will work mischief with land titles generally. We completely disagree with both the majority opinion and the specially concurring opinion of our associates. The majority, as we view it, have by their rejection of what the trial court considered and what we deem to be the true control point, to wit: the quarter section corner on the north line of section 6, 'ridden off in all directions'. They are, in our opinion, like a ship at sea without rudder or compass."

The author of the majority opinion maintains that Ida Shaw (the common grantor of the parties) conveyed said lots "* * *with reference to the north boundary line of lot 10 owned by Mrs. Newcomb * * *" which boundary had become established by acquiescence in an old fence line. On the other hand the author of the concurring opinion maintains that "* * *the side lines of Cedar Street* * *" constitute the proper control monument. (Cedar Street — one block in length—while shown on the original plat of 1888 was not opened for use until the year 1911.) We maintain that both statements are predicated upon erroneous assumptions. As a matter of fact the Ida Shaw deeds in evidence specifically state, as to the description, that the conveyance is made with reference to the plat of the Grand Avenue Addition on record in the office of the County Recorder of Maricopa County, Book 1 of Maps, page 9. Later we shall endeavor to point out in more detail the incorrectness of these assumptions and show that the government monument, supra, should be accepted as the control point for an accurate survey. Before doing this, however, there are some general observations that may well be made.

At the outset we call attention to the fact that the sole issue involved in this case is, "Where upon the ground is the true boundary line between

320   *Boundary Retracement*

lots 6 and 8 of block 31, Grand Avenue Addition?" Is it 1503 feet south of the northeast corner of the quarter section corner, as contended by appellees and as found by the court, or is it 27.35 feet north of that point, as claimed by the appellants? The only parties interested in this disputed boundary line are the appellants (plaintiffs), the record owners of lot 8, and appellees (defendants) who hold title to lot 6. The judgment of the trial court was strictly confined to a determination of this issue. This judgment, had it been permitted to stand, would not have upset incorrect boundary lines between lots owned by various other parties in the neighborhood. As between property owners to the south and the west of the lots in question, acquiescence in boundary lines or adverse possession might well be the determining factor. In any event their rights are not before us for determination in this proceeding. It is our opinion that if Ida Shaw, appellants' grantor, lost title to the south portion—actually 27.35 ft.—of lot 8 it was because of the adverse claims of appellants' neighbor on the south (Mrs. Newcomb) and not by reason of any deficiency or shortage of land in said lot 8. None of the elements of adverse possession are present as between the parties in the instant suit. It would appear appellants' remedy was to sue their grantor on her warranty rather than trying to shift to the north the true boundary lines between lots 2, 4 and 6. The trial court found (and there is no evidence to the contrary): "That there were no fences, monuments or other visible markings either along the (true) boundary line between said lots 6 and 8 * * * or along the line claimed by the plaintiffs to be the boundary between said lots at the time either the plaintiffs or defendants acquired their respective lots.

Parol Evidence and Reputation

As mentioned within the applicable cases, and discussed to some length by Hodgdon,* knowledgeable persons may offer oral evidence at to the location of boundaries and corners.

In *Thomsen v. Keil*,[†]

The plaintiff contending that the north quarter corner of section 33 is a "lost" corner his surveyor established it at a point which, if correct, would necessitate a reversal of the judgment. The surveyor of the defendant took the view that the north quarter corner is not a "lost" corner, and established it at the point which the trial court adopted as the original north quarter corner.

---

* Hodgman, F. *A Manual of Land Surveying*. Climax: The F. Hodgman Co., 1913.
† 48 Nev. 1, 226 P. 309, Nev., 1924.

The Lost Corner

A "lost" corner is defined as follows:

> "A lost corner is a point of a survey whose position cannot be determined, beyond reasonable doubt, either from original traces or from other reliable evidence relating to the position of the original monument, and whose restoration on the earth's surface can be accomplished only by means of a suitable surveying process with reference to interdependent existent corners.' Clark on Surveying and Boundaries, § 376, quoting from rules of the Department of Interior."

In the same section, the author says:

> "A corner should not be regarded as lost until all means of fixing its original location have been exhausted. It is much more satisfactory to so locate the corner than regard it as "lost" and locate by proportionate measurement.'"

At section 329, the author says:

> "The surveyor should not treat a corner as lost until he has exhausted all means of fixing its location aside from the determination thereof, by measurement thereof to other corners.'"
>
> Counsel for appellant says that the rules of the Department of Interior relative to the establishment of corners as originally established pursuant to the government survey must control. This would certainly be true if the quarter corner in question were upon the surveyed public domain; but, since it is upon land privately owned, and for which patent has been issued, we are not prepared to admit the contention, but, since the case seems to have been tried upon the theory urged, we will in disposing of it be guided by the rule invoked.
>
> The trial court no doubt kept in mind the admonitions above quoted against considering a corner as lost. Indeed, we think it would have violated the rule invoked by appellant had it concluded that the corner in question is "lost." In the light of the rule quoted we must determine if the trial court could, beyond reasonable doubt, either from original traces or from other reliable evidence relating to the original position of the corner, determine its position.

A survey is not the only means of proving a boundary line; any witness who knows the fact may be competent to prove the existence of a marked boundary line. *Bolton v. Lann*, 16 Tex. 96 (1856).

Hodgman* had this to say about parol evidence. "Lastly (after attempting other proven procedures) I shall consider the evidence of living persons. Conceding all men to be equally honest in their evidence, there is a vast

---

* About corners, paper delivered to the *Second Annual Meeting of the Michigan Association of Surveyors and Civil Engineers*, Lansing, January 11–13, 1881.

deal of difference among them with regard to their habits of observation and their ability to determine localities. Some have an exceedingly acute sense of locality, if we may so call it, and can determine very accurately the position of any object which they have been accustomed to see; while others seem to have little or no capacity of that sort. In twelve years' almost constant practice in making this class of investigations, I have found many men who would describe accurately the sort of monument used to perpetuate a corner, and who would tell you that they could put their foot on the very spot to look for it; but when the trial came I have found but few of them who would locate the point within several feet, unless they had some object near at hand to assist the memory, and even then they would frequently fail."

> It may happen where a corner post has been destroyed that its location can be more nearly determined by the testimony of persons who were familiar with it when standing, and can testify to its relations to other objects in its vicinity, than in any other way. But the surveyor in receiving this testimony should ascertain, as far as possible, what are the habits of accurate observation, and the memory of localities possessed by the person testifying, in order to know how much weight to give his testimony.

Going back to *Galbraith v. Parker*,* "The rule seems to be equally well settled that where the original United States monuments indicating the location, upon the ground, of corners, have disappeared or have been lost or obliterated, and there is no evidence or proof as to the spot of their original placing, the plat and field-notes and the calls therein are determinative of the rights of the parties in disputed boundary questions." *Stangair* v. *Roads*, 41 Wash. 583, 84 P. 405; *Washington Rock Co.* v. *Young*, 29 Utah 108, 110 Am. St. Rep. 666, 80 P. 382; *Ogilvie* v. *Copeland*, 145 Ill. 98, 33 N.E. 1085; *Read* v. *Bartlett*, 255 Ill. 76, 99 N.E. 345.

"When excesses or deficiencies occur, the rule laid down by the Land Department is that they shall be apportioned and the inaccuracies in the original surveys must be equally distributed." In the case of *Wescott* v. *Craig* (Colo.), 151 P. 934, the rule is stated to be as follows:

> "Where on a line of the same survey, between remote corners, the whole length of which line is found to be variant from the length called for, in re-establishing lost intermediate monuments, as marking subdivisional tracts, we are not permitted to presume merely that a variance arose from the defective survey of any part; but we must conclude, in the absence of circumstances showing the contrary, that it arose from the imperfect measurement of the whole line, and distribute such variance between the several subdivisions of such line, in proportion to their respective lengths." 9 Cyc. 974.

---

* 17 Ariz. 369, 153 P. 283, Ariz., 1915.

The Lost Corner

The editor's note to *Caylor* v. *Luzadder*, 45 Am. St. Rep. 183 (137 Ind. 319, 36 N.E. 909), says:

> "If lines of a survey are found to be either shorter or longer than stated in the original plat or field-notes, the causes contributing to such mistake will be presumed to have operated equally in all parts of the original plat or survey, and every lot or parcel must bear the burden or receive the benefit of a corrected resurvey, in the proportion which its frontage, as stated in the original plat or field-notes, bears to the whole frontage as there set forth. *Pereles* v. *Magoon*, 78 Wis. 27, 23 Am. St. Rep. 389, and note [46 N.W. 1047]. This rule applies to government surveys. *James* v. *Drew*, 68 Miss. 518, 24 Am. St. Rep. 287, [9 So. 293].'"

The rule is stated in 4 R.C.L. 115, as follows:

> "When division lines are run splitting up into parts larger tracts, it is occasionally discovered that the original tract contained either more or less than the area assigned to it in a plan or prior deed. Questions then arise as to the proper apportionment of the surplus or deficiency. In such cases the rule is that no grantee is entitled to any preference over the others, and the excess should be divided among, or the deficiency borne by, all of the smaller tracts or lots in proportion to their areas. The causes contributing to the error or mistake are presumed to have operated equally on all parts of the original plat or survey, and for this reason every lot or parcel must bear its proportionate part of the burden or receive its share of the benefit of a corrected resurvey. This rule for allotting the deficiency or excess among all the tracts within the limits of the survey may be applied where the original surveys have been found to have been erroneous, or where the original corners and lines have become obliterated or lost."

The case of *Tolson v. Southwestern* Improvement Association* also addressed the problem:

> Where there is an excess of land in one section and a scarcity in another, caused by a deflection from the true course in running the dividing line, the excess is not to be carried anywhere, but is to be left where it falls. 23 Ark. 710. The rule that all fractions must be thrown on the north and west where the surveys are closed is sound, but not applicable in this case because (1) it is instruction to government deputy surveyors, and not addressed to county surveyors, who are bound by law to conform their surveys to the original. Kirby's Dig. § 1136. (2) The excess out of which this suit has grown is not due to an irregular length of the township. Land Laws U. S. 511; 23 Ark. 710.

---

* 97 Ark. 193, 133 S.W. 603, Ark., 1911.

324   *Boundary Retracement*

In the Missouri case of *Hale v. Warren*,* the surveyor testified about the shortcomings of the rule: [he] stated that double proportionate measurement "is the method of last resort when all other methods of surveying have completely been exhausted ..." and it was employed if the original corner has been lost. He also related that a corner is considered lost if it is "lost beyond a reasonable doubt" and cannot "be monumented by acceptable evidence or testimony." He further testified that relocating corners using double proportionate measurement almost never "results in the corner being placed exactly where the old land surveyor put it in the original records," but, nevertheless, is the statutorily required method for re-establishing lost corners.

It should be kept in mind, however, at all times, that the goal of any procedure is to place the corner where the original surveyor placed it or would have in the case of an unmarked corner.

> Lines actually run and marked on the ground are the best evidence of the true location of a survey, and these may be proven by any evidence, direct or circumstantial, competent to prove any other disputed fact. (5 Cyc., 962.)

Furthermore, it is a fundamental principle of law that boundaries are to be located on a resurvey where the original surveyor ran the lines and called for them to be located in his field notes.[†]

Newly Discovered Evidence

Every so often, a corner is found after the fact, after being given up as lost and measures taken, or, rarely, after litigation and a decision rendered. There are two cases, one within the PLSS and one outside of the PLSS, that are noteworthy with regard to bona fide evidence discovered "after the fact."

Back in 1895, the Colorado court in the case of *Snider v. Rinehart*,[‡] the monument or stone marking the north corner of sections 31 and 32 not having been found, it was assumed to be a lost corner, and, there being a shortage or deficiency on the north line of the two sections, the shortage was apportioned, and the corner established, and by such apportionment the caverns in question were shown to be in section 32.

> But shortly after this trial, and some time during the month of June, 1888, plaintiff alleges that he by accident discovered the stone marking such

---

* 236 S.W.3d 687, Mo. App. S.D., 2007.

† *Thatcher v. Matthews*, 101 Tex. 122, 105 S.W. 317, 1907; *Bolton v. Lann*, 16 Tex. 96, 1856; *Falby v. Booth*, 16 Tex. 564, 1856.

‡ 20 Colo. 448, 39 P. 408, Colo., 1895.

The Lost Corner 325

corner, in position, where it was placed by the government surveyor at the time of making the original survey, in 1871; and avers that the country around and about said section corner is wild, broken, and mountainous, and very precipitous, the surface of the ground being covered with or made up of broken fragments of stone and rock slag, and that all efforts theretofore made to find the corner were unavailing; and that he had, by the exercise of diligence, been unable to discover its whereabouts in time to produce evidence of its location at the former trial; that by means of such corner the true boundary between sections 31 and 32, as established and subdivided by the United States government, survey, could be readily determined; and that such boundary line shows each and every part of the caverns in question to be entirely and wholly within the boundaries of section 31.

To obtain a new trial in equity on the ground of newly-discovered evidence, the complainant must show that the evidence was not discovered in time to be used in the legal proceeding. If discovered in time to have been presented upon a motion for a new trial in the legal action, relief will be denied in equity." 1 High, Inj. § 116; Ferrell v. Allen, 5 W.Va. 43; 3 Pom. Eq. Jur. § 1365; Long v. Smith, 39 Tex. 161.

The United States statutes relating to the survey of the public lands provide, in section 2395, that such "lands shall be divided by north and south lines run according to the true meridian, and by others crossing them at right angles, so as to form townships of six miles square. * * * The township shall be subdivided into sections, containing, as nearly as may be, six hundred and forty acres each, by running through the same, each way, parallel lines at the end of every two miles; and by marking a corner of each of such lines, at the end of every mile. The sections shall be numbered respectively, beginning with the number one in the northeast section and proceeding west and east alternately through the township with progressive numbers till the thirty-six be completed. * * * Where the exterior lines of the townships which may be subdivided into sections or half sections exceed, or do not extend six miles, the excess or deficiency shall be specially noted, and added to or deducted from the western and northern ranges of sections or half sections in such township, according as the error may be in running the lines from east to west, or from north to south."

Court refused to grant a new trial based on newly discovered evidence, citing the death of two of the parties and the length of time elapsed constituting laches............................

Dissenting Opinion:

The circumstances under which equity will interpose because of newly-discovered evidence are summed up by Black, in his work on Judgments (volume 1, § 386), as follows: "(1) The evidence must have been discovered since the trial. (2) It must be evidence that could not have been discovered before the trial by the plaintiff or defendant, as the case may be, by the exercise of reasonable diligence. (3) It must be material in its object, and such as ought, on another trial, to produce an opposite result on the merits. (4) It must not be merely cumulative, corroborative, or collateral."

Are these conditions met by the facts found by the court below? The "Snider Monument," as it is termed, is found to be the true government corner as established by the original government survey, and in place as originally located. Plaintiff used reasonable diligence to discover it before the trial at law. Adopting it as the true north corner for sections 31 and 32, a line correctly run therefrom to the south township line between these sections places the caverns in question in section 31. If all these things be true,—and we must assume they are on this appeal,—it would seem to admit of but little question that the appellant exercised the requisite diligence, or as to the sufficiency of the newly-discovered evidence, if verified to the satisfaction of the jury, to produce a different and decisive result on another trial.

But it is said that if it be conceded that the newly-discovered stone is properly authenticated, and admitted to be where it was originally placed by the government surveyor, it appearing that it was incorrectly placed, as shown by the field notes of the original survey, it will be inequitable to give appellant an opportunity to utilize it, even if a true corner, by awarding him another trial. Suffice it to say, the survey as made and marked upon the ground, whether incorrectly or not, fixed the boundary line between these sections, and it is not the province of the courts to correct government surveys of public land, or establish lines contrary to such surveys, however, incorrect they may be. As was said in Cragin v. Powell, 128 U.S. 691, 9 S.Ct. 203: 'Whether the official survey * * * is erroneous * * * is a question which was not within the province of the court below, nor is it the province of this court to consider and determine. The mistakes, and abuses which have crept into the official surveys of the public domain form a fruitful theme of complaint in the political branches of the government. The correction of these mistakes and abuses has not been delegated to the judiciary. * * * That the power to make and correct surveys of the public lands belongs to the political department of the government, and that, while the lands are subject to the supervision of the general land office, the decisions of that bureau in all such cases, like that of other special tribunals upon matters within their exclusive jurisdiction, are unassailable by the courts, except by a direct proceeding; and that the latter have no concurrent or original power to make similar corrections,—if not an elementary principle of our land law, is settled by such a mass of decisions of this court that its mere statement is sufficient. * * * The reason of this rule, as stated by Justice Catron in the case of Haydell v. Dufresne, is that 'great confusion and litigation would ensue if the judicial tribunals, state and federal, were permitted to interfere and overthrow the public surveys on no other ground than an opinion that they could have the work in the field better done, and divisions more equitably made, than the department of public lands could do.' 17 How. 30.' Lands are granted by the government according to the official survey, and by his patent from the government the appellant acquired title to the land conveyed thereby as described and designated by such survey, and the locus of his land is to be ascertained by reference to it

The Lost Corner

and the original landmarks placed on the ground by the government surveyor. It is equitable that this should be done in this case.

It is undisputed that about June 1, 1881, appellant discovered the caverns in question, and took immediate steps to pre-empt and acquire title to that portion of section 31 which, according to the government survey, as actually made, included them; and, whether correct or not, the location of the E. 1/2 of the N.E. 1/4, and the N.E. 1/4 of the S.E. 1/4, not only must but in justice ought to be determined by such survey, and not by the survey upon which the lines were established at the last trial, based, as it was, upon the theory that the true corner was lost. The cases are numerous which sustain the proposition that monuments placed upon the ground by the government surveyor control, when a discrepancy exists between them and the courses and distances as given in the field notes, and I do not find any case in which it is held that the extent of the discrepancy changes this rule. Among them are the following: Mayor of Liberty v. Burns, 114 Mo. 426, 19 S.W. 1107, and 21 S.W. 728; Knight v. Elliott, 57 Mo. 317; Climer v. Wallace, 28 Mo. 556; Bruckner's Lessee v. Lawrence, 1 Doug. (Mich.) 19; Nesselrode v. Parish, 59 Iowa 570, 13 N.W. 746; Johnson v. Preston, 9 Neb. 474, 4 N.W. 83; Thompson v. Harris (Neb.) 58 N.W. 712; Ogilvie v. Copeland (Ill.) 33 N.E. 1085; Beardsley v. Crane, 52 Minn. 537, 54 N.W. 740; Hess v. Meyer, 73 Mich. 259, 41 N.W. 422; George v. Thomas, 16 Tex. 74; Pollard v. Shively, 5 Colo. 309; Cullacott v. Mining Co., 8 Colo. 179, 6 P. 211.

Why should the mistake of the government, in disposing of its land upon an incorrect survey, be invoked to destroy the title of a bona fide purchaser in actual possession? And why is not such a purchaser's claim to the land entitled to recognition and protection in equity as well as at law? It seems to me that the rule invoked by appellees, and announced by the chief justice, that the mistake of the government surveyor should be visited upon the unoffending head of appellant, is somewhat in the nature of a vicarious punishment.

A more recent case is *Rautenberg v. Munnis*,[*] whereby the New Hampshire court stated, "This is a motion by the plaintiffs for a new trial in an action to determine a boundary line. The case has twice been before this court."

The motion is founded upon the discovery of new evidence, in the form of a plan dated 1902, which the plaintiffs allege "clearly indicates that the boundary line * * * is as claimed by the petitioners." The Trial Court found that the parties were not at fault in failing to discover the plan prior to the original trial, but also found that the plan "only confuses the issue and falls short of the requirements of RSA 526:1." The questions raised by the plaintiffs' exception to denial of this motion were reserved and transferred by the Presiding Justice.

The findings which are prerequisite to the granting of a new trial under RSA 526:1, on the ground of newly discovered evidence, have

---

[*] 109 N.H. 25, 241 A.2d 375, 1968.

328   *Boundary Retracement*

long been clearly defined. A new trial will be considered to be equitably required only when the following conditions are met: (1) that the moving party was not at fault for not discovering the evidence at the former trial; (2) that the evidence is admissible (*Small v. Chronicle & Gazette Publishing Company*, 96 N.H. 265, 267, 74 A.2d 544), material to the merits, and not cumulative; and (3) that it must be of such a character that a different result will probably be reached upon another trial. *McGinley v. Maine Central Railroad Co.*, 79 N.H. 320, 109 A. 715; *Haney v. Burgin*, 106 N.H. 213, 218, 208 A.2d 448.

The issue presented by the motion is one of fact for the Trial Court, and its decision is binding in this court unless it can be said to conclusively appear that a different result is probable, so that the Trial Court's conclusion is clearly unreasonable. *Cf. McGinley v. Railroad*, supra, 79 N.H. 322, 109 A. 715.

As appears from the opinion in *Rautenberg v. Munnis*, 108 N.H. 20, 226 A.2d 770, supra, the dispute centers about a common boundary line which constitutes the plaintiffs' northerly boundary and the defendants' southerly line. The westerly terminus of the line is not in dispute. The easterly terminus was described in the deed to the plaintiffs in 1950 as a "pointed boulder at the shore with an iron pipe driven on the shore side." This they claimed at the former trial was at a location then referred to as "point D." The defendants, whose predecessor in title was grantee under the first conveyance out by Jones, the common grantor of the parties, claim that the easterly terminus was an "iron set in a large boulder near high water mark at the shore" as specified in the deed from Jones to Flanders. Their evidence was that this point, called "point K" was marked by a drill hole in such a boulder, in which the "iron" had never been set. Point K was about 150 feet southerly along the shore from point D. In finding point K to be the true bound, the Court at the first trial found "persuasive" the similarity between the drill hole in the boulder at that point, and the drill hole in the ledge at the northwest corner of the defendants' tract, which was an unquestioned monument there.

The plan which the plaintiffs now offer as newly discovered evidence is a blueprint entitled "Plan of Broad View Park on Shore of Lake Winnipesaukee, Alton, N.H." It bears the date 1902 and the name L. E. Scruton, C. E., who was an engineer, now deceased. The plan shows 97 shore lots, number 21 of which includes "Rum Point," and bears the name "Flanders" which appears to have been partially erased. This lot the plaintiffs contend is the lot described in the deed from Jones to the defendants' predecessor Flanders which was dated December 2, 1901 and recorded in October, 1903. It is not disputed that the plan was never recorded, and is not referred to in any deed involved in the litigation.

A comparison of the plan with the description in the Flanders deed discloses a number of discrepancies. While both indicate an iron hub as a monument at the northwest corner and the same measurement southwesterly to a "spotted pine," the plan indicates a "sp. Birch" at the southerly end of a 111.3 foot measurement on the westerly line, while the deed refers to no such monument or measurement. The total length of the westerly boundary is the same, but the deed gives an "iron hub set in the

The Lost Corner

329

ground" as the southerly terminus, while the plan shows no monument at the southwest corner. At the disputed southeasterly corner, the plan shows "an iron hub" by the shore line, while the deed refers to an "iron set in a large boulder near high water mark."

The adjoining lot, shown on the plan as number 22, in the locus of the plaintiffs' land, bears no resemblance to the tract conveyed by Jones in 1934 to the plaintiffs' predecessor, Rollins, and in 1950 by him to the plaintiffs. The plan shows a lot measuring 180 feet along the road on the west; 261 feet on the south on a course of north 79 30 west, and 100 feet on the east, across the shore line to the "iron hub," also marking the southeasterly corner of Lot 21. Proceeding southerly from Lot 22, Lots 23 through 31 all appear to be divided by boundaries running north 79 30 west, giving each lot a width from north to south of approximately 100 feet. As to these lots the plan shows no monuments of any description.

Thus the plan is significant only as it bears upon the lot now owned by the defendants, first described in the Flanders deed of 1901. It fails to support that deed's description of an "iron hub set in the ground" at the southwest corner of the tract, but confirms the deed description of the common boundary as 250 feet in length. It indicates at the southeast corner an "iron hub" at the water mark, rather than an "iron set in a large boulder near high water mark" as specified by the Flanders deed or a "pointed boulder ... with iron pipe driven on shore side" as claimed by the plaintiffs. It gives no area for the lot, and no course for the disputed line.

As argued by the defendants, the course of the disputed line on the plan appears to run approximately south 63 or 64 degrees east striking the shore line fifty feet or more below the bound claimed by the plaintiffs. The plaintiffs offer to prove that "after allowance for changes in the magnetic field over the intervening years" the line on the plan is only four degrees below the line claimed by them, but is twenty-nine degrees above the line found by the Trial Court. In any event it must be compared with the course of "south about 73 degrees east," as given in the deed to the plaintiffs, or south 74 degrees 40 minutes east, as surveyed by their surveyor to the point D; and similarly, with south 44 degrees and 22 minutes east (or nearly due southeast) which is the course of the line to point K, as claimed by the defendants.

It is apparent, however, that as evidence of the location of the southeasterly end of the common boundary line the plan would lend dubious support to the plaintiffs' claim. The plan's designation of an "iron hub" at the northwest and southeast corners only, tends to confirm the Trial Court's previous finding of similarity in the two drill holes located at these corners where no iron hubs were found. The deed given by Jones to Rollins in 1934, which referred to the 1901 deed to Flanders, and described the southeasterly corner as a "point * * * *supposed to have been marked* by an iron set in a large boulder near high water mark" (italics supplied), would continue to cast doubt upon the accuracy both of the plan and of the plaintiffs' contention. The Scruton map fails to resolve other discrepancies between the various deeds and the claim of the plaintiffs. The absence of any reference to it in the conveyances, and of any record of it in the registry serves to further detract from its authority. The likelihood of acceptance of its location of the disputed line in preference to both of the claims presented at the trial appears to be minimal."

The finding and ruling of the Trial Court that the plan "only confuses the issue and falls short of the requirements of RSA 526:1" cannot be held to be plainly unreasonable. "The petition was heard by the judge who presided at the trial, and it would be a very clear case which would permit his conclusion of the probable effect of the new evidence upon another trial to be classed as unreasonable." *McGinley v. Railroad*, supra, 79 N.H. 320, 322, 109 A. 715.

Not finding the footsteps may result in not finding the corner, or stated another way, may be the reason the corner was not found. Not finding the corner, and applying the accepted rule, will likely result in monumenting an incorrect position. Then, later, when the true corner is discovered, and titles have been passed based on one or more incorrect surveys, there may at the least be clouded titles, at the worst, disputes followed by subsequent litigation.

Not Getting It Right Only Invites Future Problems

In this example, a government surveyor encountered four different opinions of the location of a corner, represented by four markers, all of which purported to mark the same corner (an extreme example of a "pincushion corner."). The government surveyor proceeded to locate four acceptable corners, and marked the problem corner by proportioning. Returning at a later time, for a different reason, this same surveyor by accident discovered the original corner, a distance away. Five markers now exist in the vicinity of the corner, none of which is correct. The problems that resulted, however, are that deed descriptions have been rewritten reflecting the various surveys. This may have clouded otherwise clear titles, by (1) attempting to convey land without a basis of title, and (2) omitting remainder parcels that should have been included (Figure 11.1).

Obviously, apportionment was incorrect since the corner was not actually lost even though several people thought it was; agreements are inappropriate since the corner could be, and subsequently was, found (all along, the means were within reach, as the court stated in *Williams v. Barnett*).

One of the modes by which such lost lines and corners may be restored is fully discussed in the case of *McClintock v. Rogers*, 11 Ill. 279 (1849), and perhaps indicates one of the most satisfactory means of restoring lost corners.

The court stated, "The rights of the parties in the land in controversy, must be determined by the original survey, made under the authority of the government, and returned to the office of the surveyor general; and according to

The Lost Corner

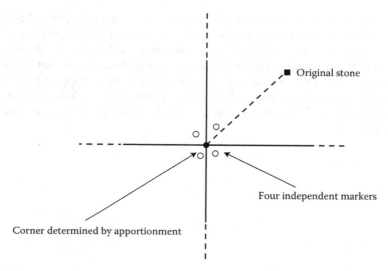

FIGURE 11.1
Diagram of foregoing problem situation.

that survey, the disputed township line was a straight one, running through the entire length, from corner to corner, and it can not now be varied to suit the interests or caprice of purchasers. To permit it to be done would be to annul the authority of all the public surveys; obliterate the lines of demarcation between the property of man and man, and open wide the door to confused, harassing, and endless litigation."

AUTHOR'S NOTE: It seems that it would behoove the retracement surveyor to do whatever it takes to find the *corner*. Failing to do this results in (1) relying on second best (proportionate measurement) which by its very nature cannot be a suitable substitute, even though it may well be the only remedy, and (2) taking a calculated risk that the corner will not be found at some point in the future. If the latter takes place, there may be a maze of titles and surveys to unravel and correct, likely necessitating litigation to clear titles.

AUTHOR'S ENDING NOTE: If you find the corner, your troubles are over. If you don't find the corner, your troubles have only begun; what then: bearings and distances? This is where surveyors are likely to disagree, due to different measurements, different points of beginning, different adjustment routines, different circumstances altogether (not to mention the inherent and cumulative errors in the current work, many of which cannot be identified, although surveyors will argue over whether one (usually theirs) survey is better than another (the other person's). Their arguments are usually based on closure, equipment, and/or computational methods, all of which are less than 100% reliable). Occasionally the argument will be over whether the corner used to start from is an, or the, appropriate corner. This process is a likely

332   *Boundary Retracement*

source of pincushion corners for which none of the subsequently set markers may be correctly marking the true corner. Boundary line agreement? This is where parties and surveyors violate the statute of frauds unless there is a legitimate dispute under the definition, causing chaos from that point forward unless, by some phenomenal stroke of luck, they happen to set their marker at the *precise* location of the true corner. Hold a fence, origin and purpose unknown? The discussion here is about a corner, not a monument or a marker.

12

Nonfederal Rectangular Surveys

Most of the northeastern United States and in scattered portions southerly down the eastern seaboard including Georgia was divided and lotted in rectangular fashion. Some areas were done in a formal manner, while others made accommodations for topography and resorted to other systems, for example, longlots, the French arpent system along water bodies, primarily along rivers.

Rectangular systems persist in western Kentucky in what is known as the Jackson Purchase, and in Tennessee, where at one time the entire state was divided in rectangular fashion by North Carolina, overseeing the land grants soon after Tennessee became a state. Most of Tennessee has since reverted to a metes-and-bounds system, although a few rectangular areas remain.

Texas, as well, with its own General Land Office (unrelated to the US GLO), has an abundance of rectangular lots and, along rivers, longlots.

When dealing with what is called "non-federal rectangular surveys," what is meant are those arrangements of rectangular lots, laid out by survey, usually mapped, often marked, and granted as such. They may occur singly as an individual grant, or in multiples such as with proprietary divisions and subdivisions of townships into smaller lots for individual acquisition. One of the biggest problems today is, represented by numerous decisions, the resulting maps are merely representative of the work done on the ground, and there is often found to be a difference between the two. There are several reasons for this. In short, the map merely shows the configuration of the lots, the placement of the ranges, and sequence of numbering in order to coordinate with the original grant, or any land descriptions in title documents.

Rectangular systems have long preceded the design and implementation of the PLSS. As the beginnings of American law relied heavily on European decisions and concepts, so did the layout and development of the wilderness of the New World encountered by Europeans. The Roman Empire, once encompassing much of Western Europe, developed a rectangular system of land and parcel organization known as the *Roman centuriatio*, or land survey. This has been characterized as "perhaps the oldest surviving form of land use survey in the Western World."[*] Present day Tunisia, established after the fall of Carthage (140 BC), began a period of complete land survey and

---

[*] Kish, G. Centuriatio: The Roman rectangular land survey. Surveying and mapping, *American Congress on Surveying and Mapping*. June, 1962.

333

334   *Boundary Retracement*

assignation of individual land parcels. Similar surveys have been found in parts of Italy, the Po lowlands, and in southern France.*

Studying this part of history, one will find a system of land utilization that is simple and straightforward, rational and working efficiently enough for it to survive, unchanged, for over twenty centuries. One will also find the layout of fields, roads, drainage ditches, and nearly all other works of man on the landscape accommodated in the rectangular grid. Roman maps demonstrate widespread use of the pattern throughout much of Western Europe, and elsewhere where the Roman Empire had exerted its influence.

Rectangular grants, lots, and parcels were often the preferred configuration. They were simple to create and lay out, and even simpler to divide. The only obstacle to complete rectilinearity of parcels is topography, where rivers, lakes, and mountain crests have become the preferred natural boundary. Even so, where one boundary of a grant was a non-straight component, usually the remaining boundaries (often three to complete roughly a square or rectangle) were straight, sometimes influenced in orientation by the natural boundary.

In short, it would seem that people tend to prefer working in somewhat of an orderly fashion, such as with a grid.

Beginnings in the New World

The first rectangular township laid out six miles square according to the cardinal directions, was Bennington, Vermont, which was chartered in 1749 (Figures 12.1 and 12.2).†

In America, especially in the Northeast, many of the towns were granted in rectangular fashion, then subsequently subdivided into squares, rectangles, and parallelograms. They were surveyed and mapped, although not all of the maps have survived or are otherwise unavailable or are in unknown locations. A few copies of originals, however, have survived and may suffice to illustrate a particular subdivision. Granting of lots, then, were usually done in a lottery, with lots designated by number, and recorded in the proprietors' records. Sometimes this is the source of title, and the origin of the original survey with its data.

One of the big mistakes people make is attempting to search for an original grant, or the original deed to a tract and hence the original survey, by searching in the usual repository, such as the registry of deeds. It is unlikely, except in a few rare cases, that the original title documents will be located there. In most instances, the documents will be found in the proprietors' records. The

---

* Ibid.
† Truesdell, W.A. The Rectangular System of Surveying, 1908.

Nonfederal Rectangular Surveys

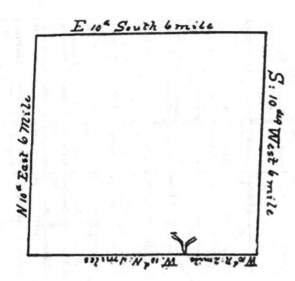

FIGURE 12.1
Original perimeter survey, Bennington, Vermont.

challenge then is to locate those records. If original town records survive, that may be the place to begin, and the very early records may include the proprietors' records of their transactions of business, including the layout of roads and the granting of individual lots. With a detailed report, in all likelihood the original grantee will be listed.

This became an issue in the New Hampshire courts, whereby we find the following early decision:

> The original proprietors of townships in this state are regarded as corporations, and are subject to the same general rules and regulations, and are invested with similar powers of corporations.
>
> Such proprietors may convey, or make partition of lands by deed, or by vote of the proprietary.*

Similar situations are found in other proprietary colonies, or where proprietorships of any sort were established. Where they owned large tracts of land, which was the usual case, they, with the aid of their surveyors, subdivided it by some scheme, and subsequently conveyed it.

Many original lots were conveyed only by lot and range designation, for example, T2 R6, a type of designation later followed in the PLSS. It is therefore essential to locate a copy of the original lotting plan, in order to learn the scheme of the particular lotting(s). Not all lots are of the same size, or even have the same orientation. Some contain exceptions for future roads, mill sites, and prior conveyances. Lot numbering can be a challenge

---

* *Atkinson v. Bemis*, 11 N.H. 44, N.H., 1840.

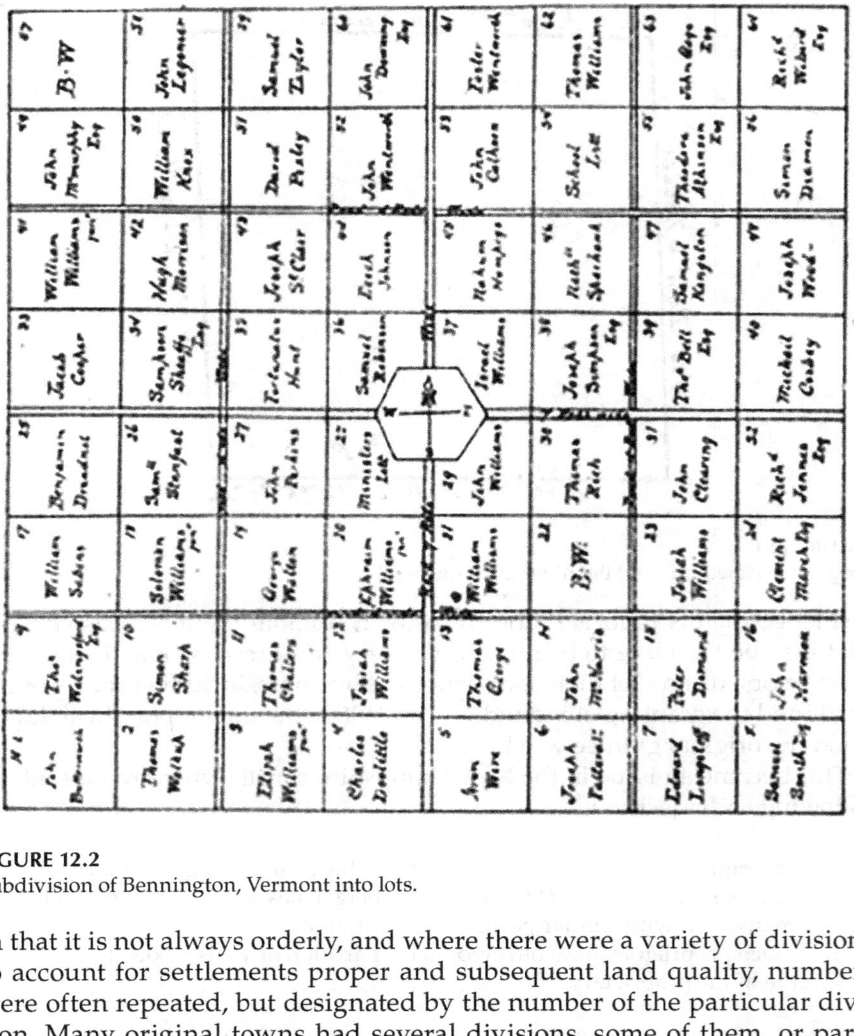

FIGURE 12.2
Subdivision of Bennington, Vermont into lots.

in that it is not always orderly, and where there were a variety of divisions to account for settlements proper and subsequent land quality, numbers were often repeated, but designated by the number of the particular division. Many original towns had several divisions, some of them, or parts thereof, later became new towns themselves. It is necessary to know the history of the particular lotting, otherwise it is very easy to become confused, or find oneself in the wrong part of town, or even the wrong part of the county.

Lots themselves were often further divided, where in present day we find relatively few original lots still intact, never having been subdivided. Only in very remote areas will original lots still be found intact without any outsales or subdivisions, areas such as northern Maine, New Hampshire, Vermont, and western New York, most of which became vast holdings by large timber companies. Scattered throughout the East will be found other remote areas with the same situation, parts of the Appalachian chain including

Nonfederal Rectangular Surveys

Pennsylvania, West Virginia, North Carolina, and Georgia, along with a few other isolated localities.

An original lot could be divided in a number of ways. Using the typical 100-acre lot as an example, two basic subdivisions are readily found.

Division by Half

Lots were often divided in half, or by other fractions. They were generally divided such that the "halfs" were east and west or north and south. Occasionally some other combination is found (Figure 12.3).

Division by Area

The alternative is to sell by acreage, such as so many acres from the east or west end, or abutting either the north or south boundary. Considering the theoretical acreage of the whole, removing the lots sequentially results in a remainder parcel which, if the lot was too large, would be larger than stated, whereas if the original lot was deficient in acreage, the remainder would be smaller than stated (Figure 12.4).

The common problem that arises here is, considering the lot was originally laid out as 100 acres, many believed that these two descriptions result in the same acreage. Since original lots tend, by design, to overrun, this often cannot be true. In the first instance, a half lot is simply one-half the entire acreage, whatever the total acreage is. To determine the area of a half lot,

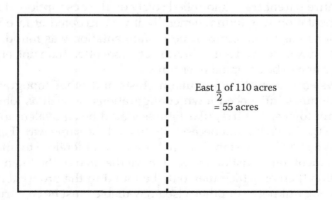

FIGURE 12.3
Dividing the original lot, 110 acres thought to be, and described as, 100 acres simultaneously.

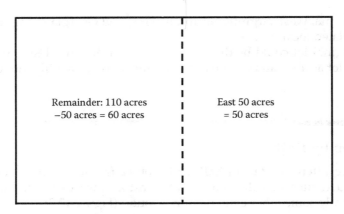

FIGURE 12.4
Dividing the original lot, 110 acres thought to be, and described as, 100 acres, sequentially.

the entire lot would have to be measured. In contrast, conveying 50 acres from what is believed to be a 100-acre lot (probably believing that 50 acres is actually a one-half even though it does not say that), leaves a remainder of whatever is left after 50 acres is removed from a part of the lot. If the original lot was not 100 acres by survey, being (likely) more than 100 acres, or in a few cases less than 100 acres, the remainder is whatever remains from the original lot after 50 acres has been conveyed away. Generally the remaining parcel will be found to be later conveyed as 50 acres, whereas it is more likely, in reality, more or less than 50 acres, and not exactly the 50 acres as believed. Again, the entire lot must be measured in order to remove the outsale (in this case 50 acres) leaving a remainder of something left to be determined, regardless of what the conveying document specifies.

That raises the second common problem with this type of conveyancing, that few retracement surveyors, especially in the past, took the time to survey the entire parent tract, and relied solely on the description. If some type of monumentation was found, generally it was accepted at face value, and not questioned as to its origin. If no monumentation was found, some creative solution was derived and markers set, more often than not, erroneously, but likely came to be accepted over time.

River systems, as well as mountain crests and other topographical features, sometimes influenced town configurations, as well as lot configurations within towns. The irregular figures would be equivalent to fractional sections in the PLSS, but not necessarily treated the same way (Figure 12.5).

Each town is different, standing on its own, and therefore to understand a lot arrangement, one must of necessity have the map of the town, or at least a description thereof, which may only be found in the proprietors' records. Those that are unknown or otherwise unavailable must be constructed from available data in the form of written descriptions and lines of occupation on the ground.

Nonfederal Rectangular Surveys

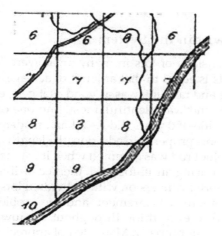

FIGURE 12.5
Lot configuration influenced by a river.

In northern Maine, for instance, many of the townships were created 6 miles square and oriented with the cardinal directions. They were also given a Lot and Range designation, surveyed and numbered westerly from the east line of the state, which is the boundary between Maine and New Brunswick, or the United States and Canada. However, not all of these towns have been further subdivided into lots, only some of them have been.

The Kentucky case of *Allen v. Blanton** is an early illustration of a rectangular (square) lot:

> Centers 1200 acres to join B's N.W. line the whole length thereof, on the waters of Glenn's creek, to lie between that creek and the first creek below it that runs into the Kentucky to extend northwestwardly so as to have the dividing ridge nearly in the middle of the survey.
>
> B had obtained a certificate for a settlement and pre-emption—and the preemption warrant had been assigned away and entered in the name of M before C's entry—C shall adjoin the line of the settlement of B not the line of M assignee of B.
>
> B's claim of 400 acres to be surveyed in a square to the cardinal points including his improvement at the intersection of the diagonals.
>
> C to adjoin B on the N. and W. the whole length of those lines—the diagonal of B's 400 acre survey to limit the base of C's 1200 acres—lines parallel to the general course of the ridge to be extended from each end of this base, until a line at right angles to these parallel projecting lines will give the quantity.

Note that B's 400 acres is to be a square laid out according to the cardinal points.

---
* 5 Ky. 523, Ky. App., 1812.

Idiosyncrasies within the System

The common dimensions of lots in many areas were 100×160 rods, since anything $\times 160$ rods is equal to that amount of acreage, so ease of computation, conveyancing and taxation was as good as it gets, even though not that accurate. But since land was plentiful, mostly no one cared. Unfortunately, surveys in the past 100–150 years did less than an adequate job and did not divide many of the lots properly. In addition, the length of the rod varied and occasionally a 16.5-foot rod was assumed when it was actually 18 foot. A few towns were laid out using an 18-foot rod, then later divided with a 16.5-foot rod. Finding the shorter conversion with individual lots leads one to believe that it was the standard measurement, and the problem arises when that standard is applied to everything throughout the town, and finding that, overall, things do not fit properly. Many do not appreciate this problem, and either set new markers, or give their clients poor advice. The chances are that a reasonable explanation exists if one searches for it, rather than just assuming that predecessors were less than careful in what they did.

Also, retracement surveyors, at least in the past, were too quick to apportion excess and deficiency in measurements, believing it was the quickest solution. Sometimes they used the wrong correction factor and in reality there was no excess or deficiency, other times they failed to find original monumentation, which, legally, would control the location of the grant, and other times they failed to take into account the usual allowances for rugged terrain, sag of chain, variations in temperature, and therefore produced an incorrect survey. Others, following in their footsteps, tended to accept the previous retracement without question or investigation, perpetuating any problems created by the first retracement surveyor. In some parts of the country, that practice persists to this day.

Bounding Descriptions

Those descriptions categorized as "bounding," or "abounding," or "bounded" are those which call for abutting parcels on all sides, or some of the sides. They easily lead one to believe that the resulting parcel is either rectangular or square. For example,

Bounded south by the highway, east by [land of] Brown, north by [land of] Smith, and west by [land of] Jones. The description may or may not recite an area of the parcel. The reader may easily be misled by this type of description, in that (1) it is probably not truly rectangular, and (2) the area figure may be merely an estimate. These are usually undeserving to be categorized as rectangular parcels, for most times they are not. Tracing the title back to

Nonfederal Rectangular Surveys

its origin generally results in a better description, ideally one with measurements based on an original survey.

Overcoming the lack of information with this type of description is discussed in Chapter 2, Title Issues. Also stressed is the incorrect procedure of relying on some combination of evidence, usually fences, that approximates the acreage call, which both courts and retracement surveyors know is the least reliable call of all.

AUTHOR'S NOTE: The author has always been under the belief that, based on original instructions, the original rectangular lots were surveyed, that is, measured, and likely, marked. If not marked, or otherwise surveyed, they may either fall under the category of protracted lots (drawn on paper only) or else subject to the rule laid down in *Larsen v. Richardson*, previously discussed.

Proprietors' Records

Another reason for examining proprietors' records is not only to find a description of the original plan, but also information in the records that is not likely to be found elsewhere. This record will also provide the first grantee of the parent tract which can be used as an element in the chain of title. Very much like field notes, proprietors' records contain information not found on plans, or in the deeds, or other title documents.

If the lots were not individually surveyed and marked, so that they only appear on paper (i.e., the plan—if available [many are not]) then the lots would necessarily be categorized as protracted lots (see Chapter 4 for a detailed discussion of this category). If tempted to undertake a retracement survey based on solely the plan and/or written description, the surveyor would be well-advised to be certain that no original survey or locations, of any kind, existed. Obviously, if there had been original survey work done, if only in part, the rules and requirements for following the footsteps apply.

Four decisions from the Vermont court provide excellent insight into the use and value of proprietors' records.

> That a line made by the proprietors in the first division of the town into lots above thirty years ago, is conclusive as a division of such lots; but a line run by the agreement of two owners of the lots is not thus conclusive; but may be weighed by the jury in connection with other testimony tending to confirm it as a division.

> *William White v. Dudley Everest*, 1 Vt. 181 (1828).

> It appeared from the proprietors' records, that the proprietors voted to divide the land in town into lots of 100 acres each, and that the committee appointed for that purpose, made and reported such survey

accordingly, and that the same was accepted and recorded. *Held*, that the presumption arising from the evidence afforded by said record was, that such a survey and division was made as was stated in said report, and that the burden of proof was upon the party who claimed a different survey and division.

Beach v. Fay, 46 Vt. 337 (1874).

The description of a township lot by reference to its number is a description in legal effect according to the lines of such lot as surveyed and established in the original division of the town, as definite as if the lines were given, and such description will determine the extent of the land.

The actual location on the ground of original township lot lines will control if ascertainable, but, when such lines have never been surveyed, or when their location on the ground cannot be ascertained, resort may be had to the lines of adjacent lots to determine their location.

In the absence of marks on the ground, a range line will be presumed to be a straight line; so that, where there were no marks on the ground of an original range line across disputed land, its location was determined prima facie by extending the range line from a point where its actual location could be ascertained.

Silsby & Co. v. Kinsley, 89 Vt. 263, 95 A. 634 (1915).

Actual location upon ground of original lot lines will control, if capable of being ascertained.

Town plan and description of lots in field book, *held* admissible on offer to show that lot lines of certain lots, as actually surveyed and marked on ground, were substantially as described in field book.

Presumption is that committee, chosen by proprietors to lay out and survey lots under charter giving committee authority to make allowances called for by it in their layout and survey, made all allowances permitted by law, and that same are included in lines of lots as shown by town plan and described in field book.

Where description in deed of lot is by reference to its number, lot lines, if surveyed upon ground, serve as monuments in fixing boundaries.

When lines have never been surveyed, or, if surveyed, their location upon ground cannot be ascertained, but lines have been actually run and marked upon ground, and have been recognized as correctly located for more than fifteen years by all parties in interest, such actual lines and monuments, marked upon ground, constitute survey, and will control courses and distances named in the original layout.

All lands are supposed to be actually surveyed, and where deed describes lot by its number, intent is to convey land according to that actual survey.

In such action, plaintiff, in measuring lots according to their descriptions in field book, held confined to courses and distances of lot lines as actually given in book, plus an allowance of one unit in thirty for swag in chain.

Nonfederal Rectangular Surveys

> Adjoining lot owners may by recognition and acquiescence in marked line as true dividing line between such lots establish division line of their lands, which will be binding upon them and their privies.
>
> *Neill v. Ward*, 103 Vt. 117, 153 A. 219 (1930).

Other States

As noted earlier, rectangular lots, of some sort, are found in all of the eastern states, and elsewhere. Some were among the earliest grants, and when occurring in the PLSS, they were honored as superior title, and rectangular lottings were designed around them. Mineral surveys, donation land grants, townsites, homestead entry surveys, and others are examples. In Florida, for example, approximately 120 Spanish land grants were made prior to the PLSS being established. Similar examples are found in most, if not all, other states.

Much of the western part of New York was divided in rectangular fashion when large land grants were subdivided. Rectangular and square towns were surveyed and marked, later to be further divided into squares and rectangles.

Areas of rectangular lots are commonly found in Pennsylvania, and parts of Ohio, especially in those land grants that predate the PLSS, such as the Connecticut Western Reserve, and some large private land grants.

Kentucky: The Jackson Purchase

The **Jackson Purchase** is a region in Kentucky bounded by the Mississippi River to the west, the Ohio River to the north, and Tennessee River to the east. Although technically part of Kentucky at its statehood in 1792, the land did not come under definitive US control until 1818, when Andrew Jackson purchased it from the Chickasaw Indians. Natives of Kentucky generally call this region *The Purchase*.

Jackson's purchase also included all of Tennessee west of the Tennessee River. However, in modern usage the term refers only to the Kentucky portion of the Jackson Purchase. The southern portion is simply called *West Tennessee*.

In 1820, the Kentucky General Assembly passed legislation that declared the public land system would be used for mapping the region by townships, ranges, and sections.

Kentucky land patents are divided into nine major groupings, each of which traces its origin to Acts of the Virginia or Kentucky General Assembly. In all instances the grantor is either the state of Virginia or the Commonwealth of Kentucky, and the grantee is the person or persons who receive the Governor's grant finalizing the patenting transaction.

344   *Boundary Retracement*

A few areas east of the Tennessee River are mapped by range-township; the rest of the state is mapped by metes-and-bounds. The Kentucky Land Office has patent records for the following Tennessee counties: Sumner, Smith, Robertson, Macon, Montgomery, Stewart, Jackson, Claiborne, Clay, Fentress, Pickett, Scott, and Campbell.

Tennessee Surveyors Districts

Early Tennessee law provided for Surveyors Districts, rectangular surveys, and appointed surveyors. The following is an excerpt from the early law.

1819—Chapter 1: An act making provision for the adjudication of North Carolina land claims, and for satisfying the same, by an appropriation of the vacant soil south and west of the Congressional Reservation Line, and for other purposes.

Sec. 1: Surveyor instructions and chain carrier oath

Sec. 2: Descriptions of districts 7–13

Sec. 3: Each of the surveyors by this act appointed and recognized, shall without delay cause his district to be divided by lines running parallel with the southern boundary line of the State, and by others crossing them at right angles, so as to form sections of five miles square, as near as may be, unless the exterior boundaries of his district may render it impracticable, and then, this rule shall be departed from, no farther than such particular circumstances may require. The corners of the sections shall be marked with progressive numbers from the beginning; each distance of a mile between the said corners shall also be distinctly marked, with marks differing from those of the corners; and the lines of the section distinguished by marks differing from other lines agreeable to instructions hereinafter given.

Sec. 4: It shall be the duty of the surveyors respectively to be caused to be marked on a tree, near each corner made as aforesaid, and within each section, the number of such section; and they shall carefully note in their respective field books, the names of the corner trees marked, and the number so made, together with all water courses and public roads over which the line he runs shall pass; the quality of the land, and the mountains or other remarkable objects, touched or crossed by a line or lines of the sections; and make return thereof to the principal surveyor, who shall therefrom make out a correct map of his district; designating the water courses, public roads and mountains, together with the division of his district into sections, and the surveys of appropriated lands, which may have heretofore

Nonfederal Rectangular Surveys

been granted or located according to law, and which may hereafter be surveyed agreeably to provisions herein after pointed out. The whole plan or map of the district shall be platted by a scale of 160 poles to the inch, the number of the sections shall correspond with the number directed to be marked on the trees with a sufficient margin, on which he shall distinctly mark the quality of the lands upon each line, distinguishing the same by colors descriptive of the quality; one fair plant of which shall always be open in his office for the inspection of any person who may have interest in obtaining a knowledge of the same; and one fair plat thereof he shall deposit in the office of the Secretary of State.

The French System of Longlots

Louisiana, settled by the French, contains an abundance of long, rectangular lots, most of which front on rivers, such as the Mississippi. The lots, also known as **arpent lots** (from the French system of measurement, both for distance and for area), also in some areas loosely known as **river lots**, are not confined to Louisiana however, and are found in many eastern locations, particularly in New England states, and most of the original colonies. Michigan and Texas also have many of these types of lots, the former having a designated area known as **The French Lots.**

This concept lent itself very well to land division since there was a lack of roads for access. The theory behind this concept is that the lots would front on a water body, providing access, then extend back into the upland region, providing bottom land, intermediate land for growing crops, and back land as a supply of wood.

Measurement systems in the various locations where longlots are found vary widely. More than one definition of the arpent was used depending on the particular location, rods used in some locations, and the vara being the preferred unit of measurement in Texas and parts of the Southwest (Figure 12.6).

Louisiana

Louisiana was settled under several different governments, which resulted in the French longlots, the square leagues of the Spanish, the irregular subdivisions in the Florida parishes which developed first under British control, then the American rectangular survey which encompasses about 85% of the state. The French longlot system was a natural fit to the levees and bayous of

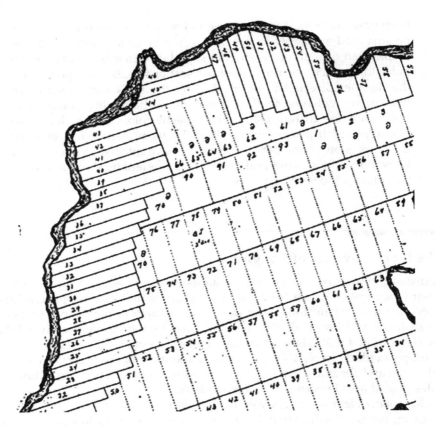

FIGURE 12.6
Example of longlots.

the southern part of Louisiana. Fronting on the stream and river banks and extending inland toward the back swamp land of the floodplain, provided an owner with a proportionate share of the best soils near the river, and the wetter soils toward the back. In addition, there was access to the water body for fishing and transportation. Marking the longlots was simple, since at first only the corners at the water were marked and directions given for the sidelines.

Many of the longlots were not surveyed when granted, but became fixed when American surveyors laid out townships and sections. The French were resistant to the American surveying of rectangles, believing that it made no sense to lay off squares that bore no relation to the usable land and might not have access to the streams. Eventually the American surveyors were authorized to survey land along streams in longlots of 160 acres.

West and northwest of the French areas, the Spanish introduced another type of land grant, known as the sitio. These grants were usually square, but often not oriented according to the cardinal directions and often a league on a side.

Nonfederal Rectangular Surveys

Texas

Texas is included within the states known as the metes-and-bounds states, although a number of original surveys and conveyances were made based on one or more rectangular schemes.

Texas lent itself to certain aspects of Spanish land division, similar to the Louisiana system of longlots.

The **sitio** grants discussed earlier in Louisiana actually were made much earlier in Texas. The unit of land measure used by the Spaniards was the **square league**, equivalent to 25 million square **varas**,* or 4428 acres. A league was the standard **sitio de granado mayor** for the raising of cattle and horses, a half-league (half a league on each side) could also be made. The phrase "one league to each wind" was coined as the surveyor's description of a square four-league tract more or less oriented with the compass.

Longlots, first laid out in 1731 as irrigated plots or **suertes**, were also used in Texas in Spanish layouts, but in a variety of forms. The river lots were used in the nonirrigated land along the lower Rio Grande River and its southern tributaries. In 1767, the policy was to grant original settlers two leagues of grazing land and twelve **caballerias** (400 acres) of cultivable land. More than 300 of the great longlots, known as **porciones**, were laid out along the Rio Grande in the 1760s, the largest of which was five miles wide by 17 miles long, containing well over 10 leagues of land.

The State of Texas

Texas was admitted to the Union in 1845, however there was no change in their land system. The public lands continued to belong to Texas and at the disposal of its government. Land transfer was by farm-size plots of 80–1280 acres, through the sale of scrip, preemption, homesteads, and veteran's bounties and donations. In addition, more than 30 million acres were granted to railroads, along with 50 million acres in support of schools, the university, and other educational endeavors. Railroads were granted 8, 16, or 20 sections per mile as encouragement. They were required to survey the lands into a checkerboard pattern, receiving the odd-numbered squares. The even-numbered squares were turned over to the public school fund, resulting in the surveys automatically divided between the railroads and the schools.

When Texas was annexed into the United States, it kept control of all of its public lands. As a result, Texas is the only US state to control all of its own public lands, unlike the PLSS states. Today, the head of the General Land Office is the Commissioner, an elected official. The office's main function

---

* The vara is a variable measure depending on the area of Spanish settlement; reported to be 33.33 inches in Texas.

348 *Boundary Retracement*

is to manage the public lands of Texas. When instituted, the GLO was overseen by a Surveyor-General, under whose instructions were deputy surveyors. Their task was to lay off assigned portions of territory, into sections, ranges, and townships, make maps thereof, and return them to the surveyor-general.

With the variety of granting systems in Texas under several different forms of government, timing is critical along with the history of the formation of the state, in order to understand the origin and development of a tract of land. More complicated than most states, the history of Texas serves as an important example of the importance of the history of a territory and the changes in its laws and policies over time. This philosophy may be applied to any of the states outside, and a few within, the PLSS, although Texas is no doubt more complex than most.

Longlots in Other Areas

Longlots have been found, primarily along rivers, in a variety of locations, primarily settled by the French. Northern Maine, New Hampshire, Vermont, Ohio, and Michigan have sizeable areas laid out in longlot fashion.

Lottings in Georgia

Georgia, as a colony, began with a headright system, but changed to rectangular lots laid out and covering large areas. Georgia held seven lotteries from 1805 to 1832, disposing of these lots.

Military Grants

As payment for military service (in many states) or as an enticement to become part of the militia (a few states, including Texas) military bounties were granted by the sovereign, which could have been a colony, a state, a republic, or other entity. Large grants were made to officers of the Seven Year's War (French and Indian War), the Revolutionary, War and the Civil War. Many of these were rectangular in form, at least in part.

Special Grants

Other grants for special purposes are found, some of them as large as a township, others smaller, but often rectangular, at least in part. Grants have been found to schools, the church, and the ministry, prisoners taken to Canada during the French and Indian War, mill sites, ferry landings, and other purposes. Today, many of these titles have not been completely resolved, and mostly the original surveys of the parcels are not easy to find. Chapter 14 contains a section of historical research, presenting modern problems traceable back to original titles in early times.

In Summary

There was relatively little granting of indiscriminate tracts of land when it came to the original creation of title and original surveys. Large areas of rectangular lottings of various sizes, shapes, and orientation dominated the landscape throughout the eastern part of the US, followed in 1785 with the Public Land System containing mostly square townships, sections, and lots.

The rectangular system of the Romans maintains its integrity, and to understand the concepts and how they were employed in practice is to understand the very underpinnings of beginning titles and original surveys in the US. Several treatises have touched on the plans and concepts over the years, especially the transition from early surveys into the Public Land Survey System. Rectangular systems are a history and study unto themselves.

13

Resolving Overlapping Grants

Seniority of Title

Many overlapping title situations have been resolved by determining which party holds the senior title. It has long been an established legal principle that the lines of the senior survey control, particularly where the junior is bounded with express reference to the elder.* There can be no overlapping of acreage where a junior survey expressly calls to begin at the senior.†

Several courts and prominent authors have shared insight and opinions regarding the surveying and location of original grants. Original lines often dictate the location and direction, even length, of subsequent lines, since parallelism and right angles, sometimes according to the cardinal directions, are the preferred methods of lot layout. Since all measurements contain errors of some sort, courts have taken that into account and allowed for it, successfully. As reviewed in Chapter 7, the federal court, in the West Virginia case of *Fay v. Crozer*,‡ quoted from Hutchinson on Land Titles, stating, "The result of this loose, cheap and unguarded system of disposing of public lands was that in less than 20 years after the adoption of the system§ nearly all of them were granted, the most part to mere adventurers, in large tracts or bodies, containing not only thousands, but in many cases hundreds of thousands, of acres in one tract. Often the grantees were nonresidents, and few of them ever saw their lands or expected to improve or use them for purposes other than speculation. The entries and surveys under warrants so cheaply and easily obtained were often made without reference to prior grants, thus creating interlocks, thereby covering land previously granted, so that in many instances the same land was granted to two or more different persons. Sometimes upon one survey actually located others were laid down by protraction—constructed on paper by the surveyors, without ever

---

* *Shackelford v. Walker*, 156 Ky. 173; 160 S.W. 807.
† *Stanolind Oil & Gas Co. v. State*, 101 S.W.2d 801.
‡ 156 F. 486, S.C.W. Va., 1907.
§ The "system" referred to was the institution of mandatory taxation and military service as obligations to the government and, when breached, the state may in turn seize the delinquent land.

351

going upon the lands, thus creating on paper blocks of surveys containing thousands of acres, none of which were ever surveyed or identified by any marks or natural monuments."

Overlapping grants mostly came about because of the lack of surveys and locations on the ground. Occasionally a person would locate their lot and, being less than satisfied with its location or its lack of quality, would move it to a more desirable location. It was sometimes possible to do that with "pitched lots," since the responsibility of the recipient was to locate their lot at a place that had not been previously granted, then make a return of its location in the pitch book so that everyone from that point forward could view what was spoken for, and therefore what was remaining. Locations with a lack of surveys and proper locations, overlapping grants in some areas seemed to be in abundance. They were found to be commonplace in Kentucky, Tennessee, and West Virginia. The following cases illustrate their treatment by the court system.

Proper retracement, that is following the original footsteps, will often lead back to original rectangular lottings which were usually surveyed (laid out and monumented). While some states have more of an abundance than others, less than adequate conveyancing and erroneous descriptions, surveying, and protraction have led to the overlapping of original grants. They have also led to gaps between grants, resulting in parcels left behind for which independent research is necessary. These will likely be found to remain in the original grantor or their heirs or successors in interest, but occasionally are lodged someplace else within the chain of title due to misdescription, incorrect retracements, and, occasionally, parcels deliberately left behind for any one of a variety of reasons.

Rectangular lottings were extremely common in Maine (part of Massachusetts), almost all of New Hampshire and Vermont, much of New York, much of Georgia, and a lot of Florida. No state east of the Mississippi River was exempt from rectangular lotting. As previously noted, nearly the entire state of Tennessee was originally lotted in rectangular fashion, but later reverted to the strict metes-and-bounds system of description. Then came the PLSS in 1785, whereby the remainder the US was lotted in rectangular fashion, excepting grants already made, such as the Spanish Grants in Florida, Ranchos in the Southwestern states, and others, also found in all of the eastern states.

As noted in Chapter 12, North Carolina granted the entire state of Tennessee according to several grids and the western part of Kentucky, the Jackson Purchase, was, and remains today, a rectangular lotting.

Many of the disputes were tried early on, some as time went on, and occasionally, a few in recent times. Four fairly recent cases illustrate the problems that are still being encountered: Bruker, Dolphin Lane, Ski Roundtop, and Barton, all discussed in Chapter 2.

We state that we must follow the original (surveyed) line, but what if two adjoining surveys either overlap, or leave a gap. Ski Roundtop had a missing

Resolving Overlapping Grants

title. The recent Tennessee Barton case talked about "no-man's Land." Senior vs. junior surveys and titles provide an answer, but what if they both have the same date, or subsequent overlapping occupation has crept in in the meantime? Dolphin Lane said the past must be explored to understand the present.

And *Wade v. McDougle** stated: "on the strength of paper title, the plaintiff showed no title; for to recover on paper as per se giving superior title, it must trace back to the state, unless the title of the contestants come from a common grantor. Ronk v. Higginbottom, 54 W.Va. 137, 46 S.E. 128."

One of the easiest ways to resolve overlapping titles with original grants, and therefore select the appropriate boundary line, is to review the appropriate decisions to understand the courts' determinations and the reasoning behind them. Because of the conditions at the time, and the less precise instrumentation and methodology, the arguments and disagreements are in abundance. Even if an original boundary does not apply to a situation, it may have had some past influence in subsequent conveyancing and boundary creation, thereby having relevance to the overall situation from its time of creation to present day.

Original Grants

Concerning original grants, guidance may be found in the early decisions of some of the colonial states, notably Kentucky and Tennessee.

One of the earliest reported cases was that of *Zachariah Herndon v. James Hogan.*[†]

> The plaintiff, on the 29th day of April, 1870, made the following entry with the survey of Kentucky county, to-wit:
>
> Zachariah Herndon, assignee of James Riddle, enters 100 acres of land, by virtue of a military warrant, in Kentucky, on the north side of Kentucky river, at Sweed's ford, and running north for quantity.
>
> The defendant, on the 4th day of December, 1782, made the following entry with the surveyor of Fayette county, to-wit:
>
> James Hogan, assignee, etc., enters on a military warrant, No. 818, fifty acres of land on Kentucky river at the mouth of Hickman creek, including the creek, and running down the river for quantity.
>
> Both entries being surveyed, the plaintiff, on the 30 day of March, 1785, entered the following caveat, to-wit:
>
> Let no grant issue to James Hogan, or his assigns, for 50 acres of land, surveyed for him on military warrant on the Kentucky river, on the north side thereof, at the mouth of Hickman creek, in Fayette county;

---

* 59 W.Va. 113, 52 S.E. 1026, W. Va. 1906.
† 2 Hughes 3, Ky, 1786.

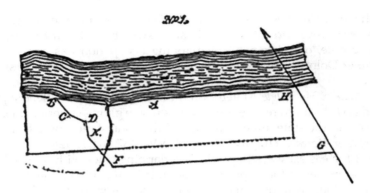

FIGURE 13.1
Diagram from the reported case of *Herndon v. Hogan*.

Zachariah Herndon claiming the same as being included in an entry made for him, prior to the said Hogan's, of 100 acres, made on military warrant, on and binding on the north side of the Kentucky river at Sweed's ford, which will be proved to be the same place; made in Kentucky county.

The annexed plat, No. 1, was returned by virtue of an order made in the cause (Figure 13.1):

A represents an old ford, and is the beginning of the plaintiff's survey, which is designated on the plat by the letters A B C D E F G H, and black lines binding on the Kentucky river. The defendant's survey is designated by dotted line, and binding on the river. The water courses are the Kentucky river and Hickman creek.

The following facts were found by a jury:

First. That there was a place on the Kentucky river called by the name of the Sweed's ford, both before, and at the time, and shortly after the plaintiff's making his entry, though it was known to but few.

Secondly. That the said ford is ten poles below the mouth of Hickman creek.

Thirdly. That the said ford acquired its name about the year 1771, and that there is no other place proved to the jury to have had that name.

Fourthly. That the ford of Kentucky river, below and near the mouth of Hickman creek, acquired the name of the mouth of Hickman about the fall 1775, and has generally been known by that name ever since.

The jury, in addition to this, found the dates of the two entries, and gave their opinions, as a *fact*, that they were neither of them surveyed agreeably to location.

BY THE COURT.—The plaintiff's entry is not sufficient in law to entitle him to the land in dispute. Caveat dismissed with costs.

The case of *Thomas Hinton, heir-at-law of Joseph Hinton, deceased, by William Morrice, his next friend, v. The Heir of William Stewart, deceased* follows shortly after.

Resolving Overlapping Grants   355

On a Caveat

On the 7th day of May, 1785, the plaintiff entered the following caveat, to-wit:

Let no grant issue to Wm. Stewart, his heirs or assigns, for 1,000 acres of land, lying and being in Fayette county, on the waters of Glenn's creek, it being a pre-emption granted to Thomas Glenn, deceased, and assigned by David Glenn, heir-at-law of the said Thomas, to the said William Stewart; because Thomas Hinton, heir-at-law of Joseph Hinton deceased, by William Morrice, his next friend, claims the same by an older improvement, for which the commissioners for the district of Kentucky granted a certificate to the said Joseph Hinton.

Joseph Hinton, on the 4th day of February, 1780, obtained from the commissioners the following certificate, to-wit:

Joseph Hinton this day claimed a pre-emption of 1,000 acres of land, at the state price, in the district of Kentucky, on account of marking and improving the same in the year 1776, lying on a fork of Glenn's creek, that heads toward Leestown, joining the lands of Cyrus McCracken, to include his cabin; satisfactory proof being made to the court, they are of opinion that the said Hinton has a right to a pre-emption of 1,000 acres of land, to include the said improvement, and that a certificate issue accordingly.'

And on the 30th day of May, in the year 1783, entered his pre-emption warrant with the county surveyor in the following words, to wit:

Joseph Hinton enters 1,000 acres of land, on a pre-emption warrant, No. 1,172, on the north fork of Glenn's creek, that empties into the said creek about two miles from the mouth of the creek, to include his improvement, the survey to be run so as to include the improvement a half mile on the west, and three-quarters of a mile on the south, and three-quarters on the east, and to extend north for quantity, at right angles, running the lines east, west, north and south.

The heirs of Thomas Glenn, on the 22d day of April, in the year 1780, obtained from the commissioners the following certificate, to-wit:

The heirs of Thomas Glenn, deceased, by David Glenn, this day claimed a pre-emption of 1,000 acres of land, at the state price, in the district of Kentucky, on account of marking and improving lands, in the year 1774, lying on the first fork of Glenn's creek, about two miles above the forks, including a small sinking spring improvement made by David Glenn. Satisfactory proof being made to the court, they are of opinion that the said heirs, etc., are entitled to the pre-emption of 1,000 acres of land, to include the above location, and that a certificate issue accordingly.

And on the 28th day of April, in the year 1783, Wm. Stewart entered the pre-emption warrant in the following words, to-wit:

William Stewart, assignee of the heirs of Thomas Glenn, enters 1,000 acres of land on a pre-emption warrant, No. 1,233, beginning on the north, or Blackford's fork of Glenn's creek, to begin on the upper side of said fork, on the course south 67 west, from a small sinking spring and improvement, and to run south 67 east, twice the width of the survey in

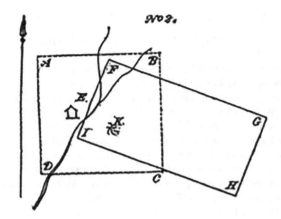

FIGURE 13.2
Plat from the case of *Hinton v. Stewart's Heir*.

length; then off at right angles north-westwardly and southwardly so as to include the quantity.

The annexed connected plat, No. 2, was returned in this cause, of which the following is an explanation (Figure 13.2):

A B C D, Thomas Hinton's pre-emption of 1,000 acres, as assignee of Joseph Hinton, as surveyed. E, Joseph Hinton's cabin. F G H I, Wm. Stewart's pre-emption of 1,000 acres, as assignee of the heirs of Thomas Glenn, as surveyed. K, the spring called for in the defendant's certificate.

The following facts were found by a jury:

First. That the spring in the survey of the defendant is the same granted by the court of commissioners to the heirs of Thomas Glenn.

Second. That the cabin in the plaintiff's survey is the same granted by the commissioners to Joseph Hinton.

Third. That the defendant did begin his survey, on the upper side of Blackford's fork, in a course south 67 west, in a direct line from a small sinking spring and improvement.

Fourth. That there was a division line established between Hinton and David Glenn, as agent for Thomas Glenn, and that the defendant has not run agreeably to the conditional line.

Fifth. That the division line was to run with the creek.

BY THE COURT.—The defendant ought to have surveyed his pre-emption on the east side of Glenn's creek. Judgment for the plaintiff for all the land in the defendant's survey which is on the west side of Glenn's creek, and is, also, within the bounds of the plaintiff's own survey. Order of survey, etc.

The case of *William Eagan v. Samuel Hinch heir-at-law of John Hinch, deceased, and John Jack, heir-at law of Samuel Jack deceased* is a relatively simple dispute between two titles.

Resolving Overlapping Grants 357

On Chancery

The complainant stated in his bill, that on the 26th day of April, in the year 1780, he obtained from the court of commissioners the following certificate, to-wit:

William Eagan this day claimed a pre-emption of 1,000 acres of land at the state price, in the district of Kentucky, on account of marking and improving the same in the year 1775, lying on a branch that runs into Greer's creek, about three-quarters or one mile west of Joseph Conway's, to include his improvement. Satisfactory proof being made to the court, they are of opinion that the said Eagan has a right to a pre-emption of 1,000 acres of land to include the above location, and that a certificate issue accordingly.

And that on the same day a certain Joseph Conway also obtained a certificate for a pre-emption of 1,000 acres of land from the said commissioners, in the following words, to-wit:

Joseph Conway this day claimed a pre-emption of 1,000 acres of land, at the state price, in the district of Kentucky, on account of marking and improving the same in the year 1776, lying on a small branch of Greer's creek, adjoining the lands of Isaac Greer and Isaac Power, to include his improvement. Satisfactory proof being made to the court, they are of opinion that the said Conway has a right to a pre-emption of 1,000 acres, to include the above location, and a certificate issue accordingly.

And that after the certificates were both issued, it was then agreed upon between the said Joseph Conway and the complainant, that a division line should be run between their respective improvements, at an equal distance from each, which line should be the boundary of each of the said claims.

In consequence of which agreement, having obtained a pre-emption warrant, he entered the same with the county surveyor on the 30th day of May, 1783, in the following words, to-wit:

William Eagan enters 1,000 acres of land on a pre-emption warrant, No. 937, on a branch of the north fork of Greer's creek, to include his improvement, and running half way to Joseph Conway's improvement that he obtained a pre-emption for, with a square line, and to extend as far on the one side of the course between the two improvements as on the other, and the survey to be four square.

He further stated that after the agreement aforesaid the said Conway sold his claim to John Hinch and Samuel Jack, who are both since deceased; land of whom the defendants are heirs-at-law; and at the time of selling the said pre-emption right, informed the said Hinch and Jack of the agreement aforesaid.

That the said Hinch and Jack having obtained a pre-emption warrant, entered the same with the county surveyor, on the 28th day of December, 1782, in the following words, to-wit:

Samuel Jack and John Hinch, assignees of Joseph Conway, enter 1,000 acres of land on a pre-emption warrant, No. 921, on the north fork of Greer's creek, joining the pre-emption of Eagan on the west, and joining Briscoe's or Dolan's on the north, and Fields' to the south, and running east and south-east for quantity.

Caused the same to be surveyed in the manner described in the connected plat, which is contrary to the agreement aforesaid, and obtained a patent of elder date than that obtained by him, on his survey, made strictly agreeably to his entry, which was made in conformity to the said agreement; and prayed for a conveyance.

John Jack, one of the defendants, who was of full age, admitted by his answer, that he had heard Joseph Conway say there was such an agreement as stated by the complainant, and said that he believed the complainant's claim was a just one, and that he was willing to convey.

The other defendant, being an infant, answered by guardian, and prayed that the complainant might be put on the proof of the agreement.

The annexed connected plat, No. 22, was returned in this cause, of which the following is an explanation (Figure 13.3):

1 2 3 4, William Eagan's pre-emption according to survey. 5 6 7 8, Hinch and Jack's heirs, assignee of Conway, do. 3 9 5 10, the interference. A, Eagan's improvement. H, and I, Hinch and Jack's or Joseph Conway's improvement.

The question of fact was submitted to the court. There were two depositions, William Crow's and John Sellers'.

Crow was present when a conversation took place between the complainant and John Hinch, deceased, in the course of which, the complainant told Hinch, he had run more than half way between the two improvements, and Hinch told him not to bring suit, for he would give it up.

And Sellers heard John Hinch say, he would not stand a suit about it.

There may have been other testimony also produced at the trial, but if there was, no memorandum of it is preserved.

BY THE COURT.—The only material questions arising in this suit, which have been submitted to the court, seem to be:

First. Was a dividing line between the two pre-emptions agreed on by the original claimants, Eagan and Conway to intersect at the half

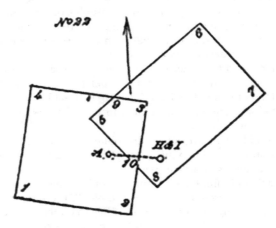

FIGURE 13.3
Plat in the case of *Eagan v. Hinch*.

Resolving Overlapping Grants

way point, a line which would pass through the center of their improvements, and to run at right angles thereto?

Second. Was this dividing line made known to, and acquiesced in by Hinch and Jack, who purchased the said Conway's pre-emption? Both points having been proved in the affirmative, to the satisfaction of the court, they are of opinion that the defendants Hinch and Jack are bound thereby; and therefore that the complainant Eagan recover from the said defendants all the land contained within their survey contrary to the said agreement, and the same is decreed accordingly.

Recent Decisions

It should be kept in mind that all titles may be traced back to an original grant. Some of them have lines coincident with an original grant boundary. The solution sought may be at the very beginning (creating) document, in identifying the basis of the problem, or discovering the solution to the problem. The old adage well applies here, "if you don't know what you started with, you can't know what you end up with." Or as the New York stated in the Dolphin Lane case, "the past must be explored to understand the present."*

In recent times, the Maine case of *Gammon v. Verrill*† is the determination of the location of an original range line, which dispute created what appeared to be an overlap of titles, but when resolved, demonstrated that it was, in fact, not such. Once again, it was a matter of following the original footsteps, and determining the correct location of the line separating the two titles, which was established sometime after 1796.

Gammon brought the action to settle a dispute about the location of the east–west boundary between their two parcels of land. The land in dispute is a triangular-shaped parcel containing approximately 36 acres. Gammon owns lot 1 range 6, which is directly south of the defendants' property, lot 1 range 7, in the Town of Bethel. The deeds in both parties' chain of title refer to the common boundary only as the range line between range 6 and range 7. Thus, the range line is the legal definition of the boundary.

The expert witnesses agree that a 1949 survey marker, located at the southwest corner of defendants' property and the northwest corner of Gammon's property, marks the location of the range line on the western side of the parcels (point 0) (See Figure 13.4). The controversy concerns the course of the range line east of point 0 as it crosses lot 1. Gammon claims that the range line bears approximately east 10° north from point 0 to the Bethel–Rumford town line. The range line asserted by Gammon has the same bearing as the range line of the adjacent lot and the southern boundary of the

---

* *Dolphin Lane Associates, Ltd. v. Town of Southampton,* 339 N.Y.S.2d 966, 72 Misc.2d 868, 1971.
† 600 A.2d 832, Me., 1991.

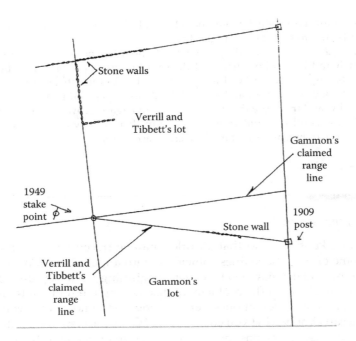

FIGURE 13.4
The included diagram in the case of *Gammon v. Verrill*.

Town of Bethel, and is parallel with the range line between ranges 7 and 8. Defendants claim the range line bears approximately east 10° south from point 0 to a post marked 1909 on the Bethel–Rumford town line. The lower court agreed with the defendants, finding that the boundary between the defendants' and the plaintiff's property ran from point 0 to the 1909 survey marker. Because the evidence provided neither original surveys nor original monuments, the court relied solely on usage in locating the range line. The trial court's location of the range line was clearly erroneous.

In boundary disputes, what the boundaries are is a question of law, but where the boundaries are on the face of the earth is a question of fact. *Liebler v. Abbott*, 388 A.2d 520 (Me.1978) (quoting *Rusha v. Little*, 309 A.2d 867 (Me.1973)); *Perkins v. Jacobs*, 124 Me. 347, 129 A. 4, (1925); *Abbott v. Abbott*, 51 Me. 575 (1863). This case does not involve a dispute as to the legal boundary in a deed; rather, it involves the factual question of where the boundary is located on the ground.

The court stated, "Although the evidence provided no original surveys or monuments, the court could have determined the approximate location of the range line based in part on the Original Plan of the Town of Bethel. The original plan is not drawn to scale, but it lays off the range lines. Several town plans were introduced into evidence, none of them identical, but they all have one feature in common: the range lines are parallel to the southern

Resolving Overlapping Grants

boundary of the town. The range line, as located by the court, deviates 21° from the bearing of the southern boundary of the town. The court erroneously discounted the usefulness of the original plan so far as that plan established that the range lines are parallel to, and on the same bearing as, the southerly boundary of the town."

> The southerly boundary of the Town of Bethel, as established by the Massachusetts General Court in 1796, was east 20° north. See 1796 Mass. Acts ch. 3. The declination for Oxford County has increased 8° 15' west since 1796.* Thus the southern boundary of Bethel today bears east 11° 45' North. Gammon's surveyor testified that the course of the range line is east 11° north.† Correcting the southerly boundary of Bethel for the change in magnetic declination, the bearing of the surveyor's range line is within 1° of the bearing of the town's southerly boundary.
>
> It is clear from this record that the range line must be approximately parallel to the south line of the town. The evidence including the survey locating the range line west of point 0, aerial photographs and defendant's own exhibit showing the range line between range 7 and 8 all establish that the range lines are parallel. Moreover, both parties' surveyors testified that the range lines in the Town of Bethel are parallel to its southern boundary. Evidence of usage as well as evidence of the marker erected on the east line in 1909 are insufficient, as a matter of law, to overcome the undisputed evidence of the bearing of the range line. Applying that bearing from point 0, the range line can be located without resort to evidence of usage. This would establish the range line where Gammon asserts it is located.

Another case tried not long after the above case was that of *DuPont v. Randall*‡ which involved a boundary dispute where the legal description of the common boundary was the line between Township Lot 156 and Township Lot 157 on Herrick's Plan of Etna. The judge directed the jury to locate the line on the face of the earth choosing either the line shown on the Duponts' surveyor's plan or the rock wall claimed by the Randalls. The jury found the boundary to be Randalls' stone wall.

The parties did not dispute the legal boundary; they disputed the location of that boundary and in such a case a factual finding will not be disturbed on appeal unless it is clearly erroneous. Because all of the evidence, including aerial photographs, show the rock wall generally deviates from the bearing of the township lot lines by 15 degrees, that wall cannot, as a matter of law, correctly locate the boundary. See *Gammon v. Verrill*, 600 A.2d 832 (Me.1991).

---

* The parties stipulated that the magnetic declination of Oxford County in 1796 was 9°02' west and is currently 17°17' west.

† Gammon's surveyor had previously located the range line west of point 0 while working on a survey for a property owner adjacent to Gammon's lot. This range line has the same bearing as the range line claimed by Gammon.

‡ 648 A.2d 437, Me., 1994.

362   *Boundary Retracement*

The jury in this case plainly reached the wrong decision. The trial court properly allowed the jury to reach a verdict, thereby avoiding the need for a second trial had we disagreed with the court's entry of a judgment as a matter of law. Having reviewed the record, however, we agree with the trial court that as a matter of law the wall cannot correctly locate the boundary.

Both these cases illustrate the importance of the original subdivision, i.e., the original layout of the town lots and their creation of their boundaries which, upon conveyance, became title lines which demanded proper retracement.

While an entire treatise could be written on this subject, a few principles supported by several examples may suffice to illustrate the point, and offer suggestions for resolution.

For example, the case of *Swift Coal & Timber Co. v. Sturgill*, et al.* provides some insight:

> The only question in this case is the proper location on the ground of a patent for 12,000 acres of land issued in 1846 to Isom Stamper. The effort, therefore, is to relocate the described boundary as did the surveyor in making the survey upon which the patent issued. The land is thus described in the patent:
>
> Beginning on three chestnuts and chestnut oak near the head of Pigeon fork; thence running the dividing ridge between Turkey creek and the Line fork to the Defeated branch; thence the dividing ridge between the Defeated branch and Turkey creek, S. 10° W. 68 poles to a chestnut; thence S. 10° E. 98 poles to a chestnut oak sarvis; thence S. 68° W. 125 poles to a stake; thence S. 5° E. 100 poles to a stake; thence S. 68° W. 2,000 poles to a stake; thence S. 40° W. 320 poles to a stake; thence N. 1,328 poles to a stake; thence N. 77° E. 1,900 poles to a stake; thence S. 33° E. 600 poles to the beginning.
>
> Upon the face of the patent it would seem that the described boundary had 11 sides, the first being the dividing ridge between Turkey creek and Line fork, the second the dividing ridge between Defeated branch and Turkey creek, and the other nine described by courses and distances; but this is not true, as is indicated by the plat filed with the certificate of survey, which shows only nine sides to the boundary, and as is conclusively proved by uncontradicted evidence that the first course and distance call in the patent, "S. 10° W. 68 poles," is from the beginning corner to a chestnut, the second corner, and that the line described as "S. 10° E. 98 poles," apparently the fourth line, is in reality, as surveyed and intended, the second line to the chestnut oak and service, the third corner. To thus survey these two lines and follow the other called-for courses and distances from the third corner around to the beginning corner will cause the patent to close, and as thus described it will contain approximately 12,000 acres of land as in the patent it was said to contain. Both parties to this litigation agree that so much is true, from which it

---

* 188 Ky. 694, 223 S.W. 1090, Ky. App., 1920.

Resolving Overlapping Grants 363

results that, upon a consideration of the patent, the primary evidences upon which it was issued (the certificate of survey and original plat), and the uncontradicted evidence that the first two lines described by courses and distances, the one to a chestnut, and the other to a chestnut oak and service, are in reality the first two lines of the boundary, and that the two ridges described must be rejected as independent and separate lines of the survey.

As to what effect, if any, must be given to these two described ridges is the real question involved in this case, as it was in two cases in the federal Circuit Court of Appeals involving the same patent; the one, *Mineral Development Co. v. Tuggle*, 151 F. 450, 81 C.C.A. 34, and the other, *Kentucky Coal Lands Co. v. Mineral Development Co.*, 219 F. 45, 133 C.C.A. 151. In both cases these described ridges were rejected as independent and separate lines of the boundary, and were held to be merely ancillary descriptions of the beginning lines of the survey, also described by courses and distances, but nevertheless of potency and to be considered, in connection with other relevant evidence, in locating the beginning lines. To this extent we concur in the conclusions reached in these cases, but this is as far as they agree or are relevant here. We have not the evidence before us upon which the Circuit Court of Appeals in the first case rejected entirely the ridges in locating the land in accordance with the plat, and the courses and distances in the patent, and in the second case held that the evidence was sufficient to carry to a jury the question of the proper location of the patent.

We come, therefore, to a consideration of the evidence in this case to determine what is the proper location of a patent which is ambiguous, because of the fact at least two, and possibly five, of its lines are twice separately and differently described. Some pertinent and controlling facts are thoroughly established. These are:

(C) The surveyor actually surveyed only the first two of the nine boundary lines he described. These two lines he marked by establishing and marking the three corners connected by them, and by marking line trees as well.

(2) These lines as run and marked by him are straight lines running the general course of the first of the two ridges referred to, but not following its sinuosities.

(3) The other seven boundary lines of the patent were not actually surveyed or marked by the surveyor, but were located by him simply by protraction, after the actual surveying on the land had been completed.

(C) The ridges referred to in the patent are well-known and permanently established objects.

(C) To follow the two lines the surveyor ran and marked out on the land, and the other seven lines that he protracted, according to the courses and distances called for, will follow the general course, but not the sinuosities, of the first of the two ridges referred to a part of its distance, but will not reach the second ridge referred to by about a mile; but the figure thus formed will conform to the plat he filed with his original survey, will close, and will contain approximately the number of acres called for in the patent.

364 *Boundary Retracement*

(C) To locate the patent by following the sinuosities of the first ridge referred to until the second is reached, and following the meanderings of the second ridge without regard to the courses, but for the approximate distance of the third, fourth, and fifth lines, will reach a point on the second ridge known as Eagle Gap, near, if not at, the end of that ridge, but which is not mentioned in the patent or certificate of survey, nor indicated upon the original plat. To close the survey thus started by starting the sixth line at Eagle Gap, the courses and distances called for on the seventh and eighth lines must be entirely disregarded, the figure will not conform in appearance to the original plat, and the boundary will contain only about half as many acres as called for by the patent and survey.

A certain and necessary inference to be drawn from these facts is that the surveyor, in locating this land for patent, so far as he located it upon the land, did not follow the sinuosities of the first ridge described, and therefore did not mean, when he said "thence running" these ridges, to follow the sinuosities of either, but intended only to follow their courses in a general way, as he had done so far as he ran them. This destroys absolutely appellees' theory that the combined distances of the third, fourth, and fifth lines were to be exhausted in following the meanderings of the top of the two ridges, and that at the point of exhaustion the sixth line was to begin. Not only so, but it leaves no way, even by disregarding both the distances and calls of the seventh and eighth lines, by which the boundary may be closed, so as to form a figure resembling the original plat or containing more than about half as many acres as were intended to be, and believed had been, included.

The only evidence offered by the appellees to justify such a location was proof that along the top of the two ridges were marked trees, which as early as 1863 had the appearance of old marks, and that Isom Stamper, from 1863, and possibly earlier, had claimed, and by general reputation in the community was believed, to own all of the land on the Turkey creek side of these two ridges to the tops of same. The preponderance of evidence upon the question of the age of marks upon such trees as are marked along the top of these two ridges is that they were not nearly so old as this patent, and there is no evidence whatever that at the time of the survey any trees, except at the first, second, and third corners and some trees on these two lines, were marked by the surveyor or with his knowledge. Upon the contrary, it is conclusively established that, when he reached the third corner which he marked and called for in his survey, he quit the field and completed his work elsewhere simply by protraction. This evidence for the appellee is therefore wholly insufficient to establish an original location of the patent along the meanders of the ridges, whatever might have been its effect if a question of adverse possession were involved. We are therefore clear that the chancellor erred, upon the evidence in this case, which includes, of course, the certificate of survey and original plat, in locating the first five lines of this patent along the tortuous course of the crest of the two ridges referred to in the patent.

But learned counsel for appellees argue most earnestly that the ridges called for are known and fixed objects on the land, and must therefore

Resolving Overlapping Grants

control in the location of the patent over courses and distances, where there is ambiguity in the description and uncertainty as to the proper location. This is unquestionably one of the most thoroughly established rules for construing and locating a patent, and when applicable often controls. The trouble here is not with the rule, but its application to the facts. The patent survey, plat, and extraneous evidence all prove that these known and fixed ridges were not run, or intended to be run, by the surveyor in the original location of this patent along their crests as they meander, but were to be followed only in a general way and by straight lines, two of which he surveyed and marked, about which there is no trouble whatever; but as to the other three straight lines, which he unquestionably intended should also follow these ridges in a general way, he has left absolutely nothing for our guidance in fixing their exact location, as we must do, but their courses and distances.

When we reach the end of the fourth line we are at a point on the side of the first ridge and at least a mile from the top of the second ridge. We might possibly extend this fourth line until it reached the second ridge in accordance with an approved rule of law (if we could tell when we reached the second ridge and at what point thereon to stop) and by adding an equal distance to the opposite (seventh) line close the survey with a boundary nearly conforming to the plat, although this would include nearly, if not quite, 17,000 acres, instead of the 12,000 acres intended and believed to have been included. But the fact that we cannot tell even approximately where we should end the fourth line, so as to lay the fifth line upon the second ridge as intended, precludes the possibility of our adoption of this plan to give effect to the reference to this ridge in locating the patent. We certainly would not be justified in extending the fourth line to the top of the second ridge, a distance of about a mile, even if we could be certain when we reached the second ridge, because the surveyor began his survey some little distance (about 150 feet) from the top and on the side of the first ridge, and the lines he surveyed (the first and second) and those protracted (the third and fourth), which follow the general course of the first ridge, do not follow the top. They are near the top merely, and cross it in one instance at least, and possibly oftener. But these lines furnish no certain location for the fifth line, which we must locate certainly from the evidence before us, and this we cannot do, except by the called-for courses and distances of the patent, which, after all, furnish the only possible method of locating this patent that is reconcilable with the patent, original survey, plat, and extraneous evidence of the original location as actually made.

We are not permitted to guess where upon the second ridge it was intended to lay the fifth line, nor can it be determined accurately from the five maps and evidence in this record where the first ridge ends and the second one begins, since the two are but one continuous ridge, with Turkey creek on one side and Line fork and Defeated branch meeting about midway on the other. Consequently the second ridge considered, as only it can be, as an aid in locating the lines otherwise and more accurately described, is of no practical value whatever, and of necessity must be disregarded entirely.

The case of *Tolson v. Southwestern Improvement Association** describes an interesting twist, having excess and deficiency.

> The excess of three chains and 62 links should be thrown on the north tier of lots. 23 Ark. 710. No change having been made by the government in the Allis survey of section 6, none can be made now. 88 Ark. 37. The southwest corner of this section, established by the government surveyors, controls the location of the lines between lots 11 and 18, though the section line may be deflected from a direct course. 85 Iowa 398. Each section is independent of every other. 7 Porter (Ala.) 428. In ascertaining boundaries, in the absence of any other controlling influence, attention should be given to the figure of the survey as shown on the original plat. 107 S.W. 307.
>
> Where there is an excess of land in one section and a scarcity in another, caused by a deflection from the true course in running the dividing line, the excess is not to be carried anywhere, but is to be left where it falls. 23 Ark. 710. The rule that all fractions must be thrown on the north and west where the surveys are closed is sound, but not applicable in this case because (1) it is instruction to government deputy surveyors, and not addressed to county surveyors, who are bound by law to conform their surveys to the original. Kirby's Dig. § 1136. (2) The excess out of which this suit has grown is not due to an irregular length of the township. Land Laws U. S. 511; 23 Ark. 710.

Opinion

> HART, J. This is a suit to settle the boundary line between lots 11 and 18 in section 6, township 9 south, range 10 west, in Cleveland County, Arkansas. The section is fractional, and the north half is divided into lots numbered from 1 to 18 inclusive. According to the plat of the original survey, both lots 11 and 18 are in the western tier of lots in the north half of the section, and lot 11 contains 38.98 acres, and lot 18 contains 38.41. Lot 11 is immediately north of lot 18. John C. Barnes became owner of both these lots by mesne conveyances from the United States Government. Appellant, George L. Tolson, by purchase, became owner of all that part of lot 11 lying south and east of the St. Louis Southwestern Railroad, containing 18 acres more or less, and lying adjoining and north of lot 18. John C. Barnes by deed conveyed lot 18 to the Southwestern Improvement Association, a corporation, which caused it to be surveyed and platted as the East Addition to the town of Rison. The other appellees bought lots from the corporation immediately adjoining lot 11, and bought with reference to the map or plat made by the corporation.
>
> The suit was instituted by appellant. He does not claim title to the land by adverse possession, but the prayer of the complaint is that the court

---

* 97 Ark. 193, 133 S.W. 603, Ark., 1911.

Resolving Overlapping Grants

establish the boundary line between lots 11 and 18. The testimony on the part of appellees does not establish an adverse claim to the strip of land in controversy for the statutory period. Hence there can be no question of title by adverse possession, as was the case in *Goodwin* v. *Garibaldi*, 83 Ark. 74, 102 S.W. 706, and that class of cases; and the sole issue raised by the appeal is, what is the true boundary between the parties?

The official plat of the United States Government shows that the boundary line between lots 11 and 18 is 60 chains north of the south boundary line of the section, and runs parallel with it; and it also shows that the western boundary line of the section from south to north is 130 chains. By actual measurement this line is 133 chains and 62 links. The evidence shows that the original survey was defectively made, and that the southwest corner of this section was established three chains and 62 links too far south; and this obviously makes a difference of that amount in the length of the west line of the section as shown by the official plat, and as it exists by actual measurement. Appellant claims that he is entitled to the excess because the official plat shows that the distance north from the southwest corner of the section to the northwest corner of lot 18 is 60 chains, that this excess of three chains and 62 links should fall on lot 11; but it will also be noted that the official plat shows that the distance north and south on the west line of lot 11 is 20 chains, and that the purchases were all made with reference to the public survey.

In the case of government sections, interior lines in the extreme northern or western tiers of quarter sections, containing either more or less than the regular quantity, are to be 20 chains wide, and the excess or deficiency of measurement is always to be thrown on the exterior lots; elsewhere the assumed subdivisional corner will always be a point equidistant from the established corners. This rule, however, has no application where the original surveys are found to be erroneous, in which case the excess or deficiency is to be apportioned to each subdivision within the boundaries where the corners are lost. 5 Cyc. P. 974, and notes.

In the case of *Caylor* v. *Luzadder*, 137 Ind. 319, 45 Am. St. 183, 36 N.E. 909, the court recognized the general rule, but said: "There seems to be a well-recognized distinction between this rule as applied to original surveys, whether in the making of such surveys or in allotting the deficiency or overplus, when the correctness of such surveys is not questioned, and that where such original surveys are found to have been erroneous or the original corners and lines are wholly lost."

The following authorities which we have examined are cited as recognizing the distinction: *Bailey* v. *Chamblin*, 20 Ind. 33; *Jones* v. *Kimble*, 19 Wis. 429; *Moreland* v. *Page*, 2 Clarke (Iowa) 139; *Westphal* v. *Schultz*, 48 Wis. 75, 4 N.W. 136; *James* v. *Drew*, (Miss.) 24 Am. St. Rep. 287.

The Supreme Court of Missouri has taken the contrary view. See *Vaughn* v. *Tate*, 64 Mo. 491; *Knight* v. *Elliott*, 57 Mo. 317.

Continuing, the Supreme Court of Indiana said: "The surveyor general was not required to, and did not, locate the half-quarter posts or line, and, having surveyed the quarter, established the lines and located the corners thereof, these defined irrevocably the boundaries or limits of the quarter; the purchasers and the Government acted upon the assumption

368   *Boundary Retracement*

that the lines were correctly measured and returned by the deputy surveyor; in this all were alike deceived; the length of lines is less than that so acted upon, and, by every principle of equity, the deficiency should be borne by the several tracts in proportion to the quantities so presumed to be contained therein at the time of the purchase." Kirby's Digest, § 1136 (referring to county surveyors), is as follows: "It shall be his duty, in subdividing any section or part of a section of land originally surveyed under the authority of the United States, to make his survey conformably to the original survey." The Revised Statutes of the United States, § 2395 *et seq.*, provide in substance that all corners marked in the surveys returned by the surveyor general shall be established as the proper corners of the sections or quarter sections, which they were intended to designate, and corners of half and quarter sections not marked shall be placed as nearly as possible "equidistant from those corners which stand on the same line," and that these boundary lines as actually run and marked "shall be established as the proper boundary lines of the section, or subdivisions, for which they were intended, and the length of such lines, as returned by either of the surveyors aforesaid, shall be held and considered as the true length thereof." It follows that the original township, section and quarter-section corners, as surveyed and established by the government surveyors, must stand as established. But, as to the division of quarter sections, there is no actual survey, and the quarter-quarter corners are placed on straight lines joining the section and quarter-section corners, and midway between them, except on the last half mile of section lines closing on the north and west boundaries of the township, or on other lines between fractional sections.

The land department of the United States has sent out the following rule to be observed in the subdivision of quarter sections into quarter-quarters: "Preliminary to the subdivision of quarter sections, the quarter-quarter corners will be established at points midway between the section and quarter-section corners and between quarter corners and the center of the section, except on the last half mile of the lines closing on the north or west boundaries of a township, where they should be placed at 20 chains proportionate measurement to the north or west of the quarter-section corner." Recognizing that there may be differences in the measurements, the following is added: "By proportionate measurement of a part of a line is meant a measurement having the same ratio to that recorded in the original field notes for that portion as the length of the whole line by actual resurvey bears to its length as given in the record." Restoration of lost or obliterated corners and subdivision of sections, revision of June 1, 1909, General Land Office, pp. 22 and 23.

It follows that the line established by the court below is not the true line; for it places the whole excess upon the land of appellees. The quarter-section corner on the west side of the section is fixed by the original survey as actually made, and must stand as the true corner. Hence the excess of three chains must fall on the western tier of lots extending from there to the northwest corner of the section. The official plat shows that the distance from the quarter-section corner on the west side of the section to the northwest corner of the section is 90 chains, when by actual

Resolving Overlapping Grants 369

measurement it is 93 chains and 62 links. The official plat also shows that the west line of lot 18 is 20 chains, and that of lot 11 immediately adjoining and north of it is 20 chains. By the rule announced above lot 11 is entitled to its proportionate part of this excess of three chains and 62 links. Therefore the decree will be reversed with directions to the chancellor to enter a decree in accordance with this opinion.

14

Forensic Applications

It's often a mistake to accept something as true, merely because it is obvious.

The truth is only arrived at by the painstaking process of eliminating the untrue.

Sherlock Holmes, "Dressed to Kill" by Sir Arthur Conan Doyle

Definition

"Forensics" is a science of interest to the legal system. The retracement surveyor, through the application of a number of sciences, obviously falls under this definition in undertaking the duty of following an ancient, or even merely an older, boundary or survey line.

As outlined in the Preface, the retracement surveyor, out of necessity, becomes involved in, maybe even inadvertently becomes embroiled in, a wide variety of scientific endeavors. While the average person may believe that the application of forensic science is confined to criminal investigations, the discipline is actually any kind of investigation involving study of past events, whether related to criminal or civil investigations, that may be of interest to the court system. Unlike most other sciences, forensic science is not concerned with predicting the future, but reconstructing the past.* Throughout this book, from the discussion of techniques and procedures to examination by the court system, example after example has been about something—an event, a series of measurements, the examination of one or more records—from the past. Retracement itself is a series of complex procedures of developing conclusions about past events with the goal of reconstructing the original survey on the surface of the earth under present conditions. Doing that effectively requires an appreciation for, and an understanding of, the past. As Winston Churchill once said, "in history lies all the secrets."

---

* Dr. Zakaria Erzinçlioglu, *True Crime Scene Investigations*, 2000.

371

The Value of Historical Knowledge and Research

Not infrequently adequate in-depth research and investigation, usually, but not always, done for other purposes discloses outstanding rights created a century or more ago and never extinguished. There have been several recent examples which are noteworthy. Although these cases have been referenced previously, their significance from a forensics and historical research perspective demands their inclusion here. Too often, people attempt to circumvent the process and not examine the entire history of a chain of title. As the Maryland court aptly stated, "a requisite for valid title is a grant from the state" (or sovereign entity).

In the case of *Ski Roundtop v. Wagerman*,* the two parties in the action both claimed a part of each other's land, creating an overlap. It turned out that neither party had record title to the disputed strip. The court referred to a previous Maryland coal case, *Maryland Coal and Realty Co. v. Eckhart*,[†] which contained a historical account on the law of land patents. This case describes Maryland's authority to patent lands derived from its sovereign heritage. The original charter, from King Charles I in 1632 to Cecil Calvert, gave him exclusive power to grant lands in Maryland, with that power vesting in the State after the American Revolution. The court stated, "Although the case law in this area is sparse, it appears that a requisite for valid title to real property is an original conveyance of public land by the State. See 3 American Law of Property, § 12:16 (1952); 73B C.J.S., Public Lands, § 188 (1983); 2 Patton on Titles, § 281 (2d ed. 1957). Absent such a conveyance, one purporting to transfer an ownership interest in such property transfers nothing, and no quantity of successive transfers by deed nor the mere passage of time will metamorphose good title from void title."

The Massachusetts Land Court decided a case in 2008 concerning a way created in the eighteenth century. In *Campbell & Others v. Nickerson & Others*,[‡] an easement was established with a conveyance by deed in 1711. The instrument conveyed shares in a large parcel of land to 14 grantees, known in this case as the "proprietors." In 1714, at a proprietors' meeting, they divided the lots among themselves, and stated that they "and their heirs and assigns forever to have and receive....privilege of passing through one or another of said lots...for egress and regress."

The court determined that the rights were to apply to the lots not only as then laid out, but also to lots "as they may be [laid out] in each respective division." The judge reasoned that the proprietors intended the rights of passing over the lots to apply to future divisions of the land, and that

---

* 79 Md.App. 357, 556 A.2d 144, Md.App., 1989.
† 25 Md.App. 605, 337 A.2d 150, Md.App., 1975.
‡ 73 Mass.App.Ct. 20, 2008.

Forensic Applications 373

those rights were established in a manner comparable to the later "common scheme" doctrine.

In 1971, the New York court recognized the value of history and its relation to the original creation of title in the case of *Dolphin Lane Associates, Ltd. v. Town of Southampton.** It stated, "this action, by its very nature, involves the tracing of chains of title going back to the earliest settlement of the Town of Southampton. Thus, the history of the early settlement and development of Southampton must be examined. The past must be explored in order to understand the present." Whereupon the court proceeded to outline in some detail the evolution of the Town of Southampton, from 1639 when eight men from Massachusetts agreed to establish a permanent settlement, through the addition of 10 more known as undertakers, a listing of applicable patents, overlapping claims from the Dutch settlers of New Amsterdam as well as the Colony of Connecticut and the eventual re-patenting and subsequent confirmations thereof.

Ultimately the title was vested to all land, not in the lawful possession of some individuals, within the boundaries named in the Town of Southampton. The legal title to this land was held by the Trustees of the Freeholders and Commonalty of the Town of Southampton.

The Pennsylvania case of *Bruker v. Burgess and Town Council of Borough of Carlisle*[†] had to do with a town square having been used for more than 200 years, which the court acknowledged that such use "is sufficient to raise a conclusive presumption of an original grant for the purpose of a public square; such is an ancient and well established principle of law. Nor can it be denied that, where such a dedication has been established and the public has accepted it, there cannot be any diversion of such use from a public to a private purpose, and it is also true that, where a dedication is for a limited or restricted use, any diversion therefrom to some purpose other than the one designated is likewise forbidden."

In this case, it was found that, "The town of Carlisle was laid out in 1751 according to a plan of Thomas Penn, then Proprietor of Pennsylvania. In a letter to Richard Peters, Secretary of the Province, he wrote that in the center of Carlisle he would have a Square and that 'the Court House may be in the middle of one side and the Gaal in any place near, there may be a places I think in the middle of the Center Square for a Market, or if that will take off too much of the lots, a lot may be given at the rent of a Shilling as for the other public used for that purpose, * * *.' A public market occupied a portion of the southeast quarter of the Square as early as 1764, and from time to time thereafter new market houses were erected on the same site and markets were regularly conducted there."

---

* 339 N.Y.S.2d 966, 72 Misc.2d 868, 1971.
† 376 Pa. 330; 102 A.2d 418, Pa., 1954.

374 *Boundary Retracement*

These modern examples underscore the importance of historical consideration of any title and an evaluation of its origin. An understanding of the origin evolution of any particular state is paramount in order to fully comprehend many of the aspects of a tract of land.

The Value of a Complete and Thorough Investigation

Many people fail in their investigations for boundary evidence and early (original) surveys by not appreciating the need for the following essentials:

Sufficient time
Sufficient funds
Sufficient resources

Modern Tools

Research into the past has never been more efficient. The use of the internet, websites, many maps, and related items in digital form and related aids make the investigator's task more efficient, and likely more successful. Maps, textbooks, family and regional histories are all mostly readily available. Internet services such as tree and wood identification, conversion of magnetic bearings for declination change, and other services greatly enhance the research process and speed the journey to a reliable conclusion.

What Forensic Science Can Do Outside of the Criminal Environment

Describe methods of clear thinking.
Provide techniques for analyzing information.
Investigative processes and techniques.
Genealogy data and compilations.
Document examination.
Photographs/postcards and insight into the past.

References

American Society of Civil Engineers. *Definitions of Surveying, Mapping, and Related Terms*. Manual of Engineering Practice No. 34. New York: Headquarters of the Society. 1954.

Barker, N. *An Essay on the Cardinal Points: Being a Collection of Authorities in Explanation of the Terms, "Due North," "Due South," "Due East," and "Due West," as Applied to Land Surveying*. Bangor, ME: Printed by Samuel S. Smith. 1864.

Black, H. C. *Black's Law Dictionary*. Deluxe 3rd edition. St. Paul, MN: West Publishing Co. 1944.

Brinker, R. C. and P. R. Wolf. *Elementary Surveying*. 7th edition. New York: Harper & Row, Publishers. 1984.

Buckner, R.B. *Fundamentals of Measurement Theory and Analysis*. Jefferson, NC: Land Surveyor's Publications. 2004.

Cadastral Training Staff, Denver Service Center, US Department of the Interior, Bureau of Land Management. *Glossaries of BLM Surveying and Mapping Terms*. 2nd edition. 1980.

Dean, D. R. Jr. and J. G. McEntyre. *Establishment of Boundaries by Unwritten Methods and the Land Surveyor*. I.S.P.L.S. Surveying Publication Series 6. Purdue University, Indiana Society of Professional Land Surveyors and School of Civil Engineering. n.d.

Erzinçlioglu, Z. *True Crime Scene Investigations*. New York: Barnes & Noble Books. 2000.

Griffin, R. J. Retracement and apportionment as surveying methods for re-establishing property corners. *Marquette Law Review*, 43(4), Spring, 1960, Article 5, 484–510.

Grimes, J. S. *Clark on Surveying and Boundaries*. 3rd Edition. Indianapolis, IN: The New Bobbs-Merrill Company, Inc. 1959.

Hodgman, F. About corners. Paper read before the Michigan Association of Surveyors and Engineers, and published in the Proceedings of the Second Annual Meeting held at Lansing, January 11–13, 1881.

Hodgman, F. *A Manual of Land Surveying*. Climax, MI: The F. Hodgman Co. 1913.

I.C.S. Staff. *Mapping and City Surveying*. Scranton, PA: International Textbook Company. 1930, 1931, 1933.

Leininger, J. Institutes of Retracement-Part I. *The Nevada Traverse*, 23(2), 1996.

Orr, C. L. Vanishing footsteps of the original surveyor. *Baylor Law Review*, IV(3), Spring, 1952, 272–295.

Quimby, E.T. *A Paper on Terrestrial Magnetism designed for the Use of Surveyors*. Concord, NH: Printed by the Republican Press Association. 1874.

Robillard, W. G., D. A. Wilson, and C. M. Brown. *Evidence and Procedures for Boundary Location*. 6th edition. Hoboken, NJ: John Wiley & Sons, Inc. 2011.

Skelton, R. H. *The Legal Elements of Boundaries and Adjacent Properties*. Indianapolis, IN: The Bobbs-Merrill Company. 1930.

United States Department of the Interior, Bureau of Land Management. *Restoration of Lost or Obliterated Corners & Subdivision of Sections. A Guide for Surveyors*. 1974 Edition. Washington, DC: US Government Printing Office. 1975.

United States Department of the Interior, Bureau of Land Management. *Glossaries of BLM Surveying and Mapping Terms.* Prepared by the Cadastral Survey Training Staff, Denver Service Center. 2nd Edition. 1980.

Wilson, D. A. *Interpreting Land Records.* 2nd Edition. Hoboken, NJ: John Wiley & Sons, Inc., 2015.

Wilson, D. A. *Forensic Procedures for Boundary and Title Investigation.* Hoboken, NJ: John Wiley & Sons, Inc. 2008.

Appendix I: List of Significant Cases on Following Footsteps, by State

The following selection of cases is representative of the courts' perspective and holdings in the various states included. Nearly every state has ruled on the issue and wherever a useful, or representative, decision was found, it is included in the list. These decisions, in summary form, should serve to illustrate the standard in the particular state's system, and provide a means to understand the rules within that state.

Alabama: *Wharton v. Littlefield*, 30 Ala. 245 (Ala. 1857).

The court charged the jury, in reference to the evidence, as follows:

> ... In all Government surveys, the section lines and corners are established by the United States surveyors. The marked lines of all sections, and the corners of all sections set up, are the correct boundaries and corners, and, when ascertained, cannot be changed by any other surveyor. The United States Government has arranged their boundaries, marked their lines and corners, and declared their contents; and the purchaser of an entire section takes all the land within those limits, be it more or less than the quantity returned by the surveyor. But in the purchase of a less quantity than a section, as between the several holders of a section, the contents of such several parts must be determined by reference to the entire section; and the purchaser of a half or quarter section is entitled to one half or one fourth of whatever the section contains. The same rule of law is applicable to fractional sections; the several purchasers thereof being entitled to a *pro-rata* quantity of such fractional section, to be ascertained in the same way as if they were the purchasers of a regular section. Each section, or subdivision of a section, is independent of any other section in the township, and must be governed by its marked and established boundaries.

Alaska: No significant retracement decisions found.

Arizona: *Wacker v. Price*, 70 Ariz. 99, 216 P.2d 707 (Ariz. 1950).

> Let it be conceded that the city survey and the Jones survey are accurate surveys of the area according to the Governmental monument located at McDowell Road and 15th Avenue. We must remember that we are not here concerned with an accurate survey of this particular area. We are concerned only with accurately ascertaining, if we can, the location of the boundary lines of Grand Avenue Subdivision according to the original plat thereof and the boundary lines of the lots and streets shown by

said map or plat. It matters not how inaccurate the plat may have been or may be, property rights have vested according to that map or plat. No surveyor whether he be acting on behalf of the city or anyone else has the right to arbitrarily change property lines or move streets from the location established by the original map or plat. The city surveyor who prepared the so-called official map of the city of Phoenix and the witness Jones wholly misconceived their duty in making their surveys of this area. The function of a surveyor in a case of this kind is not to determine where the streets or the lot lines in the subdivisions should have been according to an accurate survey but to determine where they actually were, measured by the original map or plat of Grand Avenue Subdivision. In the case of *Diehl v. Zanger*, 39 Mich. 601, the court said that a long-established fence is better evidence of actual boundaries settled by practical location than any survey made after the monuments of the original survey have disappeared. That certainly would apply where no survey appears to have been made as in this case and especially where said fences for all practical purposes conform with the original plat. Justice Cooley in a concurring opinion in that case said:

Nothing is better understood than that few of our early plats will stand the test of a careful and accurate survey without disclosing errors. * * * The (city) surveyor has mistaken entirely the point to which his attention should have been directed. The question is not how an entirely accurate survey would locate these lots, but how the original stakes located them. * * * The city surveyor should, therefore, have directed his attention to the ascertainment of the actual location of the original landmarks set by Mr. Campau, and when those were discovered they must govern. If they are no longer discoverable, the question is where they were located; and upon that question the best possible evidence is usually found in the practical location of the lines, made at a time when the original monuments were presumably in existence and probably well known: *Stewart v. Carleton*, 31 Mich. 270. As between old boundary fences, and any survey made after the monuments have disappeared, the fences are by far the better evidence of what the lines of a lot actually are. * *.

The above statement of Justice Cooley is particularly apropos in the instant case. Nothing appears on the original plat to indicate either the width of lots or of streets. No governmental monuments are indicated. There is no evidence in this record that any measurement was ever taken from the Governmental monument at the intersection of McDowell and 15th Avenue in platting Grand Avenue Subdivision.

In 110 American State Reports, page 681, under the title of "Resurveys and their Purpose and Effect," the following is found: "* * * In *Diehl v. Zanger*, 39 Mich. 601, where the first survey of lots involved in litigation was made by one Campau, and a resurvey made years afterward by the city surveyor showed that the practical location of the whole plat was wrong, it was declared that a resurvey, made after the disappearance of the monuments of the original survey, *is for the purpose of determining where they were, and not where they should have been*, and that a long-established fence is better evidence of actual boundaries settled by practical location than any survey made after the monuments of the original

Appendix I: List of Significant Cases on Following Footsteps, by State 379

survey have disappeared. "Nothing is better understood," said Justice Cooley in delivering the opinion of the court, "than that few of our early plats will stand the test of a careful and accurate survey without disclosing errors. This is as true of the government surveys as of any others, and if all the lines were now subject to correction on new surveys, the confusion of lines and titles that would follow would cause consternation in many communities. Indeed, the mischiefs that must follow would be simply incalculable, and the visitation of the surveyor might well be set down as a great public calamity. But no law can sanction this course. The (city) surveyor has mistaken entirely the point to which his attention should have been directed. The question is not how an entirely accurate survey would locate these lots, but how the original stakes located them. * * * The city surveyor should, therefore, have directed his attention to the ascertainment of the actual location of the original landmarks set by Mr. Campau, and when those were discovered they must govern. If they are no longer discoverable, the question is where they were located; and upon that question the best possible evidence is usually found in the practical location of the lines, made at a time when the original monuments were presumably in existence and probably well known: *Stewart v. Carleton*, 31 Mich. 270. As between old boundary fences and any survey made after the monuments have disappeared, the fences are by far the better evidence of what the lines of a lot actually are." * * *

... In 22 American State Reports, Ancient Boundaries, page 35, we find the following: * * * For the purpose of establishing ancient boundaries, by locating calls for corners, etc., the declarations of the parties in interest, or those who assisted in making the old survey, are admissible, when such persons are unable to testify orally or by deposition, by reason of sickness or death: *Whitman v. Haywood*, 77 Tex. 557, 14 S.W. 166; *Griffith v. Sauls*, 77 Tex. 630, 14 S.W. 230. Ancient fences, used by a surveyor in his attempt to reproduce an old survey, are strong evidence of the location of the original lines, and if they have been standing for many years, should be taken as indicating such lines, even against the evidence of a survey ignoring such fences, based upon an assumed starting-point: *Beaubien v. Kellogg*, 69 Mich. 333, 37 N.W. 691, for it will not do to allow boundaries to be disturbed upon a survey made from an assumed starting-point, without proof of its being a true line, located and fixed by the original survey. * * *

Arkansas: *Dicus v. Allen*, 2 Ark.App. 204, 619 S.W.2d 306 (Ark.App. 1981).

The established rule of property is that the original US Government survey is prima facie correct and surveys must conform as closely as possible to the original government survey.

Price v. Mauch, 1 Ark.App. 348, 616 S.W.2d 738 (Ark.App. 1981).

It is a well settled rule of surveying, recognized by the courts, that the lines actually run control over maps, plats, or field notes. The actual survey

380   *Appendix I: List of Significant Cases on Following Footsteps, by State*

originally made is evidenced by the fixed boundaries then established, and the actual survey must, therefore, govern over the erroneous plat thereof.

Pyburn v. Campbell, 158 Ark. 321, 250 S.W. 15 (1923).

Pyburn *states*: It is a well settled rule of surveying, recognized by the courts, that the lines actually run control over maps, plats, or field notes. Page 537, Clark on Surveys and Boundaries. The actual survey originally made is evidenced by the fixed boundaries then established, and the actual survey must therefore govern over the erroneous plat thereof.

Buffalo Zinc & Copper Company v. McCarty, 125 Ark. 582, 189 S.W. 355 (Ark. 1916).

It is well settled that lines and corners located and marked by the return of the Surveyor General, however erroneous, cannot be collaterally attacked, but the authority to correct such errors or mistakes rests wholly in the Land Department of the Federal government. 128 U.S. 691; 17 How. 23; 197 U.S. 510; 88 Ark. 37; 114 Mo. 426; 2 S.D. 269.

California: County of *Yolo v. Nolan*, 77 P. 1006, 144 Cal. 445 (1904).

The rule as to restoring lost corners by putting them at an equal distance between two known corners has no application, if the line can be retraced as it was established in the field. The field-notes should be taken, and from the courses and distances, natural monuments or objects, and bearing trees described therein the surveyor should endeavor to fix the line precisely as it is called for by the field-notes. He should endeavor to retrace the steps of the man who made the original survey. If by so doing the line can be located, it must be done, and, when so located, it must control. It is not the business of the surveyor to speculate as to whether one government subdivision is short and the other long in acres. He is not authorized to correct what the government has done. The line as surveyed and described in the field-notes is the description by which the government sells its land. If its description makes one section contain three hundred and twenty acres and another nine hundred and sixty acres, the parties must take according to the calls of their patents.

As said in *Kaiser v. Dalto*, 140 Cal. 172, "The lines as originally located must govern in such cases. The survey as made in the field, and the lines as actually run on the surface of the earth at the time the block were surveyed, and the plats filed must control. The parties who own the property have a right to rely upon such lines and monuments."

In *Tognazzini v. Morganti*, 84 Cal. 160, it is said by this court, speaking through the chief justice: "The rights of the respective parties of course depend upon a correct relocation of the Terrill line. ... The road it crosses [speaking of the Von Schmidt line] is a road running north and south, while the road called for in Terrill's field-notes, and delineated on his

Appendix I: List of Significant Cases on Following Footsteps, by State 381

map, runs nearly east and west. ... We think this call for the course surveyed is much less certain and trustworthy than the calls for the entrance to and exit from the Canada, the crossing of the road east and west, the entrance to the Canada Verde, and the distance of the course from the adobe houses—all of which sustain the Minto survey."

It is said in *Harrington v. Boehmer,* 134 Cal. 199: "The question in all cases similar to this is, Where were the lines run in the field by the government surveyor? A government township lies just where the government surveyor lines it out on the face of the earth. These lines are to be determined by the monuments in the field."

Colorado: *U.S. v. Doyle,* 468 F.2d 633 (1972).

The original survey as it was actually run on the ground controls. *United States v. State Investment Co.,* 264 U.S. 206, 212, 44 S.Ct. 289, 68 L.Ed. 639; *Ashley v. Hill,* 150 Colo. 563, 375 P.2d 337, 339. It does not matter that the boundary was incorrect as originally established. A precisely accurate resurvey cannot defeat ownership rights flowing from the original grant and the boundaries originally marked off. *United States v. Lane,* 260 U.S. 662, 665, 666, 43 S.Ct. 236, 67 L.Ed. 448; *Everett v. Lantz,* 126 Colo. 504, 252 P.2d 103, 108. The conclusiveness of an inaccurate original survey is not affected by the fact that it will set awry the shapes of sections and subdivisions. See *Vaught v. McClymond,* 116 Mont. 542, 155 P.2d 612, 620; *Mason v. Braught,* 33 S.D. 559, 146 N.W. 687.

The actual location of a disputed boundary line is usually a question of fact. *Gaines v. City of Sterling,* 140 Colo. 63, 342 P.2d 651. "...[T]he generally accepted rule is that a subsequent resurvey is evidence, although not conclusive evidence, of the location of the original line." *United States v. Hudspeth,* 384 F.2d 683, 688 n. 7 (9th Cir.); accord, see *Ben Realty Co. v. Gothberg,* 56 Wyo. 294, 109 P.2d 455, 458, 459. And in its trespass action the burden of proving good title to the land rests on the *Government. Yakes v. Williams,* 129 Colo. 427, 270 P.2d 765; see also *Cone v. West Virginia Pulp & Paper Co.,* 330 U.S. 212, 67 S.Ct. 752, 91 L.Ed. 849.

Connecticut: No significant retracement decisions found.

Delaware: *State el rel. Buckson v. Pennsylvania R. Co.,* 228 A.2d 587 (Del. Super, 1967), citing *Moore v. Campbell,* Tex.Civ.App. 254 S.W.2d 1018 (1953).

As to boundary disputes, the primary purpose is to track the footsteps of the original surveyor, to locate the survey as it was intended to be located on the ground by him.

Florida: The two leading cases are as discussed in the text: *Rivers v. Lozeau* and *Tyson v. Edwards.*

For further reference, see *Watrous v. Morrison,* 33 Fla. 261, 14 So. 805 (Fla., 1894).

Appendix I: List of Significant Cases on Following Footsteps, by State

In the sale of land in sections or subdivisions thereof, including lots, according to the government survey, the survey as actually made controls. It is the survey as it was actually run on the ground that governs, if the monuments, corners, or lines actually established can be located or proved. Courses and distances yield to such corners or lines so long as the latter can be located, and for the reason that the latter are the fact or truth of the survey as it was actually made, while the former are but the description of the act done, and, when inaccurate, they cannot change the fact.

Georgia: *Riley v. Griffin*, 16 Ga. 141 (Ga., 1854).

This is one of the most comprehensive retracement cases. It relies on other jurisdictions with earlier decisions:

[13.] While course and distance, depending, for their correctness, on a great variety of circumstances, are constantly liable to be incorrect. Difference in the instrument used, and in the care of Surveyors and their assistants, lead to different results. (*Les see of McCay vs. Gallway*, 3 *Ham.* 282. *Thorn berry vs. Churchill and Wife*, 4 *Monroe's Ken. R.* 32. *McNeill vs. Massey*, 3 *Hawk. R.* 91. *Beard's Lessee vs. Tullot*, 1 *Cook*, 142. *Preston's Heirs vs. Benman*, 6 *Wheat.* 58.)

This doctrine is found scattered, broadcast, throughout the authorities; and I had supposed to be too well understood and established, to require to be discussed at this day.

[14.] In *Brewer vs. Gay*, (3 *Greenleaf's R.* 126,) it was held, that in ascertaining the boundaries of lots of land, where a township has been laid out, the locations of the original Surveyor, so far as they can be found, are to be resorted to; and where they vary from the proprietor's plan, the locations actually made will control the plan.

"[15.] So, in *Dodge vs. Smith*, (2 *N. Hamp. R.* 303,) the Supreme Court say," whenever, in a conveyance, the deed refers to monuments actually erected, as the boundaries of the land, it is well settled that those monuments must prevail, whatever mistakes the deed may contain, as to the courses and distances. The same principle was decided in *Brand vs. Dawny* (20 *Marten's Lon. Rep.* 159).

[16] In *Doe vs. Paine & Sawyer* (4 *Hawk's N. Rep.* 64) the Court refer to courses and distances, as pointers or guides merely, to ascertain the natural objects of boundary.

[17.] So, also, it has been held, that where a given line is exceeded in a grant, according to the courses and distances, evidence may be received, of long occupation under it, to prove the boundaries. (*Makepeace vs. Bancroft*, 12 Mass. 469. *Sargent vs. Town*, 10 Mass. 303. *Baker vs. Sanderson*, 3 *Pick. R.* 354. *Livingston vs. Ten Brocek*, 16 *Johns. R.* 23.) *Vide Davies' Abrid. of Am. L. vol.* 3, *p.* 307, where some of the early cases decided in Massachusetts, upon this subject, are collected.

[18.] And, as landmarks are frequently formed of perishable materials, which pass away with the generation in which they are made; and are often destroyed, as in the case before the Court, by the improvement of

Appendix I: List of Significant Cases on Following Footsteps, by State 383

the country and other causes, the boundary and corners may be proved by hearsay, from the necessity of the case. (*Nicholls vs. Parker*, 14 *East*. 331. 12 *East*. 62. 1 *M. & S*. 679. 1 *T. R*. 466. 5 *T. R*. 26. 2 *Ves*. 512. 6 *Peters*, 341. 4 *Day's Conn. R*. 265. 1 *Harris & McHenry*, 84, 368, 531. 2 *Hayw. R*. 349. 1 *Yates*, 28. 6 *Binney*, 59.).

Hawaii: No significant retracement decisions found.

Idaho: *Bayhouse v. Urquides*, 17 Idaho 286, 105 P. 1066 (Idaho 1909).

The purpose of a resurvey subsequent to the taking of title by purchasers and settlers is to ascertain the lines of the original survey and the original boundaries and monuments as established and laid out by the survey under which the parties originally procured their titles. (*Martz v. Williams*, 67 Ill. 306.) On such resurvey or re-established boundaries and monuments the question of the correctness of the original survey cannot enter into the matter at all, and is a matter that does not concern the surveyor, and is not a question to be ascertained by him. (*Diehl v. Zanger*, 39 Mich. 601; *Penry v. Richards*, 52 Cal. 672 (675); *Bullard v. Kempff*, 119 Cal. 9, 50 P. 780.).

Illinois: *Martz v. Williams*, 67 Ill. 306.

Section 2 of the act of 1869, under which these proceedings were had, provides the manner by which to re-establish corners and boundaries lost, destroyed or in dispute. It nowhere confers power on the commissioners to establish new corners or run new boundary lines, but simply to re-establish those once established by the United States, the original proprietors, and by which corners and boundary lines they sold and conveyed the same to the several purchasers. The duty of these commissioners, then, was to re-establish lost corners and boundaries; to make them in the same places and at the same distances as they had been originally established by the government surveyors. From the evidence in the record, and from the report and plat returned by the commissioners, we are satisfied they have misapprehended the statute, and not been governed by its plain spirit and intent. They do not seem to have paid much attention to the returns and original field notes returned by the government surveyors, which, we think, should have been their main guide in seeking to re-establish a corner or boundary line.

M'Clintock v. Rogers, 11 Ill. 279 (Ill. 1849).

It is a question of fact to be determined, how was this line originally run by the surveyors of the United States government. If that fact can be ascertained, it must determine this case. Bleeker's map of Hosick's patent held never to be conclusive in opposition to true lines, founded upon his actual survey. *Jackson v. Joy*, 9 Johns. 103. The map necessarily shows the lots of equal size, and shows their relative locality. The legislature must

384   *Appendix I: List of Significant Cases on Following Footsteps, by State*

have contemplated that field book should accompany the map. *Jackson v. Cole*, 16 Johns. 260. The actual survey must be deemed part and parcel of the description of lot. Ibid. The government might have had the surveys re-examined before the patents issued; it is now too late to attempt to correct mistakes. *Jackson v. Cole*, 16 Johns. 264. A proprietary surveyor held not to be agent of people whose lands he surveyed, but of proprietor alone. Such surveyor could not bar title by his return, which was vested by actual survey. The courses and distances on the ground are the true survey. *Lilly v. Kitzmiller*, 1 Yeates 29; *Yoder v. Fleming*, 2 Yeates 311. A line proved, although deviating from a right line, must be the boundary. *Lyon v. Ross* et ux., 1 Bibb 466. Parol evidence admitted to show that course and boundary in the survey and patent, are otherwise on the ground. *Mageehan v. Adam's* lessee, 2 Binney 109; *Wallace v. Maxwell*, 1 J. J. Marsh., 451. Same points decided *in Hall v. Powell*, 4 S. & R., 462; *Philips v. Shaeffer*, 5 S. & R., 215. Original marked trees must govern, where they can be found. If not, courses and distances must be resorted to, as the next best rule. *Sumter v. Bracy*, 2 Bay 516; *Chenoweth v. Lessee of Haskell*, 3 Peters 96. The original surveys establish rights of parties, and must govern. *May v. Baskin*, 12 Smedes and Marshall, 429, 430. The north-east and south-west corners of patent claim established, held that law would fix north-west corner at intersection of lines extended according to patent courses, from north-east and south-west corners, unless some other point had been fixed on by original survey. *Wishart v. Crosby*, 1 Marshall 381; *Thornberry v. Churchill et ux.*, 4 Monroe 95. Where there is an ascertained place of beginning, the grant must be confined to boundaries given in the deed. *Jackson v. Wilkinson*, 17 Johns. 156; *Jackson v. Wendell*, 5 Wendell 146. The corners of sections on township lines were made when the township was laid out. They became fixed points, and if their position can now be shown by testimony, these points must be retained, although not in a straight line from A to B. The township line was not run as a single sight from A to B; it was run mile by mile, and these mile points are as sacred as the points A and B. Land Laws, vol. 1, 50, 71, 119, 120. The sections on the north and west side of the township receive excess and give way in size, if there is any excess or deficiency in the township. They are to close up to the township line wherever it was placed. See act of 1800, Land Laws, vol. 1, sec. 3, p. 71.

Sawyer v. Cox, 63 Ill. 130 (Ill. 1872).

The evidence of the various surveys of these lines is meagre, indefinite and unsatisfactory in its character. No plat or field notes of any of the surveys are given in the record; nor is there any evidence from which it may be seen how the surveyors arrived at their conclusions when the lines were located. They do not seem to have started at known and recognized true interior section corners in the township, or that there was not to be found all or a portion of the original corners to section 6. For aught we can see, the lines may have been arbitrarily run, without reference to original corners or lines. When the survey was commenced on the township lines four miles east and six miles north, there is nothing

Appendix I: List of Significant Cases on Following Footsteps, by State 385

> to show that any evidence was found of the position of any other corners established by the government on these lines. It does not appear whether a single government corner was found for mile and half mile distances along either line, or that the northwest corner of the township was found, or if not, how it was established.
>
> The object of these surveys is, first, if practicable, to find the original corners established by the surveys made by the authority of the government. It is by those lines and corners the government sold and persons purchased the public lands. And when sold, the purchaser, by his patent, acquired title to all of the land embraced within the boundary lines of the tract thus purchased. When the lines and corners established can be found and identified, the purchaser acquires title to all the lands embraced within their limits. And it does not matter whether the surveys are accurate, as the boundaries when found must control the notes or plat of the survey. Hence they govern the calls for course, distance and quantity. The plats and notes of the survey are intended to represent what was done in the field, and must yield to the lines and corners when found. But when they have become obliterated and cannot be found and traced by natural or artificial monuments, they can only be relocated by the field notes and plats of the original survey. And in doing so, then resort must be had to known lines and monuments as the basis on which to survey and find where the original lines and corners were established by the government surveyors. One of the modes by which such lost lines and corners may be restored is fully discussed in the case of *McClintock v. Rogers*, 11 Ill. 279, and perhaps indicates one of the most satisfactory means of restoring lost corners.

This case stated "when the location of a lost corner is sought, by approaching it from others which are established, we must be guided by the best lights at command, however imperfect and unsatisfactory they may be. These, in the present instance, are principally the field notes. The law cannot satisfactorily determine, in all cases, whether the courses or distances shall govern, when they do not correspond, but determine, in all cases, whether the courses or distances shall govern, when they do not correspond, but that must be determined by concurring testimony, and the circumstances of each particular case. The one that convinces the judgment the most must be selected. Unless the compass is beyond the influence of disturbing causes, and the surveyor is very careful in adjusting it properly, and in noting minutely the variation at which the line is run—and we know the date of the survey—so that the increase or decrease of the variation since, can be added or deducted, no surveyor can ever feel confident that he is running even very near to the line traversed by his predecessor, and by whose minutes he is working."

Indiana: *Herbst v. Smith*, 71 Ind. 44 (1880).

> A new survey may doubtless be had, not for the purpose of establishing the corners, lines or boundaries, as an original survey; but for the purpose of re-locating or perpetuating the corners, lines or boundaries

386   *Appendix I: List of Significant Cases on Following Footsteps, by State*

established by such original survey, where they have become obscured or lost.

Iowa: *Rowell v. Weinemann*, 119 Iowa 256, 93 N.W. 279 (Iowa 1903).

It is well settled that the lines actually run by the original government surveyors become the true boundaries, and, if they can be ascertained through monuments erected by these officials, they will control; and courses, distances, measurements, plats, and field notes must all yield. *Ufford v. Wilkins*, 33 Iowa 110; *Sayers v. City of Lyons*, 10 Iowa 249; *Root v. Town of Cincinnati*, 87 Iowa 202, 54 N.W. 206. It is well known that the original surveys were faulty in many respects, and that they will not stand the test of careful and accurate retracing. It is not the purpose of such actions as this, or of any other, for that matter, to straighten out lines, or to remove unsightly crooks, however desirable such a result might be. *Rollins v. Davidson*, 84 Iowa 237, 50 N.W. 1061. Hence everything yields to known monuments and boundaries established by the government surveyors.

Cited numerous times is the case of *Moreland v. Page*, 2 Iowa 139.

Kansas: In re The Appeal of Minnie Ralston From Survey of J. Homer Austin, 115 Kan. 842, 225 P. 343 (Kan. 1924).

The primary rules for locating city plats upon the ground are, in order of precedence in application, as follows: (1) Find the lines actually run and the corners and monuments actually established by the original survey. (2) Run lines from known, established or acknowledged corners and monuments of the original survey. (3) Run lines according to courses and distances marked on the plat. (*In re* Richardson, 74 Kan. 557, 87 P. 678).

It is urged that the section line was the proper base line which the surveyor should have ascertained and from which his measurements and calculations should have been made. That leaves out of consideration the original survey as actually located upon the ground. A certain hedge fence is spoken of as having been used as a base line in early days but time has erased that mark. The monuments and marks found by the surveyor furnished reasonably good evidence in locating the original survey. In *Ayers v. Watson*, 137 U.S. 584, 34 L.Ed. 803, 11 S.Ct. 201, it was held in effect that in ascertaining the lines of lands, the tracks of the surveyor so far as discoverable on the ground with reasonable certainty should be followed and marked trees designating corners or lines on the ground should control both parties, and that where the survey of land has been actually run and measured and ascertained, monuments are referred to in it, the footsteps of the surveyor may be traced backward as well as forward and any ascertained monument in the survey may be adopted as a starting point where difficulty exists in ascertaining the lines actually run. The court remarked that:

Appendix I: List of Significant Cases on Following Footsteps, by State 387

The beginning corner does not control more than any other corner actually ascertained, and that we are not constrained to follow the calls of the grant in the order they stand in the field notes, but may reverse them and trace the lines the other way, whenever by so doing the land embraced would more nearly harmonize all the calls and the objects of the grant. (p. 599). (See, also, *Hord v. Olivari* [Texas], 5 S.W. 57; *Ocean Beach Association v. Yard*, 48 N.J.Eq. 72, 20 A. 763.)

Kentucky: *Morris v. Jody*, 216 Ky. 593, 288 S.W. 332 (Ky.App., 1926).

All of the rules of law that have been adopted for guidance in locating disputed boundary lines have been to the end that, in so doing, the steps of the surveyor who originally projected the lines on the ground may be retraced as nearly as possible. No rule that has been adopted to accomplish that end is more firmly established than that courses and distances are controlled by marked corners and fixed monuments. Here all the evidence establishes that the lines of the survey in dispute were actually projected on the ground by the surveyor, who established the boundary lines which are in dispute, just as appellant contends they should be located.

A line or corner established by a surveyor, in making a survey upon which a grant has issued, cannot be altered because the line is longer or shorter than the distance specified, or because the relative bearings between the abuttals vary from the course named in the plat and certificate of survey. So, if the line run by the surveyor be not a right line, as supposed from his description, but be found by tracing it to be a curved line, yet the actual line must govern; the visible actual boundary, the thing described, and not the ideal boundary and imperfect description, is to be the guide and rule of property. These principles are recognized in *Beckley v. Bryan*, prin. dec. 107, and Litt. Sel. Cas. 91; *Morrison v. Coghill*, prin. dec. 382; *Lyon v. Ross*, 1 Bibb 467; *Cowan v. Fauntleroy*, 2 Bibb 261; *Shaw v. Clement*, 1 Call. 438, 3rd point; *Herbert v. Wise*, 3 Call. 239; *Baker v. Glasscocke*, 1 Hen. & Munf. 177; *Helm v. Small, Hard.* 369.

Baxter v. Evett's Lessee, 23 Ky. 329 (Ky.App. 1828).

Louisiana: Rapides Parish Police *Jury v. Grant* Parish Police Jury, 924 So.2d 357 (La.App. 3 Cir. 2006).

However, when asked whether they had meandered courses and followed the footsteps on the ground of the original GLO surveyor, which methodology is required in a "retracement" survey, the Grant Parish surveyors admitted that they had not.

Maine: *Bean v. Bachelder*, 78 Me. 184, 3 A. 279 (Me. 1886).

The defendant claimed under the earlier deed, which contained the description, "Lot No. 5, in the third range in Greenfield, according to

388 Appendix I: List of Significant Cases on Following Footsteps, by State

Herrick's plan." Herrick had surveyed the south half of the town into lots and ranges, the north half having been previously surveyed into lots and ranges by another surveyor. Herrick then made a plan of the surveyings of the whole town, which plan was in the case. The defendant's lot was in the south half, that had been surveyed by Herrick. The jury were instructed, in effect, that the lines run by Herrick upon the surface of the earth, as and for the boundaries of lot 5, would still be the boundaries of that lot, if their locality could be found; that the question for them to decide was the locality upon the surface of the earth of the lines actually run by Herrick in making the survey of that lot. The instruction was correct. *Esmond v. Tarbox*, 7 Me. 61, is express authority for it. See, also, *Pike v. Dyke*, 2 Me. 213; *Williams v. Spaulding*, 29 Me. 112. The plan was merely a picture. The survey was the substance. The plan was not made to show where the lots were to be hereafter located, or how they were to be hereafter bounded. It was made as evidence of where they had before been located and bounded. The lot actually surveyed, bounded by the lines actually run, was the lot intended to be conveyed. The plan was named in the deed, rather as a picture, indicating the location and lines of the lot. Still the actual boundaries, rather than the pictured boundaries, were to be sought for. The picture might not be wholly accurate.

Maryland: *Millar v. Bowie*, 115 Md.App. 682, 694 A.2d 509 (Md.App. 1997).

We said in *Ski Roundtop, Inc. v. Wagerman*, 79 Md.App. 357, 364–65, 556 A.2d 1144 (1989):
[T]he sole controlling focus should be whether the boundaries of the original patents establish the existence of Pleasant View. These patents precede any of the deeds referred to by the Brawners. Any discussion of subsequent deeds is irrelevant.
... Then speaking, not of subsequent surveys conflicting with prior deed descriptions, as in the case subjudice, but to conflicts between two modern surveys, the Court said:
Effectuation of the intent of the original parties ... is of paramount consideration in boundary dispute cases. ... Determination of which one of the two surveys best effects the true boundaries of the disputed land as intended by the original surveyor [grantor] is a question of fact. ...
... Of course, most boundary disputes evolve from surveying mistakes or ambiguous deeds. The fact that one surveyor's interpretation of the original survey results in a tidier or neater package, however, does not suffice, of itself, to override the intent of the original surveyor.

Massachusetts: *Kellogg v. Smith*, 61 Mass. 375 (Mass. 1851).

When, in a deed or grant, a line is described as running a particular course, from a given point, and this line is afterwards run out and located, and marked upon the earth by the parties in interest, and is afterwards recognized and acted on as the true line, the line thus actually marked out and acted on is conclusive and must be adhered to, though it may be

Appendix I: List of Significant Cases on Following Footsteps, by State 389

subsequently ascertained that it varies from the course given in the deed or grant. The line thus actually marked out on the earth's surface controls the course put down on the paper. The instrument of conveyance is not understood as requiring that the line to be run shall necessarily be absolutely and precisely according to the course described, which would probably be quite impracticable, but that the line shall be fairly run, in a skillful and proper manner, and that the actual, practical result adopted and acted on, shall be conclusive upon the parties in interest.

Thus, in the case of *Missouri v. Iowa*, 7 How. 660, it appeared that in an Indian grant of land to the United States, a line was described as running a due east course from a given point. This land was afterwards run out, and located and marked, under the authority of the United States, and the line thus marked out and located had been in various ways recognized and acted on as the true line. It was held by the supreme court of the United States, that this line, thus actually located, must be adhered to, through it was found that it varied some degrees from the due east course described in the grant. This decision fully and directly sustains the instructions of the court in the case now under consideration, as the two cases are substantially alike in their material facts. There are many other cases in which the courts have maintained and confirmed the same principle. *M'Nairy v. Hightour*, 2 Overton 302; *Newsom v. Pryor*, 7 Wheat. 7; *Avery v. Baum*, Wright, 576; *Cowan v. Fauntleroy*, 2 Bibb 261; *Young v. Leiper*, 4 Bibb 503; *Buford v. Cox*, 5 J.J. Marsh. 582, 587; *Blasdell v. Bissell*, 6 Barr 258; *Thompson v. McFarland*, 6 Barr 478.

The wisdom and propriety of the rule thus established are very clearly and forcibly illustrated by the present case, in which it is settled by the verdict, that the actual line claimed by the plaintiff was located, laid out, assented to and adopted by the parties, as the dividing line and north line of the Indian reserve. No actual survey and location of the reserve is now produced, but some deeds and other instruments are produced, made at a somewhat later period, alluding to such survey. But however the actual line was established, it was, in fact, actually established by the parties, and to their satisfaction, and so remains to the present time, undisturbed, a century and a quarter from the date of the original deed. It must also be constantly borne in mind, that this was not the line of a single lot, but a line of a large territory of eight or ten or more miles in extent. There are various grants and conveyances and acts of the legislature during this long period, conforming to this actual existing line. The strongest reasons of propriety and policy, as well as the principles of law, forbid that such a line, thus established, should be disturbed after having been established and conformed to for such a length of time. The length of this well known and ancient line renders it a matter of public and general concernment, that it should remain undisturbed. What would be the consequences of breaking up a line of this extent, and of this antiquity, to which grants and conveyances and acts of the legislature have conformed, it is impossible now to foresee. The only objection now made to the actual existing line is, that it does not exactly correspond with the points of compass as given in the original Indian deed. Probably the lines, as actually established, of a large portion of the estates in the

390   *Appendix I: List of Significant Cases on Following Footsteps, by State*

commonwealth would be found to vary more or less from the points of compass as given in the original deeds. Variation of the compass, imperfection of the instrument, unskillfulness in the use of it, roughness of surface, and other causes, inevitably produce, in every instance, more or less uncertainty of result. Whether or not the parties intended to establish the line in this case, precisely according to the points of compass, as given in the original deed, does not appear. But it does appear, and has been settled by the verdict, that they did in fact establish a line satisfactory to themselves, which has remained unquestioned down to the present time, and no sufficient reason has been shown by the defendants why it should be now destroyed. The parties in interest themselves marked out on the earth's surface, where they would have the line mentioned in the deed, and there it must be; and with that line, those claiming under them, and coming more than a century after them, must be content.

Michigan: *Adams v. Hoover*, 196 Mich.App. 646, 493 N.W.2d 280 (1992).

In surveying a tract of land according to a former plat or survey, the surveyor's only duty is to relocate, upon the best evidence obtainable, the courses and lines at the same place where originally located by the first surveyor on the ground. In making the resurvey, he has the right to furnish proof of the location of the lost lines or monuments, not to dispute the correctness of or to control the original survey. The original survey in all cases must, whenever possible, be retraced, since it cannot be disregarded or needlessly altered after property rights have been acquired in reliance upon it. On a resurvey to establish lost boundaries, if the original corners can be found, the places where they were originally established are conclusive without regard to whether they were in fact correctly located, in this respect it has been stated that the rule is based on the premise that the stability of boundary lines is more important than minor inaccuracies or mistakes. But it has also been said that great caution must be used in reference to resurveys, since surveys made by different surveyors seldom wholly agree. A resurvey not shown to have been based upon the original survey is inconclusive in determining boundaries and will ordinarily yield to a resurvey based upon known monuments and boundaries of the original survey.

Moreover, in relocating lost monuments the question is not how an entirely accurate survey would have located the lots, but how the original survey stakes located them. Callaghan's Michigan Civil Jurisprudence, Sec. 14, p. 365, citing *Diehl v. Zanger*, 39 Mich. 601 (1878). The rationale behind this proposition is primarily the public's need for finality and uniformity of boundaries and land titles. As stated by Justice Cooley in Diehl:

Nothing is better understood than that few of our early plats will stand the test of a careful and accurate survey without disclosing errors. This is as true of the government surveys as of any others, and if all the lines were now subject to correction on new surveys, the confusion of lines and titles that would follow would cause consternation in many

Appendix I: List of Significant Cases on Following Footsteps, by State 391

communities. Indeed the mischiefs that must follow would be simply incalculable, and the visitation of the surveyor might well be set down as a great public calamity.

But no law can sanction this course. The surveyor has mistaken entirely the point to which his attention should have been directed. The question is not how an entirely accurate survey would locate these lots, but how the original stakes located them. No rule in real estate law is more inflexible than that monuments control course and distance—a rule that we have frequent occasion to apply in the case of public surveys, where its propriety, justice and necessity are never questioned. But its application in other cases is quite as proper, and quite as necessary to the protection of substantial rights. The city surveyor should, therefore, have directed his attention to the ascertainment of the actual location of the original landmarks ... and if those were discovered they must govern. [Id. at 605].

Minnesota: *Dittrich v. Ubl*, 216 Minn. 396, 13 N.W.2d 384 (Minn. 1944).

A resurvey that changes lines and distances and purports to correct inaccuracies or mistakes in the old plat is not competent evidence of the true line fixed by the original plat. 8 Am.Jur., Boundaries, p. 819, § 102; *Cragin v. Powell*, 128 U.S. 691, 9 S.Ct. 203,32 L.Ed. 566, following the rule of City of *Racine v. Emerson*, 85 Wis. 80, 55 N.W. 177, 178, 39 Am. St. Rep. 819, where the court stated: "* * * A resurvey must agree with the old survey and plat to be of any use it determining it. * * * Resurveys for the lawful purpose of determining the lines of an old survey and plat are generally very unreliable as evidence of the true lines. The fact, generally known and quite apparent in the records of courts, is that two consecutive surveys by different surveyors seldom, if ever, agree; and the greater number of surveys, the greater number of differences and disagreements will occur. When two surveys disagree, the correct one cannot be determined by still another survey. It follows that resurveys are of very little use in such a case as this, except to confuse it."

Mississippi: *J. R. Buckwalter Lumber Co. v. Wright*, 159 Miss. 470, 132 So. 443 (Miss. 1931).

The original survey, whether correctly made or not, is the true boundary between sections, and the surveyor must locate the original lines as run by the original surveyor. In case he misses the recognized corner, he should not divide the difference between the owners, because this does not reestablish the true boundary line. It may have the effect of giving each landowner the number of acres to which he is entitled, but it does not necessarily do so and it does not establish the true line. He should readjust his instruments, take his bearing from the true corner, and continue his survey until he correctly traces the boundary between the two landowners.

Boyd v. Durrett, 216 Miss. 214, 62 So.2d 319 (Miss. 1953).

392   Appendix I: List of Significant Cases on Following Footsteps, by State

The general rule relating to the location or re-establishment of boundary lines, where no question of adverse possession is involved, is stated in 8 Am.Jur. p. 792, Boundaries, par. 66, as follows:

In locating and running the boundary lines of lots or tracts of land of private owners, reference is to be had to the calls in the grant and to the field notes carried into the grant or the map or plan with reference to which the conveyance was made; and if there is no ambiguity the land must be located and the lines run according to the description of the conveyance. * * * Accordingly, in restoring lost lines and corners, visible and actual landmarks are to be preferred, but if they cannot be ascertained resort must then be had to courses and distances. All lands are supposed to be actually surveyed and the intention of the grant is to convey the land according to the actual survey. It is, therefore, said that the real purpose of the inquiry is to follow the steps of the surveyor on the ground, and all calls will be construed with this in mind.

Missouri: *McKinney v. Doane*, 155 Mo. 287, 56 S.W. 304 (Mo. 1900).

The court declares the law to be that the question in this case is not how would an accurate survey locate the lots in question but how did the original survey and stakes locate them. The only purpose of the evidence of the surveyors who made the recent surveys is to enable the court to locate the original boundary if possible, and not for the purpose of determining where it ought to have been, or where it would have been by an accurate survey. The original starting points and boundaries are questions of fact for the court to find from the evidence, and not only from the evidence of the surveyors, but all the other evidence in the case bearing upon these points.

Montana: *Olson v. Jude*, 316 Mont. 438, 73 P.3d 809, 815 (2003).

When surveyors use corner sections and lines to base measurements and plot tracts, it is essential that they properly identify and authenticate the original monument. *Helehan v. Ueland* (1986), 223 Mont. 228, 725 P.2d 1192. "Original corners, as established by the government surveyors, if they can be found, or the places where they were originally established, if they can be definitely determined, are conclusive on all persons owning or holding with reference thereto, without regard to whether they were located correctly or not, and must remain the true corners or monuments by which to determine the boundaries." *Vaught v. McClymond* (1945), 116 Mont. 542, 155 P.2d 612 (quoting 11 C.J.S. *Boundaries* § 11, p. 552). Moreover, in ascertaining the lines of land or in re-establishing the lines of a survey, the footsteps of the original surveyor, so far as discoverable on the ground by his monuments, should be followed and it is immaterial if the lines actually run by the original surveyor are incorrect. *Vaught*, 116 Mont. at 550, 155 P.2d at 616 (citing *Ayers v. Watson* (1891), 137 U.S. 584, 11 S.Ct. 201, 34 L.Ed. 803; *Galt v. Willingham* (5th Cir.1926), 11 F.2d 757). *See also Buckley v. Laird* (1972), 158 Mont. 483, 493 P.2d 1070.

Appendix I: List of Significant Cases on Following Footsteps, by State

393

Vaught v. McClymond, 116 Mont. 542, 155 P.2d 612 (Mont. 1945).

"A survey of public lands does not ascertain boundaries; it *creates* them." *Cox v. Hart*, 260 U.S. 427, 43 S.Ct. 154, 157, 67 L.Ed. 332. "The quarter lines are not run upon the ground, but they exist, by law, the same as the section lines." *Keyser v. Sutherland*, 59 Mich. 455, 26 N.W. 865, 867. The location of corners and lines established by the government survey, when identified, are conclusive (*Hickerson v. Dillard, Mo.App.*, 247 S.W. 801) and the true corner of a government subdivision of a section is where the United States surveyors in fact established it, whether such location is right or wrong, as may be shown by a subsequent survey. *Beardsley v. Crane*, 52 Minn. 537, 54 N.W. 740. Original monuments of survey established during a government survey, when properly identified, control courses and distances (*Mitchell v. Hawkins*, 109 Neb. 9, 189 N.W. 175; *Langle v. Brauch*, supra) and field notes and an official plat of government surveys of record will control in ascertaining locations, even though the monuments established are gone. *Slovensky v. O'Reilly, Mo.*, 233 S.W. 478. In ascertaining the lines of land or in re-establishing the lines of a survey, the footsteps of the original surveyor, so far as discoverable on the ground, should be followed and it is immaterial if the lines actually run by the original surveyor are incorrect. *Ayers v. Watson*, 137 U.S. 584, 11 S.Ct. 201, 34 L.Ed. 803; *Galt v. Willingham*, 11 F.2d 757. "In surveying a tract of land according to a former plat or survey, the surveyor's only duty is to relocate, upon the *best evidence* obtainable, the courses and lines at the same place where originally located by the first surveyor on the ground. In making the resurvey, he has the right to use the field notes of the original survey. The object of a resurvey is to furnish proof of the location of the lost lines or monuments, not to dispute the correctness of or to control the original survey. The original survey in all cases must, whenever possible, be retraced, since it cannot be disregarded or needlessly altered after property rights have been acquired in reliance upon it. On a resurvey to establish lost boundaries, if the original corners can be found, the places where they were originally established are conclusive without regard to whether they were in fact correctly located." 8 Am.Jur. Boundaries, Sec. 102, p. 819, Emphasis ours.

Nebraska: *Pallas v. Dailey*, 169 Neb. 533, 100 N.W.2d 197 (Neb. 1960).

A resurvey not shown to have been based upon the original survey is inconclusive in determining boundaries, and will ordinarily yield to a resurvey based upon known monuments and boundaries of the original survey.

Nevada: *Gray v. Coykendall*, 53 Nev. 466, 6 P.2d 442 (Nev. 1931).

It was said in *Treadwell v. Marrs*, 9 Ariz. 333, 83 P. 350, on an issue as to the location of a mining claim, "that, where the monuments are found upon the ground, or their position or location can be determined with

394 Appendix I: List of Significant Cases on Following Footsteps, by State

certainty, the monuments govern, rather than the location certificate; but where the course and distances are not with certainty defined by monuments or stakes, the calls in the location notice must govern and control." This is a salutary and well settled rule calculated to require the best evidence of the true boundaries of a claim, and to prevent the swinging or floating of claims to the detriment of subsequent locators. Of course inaccuracies or mistakes in a mining location will not invalidate the location, and in such cases monuments originally erected on the ground control the courses and distances. *Book v. Justice Min. Co. (C. C.)* 58 F. 106; *Gibson v. Hjul,* 32 Nev. 360, 108 P. 759. It is by such means that mistakes may be made known. But this applies only where the monuments or stakes can be clearly ascertained, otherwise the description in the location notice controls. *Swanson v. Koeninger,* 25 Idaho, 361, 137 P. 891; *Tiggeman v. Mrzlak,* 40 Mont. 19, 105 P. 77; *Flynn Group Min. Co. v. Murphy,* 18 Idaho, 266, 109 P. 851, 138 Am. St. Rep. 201; *Thallman v. Thomas (C. C.)* 102 F. 935; Lindley on Mines (3d Ed.) § 375; 40 C.J. 807.

New Hampshire: *Smith v. Dodge,* 2 N.H. 303 (N.H. 1820).

Whenever in a conveyance of land the deed refers to monuments actually erected as the boundaries of the land, it is well settled that those monuments must prevail, whatever mistakes the deed may contain as to the distances between the monuments.

New Jersey: No significant retracement decisions found.

New Mexico: *Pacheco v. Martinez,* 97 N.M. 37, 636 P.2d 308 (N.M. App. 1981).

A resurvey not shown to have been accurately based upon established points in an original survey is generally inconclusive in determination of boundaries between tracts. *Pallas v. Dailey,* 169 Neb. 533, 100 N.W.2d 197 (1960); *Thein v. Burrows,* 13 Wash.App. 761, 537 P.2d 1064 (1975). In *Thein v. Burrows,* the plaintiffs brought suit for a permanent injunction and sought damages for the wrongful cutting of timber on plaintiffs' lands. At trial both the plaintiff and defendant presented plats prepared by surveyors hired by each of the parties. The rule was declared in that case:

The general rule governing the determination of boundary lines by resurvey is that the intent of the new survey should be to ascertain where the original surveyors placed the boundaries rather than to determine where new and modern surveys would place them. 13 Wash.App. at 763, 537 P.2d at 1066.

New York: *Wates v. Crandall,* 144 N.Y.S.2d 211 (1955).

It is further stated that "A resurvey not shown to have been based upon the original survey is inconclusive in determining boundaries, and will ordinarily yield to a resurvey based upon known monuments and boundaries of the original survey." 11 C.J.S., Boundaries, § 61, pp. 634, 635.

Appendix I: List of Significant Cases on Following Footsteps, by State 395

North Carolina: *Cherry v. Slade*, 3 Murph. 82 (1819).

This case is the subject of considerable discussion within the text.

North Dakota: *Radford v. Johnson*, 8 N.D. 182, 77 N.W. 601 (N.D. 1898).

In a resurvey of the land which originally belonged to the United States, and which it has caused to be surveyed under its authority, such resurvey must conform to the survey made under the authority of the government, if the mounds and corners of the original government survey can be identified. If the stakes and monuments placed by the government in making the survey to indicate the section corners and quarter section posts can be found, or the places where they were originally placed can be identified, they are to control in all cases. Further, the corners established by the original surveyors under the authority of the United States cannot be altered. Whether properly placed or not, no error in placing them can be corrected by any surveyor deriving his authority from the laws of the state.

Ohio: *Sellman v. Schaff*, 26 Ohio App.2d 35, 269 N.E.2d 60 (Ohio App. 3 Dist. 1971).

The concept governing all resurveys. When an original survey has been made, it is not the plat or the metes and bounds description that is primary. The primary function of the second surveyor is to find first where the boundaries were established by the first surveyor. Only where this becomes impossible of accomplishment does the second survey turn to the courses, distances, and still-existent monuments to determine the boundaries. The essential rule governing the resurvey is to follow the steps of the first surveyor.
2 Proof of Facts 651, Boundaries, states:
A survey is the locating and marking on the ground of the land described in a grant. Once a tract has been located by survey, and its boundaries have been marked, those boundaries cannot be altered by a subsequent survey. In making a resurvey, the duty of the surveyor is merely to locate the monuments placed by the original surveyor, or, where such monuments no longer exist, the places where they originally stood.

Oklahoma: *Fellows v. Willett*, 98 Okla. 248, 224 P. 298 (Okla. 1923).

When a surveyor is called upon to locate government corners and lines, he is not employed as an arbiter of disputes between the adjoining landowners. Neither is it his province to correct mistakes in the original survey. It is his duty to locate the corners and lines as formerly established.
The original section corners as established by the government survey, or the place where they were established, if they can be definitely

396   *Appendix I: List of Significant Cases on Following Footsteps, by State*

determined, are conclusive on all present owners or holders with reference thereto, without regard to whether they were located correctly in the first instance, and must remain the true corners or monuments from which to determine the boundaries.

The foregoing is the rule of law even though the location of the section corner in the first instance is the result of an inaccurate survey.

If the section corners established by the government survey are the result of an inaccurate survey, and become obliterated, the resurvey must fix the corners where originally placed by the government, if the corners can be located by other natural objects or convincing evidence. This rule should be followed, even though to do so will awry the shape and boundary of the tract or section affected.

In making a survey or resurvey of lands the county surveyor is not authorized to take into consideration adverse possession and claims in locating lines and corners of sections or tracts of land. The duty of the county surveyor in making a resurvey of a former government survey is to locate the lines and corners of the section as fixed by the government survey without regard to adverse claims by adjoining landowners. On appeal from the surveyor's report it is not proper to try adverse claims between the parties.

Oregon: *Van Dusen v. Shively*, 22 Or. 64, 29 P. 76 (Or. 1892).

The line as actually run on the ground by Trutch, if it can be ascertained, is the line which must govern in this case, and courses and distances as given in the field-notes must yield thereto.

Goodman v. Myrick, 5 Or. 65. The location of this line is a question of fact to be ascertained from the evidence. The courses and distances as given in the field-notes are but descriptions which serve to assist in determining where the line was actually run. But where the line can be shown from the marks and blazes on the trees, or other natural monuments or calls, the courses and distances must yield to it. In cases of this kind the object is to follow in the "footsteps of the surveyor" as nearly as possible. No fixed or certain rules can be laid down by which questions of disputed boundaries can be settled, but each case must depend upon its own particular facts. The courses and distances in this case are entitled to but little weight in determining the line in dispute, as they do not correspond with the line as claimed by either party.

Trotter v. Town of Stayton, 41 Or. 117, 68 P. 3 (Or. 1902).

It will thus be seen that Gobalet, in making his survey, was not attempting to relocate the original lines, but, as instructed by the town authorities, to straighten out the streets, so that those in the original town and the subsequent additions might conform as nearly as practicable. Under such circumstances his survey could be of no value as evidence in determining the question before the court. The ruling question is the true location of the west line of Third street and the south line of Ida street.

Appendix I: List of Significant Cases on Following Footsteps, by State   397

As the plaintiff's property is described with reference to the lots and blocks as originally laid out in accordance with certain courses and distances, the point to be determined is the location of the lines so established, and no survey could be accurately based upon any other data. All subsequent surveys or resurveys are of no effect as evidence unless they tend to determine that question. *Bower v. Earl*, 18 Mich. 367; *Hale v. Cottle*, 21 Or. 580, 28 P. 901; *King v. Brigham*, 19 Or. 560, 25 P. 150; *City of Racine v. Emerson* (Wis.) 55 N.W. 177, 39 Am.St.Rep. 819; *Albert v. City of Salem (Or.)* 65 P. 1068.

Pennsylvania:

The survey and the conveyance are the distinctive and operative acts in the transmission of real property, and, where they differ from each other, one must of necessity control the other, if each stands unaided by extrinsic explanation, or the discrepancy could never be adjusted. In such an exigency, courts have always held the demarcation of the boundaries as the substantive act in fixing the limits of the grant, and the deed as evidence of them—the former the factum, and the latter the representative; and hence the rule that the survey, if unimpeachable for fraud or mistake, controls the deed, where there is a discrepancy as to the extent of the grant. 34 Pa. 195.

In *Lodge v. Barnett*, supra, it is said: "On this point there is nothing more fixed or better ascertained than the law of this state."

The courses and distances in a deed always give way to the boundaries found upon the ground, or supplied by the proof of their former existence when the marks or monuments are gone. So the return of a survey, even though official, must give way to the location on the ground, while the patent, the final grant of the state, may be corrected by the return of survey, and, if it also differs, both may be rectified by the work on the ground. One of the strongest illustrations of this rule is to be found in the instance of the surveys of the donation lands, set apart for the soldiers of the Pennsylvania line in the Revolutionary War. The law required the tract to be identified by marking the number of it upon a tree within and nearest to the northwestern corner. It was held that this number controlled all the remainder of the description in the patent, so as to wrest it entirely from its position and adjoiners, as described in the patent and general draft. *Smith v. Moore*, 5 Rawle (Pa.) 348; *Dunn v. Ralyea, 6 Watts & S. (Pa.)* 475. Chief Justice Gibson, in the former case, stated the general principle thus: "It is a familiar principle of our system, and one in reason applicable to this species of title, as well as any other, that it is the work on the ground, and not on the diagram returned, which constitutes the survey, the latter being but evidence (and by no means conclusive) of the former. * * * It is conceded that the patent may be rectified by the return of survey; and why not the return of survey by the lines on the ground, and particularly the numbered tree, which is the foundation of the whole?" In the latter case, Kennedy, J., said: "That the original lines as found marked on the ground must govern, in determining the location and extent of the survey, is a well-established rule, in general applicable

396 *Appendix I: List of Significant Cases on Following Footsteps, by State*

to all cases." * * * We know, in point of fact, that the marks made on the ground at the time of making the survey are the original, and therefore the best evidence of what is done in making it; that everything that is committed to paper afterwards in relation to it is intended and ought to be, as it were, a copy of what was done, and ought to appear on the ground, in the doing of which errors may be committed, which renders it less to be relied on than the work as it appears by the marks made on the ground. 46 Pa. 484, 485.

Hall v. Powel, 8 Am.Dec. 722, 4 Serg. & Rawle 456, Pa. 1818.

The court state the law to be, "that where lines and corners are to be found on the ground, they cannot be departed from, though there may be some variance in the courses, and if the waters are found to agree with the lines and corners returned, this is strong evidence of a survey actually made." So has the law ever been held. The question has been frequently agitated, and is now put at rest. The field notes, the original plots made by the surveyor, the survey returned, and the patent, are only evidence of the survey. The real survey, the primary evidence, is the marks on the ground. In *Yoder v. Fleming*, before Shippen and Yeates, Justices, at Nisi Prius, at Lewistown 1798, 2 Sm. L. 256, the question occurred, whether the pretensions of a party should be determined by the courses and distances expressed in the return of survey, or by the marked trees, and lines actually run; and thus was the law laid down by these judges, whose experience in questions of this nature, was greater than that of any men now living. "The natural or artificial boundaries of a survey, have uniformly prevailed, and there is absolute certainty, when a right line is followed from one corner to another; but the best instruments will vary in some small degree. For the sake of public convenience, and individual safety, all the lands comprised within certain marked lines, or proceeding from marked and known corners, will pass in a deed. Any surplus measure, or variation in the courses and distances, will not vitiate the instrument. The lines actually run on the ground, are the true survey and appropriation of the land contracted for; but the return of survey is only evidence thereof, and shall be controlled by the actual survey. This point had been frequently determined, and particularly in *Walker v. Furry & Krehl*, before Ch. J. McKean, in 1790."

Rhode Island: *Acampora v. Pearson*, 899 A.2d 459 (R.I. 2006):

A professional engineer, testified as an expert witness for the Pearsons. He opined that the proper way to resurvey a parcel is to use the best evidence currently available to physically locate the lot lines originally set by the creator of the plat.

South Carolina: No significant retracement decisions found.

South Dakota: *Titus v. Chapman*, 2004 SD 106, 687 N.W.2d 918 (S.D. 2004).

Appendix I: List of Significant Cases on Following Footsteps, by State   399

Government surveys, not surveys conducted by private individuals, create, rather than merely identify, boundaries. *Cox v. Hart*, 260 U.S. 427, 436, 43 S.Ct. 154, 157, 67 L.Ed. 332, 337 (1922). The term "original survey" refers to the official government survey performed under the laws of the federal government by its official agency. *See Id.; Block v. Howell*, 346 N.W.2d 441, (S.D.1984); Walter G. Robillard & Lane J. Bouman, Clark on Surveying and Boundaries § 4.12 (5th ed. 1976).

A subsequent survey by a private individual or non-government entity is more accurately described as a retracing or resurvey. *Block*, 346 N.W.2d at 444; *Randall v. Burk Tp.*, 4 S.D. 337, 57 N.W. 4, 10 (1893). In a retracing or resurvey, a surveyor must "take care to observe and follow the boundaries and monuments as run and marked by the original survey." *Block*, 346 N.W.2d at 444. Boundaries as established by original government surveys are unchangeable and must control disputes. *Christianson v. Daneville Tp.*, 61 S.D. 55, 58, 246 N.W. 101, 102 (1932).

Original monuments, those located by the original surveyor, mark true corners. *Lawson v. Viola Tp.*, 50 S.D. 555, 210 N.W. 979, 980 (1926). "Where the location of the original monument can be found, or can be established by evidence, such location shall be held to be the true corner, regardless of the fact that resurveys may show that it should have been located elsewhere." *Id.* (citing *Byrne v. McKeachie*, 34 S.D. 589, 149 N.W. 552 (1914); *Hoekman v. Iowa Civil Township*, 28 S.D. 206, 132 N.W. 1004 (1911); *Randall*, 4 S.D. 337, 57 N.W. 4; *Beardsley v. Crane*, 52 Minn. 537, 54 N.W. 740 (1893); *Ogilvie v. Copeland*, 145 Ill. 98, 33 N.E. 1085 (1893); *Nesselroad v. Parrish*, 59 Iowa 570, 13 N.W. 746 (1882). Where the original monument is obliterated, that is it cannot be located nor established by evidence, then a corner can be established by a new survey. *Lawson*, 50 S.D. at 558, 210 N.W. at 980 (citing *Randall*, 4 S.D. at 355, 57 N.W. at 10; *Washington Rock Co. v. Young*, 29 Utah 108, 80 P. 382 (1905)). Only upon obliteration of an original corner may a new survey be made from points that can be determined in accordance with the original surveyors field notes. *Id.* However, if the point at which an original monument was located can be ascertained by the court, the line as indicated by the government survey prevails. *Dowdle v. Cornue*, 9 S.D. 126, 127, 68 N.W. 194 (1896).

SDCL 43-18-7 provides:

In retracing lines or making the survey the surveyor shall take care to observe and follow the boundaries and monuments as run and marked by the original survey, but shall not give undue weight to partial and doubtful evidence or appearances of monuments, the recognition of which shall require the presumption of marked errors in the original survey, and he shall note an exact description of such apparent monuments.

Randall v. Burk Tp., 4 S.D. 337, 347, 57 N.W. 4, 10, 1893.

The rule is well settled that on a resurvey of land originally belonging to the United States, and which it has caused to be surveyed under its authority, such resurvey must follow the boundaries and monuments, as run and made by the original government survey, if the monuments placed by the government, in making the survey, to indicate the section corners and quarter section posts, can be found, or the places where they were originally placed can be identified.

400   *Appendix I: List of Significant Cases on Following Footsteps, by State*

Tennessee: *Wood v. Starko*, 197 S.W.3d 255 (Tenn.App. 2006).

One of the most useful retracement cases in the metes-and-bounds states. It reviews a number of previous decisions from a variety of jurisdictions. In its summary, the decision highlights many of the important points contained in this case.

Staub v. Hampton, 117 Tenn. 706, 101 S.W. 776 (1907) is emphasized, which states "While these general rules apparently have their origin in surveys reflecting government grants, such rules are equally applicable to private surveys."

Texas:

The Texas courts have issued more excellent decisions regarding retracement of original surveys than any other state. Among them are the following.

Outlaw et al. v. Gulf Oil Corporation et al., 137 S.W.2d 787 (Tex., 1940), *Taylor v. Higgins Oil & Fuel Co.*, 2 SW.2d 288 (1928), *Ballard v. Stanolind Oil & Gas Co.*, 80 F.2d 588 (Tex., 1935), *Blaffer v. State*, 31 S.W.2d 172 (Tex., 1930), and *Hart v. Gries et al*, 155 S.W.2d 997 (Tex., 1941).

The purpose of the inquiry, and the end to which all evidence is addressed, in a boundary suit is to find the footsteps of the original surveyor. *Goodson v. Fitzgerald*, 40 Tex.Civ.App. 619, 90 S.W. 898 (Tex. Civ.App. 1905)

The case was tried by all parties on the theory that the surveys were actually made on the ground, and the testimony related to the search for the footsteps of the surveyors. As we also said in the Turnbow case: "In all of the cases, from *Stafford v. King*, 30 Tex. 257, 94 Am.Dec. 304 (1867), on down, the law has been that search must be made for the footsteps of the surveyor, and that, when found, the case is solved."

Stafford v. King, 30 Tex. 257, 94 Am.Dec. 304 (1867).

This is probably the leading case in this state, regardless, it is one of the most important retracement decisions in the United States.

Utah: *Henrie v. Hyer*, 92 Utah 530, 70 P.2d 154 (Utah 1937).

It is conceded, as it must be, that the original corners as established by the government surveyors, if they can be found, or the places where they were originally established, if that can be definitely determined, are conclusive on all persons owning or claiming to hold with reference to such survey and the monuments placed by the original surveyor without regard to whether they were correctly located or not. Surveyors, in making resurveys or in searching for or relocating or re-establishing lost

Appendix I: List of Significant Cases on Following Footsteps, by State 401

or obliterated corners, may consider extrinsic and material evidence, as well as the field notes, if there is doubt or uncertainty in the field notes, for the purpose of determining the exact location of lost lines or corners of the original survey. Monuments control over courses and distances. *Washington Rock Co. v. Young*, 29 Utah 108, 80 P. 382, 110 Am. St. Rep. 666.

Vermont:

Carrying out the theory of appellants, we quote from the case of *Silsby & Co. v. Kinsley*, 89 Vt. 263, 95 A. 634, 638: "The actual location upon the ground of original lot lines will control, if capable of being ascertained; but, when such lines have never been surveyed or, if surveyed, their location upon the ground cannot be ascertained, resort may be had to the lines of adjacent lots to determine their location."

Virginia: *Moody v. Farinholt*, 158 Va. 234, 164 S.E. 258 (Va. 1932).

W. F. O'Hara, a civil engineer introduced as a witness on behalf of the plaintiff has undertaken to establish this west line of Shooters Hill, and since that west line is a part of the Northam survey, it was of course necessary for him to retrace it. He said that he examined the original, which is of record, and some older deed on which it was in a measure based. The ash tree in the center of Carter Braxton's mill pond is gone. At what is described as "a cedar at the west side of a small swamp" he found a cove or branch making into the mill pond, and the remains of a large dead cedar tree. He also found a rotten cedar stob where that line crossed the public road. Projected, this line reached the head of a fresh water stream running into the Piankatank river.

"The witness further stated that in running this western boundary line of Shooters Hill he went down a ditch a portion of the way which was at places three or four feet deep now and three or four feet wide. This was evidently an old ditch because there were large trees growing on the bottom of the ditch some of them being as much as two feet across the stump; that he got the present course of the line from 9 to 10 as shown on said blue print by the usual methods of calculating the magnetic variation since 1837."

That line in the Fauntleroy deed of 1837 runs south thirty-nine and one half degrees west. O'Hara found it to be now south forty-three degrees forty-eight minutes west, which is about what would have been expected, making allowance for magnetic variation. With allowances so made it is practically parallel with the eastern boundary of Shooters Hill, which boundary is definitely marked by monuments. By the Northam survey these two lines vary in bearing by only one and one-half degrees.

All of this violates one of the fundamentals of surveying. It neither touches nor purports to touch one of the landmarks or monuments on the Shooters Hill line, although some of them are as plain today as they

402   *Appendix I: List of Significant Cases on Following Footsteps, by State*

were in 1837. Established, they fix this line, and so established, it cannot possibly be changed by the lines of other surveys. No amount of evidence can change a right line from A to B where A and B are ascertained monuments. The ditch of itself, down which the O'Hara line runs, is enough.

Washington: *Strunz v. Hood*, 44 Wash. 99, 87 P. 45 (Wash. 1906).

Appellants correctly contend that a court or a court commissioner cannot correct the United States government surveys, or establish government corners at points other than those fixed by the government surveyors; that in any attempt to re-establish an original survey the purpose should be to follow the footsteps of the government surveyor as nearly as possible, and that when there is any variance between field notes and monuments, as set up by the United States government surveyors, the monuments must prevail. It was undoubtedly the duty of the commissioner to ascertain, if possible, where the original government monuments had been actually located and established, rather than where he might think they ought to be located or established.

West Virginia: *Billups v. Woolridge*, 80 W.Va. 13, 91 S.E. 1082 (W.Va. 1917).

The court told the jury that in endeavoring to locate the land described in the patent they should search for the footsteps of the surveyor in locating the survey upon which the patent was based. It is argued that inasmuch as the only patent offered in evidence was the patent for 777 acres, and it was junior to the patent for the 104 acres, and the former called for and was limited to the true location of the line of the 104 acre tract, the line in dispute, the instruction was misleading and prejudicial to defendant's interests. Defendant did not introduce the patent for the 104 acre tract, but without objection, the surveyors and other witnesses referred to the calls of the senior patent, and identified the call in the one as alike and coincident with the calls in the other, so far as they relate to the disputed line.

Wisconsin: City of *Racine v. Emerson*, 85 Wis. 80, 55 N.W. 177 (Wis. 1893).

A resurvey that changes lines and distances and purports to correct inaccuracies or mistakes in the old plat is not competent evidence in the case. There are only two questions: (1) Where is the true line fixed by the original plat? (2) Is the fence in question on that line? A resurvey that changes or corrects the old survey and plat can never determine the first question. A resurvey must agree with the old survey and plat to be of any use in determining it.

Resurveys for the lawful purpose of determining the lines of an old survey and plat are generally very unreliable as evidence of the true lines. The fact, generally known and quite apparent in the records of

Appendix I: List of Significant Cases on Following Footsteps, by State   403

courts, is that two consecutive surveys by different surveyors seldom, if ever, agree; and the greater number of surveys, the greater number of differences and disagreements will occur. When two surveys disagree, the correct one cannot be determined by still another survey. It follows that resurveys are of very little use in such a case as this, except to confuse it.

Pereles v. Gross, 126 Wis. 122, 105 N.W. 217 (Wis. 1905).

In resurveying a tract of land according to a former plat or survey, the surveyor's only function or right is to relocate, upon the best evidence obtainable, the corners and lines at the same places where originally located by the first surveyor on the ground. Any departure from such purpose and effort is unprofessional, and, so far as any effect is claimed for it, unlawful. To fix lines variant from the originals and according merely to his notion of a desirable arrangement of lots and streets leads naturally to confusion of claims among lot owners, and, when done by a city surveyor as a basis for occupation of land for streets, is attempted confiscation. The evidence shows that in the city survey nothing was found on the ground to show where any of the subdivisions between the points A and E were originally located; also, that the surveyor proceeded, while retaining the same number of lots, to give those lots such arbitrary width as he saw fit, with the purpose and result of making the lines of the streets on the northeast side of Water street coincide with the extended lines of the north and south streets. This resurvey is therefore wholly valueless, and not even evidentiary, unless it be found as a fact that such coincidence of street lines did in fact exist in the original survey.

Wyoming: *Bentley v. Jenne*, 33 Wyo. 1, 236 P. 509 (Wyo. 1925).

Original corners, if found, are conclusive whether correctly located or not, 9 C. J. 164. A resurvey merely traces original lines; *Bayhouse v. Urquides*, 105 P. 1066; *Steel v. Co.*, 106 U.S. 447; *Weaver v. Howatt*, 152 P. 928; *Trotter v. Stayton, (Ore.)* 68 P. 3.

Appendix 2: Lost Corners, by State

One of the earliest issues faced by the new United States Government after the conclusion of the War of Independence was the appropriate development and disposition of the public lands owned by the Federal Government. The issue was addressed in the United States Constitution under Article IV, Section 3, clause 2, which provides that "The Congress shall have Power to dispose of and make all needful Rules and Regulations respecting the Territory or other Property belonging to the United States" ensuring that the Federal Government, and not the individual States, would determine how the public lands of the United States were to be administered.

Prior to the Constitution's adoption in 1788 and while the Articles of Confederation (1781) were still in effect, the Congress passed the Land Ordinance of 1785, establishing a system for surveying and thereby demarcating the public lands for their (1) orderly disposition into new States, (2) conveyance from Federal into State and private ownership, or (3) retention for Federal administration. This "rectangular system of survey" typically describes townships of 36 square miles comprised of sections of 1 square mile (640 acres, more or less), each subdivided into quarter sections (160 acres) and quarter-quarter sections (40 acres). Under this land tenure system, each tract of land would receive a unique identifying description. Before a survey was completed, the lands were known as "unsurveyed public lands" and could not be disposed out of Federal ownership.

Since the Land Ordinance of 1785, it has been the continuous policy of the United States that land shall not leave Federal ownership until it has first been surveyed, and an approved plat of survey has been filed. After the survey, persons interested in homesteading or making other authorized land entries under the Federal public land laws could identify what lands were available for claim and entry. The corner monuments on the ground established actual on-the-ground locations for the boundaries of the lands entered, patented, and/or otherwise conveyed. This process assures the orderly disposition of the public lands and avoids confusion and contention.

Thirty of the fifty current States ("public States") were originally surveyed under this system. With very few exceptions all chains of title to privately owned land in those 30 States trace back to a Federal land patent or other grant. Land ownership and boundaries in the other 20 States, that is, the Thirteen Original States plus Hawaii, Kentucky, Maine, Tennessee, Texas, Vermont, and West Virginia, were established by other means and surveyed according to a variety of different systems and standards.

406   *Appendix 2: Lost Corners, by State*

The current *Manual of Surveying Instructions* (Manual, 2009) represents
the most recent in a series of official and binding survey instructions
dating back to 1804. The previous official Manual was issued in 1973 (the
1973 Manual), prior to which the Manual of 1947 contained the official
instructions. The Manual describes how cadastral surveys are made in
conformance with statute law and its judicial interpretation.*

The Manual is supplemented for the purposes of re-surveying and retracement surveys with a supporting official document known as *Restoration
of Lost or Obliterated Corners & Subdivision of Sections, a guide for surveyors.*
This document is frequently referred to as the "supplement to the Manual."
It has been available since its first printing in 1883, and offers guidance to
surveyors and others attempting to locate and follow the original surveys
done in accordance with the Manual. Any surveyor within the Public Land
States (the PLSS) should be familiar with both documents, keeping in mind
that original surveys were intended to be performed in accordance with the
Manual in existence at the time of the survey. In addition, surveyors outside
the PLSS will undoubtedly find both of the documents helpful from time to
time.

Some Representative Cases Involving Lost Corners

A few states, along with the federal courts, have a few decisions specifically
on the topic of lost corners that are noteworthy. Some are included in the
following.

The federal cases are included within the state where the controversy took
place. There are several decisions by the federal courts which the retracement surveyor should be familiar with.

Alabama: "Originally the lands lying in this state were the property of
the federal government. The Congress authorized the survey of these lands,
dividing them into townships, ranges, sections, and subdivisions of sections. When these surveys were made, the lines and corners were marked
by the surveyors. These marks, of whatever nature, became the 'monuments'
marking the divisions of land by which the government described the lands
when deeded by it to the citizen or purchaser. These monuments, therefore,
became a part of the purchaser's muniment of title marking the limits of the
area owned by him and were a pledge to him by his government that there
would be no change in these monuments. These corners or monuments as
established by the government, if they can be found, or the places where

---

* *Manual of Surveying Instructions for the Survey of the Public Lands of the United States.* United
States Department of the Interior, Bureau of Land Management, Cadastral Survey, 2009.

Appendix 2: Lost Corners, by State 407

they were originally established, are conclusive on all persons owning or holding land with reference thereto, without reference to whether they were located correctly or not and must remain the true corners or monuments. 9 C.J. 164 (18)b. In other words, a corner located by the original government survey, from which the lands are disposed of by the government and successive grantors, becomes the monument of that corner, and there cannot be another substituted for it so long as it exists or may be located. When, however, the monument establishing an original corner is lost, the object is always to carry into effect what was done or attempted to be done when the survey was made. *Ramsey v. Morrow*, 133 Ky. 486, 118 S.W. 296. When this is done, the survey made in accordance with law in undertaking a relocation becomes evidence from which the jury must say whether it is the corner or not. 9 C.J. 162(15)(2). It follows that the original monument, or, in case of its loss, its substitute located according to law, is the monument protected by the statute. *Cornelious v. State*, 22 Ala.App. 150, 113 So. 475 (Ala.App. 1927)."

AUTHOR'S NOTE: It is noteworthy here that the Alabama court (rectangular state) is referencing a Kentucky (metes-and-bounds state) case.

Arizona: *Galbraith v. Parker*, 17 Ariz. 369, 153 P. 283 (Ariz. 1915).

The rule seems to be equally well settled that where the original United States monuments indicating the location, upon the ground, of corners, have disappeared or have been lost or obliterated, and there is no evidence or proof as to the spot of their original placing, the plat and field-notes and the calls therein are determinative of the rights of the parties in disputed boundary questions. *Stangair v. Roads, supra; Washington Rock Co. v. Young*, 29 Utah 108, 110 Am. St. Rep. 666, 80 P. 382; *Ogilvie v. Copeland*, 145 Ill. 98, 33 N.E. 1085; *Read v. Bartlett*, 255 Ill. 76, 99 N.E. 345.

Arkansas: *Thompson v. Darr*, 174 Ark. 807, 298 S.W. 1 (Ark. 1927).

It may be said also that Woolverton was correct in his apportionment of the excess of the 80 links more than the mile which the Government field-notes showed to be the length of the south line of section 33, the apportionment being the addition of 20 links to the south line of each 40-acre tract bounded by the south line of the section.

The surveyor could not change the corners established by the Government survey, as these fixed monuments prevail over both course and distance. *Meyer v. Board of Imp. Pav. Dist. No. 3*, 148 Ark. 623,231 S.W. 12. Section 2396 of the U.S. Revised Statutes so provides.

The Supreme Court of Wisconsin, in the case of *Lewis v Prien*, 73 N.W. 654, 98 Wis. 87, said:

The unvarying rule to be followed in such cases is to start at the nearest known point on one side of the lost corner, on the line on which it was originally established; to then measure to the nearest known corner on the other side, on the same line; then, if the length of the line is in excess

408   *Appendix 2: Lost Corners, by State*

of that called for by the original survey, to divide it between the tracts connecting such two known points in proportion to the lengths of the boundaries of such tracts on such line as given in such survey.

California: *Higgins v. Ragsdale*, 23 P. 316, 83 Cal. 219 (1890).

A part of the testimony for the plaintiff was that of C. E. Uren, as follows:

I am a surveyor and civil engineer. I know the lines in controversy. I went upon the ground this spring at the request of plaintiff to make a survey to establish the corner. I sought for the missing corner, and could not find it. The north and south section corners are in place, and are easily discoverable without the aid of field-notes. Not being able to find the corner, I established it as a lost corner by dividing the distance between the two section corners equally. The line is over a mile long, while the survey calls for a mile long only. I examined carefully all the marks that would aid me. I discovered a line of blazed trees from the south stake, which were originally line-trees. I followed along the same, and at half a mile set a stake, near which and about right distance and direction was the remains of a yellow pine stump badly burned, but which I believe was the bearing-tree.

The only pine-tree which is now standing has two blazes on the west side, which are old, and made quartering just like line blazes. This tree could not be a witness-tree. I examined it very carefully. It has a large burned place on the side defendant claims the stake to be, and facing stake. There is no mark outside the burned place. Surveyors do not make bearing-marks on dead trees or burned wood.

About two hundred and fifty feet south of the pine is an oak eighteen feet east of my line with old blaze on the west side. If the line ran where defendant claims, it would be on the east side of this oak-tree. I found line trees all along my line. I also found line-trees on the north half of the line. Section lines are always originally run from the south toward the north. I ran the line through from section-post to section-post, and placed a stake exactly midway; the distance from said post to the north and south stakes being exactly 40.875 chains in each instance. The fence of defendant is 57 3/4 feet north of the stake so placed, and incloses 2.66 acres of plaintiff's land.

I was absolutely unable to find or establish the original one-quarter section stake, and there is nothing on the ground by which it or a bearing-tree can be established. I examined very carefully.

Colorado: *Gaines v. City of Sterling*, 140 Colo. 63, 342 P.2d 651 (Colo. 1959).

In *Westcott v. Craig*, 1915, 60 Colo. 42, 151 P. 934, this court said:

The method of re-establishing interior corners is fixed not only by the rules of the General Land Office, but by numerous decisions of the courts.

The court in Westcott proceeded to quote from the leading case of *Moreland v. Page*, 1856, 2 Iowa 139, what it describes as the correct rule

Appendix 2: Lost Corners, by State

of apportionment conforming to the regulations of the General Land Office:

Where on a line of the same survey, between remote corners, the whole length of which line is found to be variant from the length called for, in re-establishing lost intermediate monuments, as marking subdivisional tracts, we are not permitted to presume merely that a variance arose from the defective survey of any part; but we must conclude, in the absence of circumstances showing the contrary, that it arose from the imperfect measurement of the whole line, and distribute such variance between the several subdivisions of such line, in proportion to their respective lengths.

The Colorado court then commented:

This case has been followed in numerous decisions, and the rule laid down is identical in principle with that by which an excess or shortage in the frontage of a subdivision is apportioned among all the lots affected. We regard the rule thus announced as correct in principle, and it is supported by abundant authority.

This decision was followed as to the method of locating lost corners in *Beaver Brook Resort Co. v. Stevens*, 1924, 76 Colo. 131, 230 P. 121.

The *Manual of Instructions for the Survey of the Public Lands of the United States, 1947*, section 376, at page 298, before the trial court, states:

All lost (obliterated) quarter-section corners on the section boundaries within the township will be restored by single proportionate measurement between the adjoining section corners, after the section corners have been identified or relocated. (Wording in parenthesis added.)

C.R.S. '53, 136-1-1 provides:

Whenever a surveyor is required to make a subdivision of a section, as established by the United States survey, he shall proceed as follows: Commencing at either quarter-section corner of the section, he shall proceed to run two lines respectively, to the opposite quarter-section corners, and at the point of their intersection establish a post which shall make the interior quarter-section corner for that section. Any less subdivision shall be made by proceeding in the same manner, all interior posts being established by the point of intersection of lines, bisecting opposite sides; provided, that all fractional quarter-sections shall be so divided as to give to occupants of any part thereof the quantity of land which shall be held to appertain to such part, at any United States land office in this state.

The general rule (known as the apportionment rule) is that where a tract of land is subdivided into parts or lots, title to which becomes vested in different persons, none of the grantees are entitled to a preference over the others upon the discovery of an excess or deficiency in the quantity of land contained in the original tract. The excess or deficiency is then divided among all the lots or parcels in proportion to their areas. See 97 A.L.R. 1227-1228 et seq.; and 11 C.J.S. Boundaries §§ 124-125, pp. 737–739.

Some limitations and exceptions have inevitably grown up to invade the pure application of the apportionment rule. It has been applied when the original surveys were made at the same time and so are held to have equal standing. See *Pandem Oil Corp. v. Goodrich, Tex.Civ.App.*1930, 29 S.W.2d 877, reversed on other grounds Tex.Com.App.1932, 48 S.W.2d 606. Another court has applied it only as a last resort. *Chandler v. Hibberd, Cal. App.*1958, 332 P.2d 133. It has been limited to each block within a subdivision. *Tyner v. McDonald, Fla.*1953, 63 So.2d 504. And at least one court has refused to apply it at all where possessory rights have intervened. *Vance v. Gray*, 1911, 142 Ky. 267, 134 S.W. 181. The remnant rule has been applied to limit it and to apply to lots or parcels sold under a plan with the last grantee receiving only what is left. *Barrett v. Perkins*, 1911, 113 Minn. 480, 130 N.W. 67; *Adams v. Wilson*, 1902, 137 Ala. 632, 34 So. 831, and *Geiger v. Uhl*, supra. Yet the latter rule has been rejected and vigorously attacked by a New York court as not being what the grantor intended. *Mechler v. Dehn*, 1922, 203 A.D. 128, 196 N.Y.S. 460, affirmed without opinion in, 1923, 236 N.Y. 572, 142 N.E. 288.

Other rules have grown up to solve controversies which have arisen as to this problem under government sections such as we have here. One of these widely applied is that interior lots in the extreme northern and western tiers were to contain the legal number of acres and any excess or deficiency is always to be thrown on the exterior lots. *Keesling v. Truitt*, 1868, 30 Ind. 306; *Grover v. Paddock*, 1882, 84 Ind. 244; *Fuelling v. Fuesse*, 1909, 43 Ind.App. 441, 87 N.E. 700; *Knight v. Elliott*, supra; *Vaughn v. Tate*, supra.

A modification of the above rule that any excess or deficiency in fractional sections is to be placed upon the exterior lots is that it has no application to interior sections. See *Suit v. Hershman,*1917, 66 Ind.App. 388, 118 N.E. 310; *Hootman v. Hootman*, 1907, 133 Iowa 632, 111 N.W. 60.

Most courts recognize a distinction between the application of the rules relating to government surveys as between correct original surveys and their application where the original surveys are found to have been erroneous or the original corners and lines have been wholly lost. In the latter case the surplus or deficiency is apportioned in most states but not all. See 97 A.L.R. 1233-1234, in which the Colorado case of *Westcott v. Craig*, supra, is cited among cases from many other jurisdictions following the reasoning in *Moreland v. Page*, supra, that there should be a clear application of the apportionment rule without regard to the distinction above mentioned.

The rule which we now adopt as being fair and just and in harmony with the Acts of Congress and the regulations of the United States Public Land Office, as well as the previous decisions of this court.

United States v. Doyle, 468 F.2d 633 (10th Cir. 1972).

The procedures for restoration of lost or obliterated corners are well established. They are stated by the cases cited below and by the supplemental manual on Restoration of Lost or Obliterated Corners and Subdivisions of Sections of the Bureau of Land Management (1963 ed.). The supplemental manual sets forth practices and contains explanatory and advisory comments.

Practice 1 of the supplemental manual recognizes that an existent corner is one whose position can be identified by verifying evidence of the

Appendix 2: Lost Corners, by State

monument, the accessories, by reference to the field notes, or "where the point can be located by an acceptable supplemental survey record, some physical evidence, or testimony." Practice 2 recognizes that an obliterated corner is one at whose point there are no remaining traces of the monument, or its accessories, but whose location has been perpetuated, or the point for which may be recovered beyond a reasonable doubt, by the acts and testimony of the interested land owners, competent surveyors, or other qualified local authorities, or witnesses, or by some acceptable record evidence. Practice 3 states that a lost corner is one whose position cannot be determined, beyond reasonable doubt, either from traces of the original marks or from acceptable evidence or testimony bearing on the original position, and whose location can be restored only by reference to one or more interdependent corners.

The authorities recognize that for corners to be lost "[t]hey must be so completely lost that they cannot be replaced by reference to any existing data or other sources of information." *Mason v. Braught*, supra, 146 N.W. at 689, 690. Before courses and distances can determine the boundary, all means for ascertaining the location of the lost monuments must first be exhausted. *Buckley v. Laird*. 493 P.2d 1070, 1075 (Mont.); Clark, Surveying and Boundaries § 335, at 365 (Grimes ed. 1959); see advisory comments of the supplemental manual, supra at 10.

The means to be used include collateral evidence such as boundary fences that have been maintained, and they should not be disregarded by the surveyor. *Wilson v. Stork*, 171 Wis. 561, 177 N.W. 878, 880. Artificial monuments such as roads, poles, fences and improvements may not be ignored. *Buckley v. Laird*, supra, 493 P.2d at 1073; *Dittrich v. Ubl*, 216 Minn. 396, 13 N.W.2d 384. And the surveyor should consider information from owners and former residents of property in the area. See *Buckley v. Laird*, supra, 493 P.2d at 1073–1076. "It is so much more satisfactory to so locate the corner than regard it as 'lost' and locate by 'proportionate' measurement." Clark, supra § 335 at 365.

Florida: *City of Jacksonville v. Broward*, 120 Fla. 841, 163 So. 229 (Fla. 1935).

To support its contention appellant relies on the rule stated in 4 R. C. L. 115, as follows:

When division lines are run splitting up into parts larger tracts it is occasionally discovered that the original tract contained either more or less than the area assigned to it in a plan or prior deed. Questions then arise as to the proper apportionment of the surplus or deficiency. In such cases the rule is that no grantee is entitled to any preference over the others, and the excess should be divided among, or the deficiency borne by, all of the smaller tracts or lots in proportion to their areas. The causes contributing to the error or mistakes are presumed to have operated equally on all parts of the original plat or survey, and for this reason every lot or parcel must bear its proportionate part of the burden or receive its share of the benefit of a corrected resurvey. This rule for allotting the deficiency or excess among all the tracts within the limits of the survey may be applied where the original surveys have been found

412   *Appendix 2: Lost Corners, by State*

to have been erroneous, or where the original corners and lines have become obliterated or lost.

If the lines of a survey are "found to be either shorter or longer than stated in the original plat or field notes, the causes contributing to such mistakes will be presumed to have operated equally in all parts of the original plat or survey, and hence every lot or parcel must bear the burden or receive the benefit of a corrected resurvey, in the proportion which its frontage as stated in the original plat or field notes bears to the whole frontage as there set forth." *Pereles et al. v. Magoon et al.*, 78 Wis. 27, 46 N.W. 1047, 23 Am. St. Rep. 389, 392, note.

This rule applies to government surveys. *James v. Drew*, 68 Miss. 518, 9 So. 293, 24 Am. St. Rep. 287.

The foregoing rule is approved in *Porter v. Gaines*, 151 Mo. 560, 52 S.W. 376; *Mosher v. Berry*, 30 Me. 83, 50 Am. Dec. 614; *Marsh v. Stephenson*, 7 Ohio St. 264, 70 Am. Dec. 72; *Booth v. Clark*, 59 Wash. 229, 109 P. 805, Ann. Cas. 1912A, 1272; *Coppin v. Manson*, 144 Ky. 634, 139 S.W. 860; Clark on Surveys and Boundaries c. 180, p. 161.

Georgia: *Riley v. Griffin*, 16 Ga. 141 (Ga. 1854).

The decision has been discussed in great detail at several places within the text.

Idaho: *Craven v. Lesh*, 22 Idaho 463, 126 P. 774 (Idaho 1912).

It is clear from a review of all the evidence that said corner is an obliterated corner and not a lost one. The field notes of the government original survey show that a willow stake was set at said corner with charcoal at the bottom thereof, and that said corner was established in the manner followed by the government in establishing such corners.

Richardson v. Bohney, 19 Idaho 369, 114 P. 42 (Idaho 1911).

When public land has been surveyed by authority of the United States, and patented with reference to the boundaries as fixed by such surveys, the corners and lines so established, whether correct or not, are conclusive, and cannot be altered and controlled by other surveys. (*Billingsley v. Bates*, 30 Ala. 376, 68 Am. Dec. 126; *Climer v. Wallace*, 28 Mo. 556, 75 Am. Dec. 135; *Mayor of Liberty v. Burns*, 114 Mo. 426, 19 S.W. 1107, 21 S.W. 728; *Granby Mining Co. v. Davis*, 156 Mo. 422, 57 S.W. 126; *Arneson v. Spawn*, 2 S.D. 269, 39 Am. St. 783, 49 N.W. 1066; *Goodman v. Myrick*, 5 Ore. 65; *Jones v. Kimble*, 19 Wis. 429; *Trinwith v. Smith*, 42 Ore. 239, 70 P. 816.)

The true corner of a government subdivision of land is where the United States survey in fact established it, whether such location is right or wrong, as may be shown by a subsequent survey. (*Nesselrode v. Parish*, 59 Iowa 570, 13 N.W. 746; *Beardsley v. Crane*, 52 Minn. 537, 54 N.W. 740.)

Appendix 2: Lost Corners, by State

The question is not where the government corner should have been located, but where in fact it was located. And when it is once found, or the place of its location identified, it must control. (*Doolittle v. Bailey*, 85 Iowa 398, 52 N.W. 337.)

If the stakes or monuments placed by the government in making a survey to indicate section corners and quarter posts can be found, or the place where they originally were can be identified, they are to control in all cases. If they cannot be found, or if lost or obliterated, they must be restored upon the best evidence obtainable which tends to prove where they were originally. It is for this purpose that resurveys are made, and the lines retraced as nearly as possible. (*Hess v. Meyer*, 73 Mich. 259, 41 N.W. 422; *Washington Rock Co. v. Young*, 29 Utah 108, 110 Am. St. 666, 80 P. 382.)

Illinois: *Krause v. Nolte*, 217 Ill. 298, 75 N.E. 362 (Ill. 1905).

The principal question presented by the record in this case is whether the question of title to the strip of land in controversy was passed upon and determined by the circuit court in its decree rendered at the April term, 1904, approving a surveyors' report, made by two of the commission of three surveyors, appointed by said court at the October term, 1903, thereof to make a survey and locate the boundary line between the farms of appellant and appellee.

We are of opinion that the court erred in holding that the decree rendered in the proceeding for the establishment of the boundary line was conclusive upon the appellant upon the question of title to the strip. The act of May 10, 1901, is the same as the act of March 25, 1869, upon the same subject. Gross' St. Ill. 1871, pp. 726, 727. In construing the latter act, this court said in *Martz v. Williams*, 67 Ill. 306, that "it nowhere confers power on the commissioners to establish new corners or run new boundary lines, but simply to re-establish those once established by the United States." *Allmon v. Stevens*, 68 Ill. 89. In *Irvin v. Rotramel*, 68 Ill. 11, in speaking of this same act of March, 1869, it is said: "It is evident from this section, and, indeed, from the whole tenor of the act, its object is to provide means by which lost corners may be restored and placed where they properly belong according to the original survey." What was said in the cases thus referred to in regard to the act of March, 1869, applies to the act of May 10, 1901. The object of the latter act is merely to restore and re-establish the corners and boundary lines once established by the United States. This being so, the proceeding is not one for the purpose of establishing the title of the parties to any portion or portions of the property as to which a boundary line is to be restored. Establishing title is an entirely different matter from re-establishing and restoring lost corners of disputed boundaries.

One of the modes by which such lost lines and corners may be restored is fully discussed in the case of *McClintock v. Rogers*, 11 Ill. 279, and perhaps indicates one of the most satisfactory means of restoring lost corners.

414   *Appendix 2: Lost Corners, by State*

Indiana: *Bailey v. Chamblin,* 20 Ind. 33 (Ind. 1863).

The effect of these statutes, construed together, is to maintain the corners originally marked, if they can be found, whether the sub-divisions contain more or less than the legal quantity; but if they cannot be found, then the excess or deficiency shall be apportioned to each sub-division contained within the boundary where the corners are thus lost.

Iowa: *Williams v. Tschantz,* 88 Iowa 126, 55 N.W. 202 (Iowa 1893).

The basis of all surveys of lands in this state is the government survey. The lapse of time and improvement of the country has and will continue to obscure and destroy the monuments that mark these and later legal surveys. The preservation of these corners and boundaries is a matter of great importance, yet without a statute it could only be done by tedious and expensive litigation. The necessity for a plain, speedy, and inexpensive proceeding by which lost corners and boundaries can be established and perpetuated is beyond question. Chapter 8, as I construe it, meets this necessity. Under section 1, adjoining owners may, by agreeing upon a survey, and recording the plat thereof, permanently establish their corners and boundaries. Under the sections following, if they do not so agree, this proceeding may be had to permanently establish corners and boundaries that "are lost, destroyed, or are in dispute." With their corners and boundaries permanently and legally established, there is no room for further dispute as to rights resting thereon. The skill of the surveyor, rather than the judgment of courts and juries, is required in relocating lost corners and boundaries; hence the provision for appointing a commission of one or more surveyors. If it was a question of occupancy, the judgment of a court and jury, rather than the skill of the surveyor, would be required. Lost corners are relocated by measurements from recognized points; hence the commissioners are authorized to take evidence of persons able to identify any original government corner, or other legally established corner or corners, that have been recognized as such by the adjoining proprietors for over ten years. Thus it is that the surveyor gets his starting points from which to relocate the lost corner. If a government corner is found, that is controlling; if not, then any legally established corner, or corner that has been acquiesced in as such for over ten years.

Kansas: *Stanley v. The County Surveyor of Sheridan County,* 126 Kan. 95, 266 P. 929 (Kan. 1928).

It is the duty of the county surveyor notified under the statute (Gen. Stat. 1909, § 2272) to survey land and establish its corners and boundaries, to proceed according to the statutory rules. (*In re Martin's Appeal,* 86 Kan. 336, 120 P. 545. See, also, *Roadenbaugh v. Egy,* 88 Kan. 341, 128 P. 381.)

The pertinent statutory rule provides:

In the resurvey of lands surveyed under the authority of the United States, the county surveyor shall observe the following rules, to wit:

Appendix 2: Lost Corners, by State

First, section and quarter-section corners, and all other corners established by the government survey, must stand as the true corners. Second, they must be reestablished at the identical spot where the original corner was located by the government surveyor, when this can be determined. Third, when this cannot be done, then said corners must be reestablished according to the government field notes, adopting proportionate measurements where the present measurements differ from those given in the field notes. (R. S. 19-1422.)

Kentucky: *Ramsey v. Morrow*, 133 Ky. 486, 118 S.W. 296 (Ky.App. 1909).

One rule laid down for establishing a lost stake corner is to run the courses called for from the known corners to the intersection of the lines. *Haggan v. Wood's Heirs, Ky. Dec.* 274. But there is no hard and fast rule on the subject. The object is always to carry into effect what was done and attempted to be done when the entry and survey were made. *The rule laid down in Haggan v. Wood*, supra, cannot be applied in this case, because if the lines from the seventeenth and nineteenth corners should be extended, they would never intersect. It is clear that an error was made by the original surveyor in copying his field notes, or in entering them. The same surveyor surveyed each entry, surveying the Morrow-Dodson entry first. He had the same person as a chairman in each instance. It is quite likely that, when the Dick entry was run out until its line reached the Morrow-Dodson entry, from that point the calls of the latter were copied as the calls for the Dick survey. That was a customary and a natural way of doing such work. Consequently any error made in entering or transcribing the calls of the former survey along that line would occur also in the Dick patent. In running the line from the eighteenth corner, the stake, to the red bud and the black oak, if the course named, N. 75° E., is pursued, the natural object called for would never be reached. As the corners before the eighteenth were natural objects, and satisfactorily established, the error must be corrected somewhere between the seventeenth and nineteenth corners. By beginning at the red bud and black oak, and reversing its call to the stake, then reversing the next call, it is found that by changing the course from N. 60° E. to N. 80° E. the distance 108 poles runs out at the seventeenth corner. Ordinarily the rule is that distances yield to courses, and both to natural objects, or marked monuments. *Bryan v. Beckley, Litt. Sel. Cas.* 93, 12 Am. Dec. 276. But this is not always so, for if it be manifest from the patent as run out, that approximate certainty may be accomplished by changing the course instead of the distance, then that is to be done. *Blight v. Atwell*, 4 J. J. Marsh. 278.

Louisiana: *Houston Ice & Brewing Co. v. Murray Oil Co.*, 149 La. 228, 88 So. 802 (La. 1921).

The evidence in this case does not show where is the nearest identified original standard corner on a continuation of the west boundary of section 14, extending north. If there is an identified original standard corner

416   *Appendix 2: Lost Corners, by State*

nearby, on a continuation of that line northward, the correct way to relocate the west boundary line of section 14, and thereby to restore the two obliterated standard corners, would be to project a straight line from the identified original standard quarter corner on the west line of section 23 northward to the nearest identified original standard corner on a continuation of the supposed west boundary line of section 14. That is our interpretation of the appropriate one of the rules established by the General Land Office and approved by the courts generally. 9 C. J. p. 164 et seq. The relocation of the north boundary line of section 14 should be made by projecting a straight line from the nearest identified original standard corner towards the east (which appears from the record in this case to be the northeast corner of section 14), to the northwest corner of section 15. The point where that projected line will cross the line projected from the identified original standard quarter corner on the west boundary line of section 23, northward to the nearest identified original standard corner or quarter corner, should be the northwest corner of section 14. The relocation of the south boundary line of section 14 should be made by projecting a straight line from the identified original standard southeast corner of that section westward to the original identified standard southwest corner of section 16. The point where that projected line will cross the line projected northward from the identified original standard quarter corner on the west line of section 23, to the nearest identified original standard corner or quarter corner, should be the southwest corner of section 14. In projecting these lines, of course, measurements should be made of the actual distances between the identified original standard corners thus connected, so as to distribute or apportion any discrepancies found between the actual measurements and the measurements shown on the original government survey. In that connection, too, the field notes should be observed wherever practicable, so as to restore or relocate the lost or obliterated corners at the points where the government surveyor placed them, whether right or wrong. If the nearest identified original standard corner or quarter corner on a continuation of the line running northward from the identified original standard quarter corner on the west boundary line of section 23 is so far away as to make it impracticable to connect the two identified original standard corners or quarter corners, or if for any other reason it is impracticable to connect them with a straight line, the relocation of the west boundary line of section 14 will have to be made by some other method to be adopted by the surveyors. But the record does not disclose that the method primarily required is impracticable. Any section corner or quarter corner that is identified as having been established by an official survey of the United States government must stand as being correctly located, however plain it may appear that the location is wrong; because the government surveys cannot be changed in an action at law between individuals.

Maine: *Stetson v. Adams*, 91 Me. 178, 39 A. 575 (Me. 1898).

The plan is the picture; it is the guide; it is the finger post. But what controls is what was done really upon the surface of the earth, the lines as they were

Appendix 2: Lost Corners, by State 417

run; the stakes as they were put down. The bounds as they were made are the controlling bounds, even though they may vary from the plan, and even though the plan may have been accurate in locating them.

Where the boundaries are first—especially where the lines are first run—there they stay.

Nothing is more firmly established in this state than that, in such case, the survey must govern when its location can be shown. When it cannot be, then the plan may locate it. *Bean v. Batchelder*, 78 Me. 184, 3 A. 279.

Maryland: *Delphey v. Savage*, 227 Md. 373, 177 A.2d 249 (Md. 1962).

A monument, when used in describing land, can be defined as any physical object on ground which helps to establish location of line called for. It may be either natural or artificial, and may be a **tree**, stone, **stake**, pipe, or the like. Because the intention of the parties governs the interpretation of deeds, monuments named in deeds are given precedence over courses and distances; the parties can see the monument referred to in the deed, but would require equipment and expert assistance to find a course and distance.

Michigan: *Case v. Trapp*, 49 Mich. 59, 12 N.W. 908 (Mich. 1882).

"Complainant is owner of the S.E. 1-4 of the N.E. 1-4 of section 6, in township 11 N., of range 4 W., in Gratiot county, with some exceptions not important to this controversy. Defendant is owner of the parcel lying immediately south of this. In the year 1856 the highway commissioners of the township laid out a highway on what they supposed was the line between the two parcels, having first had a survey made for the purpose. The country was then but little settled, and complainant's land had been purchased of the government only two years before. For more than 20 years the dividing line between the two parcels was supposed to be the center of this highway, and complainant and those through whom she derived title exercised dominion over the land up to the highway without disturbance or controversy. In 1879 the county surveyor was brought upon the ground to survey out the line. The account he gives of his action is as follows:

I surveyed it in May—on the thirteenth, fourteenth and twenty-ninth days of that month—1879, at the request of a majority of the resident land-owners of said section. The chainmen were Jesse Trapp and A. Edmondson; they were sworn. I think the east and west quarter line of said section ran some six or seven rods north of the line of an old east and west road. The east quarter post was set some six or seven years ago, when I surveyed section 5 of same township; the old bearing trees were all gone, and I surveyed and measured the section line between sections 5 and 6, and set the post at proportional distances, according to law with reference to setting quarter posts where they are lost. The north and east lines of section 6 are the township lines, and the north and west subdivisions are fractional. On the west side of a township in

418   *Appendix 2: Lost Corners, by State*

the government survey they never set a quarter post, so I surveyed the west side of section 6 in order to ascertain where to set the quarter post; I set it at proportional distances, using the same method that I did on the east side. In the government survey between section 6 and section 1 of the town west adjoining said section 6, there is a jog of 1 chain and 63 links at the section corner, as shown by the plat.

The cross-examination elicited the fact that the surveyor took the word of the parties employing him that the survey was at the request of a majority of the parties interested, and that in fact the survey was ex parte and with only one set of claimants represented. This survey gave to the defendant a strip of land north of the highway, some six or seven rods in width, and he immediately claimed title and asserted a right to possession. Complainant then filed this bill to quiet her own title.

It is assumed on the part of the defence that the survey corrects an old error, and unquestionably locates the true line. We do not think there is any certainty of this. On the contrary the probabilities are all against it. If the government surveys had all been accurate and in accordance with the rules laid down for the guidance of the surveyor, there would commonly be no difficulty in locating the government subdivisions with reasonable certainty, even after the monuments had disappeared. But it is notorious that the errors in the original surveys were frequent and sometimes considerable, and they were often such as to make irregular lots and lots of different sizes in the same neighborhood, where according to the field notes and the plats they should be alike. The consequent difficulty in locating boundaries after corner posts and witness trees are gone is appreciated by every intelligent surveyor.

We have no doubt the county surveyor who made the survey of 1879 was intelligent and had the purpose to be accurate. But we have no reason to suppose the one who made the survey of 1856 was not equally trustworthy, and the circumstances were altogether more favorable to a correct result at that time than when the last survey was made. The land was then but recently located; it is highly probable that corner posts and witness trees remained; if they were gone, there might be disinterested persons in the neighborhood who had seen them and could assist in determining their location, and there was then no existing dispute and litigation to interest parties in leading the surveyor astray. In 1879, on the other hand, the witness trees were gone, and the survey was made at the instance and with the assistance of one party in interest, and without the knowledge, so far as appears, of the other. When under such circumstances the two surveys are found to disagree, there should be no hesitation in saying that the probabilities favor the earlier one, especially when it has remained unquestioned for many years.

Hess v. Meyer, 88 Mich. 339, 50 N.W. 290 (Mich. 1891).

In the circular referred to the method established by the department of the interior for the restoration of corners on base and correction lines is as follows: 'Run a right line between the nearest existing corners on such

Appendix 2: Lost Corners, by State 419

line, whether base or correction line, which corners must, however, be fully identified, and at the point proportionate to the distance given in the field-notes of the original survey establish a new corner. This point should be verified by measurements to the nearest known corners north or south of the base or correction line, or both.

Minnesota: *Ferch v. Konne,* 78 Minn. 515, 81 N.W. 524 (Minn. 1900).

Rule for Locating Lost Corner.

In such a case, there being no surplus to distribute, the actual location of a lost closing corner between two sections lying south of the standard parallel is to be determined by measuring east or west, as the case may be, from the nearest standard corner, the distance shown by the government field notes. These notes and the original plats control in such cases.

According to the rules of the general land office in force when Moyer made his survey (circular of 1883, and really not changed by the circular of 1896), lost or obliterated corners along a standard parallel or correction line are restored by proportionate measurements on the line, conforming as nearly as possible to the field notes, and joining the nearest identified original standard corners on opposite sides of the missing corner or corners, as the case may be, and then applying the rule:

As the original field-note distance between the selected known corners is to the new measure of said distance, so is the original field-note length of any part of the line to the required new measure thereof. The sum of the computed lengths of the several parts of a line must be equal to the new measure of the whole distance.

Missouri: *Woods v. Johnson,* 264 Mo. 289, 174 S.W. 375 (Mo. 1915).

So long as the monuments placed upon the earth's surface by the United States surveyors can be identified, there are no lost corners, and they control, no matter what more recent surveys, by courses and distances disclose. *Jacobs v. Moesley,* 91 Mo. 464; *Knight v. Elliott,* 57 Mo. 317. Measurements must yield to the true location controlled by the Government's monuments, as indicated in the records. *Brown v. Carthage,* 128 Mo. 10. A Government corner must prevail over distances in the ascertainment of the true line. *Mining & Smelting Co. v. Davis,* 156 Mo. 422; *Campbell v. Clark,* 8 Mo. 553. Government corners must prevail. *Major v. Watson,* 73 Mo. 661; *Carter v. Hornback,* 139 Mo. 238; *Climer v. Wallace,* 28 Mo. 556; *Mayor of Liberty v. Burns,* 114 Mo. 426. Field notes of the Government surveyors will control in ascertaining location of corners and boundary lines, even when the established corner by the surveyor cannot be found. *Bradshaw v. Edelin,* 194 Mo. 640. Government surveys are conclusive. *Frederitzie v. Boeker,* 193 Mo. 229. Conflict between plat and actual survey is controlled by the survey. *McKinney v. Doane,* 155 Mo. 289. Original corners as established by the government surveyors, if they can be found, or the places where they were originally established, if they can be definitely determined, are conclusive, without regard to

420   *Appendix 2: Lost Corners, by State*

whether they were located correctly or not. 5 Cyc. 873. The general duty of a surveyor in instances like this are plain enough; he is not to assume that a monument is lost until after he has thoroughly sifted the evidence and found himself unable to trace it, and exhausted every recourse to which surveyors have access, as the original Government surveys must govern. *Stewart v. Carlton*, 31 Mich. 270; *Diehl v. Zanger*, 39 Mich. 601; *Dupont v. Sparring*, 42 Mich. 492.

Montana: *Vaught v. McClymond*, 116 Mont. 542, 155 P.2d 612 (1945).

An obliterated quarter section corner should be fixed as originally located. Lost quarter section corners, in regular interior sections should, under the generally accepted rules, be relocated on a straight line between the section corners and equidistant therefrom, ***. 11 C.J.S., Boundaries, § 13, p. 556.

Nebraska: *State, Bd. of Educational Lands and Funds v. Jarchow*, 219 Neb. 88, 362 N.W.2d 19 (Neb. 1985).

The boundaries of the public lands established by the duly appointed government surveyors, when approved by the Surveyor General and accepted by the government, are ... held and considered as the true corners ... and the restoration of lines and corners of said surveys and the division of sections into their legal subdivisions shall be in accordance with the laws of the United States, [and] the circular of instructions of the United States Department of the Interior, Bureau of Land Management, on the restoration of lost and obliterated section corners and quarter corners....

Both parties agree that exhibit 6, which is Chapter V: Restoration of Lost or Obliterated Corners, of the Manual of Instructions for the Survey of the Public Lands of the United States 1973, contains the instructions which must be followed in a case such as this.

At this point it is necessary to set forth several applicable rules, which have been paraphrased from the manual.

5-1 In restoring the lines of a survey, the purpose is not to correct the original survey, but to determine where the corner was established in the beginning.

5-5 An existent corner is one whose position can be located by an acceptable supplemental survey record, physical evidence, or testimony of one or more witnesses who have a dependable knowledge of the original location.

5-9 An obliterated corner's location may be recovered if proven beyond a reasonable doubt by acts and testimony of interested landowners. A position that depends upon the use of collateral evidence can be accepted only as duly supported through relation to known corners, natural objects, or unquestionable testimony.

5-10 A corner is not considered lost if its position can be recovered satisfactorily by means of the testimony and acts of witnesses having positive knowledge of the precise location of the original monument.

Appendix 2: Lost Corners, by State

The restoration of lost corners

5-20 A lost corner is a point of a survey that cannot be determined beyond a reasonable doubt from acceptable evidence or testimony concerning the original position, and whose location can be restored only by reference to one or more interdependent corners.

5-21 The rules for restoration of lost corners should not be employed until all original and collateral evidence has been developed. The surveyor will then turn to proportionate measurement, which is always employed to relocate a lost corner unless outweighed by conclusive evidence of the original survey.

5-24 Proportionate measurement is one that gives equal weight to all parts of the line. The excess or deficiency between two existent corners is so distributed that the amount given to each interval bears the same proportion to the whole difference as the record length of the interval bears to the whole record distance. After the proportionate difference is added to or subtracted from the record length of each interval, the sum of the several parts will equal the new measurement of the whole distance.

Nevada: *Thomsen v. Keil*, 48 Nev. 1, 226 P. 309 (Nev. 1924).

A "lost" corner is defined as follows:

A lost corner is a point of a survey whose position cannot be determined, beyond reasonable doubt, either from original traces or from other reliable evidence relating to the position of the original monument, and whose restoration on the earth's surface can be accomplished only by means of a suitable surveying process with reference to interdependent existent corners.' Clark on Surveying and Boundaries, § 376, quoting from rules of the Department of Interior.

In the same section, the author says:

A corner should not be regarded as lost until all means of fixing its original location have been exhausted. It is much more satisfactory to so locate the corner than regard it as "lost" and locate by proportionate measurement.

At section 329, the author says:

The surveyor should not treat a corner as lost until he has exhausted all means of fixing its location aside from the determination thereof, by measurement thereof to other corners.

Counsel for appellant says that the rules of the Department of Interior relative to the establishment of corners as originally established pursuant to the government survey must control. This would certainly be true if the quarter corner in question were upon the surveyed public domain; but, since it is upon land privately owned, and for which patent has been issued, we are not prepared to admit the contention, but, since the case seems to have been tried upon the theory urged, we will in disposing of it be guided by the rule invoked.

420

Appendix 2: Lost Corners, by State

The trial court no doubt kept in mind the admonitions above quoted against considering a corner as lost. Indeed, we think it would have violated the rule invoked by appellant had it concluded that the corner in question is "lost." In light of the rule quoted, we must determine if the trial court could, "beyond reasonable doubt, either from original traces or from other reliable evidence relating to the original position of the corner, determine its position."

New Mexico: *Velasquez v. Cox*, 50 N.M. 338, 176 P.2d 909 (N.M. 1946).

> It appears that the main question in this case is as to the true location on the ground of the line between the quarter-section corner on the north boundary of Section 13 and the quarter-section corner on the south boundary of said section. The first step is to determine, if possible, where the quarter-section corner stones were set by the United States Deputy Surveyor in the year 1880.

The foregoing excerpts from the testimony of the expert witnesses indicate the different theories as to the surveys. Plaintiffs' attorney, while admitting that defendant's Exhibit 2 was admissible as a public record (N.M.S.A. 1941, 19-101(44); 11 C.J.S., Boundaries § 111, p. 707), strenuously maintains that Mr. Harvey, plaintiffs' engineer, was the only competent witness to testify as to the location of corners of the public surveys; and that he located the median line dividing Sec. 13 into east and west halves in accordance with Circular 1452 United States Department of the Interior General Land Office, "Restoration of *Lost or Obliterated Corners and Subdivision of Sections.*" *Cordell v. Sanders*, 331 Mo. 84, 52 S.W.2d 834, 838.

The circular 1452, supra, states general rules about which there is no dispute, which we quote:

> First. That the boundaries of the public lands, when approved and accepted, are unchangeable.
>
> Second. That the original township, section, and quarter-section corners must stand as the true corners which they were intended to represent, whether in the place shown by the field notes or not.
>
> Fourth. That the center lines of a section are to be straight, running from the quarter-section corner on one boundary to the corresponding corner on the opposite boundary.
>
> Sixth. That lost or obliterated corners are to be restored to their original locations whenever it is possible to do so.

On the other hand, defendant insists that the quarter-section corners are to be restored to their original locations regardless of the distance or direction from the re-established corners on the east line of Section 13. Defendant insists that there is substantial evidence to support his theory that the obliterated quarter-section corners on the north and south lines of Section 13 were originally located as indicated by the map of the hydrographic survey made

Appendix 2: Lost Corners, by State

in the year 1938 (Defendant's Exhibit 2), which plaintiffs' attorney admits corresponds approximately with the east boundary fence of defendant and defendant points out that the quarter-section corner on the north boundary of Section 13 harmonizes with the quarter-section corners on the north boundaries of the two sections to the west of Section 13 in which there are irrigated fields belonging to other parties. He dwells upon the facts that 60 years has elapsed since the patent to the defendant's land was issued by the United States to the homesteader, Martinez, and the fence in question was built by him; that 49 years elapsed before the location of the boundary fence was questioned; that Martinez, the homesteader, was required under the homestead law to reside 5 years on his homestead, and that was his place of residence in the year 1880 when the township was surveyed; that it is probable that he knew where the quarter-section corners on the north and south boundaries of Section 13 were placed by the U.S. Deputy Surveyor in 1880, the north quarter-section corner being a corner of his land, and that he built his east boundary fence line in conformity therewith; that he planted an orchard and made other valuable improvements on the disputed strip. Plaintiffs' land being public domain at that time, Martinez had the choice of entering those subdivisions on which the plaintiffs now claim two-thirds of the irrigated lands on the Martinez homestead—defendant's farm—are located. The non-irrigated lands are of small value—taxed on a valuation of $1 per acre.

Plaintiffs' witness and engineer, Harvey, expressed the belief that corners within the Township 30 N., R. 8 W., were never set. In other words, the land lines were never run, nor stones placed for the corners by the United States Deputy Surveyor, and that the field notes, which were accepted and approved by the government were fictitious, or what is sometimes called an "office survey." He testified that he found one original corner, the southwest corner of Section 13. Defendant claims that quarter-section corners on the north boundaries of Sections 14 and 15 were found, but Mr. Harvey states that he had never run across such large discrepancies as indicated by the hydrographic map. There are such discrepancies in the public surveys. *Canavan v. Dugan, etc.*, 10 N.M. 316, 62 P. 971, 11 C.J.S. Sec. 51, Boundaries, pp. 612–615.

In the case of *Byrne v. McKeachie*, 34 S.D. 589, 149 N.W. 552, 553, the court said: "No matter how erroneously the work of the government surveyor may have been done, and no matter how far out of its proper location a government corner may have been established, if such location can be fixed it must control. *Hoekman v. Iowa Civil Twp.*, 28 S.D. 206, 132 N.W. 1004."

And in this same case is quoted from the earlier case of *Wentzel v. Claussen*, 26 S.D. 89, 127 N.W. 621, 622, as follows: "Where people for 20 or more years have recognized lines as the true boundaries throughout a whole township, there must be most satisfactory proof that the government corners have become absolutely 'lost,' as distinguished from 'obliterated,' before it will be allowed the township authorities or private parties to institute a new survey and locate corners throughout a township at points clearly not where the original corners were located."

424   *Appendix 2: Lost Corners, by State*

In the case of *Bentley, et al. v. Jenne*, 33 Wyo. 1, 236 P. 509, the court made reference to what is termed an independent survey made by the government by which the township was platted into a number of lots. The homesteads that had been patented were given lot numbers, and the court said: "Counsel for appellant have devoted a great deal of their argument to this subject, but we fail to see the importance thereof. We concede that a resurvey cannot disturb the title which parties have acquired up to the time that it is made. We do not, however, understand that respondents claim any rights under any new survey. The only purpose, apparently, of introducing any evidence in regard thereto was to settle all future questions as to the actual location and description of the Hamilton land, and to locate and describe it in accordance with the new survey."

In Circular No. 1452 of the Department of the Interior, supra, it is stated:

> 1006. The act of Congress approved September 21, 1918, entitled "An act authorizing the resurvey or retracement of lands heretofore returned as surveyed public lands of the United States under certain conditions" provides authority for the resurvey by the Government of townships theretofore held to be ineligible for resurvey by reason of the disposals being in excess of fifty per centum of the total area thereof. * * *
>
> 1011 * * * The independent resurvey is one which makes a new subdivision and new lottings of the vacant public lands, which are designed to supersede the original survey. Provision is made for the segregation of individual tracts of privately-owned lands, entries, or claims that may be based upon the original plat, when necessary for their protection, or for their conformation to the regular subdivisions of the resurvey if that may be feasible, or for an amendment of the entry or patent after an adjudication of the rights that may be involved.

These independent surveys, where they do not conform to original lines and corners, are not binding on owners of patented lands, but often enable men of good will to avoid expensive litigation by accepting the good offices of the federal officials in charge of the resurvey of public lands.

It is stipulated that the lands of plaintiffs were patented in the year 1908 and 1922 to homesteaders. These parties and their successors in interest lived on these homesteads and farmed up to the fence line in question for more than 10 years before they sold to the plaintiffs, and the plaintiffs after they discovered what they thought was the error in the original survey or location of the fence line waited 11 years before filing this suit.

There are arguments in the briefs on acquiescence and adverse possession. We have lately discussed these subjects in the cases of *Thurmond v. Espalin*, 50 N.M. 109, 171 P.2d 325; *Lovelace v. Hightower*, 50 N.M. 50, 168 P.2d 864; *Christmas v. Cowden*, 44 N.M. 517, 105 P.2d 484; *Rodriguez v. La Cueva Ranch Co.*, 17 N.M. 246, 134 P. 228; *Ward v. Rodriguez*, 43 N.M. 191, 88 P.2d 277, 279.

In *Ward v. Rodriguez*, we said: "We recognize the rule to be that the government has the right to re-survey public land, as corrective and as a retracing,

Appendix 2: Lost Corners, by State

but such survey will be construed to have and follow the lines of the original U.S. survey where it would affect bona fide private rights held under such original survey. U.S.C.A. Tit. 43, § 772; *Cragin v. Powell*, 128 U.S. 691, 9 S.Ct. 203, 32 L.Ed. 566; *Lane v. Darlington*, 249 U.S. 331, 39 S.Ct. 299, 63 L.Ed. 629."

In other words, what we were attempting to make plain was that no resurvey by the government can change the lines of the original survey to the prejudice of private rights acquired in good faith in reliance on the integrity of the original survey.

But we rest our conclusion herein on the evidence as to the true location on the ground of the dividing line between the lands of the plaintiffs and defendant. Before the trial court could consider the accuracy of the work of plaintiffs' engineer, he was called upon to determine, if possible, where the original quarter-section corners on the north and south boundaries of Section 13 were actually established by the government survey in the year 1880. Twitchell's Leading Facts of New Mexican History says that the irrigation of the lands in San Juan County was commenced about the year 1877. It is probable that there were settlers in Twp. 30 N., R. 8 W., in 1880, and that the Deputy U.S. Surveyor established some corners in the settlements, although the belief of the witness Harvey that it was in the main an "office survey" may be true. The map of the hydrographic survey, defendant's Exhibit 2, is some evidence that the dividing line claimed by defendant as evidenced by his fence harmonizes with the lines recognized by the landowners of irrigated lands in the two sections immediately west of Section 13.

North Dakota: *Nystrom v. Lee*, 16 N.D. 561, 114 N.W. 478 (N.D. 1907).

> It is conceded that, in all cases where the location of the original monuments can be ascertained, they must control, and that the object of the resurvey is to locate, if possible, the original monuments. It is also well established that, in proving the location of lost or obliterated monuments, any evidence may be used which tends to establish the location of such monuments. 5 Cyc. 956. The testimony of the witnesses who have seen them and remember their location is competent. The use of the field notes made on the original survey is likewise competent, and, when used, they have the force of a deposition made by the surveyor. Although they would have been competent evidence, we do not consider the field notes as necessary evidence in this case. *White v. Amrhien*, 14 S.D. 270, 85 N.W. 191; *Randall v. Burk Township*, 4 S.D. 337, 57 N.W. 4; *Radford v. Johnson*, 8 N.D. 182, 77 N.W. 601; *Dowdle v. Cornue*, 9 S.D. 126, 68 N.W. 194; *Neary v. Jones*, 89 Iowa 556, 56 N.W. 675.
>
> Rule No. 147 of the Manual of Surveying Instructions, issued by the United States General Land Office, requires that in making surveys of public lands quarter section corners, both upon the meridional and latitudinal section lines, be established at points equidistant from the corresponding section corners, except (here follow exceptions which have no application to interior sections). The instructions issued by the General Land Office March 14, 1901, regarding the re-establishment of interior

426 *Appendix 2: Lost Corners, by State*

quarter section corners, say: "The missing quarter section corner must be re-established equidistant between the section corners marking the line according to the field notes of the original survey." We take this to mean that, when the section corners on any side of the section are found or located, the quarter corner must be placed midway between them. Conversely, it must be true that, when the northwest corner of an interior section and the quarter corner on the north line are found or located, and they are found to be one-half mile apart, the northeast corner of the section must at least be presumed to be one-half mile east of the quarter corner, and the same rule must apply to the southwest corner of the section, when, as in this case, the northwest corner and the west quarter corner are one-half mile apart. If this rule is correct, as we think it must be, then, when the surveyor ran the east line one-half miles south from the northeast corner which he had established, and a line between that point and the quarter monument on the west line was found to be one mile in length, it must at least furnish prima facie evidence of the location of the original corners and boundaries. We think the evidence on this point was competent, and that the plaintiff did not have to resort to every known test to ascertain the correctness of the survey. It is held in several states that lost quarter posts should be relocated at equal distances between the section corners. *Hess v. Meyer*, 73 Mich. 259, 41 N.W. 422; *Frazier v. Bryant*, 59 Mo. 121; *Knight v. Elliott*, 57 Mo. 317; *Lemmon v. Hartsook*, 80 Mo. 13. A line marked part of the distance must be followed in the same direction for the whole distance unless there is some marked corner to divert it. *Thornberry v. Churchill*, 4 T.B. Mon. 29, 16 Am. Dec. 125; *George v. Thomas*, 16 Tex. 74, 67 Am. Dec. 612, and cases cited. Where boundaries and termini are is a question for the jury. *Doe v. Paine*, 11 N.C. 64, 15 Am. Dec. 507; *Comegys v. Carley*, 3 Watts 280, 27 Am. Dec. 356.

Ohio: *Sellman v. Schaaf*, 26 Ohio App.2d 35, 269 N.E.2d 60 (Ohio App. 3 Dist. 1971).

This is by far one of the most important cases relative to retracement work. It references many additional decisions, from a variety of jurisdictions.

Oklahoma: *Overton v. Leonard*, 79 Okla. 219, 192 P. 221 (Okla. 1920).

The law applicable to the case, section 1711, Revised Laws 1910, is as follows:

The resurvey and subdivision of lands by county surveyors shall be according to the laws of the United States, and the instructions issued by the officers thereof in charge of the public land surveys, in all respects; and in the subdivision of fractional sections, bounded on any side by a meandered lake or river, or the boundary of any reservation of irregular survey, the subdivision lines running toward and closing upon the same shall be run at courses in all points intermediate and equidistant, as near as may be, between the like section lines established by the original survey.

Appendix 2: Lost Corners, by State

The regulations of the General Land Office in regard to the restoration of lost or obliterated corners promulgated June 1, 1909, contain the following provision:

The method now in practice, where regular conditions are found, requires section lines to be initiated at the corners of the south boundary of the township, and to close on existing corners on the east, north and west boundaries of the township, except that when the north boundary is a base line or standard parallel, new corners are set therein, called closing corners.

The United States statutes that regulated the method for locating the half and quarter section corners are (sections 4804, United States Compiled Statutes 1916; section 2396, R. S.) as follows:

First. All the corners marked in the surveys, returned by the Surveyor General, shall be established as the proper corners of sections, or subdivisions of sections, which they were intended to designate; and the corners of half and quarter sections, not marked on the surveys, shall be placed as nearly as possible equidistant from two corners which stand on the same line.

Second. The boundary lines, actually run and marked in the surveys returned by the Surveyor General, shall be established as the proper boundary lines of the sections, or subdivisions, for which they were intended, and the length of such lines, as returned, shall be held and considered as the true length thereof. And the boundary lines which have not been actually run and marked shall be ascertained, by running straight lines from the established corners to the opposite corresponding corners; but in those portions of the fractional townships where no such opposite corresponding corners have been or can be fixed, the boundary lines shall be ascertained by running from the established corners due north and south or east and west lines, as the case may be, to the water course, Indian boundary line, or other external boundary of such fractional township.

Third. Each section or subdivision of section, the contents whereof have been returned by the Surveyor General, shall be held and considered as containing the exact quantity expressed in such return, and the half sections and quarter sections, the contents whereof shall not have been thus returned, shall be held and considered as containing the one-half or the one-fourth part, respectively, of the returned contents of the section of which they may make part.

Section 4805 (R. S. § 2397) is as follows:

In every case of the division of a quarter section the line for the division thereof shall run north and south, and the corners and contents of half-quarter sections which may thereafter be sold, shall be ascertained in the manner and on the principles directed and prescribed by the section preceding, and fractional sections containing one hundred and sixty acres or upwards shall in like manner as nearly as practicable be subdivided into half-quarter sections, under such rules and regulations as may be prescribed by the Secretary of the Interior, and in every case of a division of a half-quarter section, the line for the division thereof shall run east and west, and the corners and contents of quarter–quarter

428

Appendix 2: Lost Corners, by State

sections which may thereafter be sold, shall be ascertained, as nearly as may be, in the manner, and on the principles, directed and prescribed by the section preceding; and fractional sections containing fewer or more than one hundred and sixty acres shall in like manner, as nearly as may be practicable, be subdivided into quarter–quarter sections, under such rules and regulations as may be prescribed by the Secretary of the Interior.

The regulation of the General Land Office in regard to the re-establishment of the quarter section corners on closing section lines between fractional sections contained the following provision:

Re-establishment of Quarter Section Corners on Closing Section Lines between Fractional Sections—this class of corners must be re-established proportionately, according to the original measurement of 40 chains from the last interior section corner. If the whole measurement does not agree with the original survey, the excess or deficiency must be divided proportionately between the two distances expressed in the field notes of original survey. The section corner started from and the corner closed upon should be connected by a right line, unless the retracement should develop the fact that the section line is either a broken or curved line, as is sometimes the case.

The surveyor locating the closing section corners found the distance between the corners to be 5,326.4 feet instead of 5,280 feet, or an excess of 46.4 feet. The question presented is: How should he divide the excess?

The rules and regulations of the General Land Office provide:

If the whole measurement does not agree with the original survey the excess or deficiency must be divided proportionately between the two distances expressed in the field notes of the original survey.

In fixing the line or corner between the lots, the county surveyor should have followed the rules and regulations of the General Land Office and the statute of the United States, and this portion of the section being divided in lots, instead of half or quarter sections, he should make the width of the lots all the same.

The Supreme Court of the United States, in the case of *St. Paul & P. R. R. Co. v. Schurmeier*, 7 Wall. 272, 19 L.Ed. 74, says:

Provision was made by the Act of February 11, 1805, that townships should be "subdivided into sections, by running straight lines from the mile corners, marked as therein required, to the opposite corresponding corners, and by marking on each of the said lines intermediate corners, as nearly as possible equidistant from the corners of the sections on the same." Corners thus marked in the surveys, are to be regarded as the proper corners of sections, and the provision is that the corners of half and quarter sections, not actually run and marked on the surveys, shall be placed, as nearly as possible, equidistant from the two corners standing on the same line. 2 Stat. at L. 313. Boundary lines actually run and marked on the surveys returned, are made the proper boundary lines of the sections or subdivisions for which they were intended, and the second article of the second section provides that the length of such lines, as returned, shall be held and considered as the true length thereof. Lines intended as boundaries, but which were not actually run and marked,

Appendix 2: Lost Corners, by State

must be ascertained by running straight lines from the established corners to the opposite corresponding corners; but where no such opposite corresponding corners have been, or can be fixed, the boundary lines are required to be ascertained by running from the established corners due north and south, or east and west, as the case may be, to the water course, Indian boundary line, or other external boundary or such fractional township. * * *

Meander lines are run in surveying fractional portions of the public lands bordering upon navigable rivers, not as boundaries of the tract, but for the purpose of defining the sinuosities of the banks of the stream, and as the means of ascertaining the quantity of the land in the fraction subject to sale, and which is to be paid for by the purchaser.

Other cases following the same rule of reasoning are as follows: *Jefferis v. Land Co.,* 134 U.S. 178, 10 S.Ct. 518, 33 L.Ed. 872; *Ivy v. Parker,* 18 Ariz. 503, 163 P. 258; *Tolson v. S.W. Improvement Ass'n,* 97 Ark. 193, 133 S.W. 603; *Brooks v. Stanley,* 66 Neb. 826, 92 N.W. 1013; *Hootman v. Hootman,* 133 Iowa, 632, 111 N.W. 61; *Gerke v. Lucas,* 92 Iowa, 79, 60 N.W. 538; *Edinger v. Woodke,* 127 Mich. 41, 86 N.W. 397; *Miller v. Topeka Land Co.,* 44 Kan. 354, 24 P. 420; *McAlpine v. Reicheneker,* 27 Kan. 257; *Booth v. Clark,* 59 Wash. 229, 109 P. 805, Ann. Cas. 1912A, 1272; *Andrews v. Wheeler,* 10 Cal.App. 614, 103 P. 144; *Pereles v. Magoon,* 78 Wis. 27, 46 N.W. 1047, 23 Am. St. Rep. 389.

The surveyor, instead of following the statute, attempted to fix the dividing lines between the lots by the use of certain meander corners, established by the government. In this the surveyor was in error.

The defendants in error introduced over the objection of plaintiff in error certain evidence of other surveyors contending that the closing corners established by the surveyor were not correctly located. This was not properly before the trial court for the reason the portion of the findings of the survey establishing the corner was not appealed from by defendant nor the plaintiff, nor did either side take any exception to that portion of the report of the surveyor fixing and establishing the corners.

The surveyor after establishing the corners used an improper method of locating the dividing line between the lots, and did not make the lots of equal width. This was error, and the district court committed error in not so modifying said report of the surveyor.

The judgment of the court will therefore be reversed and remanded, with instructions to the district court to modify the report of the surveyor and establish the subdivision lines by dividing the section in four equal parts, that is, the width of each lot to be the same, and the line between lots 1 and 2 is established at 1,331.6 feet west of the east line of said section.

Oregon: *Kincaid v. Peterson,* 135 Or. 619, 297 P. 833 (Or. 1931).

Section 27-1820, Oregon Code 1930, provides that the county surveyor, in the resurvey of lands surveyed under authority of the United States, shall observe the following rules: '(1) Section and quarter-section corners, and all other corners established by the government survey, must

430 *Appendix 2: Lost Corners, by State*

stand as the true corners; (2) they must be re-established at the identical spot where the original corner was located by the government survey, when this can be determined; (3) when this cannot be done, then said corners must be re-established according to the government field notes, adopting proportionate measurements where the present measurements differ from those given in the field notes.' In the establishment of the lost corner, the surveyors who testified for the plaintiff obeyed the above statutory rules. Their lines were run from known and established corners as made in the original survey.

South Dakota: *Hanson v. Township of Red Rock in Minnehaha County,* 4 S.D. 358, 57 N.W. 11 (S.D. 1893).

Established such corner monuments at points other than those indicated by the government field notes. The rule that fixed monuments will control courses and distances only prevails when the boundaries are fixed and known, and unquestioned monuments exist; and where the boundary is not fixed and known, and the location of the monuments themselves is uncertain, or left in doubt by the evidence, then courses and distances will be considered in fixing the boundaries. The rule is well stated by the supreme court of Iowa in *Yocum v. Haskins,* 81 Iowa, 436, 46 N.W. 1065. In that case the court says: "It is a well-established rule that when boundaries are fixed and known, and unquestioned monuments exist, and neither courses, distances, nor computed contents correspond with the monuments, the monuments govern. *Pernam v. Wead,* 6 Mass. 131; *Nelson v. Hall,* 1 McLean, 518. When the boundary is not fixed and known, but is in dispute, courses, distances, and contents may be considered, in fixing and knowing the true boundary. When, as in this case, the dispute is as to which of two points is the established corner, and one point is where such corners are usually established, and such as to give to each owner the quantity of land purchased, and the other is remote, and gives to some more, and others less, than the quantity of land purchased, it will surely require less evidence to convince the mind that the former is the true line than that the latter is." The rule is also clearly stated by Chief Justice Parsons in *Pernam v. Wead,* 6 Mass. 131. The court says: "When the boundaries of land are fixed, known, and unquestionable monuments, although neither courses nor distances nor the computed contents correspond, the monuments must govern." With respect to courses, from errors in surveying instruments, variation of the needle, and other causes, different surveyors often disagree. The same observations apply to distances arising from the inaccuracies of measures, or of the Section 2395, Rev. St. U.S. provides as follows: "The public lands shall be divided by north and south lines run according to the true meridian, and by others crossing them at right angles, so as to form townships of six miles square, unless where the lines of an Indian reservation, or of tracts of land heretofore surveyed or patented, or the course of navigable rivers, may render this impracticable; and in that case this rule must be departed from no further than such particular circumstances require." It is quite clear from the provisions of this section that all

Appendix 2: Lost Corners, by State 431

township lines are required to be straight lines connecting the township corners, and that all section and quarter section corners established by the government surveyor in establishing township lines should be made to coincide with such township lines. Therefore, when the government surveyor, in the field notes returned by him to the government, shows that such section and quarter section corners are established on such straight lines between the township corners, and fixes their location by courses and distances, these field notes are to be accepted as presumptively correct, and can only be overcome by the most clear and satisfactory evidence that such surveyor party measuring and computations are often erroneous. But fixed monuments remain. About them, there is no dispute or uncertainty; and what may be uncertain must be governed by monuments, about which there is no dispute." The law of this state as to making resurveys is as follows: Comp. Laws, § 694: "In retracing lines or making any survey he [the surveyor] shall take care to observe and follow the boundaries and monuments as run and marked by the original survey, but shall not give undue weight to partial and doubtful appearances or evidences of monuments, the recognition of which shall require the presumption of marked errors in the original survey, and he shall note an exact description of such apparent monuments." Therefore, while the surveyor, in making a resurvey, is required to take care to observe and follow the boundaries and monuments as run and marked by the original survey, he is also required not to give 'undue weight to partial and doubtful appearances or evidences of monuments, the recognition of which shall require the presumption of marked errors in the original survey.'

The duty of a surveyor in relocating lost corners on township lines, when the monuments claimed to be government monuments are disputed, and not clearly established, is to establish them on a line coinciding with the township line at the points indicated by the government field notes; that is, on a straight line connecting known and undisputed government monuments on such township line. The law governing this class of cases is so clearly stated by the supreme court of Michigan in *Hess v. Meyer*, 41 N.W. 422, 73 Mich. 259, (decided in 1889,) that we quote quite largely from the opinion. Mr. Justice Champlin, speaking for the court, says: "The exterior lines of the township are entirely independent of the interior subdivisions, and are to be made by different surveyors; the regulations of the department of the interior, which have the force of law, not allowing the same surveyor who ran the exterior lines of the township to subdivide it." Township lines are required to be straight lines a distance of 480 chains. Therefore, when any two known monuments are found to exist on such line, a right line between these monuments would represent the location of the town line; and although the section corners on an east and west town line may, through error in the chainmen, be located and placed by the government surveyor either east or west of where they properly should have been placed, and must so remain, there is no such liability to error as to placing them either north or south of the proper place on such line. They are not dependent upon section corner or quarter posts placed when the interior of the sections

432   *Appendix 2: Lost Corners, by State*

are surveyed, because, as before stated, they are placed in position anterior to, and independent of, such interior surveys. It follows that, where a section corner on an east and west township line is lost, the proper method would be to run a straight line from the nearest known monument on the town line on either side of the lost corner or corners, and replace the post, according to the field notes of the government survey, upon the straight line connecting the two known corners. Such town line cannot be swerved from a right line by measuring from a known quarter corner north of the line to one south of the line, and dividing the distance. To do so would make the survey of the town line subordinate to the survey of the subdivision of the township, when the contrary is not only the rule, but the fact. The northeast corner of section 3, being a corner on the town line as originally surveyed, is lost. In relocating it, the surveyor, whose survey is relied upon by the plaintiff, as it was evidently by the jury, measured from the quarter section post north to a quarter section corner south of the town line—both being known monuments—and divided the distances equally, and located the section corner equi-distant from the two quarter posts. That would be right, provided it coincided with a right line run between the nearest two known monuments on the government survey of the town line. * * * In case of lost section corners on town or range lines, the lines should be resurveyed between the nearest known government monuments (on the town line) on either side of the lost corner or corners, and the section corners relocated on a straight line between such monuments at the distance indicated in the field notes, and the lost quarter posts at equal distances between the section corners.

Utah: *Moyer v. Langton,* 37 Utah 9, 106 P. 508 (Utah 1910).

"'The rule for determining lost corners of a survey, when some remain, is to run the lost lines according to the courses and distances in the survey, unless the lines so run do not close the survey with the corners remaining, in which case the courses in the survey must be followed and the distances disregarded, and if the survey cannot then be made to close, the courses themselves must be deviated from. And where vestiges of an ancient boundary are to be seen new posts should be fixed, but they must be placed where the former limit or fence stood.' (5 Cyc., 872, together with notes and authorities there cited.)"

Washington: *King v. Carmichael,* 45 Wash. 127, 87 P. 1120 (Wash. 1906).

The original corner was lost and could not be found, and that its location must be determined according to the rules laid down by the Interior Department for the establishment of lost corners. A dividing line had been run by the county surveyor according to these rules, apportioning the shortage between the two quarter sections.

The case presents a question of fact only. If the place of location of the original corner has been preserved, then unquestionably it marks one

Appendix 2: Lost Corners, by State 433

of the points from which the dividing line between the quarter sections must be run, no matter at what point on the line it is found. On the other hand, if the corner is lost, then it must be established according to the rule followed by the county surveyor, or by some rule equally satisfactory and which will produce an equitable result.

Heybrook v. Index Lumber Co., 49 Wash. 378, 95 P. 324 (Wash. 1908).

If the corner on the north was actually lost, the line through the section should have been run from the known quarter corner to a post on the north line half way between the section corners, and this regardless of the magnetic variation reported by the government surveyor.

Wisconsin: *City of Racine v. Emerson*, 85 Wis. 80, 55 N.W. 177.

This is one of the most important retracement cases, referenced by numerous courts and discussed in some detail within the text at appropriate locations.

Wyoming: *Carstensen v. Brown*, 32 Wyo. 491, 236 P. 517 (Wyo. 1925).

field notes are the best evidence when corners are lost, *Galbraith v. Parker*, 153 P. 283; *Stangair v. Roads*, 41 Wash. 583; *Reed v. Bartlett*, 255 Ill. 76.

Final Thoughts

The safest way to ascertain boundaries is to compare the grants with the marks and natural objects on the ground; that the proper method for ascertaining the boundaries of a grant is to find, if possible, the lines which the surveyor surveyed on the ground; that the boundaries of a grant as actually surveyed are the limits of grantee's right; and that the calls which are most certain, about which there is less probability of mistake or inaccuracy, are to prevail.

Richardson et al. v. Schwoon, et al., 3 Tenn. App. 512, 1925

In suit involving boundary question, search must be made for the footsteps of the original surveyor and, when found, the case is solved.

Hart v. Greis et al., 155 S.W.2d 997, 1941.

Table of Authorities

Cases

A. B. Moss & Bro. v. Ramey	133
Abbott v. Abbott	320
Acampora v. Pearson	354
Adams Express Co. v. Ohio State Auditor	120
Adams v. Hoove	346
Adams v. Wilson	364
Affleck v. Morgan	49
Akin v. Godwin	26, 32, 34, 35, 37, 40, 277
Albert v. City of Salem (Or.)	353
Albertson v. Chicago Veneer Co.	120, 125
Alexander v. Hill	121
Alexander v. Lively	275
Allely v. Fickel	141
Allen v. Duvall	74
Allen v. Kingsbury	184
Ames v. Hilton	111
Anderson v. Johanesen	142
Andrews v. Wheeler	382
Ansley v. Graham	276
Antone v. Hoffman	167
Appeal of Moore	48
Armijo v. New Mexico Town Co.	73
Arms v. City of Owatonna	287
Arneson v. Spawn	367
Asher v. Fordson Coal Co.	127, 265
Asher v. Howard	119
Ashley v. Hill	338
Atwood v. Willacy County Navigation District	165
Avery v. Baum	345
Ayers v. Watson	39, 78, 128, 131, 217, 276, 343, 349
B. H. Bassett v. W. H. Rigell	163
Bailey v. Chamblin	327, 368
Bailey v. Look	82
Baker & Sons v. Sherman	228
Baker v. Glasscocke	86, 344
Baker v. Talbott	82
Baker v. Sanderson	339
Ballard v. Stanolind Oil & Gas Co.	356
Banks v. Talley	86
Barnes v. Wingate	167

Barrett v. Perkins	*364*
Bartholomew v. Murry	*69*
Bartlett Land, etc. Co. v. Saunders	*86, 196*
Bass v. Mitchell	*64*
Baxter v. Evett's Lessee	*86, 121, 344*
Baybutt Construction Corp. v. Commercial Union Ins.	*112*
Beach v. Fay	*307*
Bean v. Bachelder	*88, 344*
Bean v. Batchelder	*282, 371*
Beardsley v. Crane	*96, 142, 296, 349, 355, 367*
Beaubien v. Kellogg	*336*
Beaver Brook Resort Co. v. Stevens	*364*
Beckley v. Bryan	*86, 117, 219, 220, 225, 344*
Beckman/Tillman v. Bennett	*41*
Bell v. Morse	*184*
Bellows v. Copp	*183, 185*
Ben Realty Co. v. Gothberg	*338*
Bentley v. Jenne	*99, 359*
Berkley v. Bryan &c	*81*
Billingsley v. Bates	*367*
Billups v. Woolridge	*358*
Bishop v. Johnson	*31, 34, 277*
Bittner v. Walsh	*35*
Black v. Sprague	*231, 278*
Blaffer v. State	*356*
Blake v. Pure Oil Co.	*165*
Blasdell v. Bissell	*346*
Blight v. Atwell	*370*
Block v. Howell	*94, 95, 355*
Bolton v. Lann,	*167*
Bond v. Middleton	*61, 166, 168*
Book v. Justice Min. Co.	*350*
Boon v. Hunte	*135*
Booth v. Clark	*367, 382*
Booth v. Upshur	*104*
Bower v. Earl	*353*
Bowman v. Farmer	*191*
Boyd v. Durret	*348*
Boyd v. Durrett	*48*
Bradford v. Pitts	*16, 44*
Bradshaw v. Edelin	*285, 373*
Bramblet v. Davis	*114, 116*
Breck v. Young	*192*
Brockman v. Rose	*121*
Brooks v. Tyler	*174, 175, 190*

Table of Authorities

Brown v. Carthage	*285, 373*
Brown v. Clark	*185*
Brown v. Clements	*132*
Brown v. Eubank	*166*
Brown v. Lyons	*259*
Brown v. McKinney	*256*
Bruce v. Morgan	*121*
Bruce v. Taylor	*118, 121, 219*
Bruckner's Lessee v. Lawrence	*296*
Bruker v. Burgess and Town Council of Borough of Carlisle	*331*
Bruker v. Burgess and Town Council of Carlisle	*57*
Bryan v. Beckley	*117, 179, 206, 370*
Bryant v. Strunk	*121, 122, 126, 265*
Buckley v. Laird	*39, 283, 349, 366*
Buffalo Zinc & Copper Company v. McCarty	*337*
Buford v. Cox	*346*
Buie v. Miller	*69*
Bulor's Heirs v. James M'Cawley	*228*
Burton v. Duncan	*58, 136*
Byrne v. McKeachie	*96, 355, 377*
California v. Deseret Water, Oil & Irr. Co.	*133*
Campbell & Others v. Nickerson & Others	*330*
Campbell v. Branch	*200*
Campbell v. Carruth	*73*
Campbell v. Clark	*285, 373*
Campbell's Executors v. Wilmore	*159*
Canavan v. Dugan, etc.	*377*
Cardigan v. Page	*188, 190*
Carey v. Clark	*288*
Carstensen v. Brown	*260, 386*
Carter v. Beals	*185*
Carter v. Hornback	*285, 373*
Case v. Trapp	*371*
Caylor v. Luzadder	*292, 326*
Chandler v. Hibberd	*364*
Chapman & Dewey Lbr. Co. v. St. Francis Levee Dist	*132*
Chapman v. Polack	*231*
Chapman v. Pollack	*277*
Chapman v. Twitchell	*273*
Chenoweth v. Lessee of Haskell	*341*
Cherry v. Slade	*224, 351*
Christian v. Gernt. et al.	*75, 281*
Christianson v. Daneville Tp.	*95, 355*
Christmas v. Cowden	*378*
City of Jacksonville v. Broward	*366*

City of North Mankato v. Carlstrom	53, 243, 287
City of Pompano Beach v. Beatty	26
City of Racine v. Emerson	37, 40, 287, 347, 353, 358, 386
Clark v. Moore	274
Clark v. Northy	184, 190
Cleaveland v. Flagg	183
Climer v. Wallace	142, 285, 296, 367, 373
Clough v. Bowman	184
Clough v. Sanborn	183
Cody v. England	203
Colby v. Collins	183
Coleman County v. Stewart	123, 124
Combs v. Jones	217, 265
Combs v. Valentine	218
Combs v. Virginia Iron, Coal & Coke Co.	218
Comegys v. Carley	380
Cone v. West Virginia Pulp & Paper Co.	338
Cook v. Babcock	184
Cook v. Combs	184
Coppin v. Manson	367
Corbett v. Norcross	190, 191
Cordell v. Sanders	285, 376
Corlies v. Little	224
Cornelious v. State	362
Cornett v. Kentucky River Coal Co.	218
County of Yolo v. Nolan	284, 337
Cowan v. Fauntleroy	86, 344, 345
Cox v. Finks	159
Cox v. Hart	95, 142, 349, 355
Cragin v. Powell	37, 41, 101, 132, 262, 295, 347, 378
Craig v. Russell	33
Cram v. Ingalls	185
Craven v. Lesh	367
Creech v. Johnson	115, 119, 258
Crockett v. Andrews	91
CSX Hotels, Inc., dba the Greenbrier Resort, a West Virginia corporation, v. City of White Sulpher Springs	90
Curtis v. Aaronson	224
Curtis v. Francis	184
Dagget v. Wiley	273
Dallas Borough Annexation Case	72
Daniel v. Florida Industrial Co.	141
Darling v. Crowell	190
Davenport v. Bass	158
Davies v. Craig	227

Table of Authorities

441

Davis v. Curtis	265
Dawes v. Prentice	184
Dean v. Erskine	184
Delphey v. Savage	371
Desha v. Erwin	143
Dicus v. Allen	336
Diehl v. Zanger	18, 49, 222, 265, 284, 285, 287, 335, 340, 347, 374
Dimmitt v. Lashbrook	42
Dittrich v. Ubl	37, 40, 283, 347, 366
DiVirgilio v. Ettore	84
Dobson v. Finley	274
Doe v. Hildreth	157
Doe v. Paine	380
Dolphin Lane Associates, Ltd. v. Town of Southampton	57, 320, 330
Doolittle v. Bailey	367
Dowdle v. Cornue	96, 355, 379
DRED v. Dow Sand & Gravel	243, 246
Duff v. Fordson Coal Co.	127
Dugger v. McKesson	261
Duncan v. Hall	202, 203, 221
Dundalk Holding Co. v. Easter	56
Dunn v. Ralyea	38, 216, 353
DuPont v. Randall	321
Dupont v. Sparring	265, 284, 285, 374
Eagan v. Hinch	319
East Texas Pulp and Paper Co. v. Cox	166, 167
Edinger v. Woodke	382
Edmunds v. Griffin	185
Emery v. Fowler	112
Enfield v. Day	202
Esmond v. Tarbox	282, 344
Everett v. Lantz	338
Fagan v. Walters	79
Farr v. Woolfolk	245, 258
Fay v. Crozer	107, 222, 313
Fellows v. Willett	143, 352
Ferch v. Konne	373
Ferguson v. Bloom	267, 268
Ferrell v. Allen	294
Finberg v. Gilbert	71, 165
Findlay v. State	167
Flagg v. Thurston	184
Flynn Group Min. Co. v. Murphy	350
Foard v. McAnnelly	254

442
Table of Authorities

Fordson Coal Co. v. Napier	127, 265
Fordson Coal Co. v. Osborn	265
Fore v. Gilliam	265
Forsaith v. Clark	185, 186
Foster v. Duval County Ranch Co.	111
Frazier v. Bryant	380
Frederick W. Kretzer, et al., v. City of White Sulphur Springs	90
Froscher v. Fuchs	26, 33, 35, 36, 262, 277
Fuelling v. Fuesse	365
Furbush v. Goodwin	184
Gage v. Gage	183
Gaines v. City of Sterling	338, 363
Galbraith v. Lunsford	253
Galpin v. Atwater	184
Galt v. Willingham	39, 78, 349
Gammon v. Verrill	320, 321, 322
Garrett v. Cook	243, 244, 246
Gates v. Asher	166
Gazzam v. Lessee of Phillips	132
Geiger v. Uhl	364
George v. Thomas	296, 380
Gerke v. Lucas	382
Gibson v. Hjul	350
Gibson v. Universal Realty Co.	166
Gilmer v. Young	203
Givens v. U.S. Trust Co.	120, 125
Glenn v. Whitney	44, 255
Goff v. Avent	73
Goff v. Lowe	219
Goodman v. Myrick	43, 352, 367
Goodson v. Fitzgerald	165, 168, 356
Goodwin v. Garibaldi	326
Granby Mining Co. v. Davis	367
Graves v. Amoskeag Co.	185
Gray v. Coykendall	350
Griffith v. Sauls	336
Grover v. Paddock	365
Guise v. Shuman	35
Gulf Oil Corp. v. Marathon Oil Co.	112
Gulf Production Co. v. Spear	217
Gunter v. Mfg. Co.	274
Hagerman v. Thompson	62, 221
Hagerman, et al. v. Thompson, et al.	22
Hagerman. v. Thompson	210

Table of Authorities 443

Hagey v. Detweiler	257
Haggan v. Wood	369
Haggan v. Wood's Heirs, Ky	369
Hagins v. Whitaker	219
Hale v. Cottle	353
Hale v. Warren	286, 293
Hall v. Davis	192
Hall v. Powel	354
Hall v. Powell	341
Halstead v. Aliff	86, 196
Hancock v. Bennett	166
Hannah v. Pogue	244
Hansford v. Chesapeake Coal Co.	126
Hanson v. Township of Red Rock in Minnehaha County	383
Hanstein v. Ferrall	274
Hardee v. Horton	134
Harris v. Lavin	219
Harry v. Graham	118, 203, 274
Hart v. Greis et al.	104, 387
Hart v. Gries	356
Harvey v. Mitchell	183
Haydel v. Dufresne	262
Haydell v. Dufresne	295
Hayes v. Hanson	188
Helehan v. Ueland	348
Helm v. Small	86, 344
Henrie v. Hyer	46, 49, 195, 206, 357
Herbert v. Wise	86, 344
Herbst v. Smith	342
Herndon v. Hogan	315
Hess v. Meyer	142, 195, 206, 296, 368, 373, 380, 385
Heybrook v. Index Lumber Co.	386
Higgins v. Ragsdale,	138, 362
Highway Properties v. Dollar Sav. Bank	74
Higuera's Heirs v. United States	128, 131, 221
Hinton v. Stewart's Heir	317
Hite v. Graham	206
Hodges v. Sanderson	141
Hoekman v. Iowa Civil Township	96, 355
Hoekman v. Iowa Civil Twp.	377
Hoffman v. Beecher	104
Hogg v. Lusk	218
Holloway's Unknown Heirs v. Whatley	69
Home Owners' Loan Corp. v. Dudley	255
Homer v. Cilley	187

444 Table of Authorities

Hootman v. Hootman	365, 382
Hord v. Olivari	343
Hoskins Heirs v. Boggs	83, 258
Houston Ice & Brewing Co. v. Murray Oil Co.	370
Howell v. Ellis	217
Howes v. Wells	218
Humble Oil & Ref. Co. v. State.	107
Hunt v. Barker	170
in re	135
In re Lynch's Estate; In re Duffy	69
Ivalis v. Curtis v. Harding	136, 262
J. R. Buckwalter Lumber Co. v. Wright	348
J.R. Buckwalter Lumber Co. v. Wright	266
J.W. Robinson v. Elias Mosson et al.	64
Jackson v. Cole	340
Jackson v. Douglas	254
Jackson v. Joy	340
Jackson v. McConnell	255
Jackson v. Stoats	190
Jackson v. Wendell	341
Jackson v. Wilkinson	341
Jacobs v. Moesley	264, 284, 285, 373
Jacobs v. Moseley	254
Jahnke v. McMahon	254
Jakeway v. Barrett	192
James v. Drew	292, 327, 366
Janke v. McMahon	242, 243
Jarvis v. Swain	274
Jefferis v. East Omaha Land Co.	132
Jefferis v. Land Co.	382
Jenks v. Morgan	183
Johnson v. M'millan	206
Johnson v. Westrick	62, 98
Jones v. Burgett	64, 80
Jones v. Hamilton	121
Jones v. Kimble	21, 327, 367
Jones v. Merrimack Co.	185
Jones v. Pashby	253
Jordan v. Deaton	255
Justice v. McCoy	205
Kahn v. Delaware Securities Corporation	37, 40
Kaiser v. Dalto	44, 337
Kashman v. Parsons	66
Keesling v. Truitt	365
Kellogg v. Smith	183, 192, 223, 345

Table of Authorities

Kelsey v. Lake Childs Co.	37, 143
Kennedy v. Oleson	140
Kentucky Coal Lands Co. v. Mineral Development Co.	323
Kentucky River Timber & Coal Co. v. Morgan	121
Kentucky Union Co. v. Hevner, et al.	107
Kerr v. Fee	87
Kimms v. Libby	254
Kincaid v. Peterson	383
King v. Brigham	353
King v. Carmichael	386
Kip v. Norton	255
Kirby Lumber Co. v. Adams	167
Kirby Lumber Corp. v. Lindsey	166, 167
Kirch v. Persinger	142
Knapp v. Marlborough	184
Knight v. Elliott	264, 284, 285, 296, 327, 365, 373, 380
Knox v. Silloway	183
Knupp v. Barnard	266
Krause v. Nolte	368
Laflin Borough v. Yatesville Borough	56
Lampson Lumber Co. v. Caporale	69
Lane v. Darlington	378
Lane v. James	188
Langle v. Brauch	143, 349
Larsen v. Richardson	43, 102, 134, 159, 264, 306
Lawler v. Rice & Goodhue County	142
Lawson v. Viola Tp	96, 355
LeCompte v. Lueders	37, 40
Lemmon v. Hartsook	380
Lessee of Alshire v. J.R. Hulse	229
Lessee of Brown v. Clements	132
Lessee v. Walker	128, 131, 231
Lewen v. Smith	273
Lewis v. Ogram	254
Liddon v. Hodnett	142, 255
Liebler v. Abbott	320
Lillis v. Urrutia	86
Lilly v. Kitzmiller	341
Lilly v. Marcum	86
Lind v. Hustad	254
Linney v. Wood	64, 165, 168
Linscott v. Fernald	184
Lippincott v. Souder	224
Livingston v. Ten Brocek	339
Lloyd v. Benson	81, 211

Lodge v. Barnett	104, 353
Long v. Smith	294
Loring v. Norton	80, 190
Lovelace v. Hightower	378
Lugon v. Closier	75, 271
Lugon v. Crosier	76, 80, 282, 286
Lund v. Parker	185
Lyon v. Ross	85, 86, 341, 344
M. K. & T. Ry. Co. v. Merril	133
M'Clintock v. Rogers	48, 85, 172
M'Iver v. Walker	190
M'Iver's Lessee v. Walker	131, 175, 206
Maddox Bros. & Anderson v. Fenner	64, 123
Mageehan v. Adam's lessee	341
Major v. Watson	264, 280, 284, 285, 373
Manufacturer's National Bank of Detroit v. Erie County Road Commission	99
Marsh v. Stephenson	367
Martin v. Carlin	139
Martin v. Hall	258
Martin v. Lopes	244
Maryland Coal and Realty Co. v. Eckhart	56, 330
Mason v. Braught	195, 206, 283, 338, 366
Masterson v. Ribble	71
Matador Land & Cattle Co. v. Cassidy-Southwestern Commission Co.	64
Matthews v. Parker	80
May v. Baskin	48, 341
Mayor of Liberty v. Burns	285, 296, 367, 373
McAlpine v. Reicheneker	382
McAndrews, etc. Co. v. Camden Nat. Bank	84
McClintock v. Rogers	272, 299, 342, 368
McCourry et al. v. McCourr	177
McCoy v. Galloway	273
McCullough v. Absecon Beach Co.	104, 224
McEwen v. Den	203, 204
McIver v. Walke	174
McKee v. Bodley	206
McKinney v. Doane	82, 285, 348, 374
M'Clintock v. Rogers	103, 340
Md. Construction Co. v. Kuper	56
Mechler v. Dehn	365
Melcher v. Flanders	185
Mello v. Weaver	242
Mellor v. Walmesley	140

Table of Authorities

Mellow v. Weaver	243
Melvin v. Marshall	187
Mercer v. Bate	228
Mercer v. Bate	84, 270, 271
Mercer v. Bate, 4 J.J. Marsh. 334	122, 225
Meyers v. Johnson	258
Miles v. The Pennsylvania Coal Co.	256
Millar v. Bowie	345
Miller v. Southland Life Ins. Co.	166
Miller v. Topeka Land Co.	382
Miller v. White	142
Milligan v. Milligan	82
Miner v. Brader	239
Mineral Development Co. v. Tuggle	323
Mining & Smelting Co. v. Davis	285, 373
Minot v. Brooks	183
Missouri v. Iowa	345
Mitchell v. Hawkins	349
M'Nairy v. Hightour	345
Mock v. Astley	267
Moody v. Farinholt	357
Moore v. Campbell	338
Moreland v. Page	327, 343, 363, 365
Morgan v. Renfro	121
Morris v. Jody	343
Morrison v. Coghill	86, 344
Morse v. Rogers	200
Mosher v. Berry	367
Moyer v. Langton	385
Muldoon v. Sternenberg	166
Myrick v. Peet	139, 170, 221, 253
National City v. California Water and Telephone Company	69
Neary v. Jones	379
Neill v. Ward	194, 308
Nelson v. Hall	384
Nesselrode v. Parish	296, 367
Nesselrode v. Parrish	142, 355
Newfound Management v. Sewer	263
Newsom v. Pryor	165, 345
Newsom v. Pryor's Lessee	165
Nichol v. Lytle's Lessee	254
Nixolin v. Schnerderline	278
Normanoch Ass'n v. Deiser	21
Norwood v. Crawford	203, 274
Nutting v. Herbert	184

Table of Authorities

Nystrom v. Lee	379
Ocean Beach Association v. Yard	343
Ogilvie v. Copeland	96, 142, 292, 296, 355, 362
Olson v. Jude	39, 41, 348
Osborn v. King	35
Osteen v. Wynn	111, 245, 257
Outlaw v. Gulf Oil Corporation	87, 140, 356
Overstreet v. Dixon	87
Overton v. Leonard	380
Owen v. Foster	190
Oxford v. White	73
Pacheco v. Martinez	350
Pallas v. Dailey	350
Pandem Oil Corp. v. Goodrich	364
Parker v. Brown	185
Parkhurst v. Van Cortland	184
Parran v. Wilson	66, 79
Peacher v. Strauss	53, 60, 72, 112, 211
Pearson v. Baker, 34 Ky. 321	117, 220
Peaslee v. Gee	183, 184
Pennington v. Flock	73
Penry v. Richards	340
People of the State of California v. Thompson & Gray, Inc.	284
People v. Holmes	140
Pereles et al. v. Magoon et al.	366
Pereles v. Gross	98, 359
Pereles v. Magoon	292, 383
Perich v. Maurer	288
Perkins v. Jacobs	320
Pernam v. Wead	384
Pernam v. Weed	190
Perry v. Buswell	18
Perry v. Pratt	84
Peterson v. Johnson	256
Petty v. Paggi Bros. Oil Co.	166
Philips v. Shaeffer	341
Pierce v. Richardson	188, 190
Pierce v. Schram	168
Pike v. Dyke	282, 344
Pilgrim v. Kuipers	66, 254
Pinkham v. Murray	188
Piotrowski v. Parks	111
Pitman v. Nunnelly	121
Pittsburgh etc. R. C. O. v. Beck	73
Plowman v. Mille	166

Table of Authorities

Pollard v. Shively	103, 296
Porter v. Gaines	367
Potts v. Everhart	272
Prescott v. Hawkins	184
Prescott v. Hayes	185
Preston Heirs v. Bowmar	220
Preston v. Bowmar	120
Preston's Heirs v. Bowmar	117
Price v. Earl of Torrington	140
Price v. Mauch	336
Pride v. Lunt	184
Proctor v. Hinkley	80
Puget Mill Co. v. North Seattle Imp. Co.	142
Pyburn v. Campbell	336
Quesnel v. Woodlief	159
Radford v. Johnson	351, 379
Railroad Co. v. Schurmeier	85
Ralston v. Dwiggins	288
Ralston v. M'Clurg	81
Ralston v. McClurg	122
Ramsay v. Butler, Purdum & Co.	66
Ramsey v. Morrow	361, 369
Rand v. Dodge	183, 184
Randall v. Burk Township	379
Randall v. Burk Tp	95, 231, 355, 356
Randall v. Burk Tp. of Minnehaha County	94
Randleman v. Taylor	253
Rapides Parish Police Jury v. Grant Parish Police Jury	179, 344
Read v. Bartlett	292, 362
Reel v. Walter	43
Reid v. Dunn	76, 284
Reynolds v. Bradford	168
Ricci v. Godin	82
Richardson v. Bohney	367
Riley v. Griffin	103, 190, 339, 367
Rivers v. Lozeau	22, 97, 264, 338
Roadenbaugh v. Egy	369
Rockwell v. Adams	255
Rodriguez v. La Cueva Ranch Co.	378
Rollins v. Davidson	222, 343
Ronk v. Higginbottom	315
Root v. Town of Cincinnati	342
Roth v. Halberstadt	70
Routh v. Williams	40
Rowe v. Kidd	42, 84, 108, 125, 270

450   *Table of Authorities*

Rowell v. Weinemann	*221, 342*
Ruddy v. Rossi	*133*
Runkle v. Smith	*167*
Rusha v. Little	*320*
Russell v. Dyer	*186*
S.R.H. Corp. v. Rogers Trailer Park, Inc.	*159*
Safret v. Hartman	*203*
Sala v. Crane	*128, 130, 133*
Sanborn v. Clough	*184*
Sansing v. Bricka	*167*
Savannah, Florida and Western Railway v. Geiger	*271*
Sawyer v. Cox	*341*
Sayers v. City of Lyons	*342*
Scammon v. Scammon	*186, 188*
Schraeder Min. Co. v. Packer	*254*
Schurmeier v. St. Paul, etc., R. Co.	*85*
Scott v. Thacker Coal Mining Co.	*265*
Seabrook v. Coos Bay Ice Co.	*85, 129*
Sellman v. Schaaf	*38, 41, 229, 380*
Sellman v. Schaff	*351*
Serrano v. Rawson	*231*
Shackelford v. Walker	*313*
Shaw v. Clement	*86, 344*
Silsby & Co. v. Kinsley	*307, 357*
Simmons Creek Coal Co. v. Doran	*276*
Simpkins' Adm'r v. Wells	*218*
Simpkins v. Wells	*275*
Ski Roundtop Inc. v. Wagerman	*54, 329, 345*
Slovensky v. O'Reilly	*349*
Smart v. Huckins	*68, 71, 72*
Smita v. Young	*277*
Smith v. Bodfish	*190*
Smith v. Bradley	*188*
Smith v. Dodge	*350*
Smith v. Messer	*187, 188, 190*
Smith v. Moore	*216, 353*
Smith v. Nelson	*255*
Smith v. Turner	*166*
Sneed v. Osborn	*244*
Snider v. Rinehart	*294*
South Penn Oil Co. v. Knox	*64*
Sowerwine v. Neilson	*79*
Sparhawk v. Ballard	*185*
St. Paul & P. Ry. Co. v. Schurmier	*131, 133*
Staaf v. Bilder	*40, 41*

Table of Authorities

Stack v. Pepper	203
Stadin v. Helin	80, 227
Stafford v. King	46, 104, 165, 170, 193, 195, 225, 268, 269, 356
Stanley v. The County Surveyor of Sheridan County	369
Stanolind Oil & Gas Co. et al. v. Wheeler et al.	71
State ex rel. Brayton v. Merriman	143
Stanolind Oil & Gas Co. v. State	313
State ex rel. Brayton v. Merriman	143
State el rel. Buckson v. Pennsylvania R. Co.	338
State ex rel. Cohen v. Manchin	91
State v. City of Sarasota	73
State v. Coleman-Fulton Pasture Co.	122
State v. King	64
State v. Moore	133
State v. Palacios	122, 123, 139
State v. Sullivan	166
State, Bd. of Educational Lands and Funds v. Jarchow	374
Staub v. Hampton	38, 41, 105, 216, 356
Steele v. Taylor	219
Steinherz v. Wilson	109
Stetson v. Adams	371
Stewart v. Carleton	48, 49, 237, 335, 336
Stewart v. Carlton	264, 284, 285, 374
Stewart v. Hoffman	40
Stover v. Freeman	273
Straw v. Jones	185
Strong v. Sunray DX Oil Co.	165
Strunk v. Geary	265
Strunz v. Hood	358
Suit v. Hershman	365
Sullivan v. Rhode Island Hospital Trust Co.	68
Sumner v. Sherman	188
Sumter v. Bracy	341
Suydam v. Williamson	120
Swanson v. Koeninger	350
Swift Coal & Timber Co. v. Sturgill	265, 322
Syllabus Point 1, Miners in General Group v. Hix	91
Tappan v. Tappan	183, 185
Tarpenning v. Cannon	288
Taylor v. French	188
Taylor v. Higgins Oil & Fuel Co.	268, 356
Terry v. Chandler	254
Tewksbury v. French	104
Texas International Petroleum v. Delacroix Corp	180

Thallman v. Thomas	350
Thatcher v. Matthews	71, 137, 293
The Appeal of Minnie Ralston From Survey of J. Homer Austin	343
Thein v. Burrows	350
Theriault v. Murray	81
Thomas Hinton, heir-at-law of Joseph Hinton, deceased, by William Morrice, his next friend, v. The Heir of William Stewart, deceased	316
Thomas Jordon, Inc. v. Skelly Oil Co.	165
Thomas v. Perry	159
Thompson v. Hill	88, 276
Thompson v. McFarland	346
Thomsen v. Keil	236, 290, 375
Thornberry v. Churchil	117, 341, 380
Thurmond v. Espalin	378
Tiggeman v. Mrzlak	350
Titus v. Chapman	94, 134, 355
Tognazzini v. Morganti	337
Tolson v. Southwestern Improvement Association	293, 325
Tompkins v. American Republics Corp.	69
Town v. Greer	143
Treadwell v. Marrs	350
Trinwith v. Smith	143, 367
Tripp v. Bagley	255, 256
Trotter v. Stayton	287, 359
Trotter v. Town of Stayton	352
Trustees, etc., v. Wagnon	121
Tucker v. Aiken	188
Turnbow v. Bland	47, 262, 268
Turner v. Baker	253
Tyner v. McDonald	33, 35, 364
Tyson v. Edwards	22, 26, 27, 28, 33, 41, 42, 88, 338
U.S. v. Champion Paper Inc.	157
U.S. v. Doyle	238, 283, 338
U.S. v. Gallas	229
Ufford v. Wilkins	342
Uhl v. Reynolds	258
Ujka v. Sturdevant	85
United Fuel Gas Co. v. Snyder	64
United States v. Champion Paper, Inc.	70
United States v. Champion Papers, Inc.	61, 160
United States v. Doyle	43, 283, 365
United States v. Gratiot	133
United States v. Hudspeth	338

Table of Authorities 453

United States v. Lane	338
United States v. State Investment Co.	210, 338
Urban v. Urban	68
Van Dusen v. Shively	43, 352
Vance v. Fore	231
Vance v. Gray	364
Vaughn v. Tate	327, 365
Vaught v. McClymond	338, 349, 374
Velasquez v. Cox	376
Voigt v. Hunt	254
Vosburgh v. Teator	254
Vowinckel v. N. Clark & Sons	242
W. H. Rigell v. Ed Mays	163
Wacker v. Price	287, 334
Wade v. McDougle	314
Walker v. Curtis	140
Walker v. Furry & Krehl	354
Wallace v. Maxwell	121, 341
Warcynski v. Barnycz	141
Ward v. Fuller	185
Ward v. Rodriguez	378
Wash v. Holmes,	226
Washington Rock Co. v. Young	46, 96, 292, 355, 357, 362, 368
Wates v. Crandall	351
Watrous v. Morrison	142, 254, 338
Weatherly v. Jackson	167
Weaver v. Robinett	273
Wells v. Burbank	186, 187, 188, 190
Wells v. Jackson Iron Manufacturing Co.	174
Wells v. Jackson Iron Manufacturing Company	175
Wells v. Jackson Iron Mfg. Co.	69, 180
Wells v. New York Mining And Manufacturing Co.	259
Wendell v. Blanchard	185
Wentzel v. Claussen	377
Wescott v. Craig	292
Westcott v. Craig	363, 365
Westphal v. Schultz	327
Wharton v. Littlefield	334
Whitaker v. Hall	206
White v. Amrhien	379
Whitehead v. Atchison	82
Whiting v. Gardner	128, 129
Whitman v. Haywood	336
Whitmore v. Brown & Gilley; Smallidge, et al. v. Same	70
Wightman v. Campbell	140, 141

Wildeboer v. Hack	*36, 277*

William Eagan v. Samuel Hinch heir-at-law of John Hinch,
  deceased, and John Jack, heir-at law of Samuel Jack
  deceased
	318

William White v. Dudley Everest	*307*
Williams v. Barnett	*241, 243, 299*
Williams v. Brush Creek Coal Co.	*265*
Williams v. Spaulding	*282, 344*
Williams v. Tschantz	*265, 368*
Willis v. Campbell	*26, 41*
Wilson v. DeGenaro	*69*
Wilson v. Giraud	*165*
Wilson v. Inloes	*190*
Wilson v. Stork	*283, 366*
Wise v. Burton	*231*
Wishart v. Crosby	*341*
Wm. Cameron & Co. v. Taylor	*167*
Wood v. Hildebrand	*56*
Wood v. Starko	*37, 38, 105, 356*
Woods v. Banks	*185*
Woods v. Johnson	*170, 280, 284, 373*
Wyatt v. Foster & Rafferty	*172*
Yakes v. Williams	*338*
Yocum v. Haskins	*383*
Yoder v. Fleming	*341, 354*
Yolo County v. Nolan	*142*
Young v. Leiper	*190, 345*
Zachariah Herndon v. James Hogan	*315*
Zawatsky Constr. Co. v. Feldman Development Corp.	*56*
Zeringue v. Harang	*229*
Ziebold v. Foster	*264*
Zwakhals v. Senft	*26*

Index

Note: Page numbers followed by *"fn"* indicate footnotes.

A

Abbott v. Abbott, 360
A. B. Moss & Bro. v. Ramey, 137
Abounding descriptions, 57, 340
About corners, 298–301
Absolutely appellees' theory, 364
Abutting parcels, 57, 108, 227, 340
Abutting tracts, descriptions by, 56–60
Acampora v. Pearson, 398
Accessory
    corner, 71
    monument, 75–76
Accidental error, 230
Accretion, 43
Accurate survey map, 88, 130
Acquiescence, 59, 272, 273, 281, 286
    boundary, 273
Acreage call, 341
Act for Prevention of Frauds and Perjuries, 264fn, 271fn, 278
Actual location, 342
Actual notice, 50
Actual survey, 219
    appellant, 222
    comparing measurements, 223
    custom of surveyor, 219–220
    distance deficiencies, 222–223
    legal rule, 221
    problems with area recitations, 223–224
Adams Express Co. v. Ohio State Auditor, 122
Adams v. Hoover, 390–391
Adams v. Wilson, 410
Adjacent surveys, 57
Adjoining descriptions, 58
Adverse possession, 43, 264, 272
AFFIRMED case, 20
Affleck v. Morgan, 40
Agonic line, 189

Agreements, 268
Agreement upon definite line, 275
Akin v. Godwin, 12, 19, 21, 24, 26, 29, 305
Albertson v. Chicago Veneer Co., 128
Alby v. Booth, 324fn
Alexander v. Hill, 124
Alexander v. Lively, 303
Allely v. Fickel, 146
Allen v. Blanton, 339
Allen v. Kingsbury, 198
Allmon v. Stevens, 413
Aluminum metals, 258
Ambiguities, 51–53, 64–65
    in field notes, 53–54, 180
American law, 333
Ames v. Hilton, 111
Anderson v. Johanesen, 149
Andrews v. Wheeler, 429
Angles, 9
Ansley v. Graham, 304
Appalachian chain, 336–337
Apportionment rule, 5–7, 30–31, 409–410
    Bureau of Land Management, 32–33
    conditions for rule to apply, 6
    diagram from *Tyson v. Edwards*, 21
    establishing internal lines within Rizzo's subdivision, 11
    footsteps contrary to established rule, 34–36
    plan from *Tyson v. Edwards*, 15
    plat as grid, 21
    plat redrawing, 23
    purpose of resurvey, 29
    quarter-quarter section, 7
    retracement surveys, 13
    State Road Department survey, 19
    surveying and boundaries, 28
    surveying principles, 16, 31
    surveyor roles, 9
    true meridian and base line, 10

455

Index

Area recitations, problems with, 223–224
Areas of rectangular lots, 343
Armijo v. New Mexico Town Co., 66
Arneson v. Spawn, 412
Arpent lots, 345
Artificial monuments, 74
Asher v. Fordson Coal Co., 131
Asher v. Howard, 122
Avery v. Baum, 389
Ayers v. Watson, 28, 72, 136, 236, 304, 386, 392, 393

B

Bailey v. Chamblin, 367, 414
Baker & Sons v. Sherman, 250
Baker v. Glasscocke, 387
Ballard v. Stanolind Oil & Gas Co., 400
Barrett v. Perkins, 410
Bartholomew v. Murry, 62
Bass v. Mitchell, 57
Baxter v. Evett's Lessee, 124, 387
Baybutt Construction Corp. v. Commercial Union Ins, 112
Bayhouse v. Urquides, 132, 383, 403
Beach v. Beatty, 12
Beach v. Fay, 342
Bean v. Bachelder, 82, 311, 387–388
Beardsley v. Crane, 92fn, 393, 399, 412
Bearing trees and objects, 253–255
Beaubien v. Kellogg, 379
Beaver Brook Resort Co. v. Stevens, 409
Beckley v. Bryan, 239, 387
Beckley v. Bryan and Ransdale, 245
Beckley vs. Bryan &c, 76
Bellows v. Copp, 197, 198
Bell v. Morse, 197
Bemis v. Bradley, 111
Bennington, Vermont, 334
Ben Realty Co. v. Gothberg, 381
Bentley v. Jenne, 403, 424
Bethel–Rumford town line, 360
Billingsley v. Bates, 412
Billups v. Woolridge, 402
Bishop v. Johnson, 18, 22, 305
Bittner v. Walsh, 24
Black's Law Dictionary, 87
Black v. Sprague, 253, 305

Black Walnut Bottom, 46
Blaffer v. State, 400
Blasdell v. Bissell, 389
Blazed line, 78
Blazed tree, 247
Blight v. Atwell, 415
"Block survey", 293
Block system, 293
Block v. Howell, 91fn
"Blucher State Map", 126
Blueprint, evidence discovering, 328
Blunder, 232
Bolton v. Lann, 321, 324fn
Bond v. Middleton, 53
Book v. Justice Min. Co., 393–394
Boon v. Hunter, 139
Booth v. Clark, 412, 429
Booth v. Upshur, 101
Boundaries, 31, 37–38, 40, 69, 84–85, 360
  by acquiescence, 273
  markers, 109fn
Boundary agreements, 263
  acquiescence, 273, 286
  adverse possession, 272
  agreement upon definite line, 275
  boundary by acquiescence, 273
  conditional lines, 284–286
  consentable lines, 282–284
  dispute as to location of boundary, 274–275
  doctrine grew, 268
  estoppel, 273–274
  line by accepting evidence, 265
  "Line of Occupation", 266
  meaning of uncertainty, 270–272
  parol agreement, 269, 274
  possible outcome of boundary line agreements, 264
  practical location of boundary, 275–278
  principles, 267
  recognition, 286
  requirement, 280–282
  statute law in new Hampshire, 269–270
  effect of statute of frauds, 274
  statute of frauds, 278–280
Boundary lines, 3, 69, 77, 81, 133, 269, 274–275, 289, 427

Index

457

agreements, 263, 332
Bounded descriptions, 57, 340
Bounding descriptions, 57–60, 340–341
Bounds, 9, 16, 84
Bouvier's Law Dictionary, 44
Boyd v. Durrett, 38*fn*, 391
Bradford v. Pitts, 34
Bradshaw v. Edelin, 314, 419
Brand vs. Dawny, 382
Brashears v. Joseph, 124, 237
Brass metals, 258
Brewer vs. Gay, 382
Brockman v. Rose, 124
Brooks v. Stanley, 429
Brooks v. Tyler, 186, 188
*Brown's Boundary Control and Legal
    Principles*, 99, 100, 101
Brown v. Carthage, 314, 419
Brown v. Clark, 198
Brown v. Clements, 136
Brown v. Lyons, 285
Brown v. McKinney, 283
Bruce v. Morgan, 124
Bruce v. Taylor, 124, 239
*Bruker v. Burgess and Town Council of
    Borough of Carlisle*, 373
*Bruker v. Burgess and Town Council of
    Carlisle*, 49
Bryant v. Strunk, 123, 124, 130, 292
Bryan v. Beckley, 118, 223
Bryan v. Beckley, Litt. Sel., 415
Buckley v. Laird, 28, 312, 411
*Buffalo Zinc & Copper Company v.
    McCarty*, 380
Buford v. Cox, 389
Buie v. Miller, et al., 62
Bullard v. Kempff, 132
Bulor's Heirs v. James M'Cawley, 249
Burton v. Duncan, 49–50
Byrne v. McKeachie, 399, 423

C

Caballerias, 347
*California v. Deseret Water, Oil
    & Irr. Co.*, 137
"Call" survey, 118
*Campbell & Others v.
    Nickerson & Others*, 372

Campbell's Executors v. Wilmore, 168
Campbell v. Carruth, 66
Campbell v. Clark, 314, 419
Canavan v. Dugan, 423
Cardigan v. Page, 202
Cardinal principle, 96
Cardinal rule, 2*fn*
Carey v. Clark, 318
"Carolina", 45
"Carricks Chance" 46
Carroll v. Norwood's Heirs, 243
Carstensen v. Brown, 286, 433
Carter v. Beals, 198
Carter v. Hornback, 314, 419
Case v. Trapp, 417–418
Castleman v. Pouton, 245
Caveat, 355–356
Caylor v. Luzadder, 323, 367
Center of section, 74–75
Cf. McGinley v. Railroad, 328
cf. *Tyner v. McDonald*, 21
Chaining, errors in, 210
Chain of title, 47, 100, 290, 351, 352, 359
Chancery, 357–359
Chandler v. Hibberd, 410
*Chapman & Dewey Lbr. Co. v. St. Francis
    Levee Dist.*, 137
Chapman v. Polack, 253
Chapman v. Pollack, 305
Chapman v. Twitchell, 300
Chenoweth v. LesSee of Haskell, 384
Cherry v. Slade, 244, 395
Christianson v. Daneville Tp., 91*fn*, 399
Christian v. Gernt, 310*fn*
Christian v. Gernt. et al., 310*fn*
Christmas v. Cowden, 424
Circuit Court of Appeals, 363
City of Hastings v. Foxworthy, 137
City of Jacksonville v. Broward, 411–412
City of Mankato v. Carlstrom, 267
City of North Mankato v. Carlstrom, 44,
    266, 267*fn*
City of Racine v. Emerson, 26, 30, 316, 391,
    397, 402–403, 433
City of Racine v. J. I. Case Plow Co., 261
Clark on Surveying and Boundaries, 99, 100
Clark v. Moore, 302
Clark v. Northy, 197
Cleaveland v. Flagg, 197

458   *Index*

Climer v. Wallace, 149, 314, 412, 419
Closing corner, 71
Closing error, 230, 234
Clough v. Bowman, 197
Clough v. Sanborn, 197
Cody v. England, 219
Colby v. Collins, 197
Coleman County v. Stewart, 125, 128
Collateral evidence, 411
Combs v. Jones, 236, 292
Combs v. Valentine, 237
Combs v. Virginia Iron, Coal & Coke Co., 237
Comegys v. Carley, 426
Compass, 187–188
Compass-chain, 181
Compass-tape, 181
Compensating errors, 230
Compromise line, 79
Conditional corner, 286
Conditional lines, 78, 284–286
Cone v. West Virginia Pulp & Paper Co., 381
Congress, The, 406–407
Consentable line, 78
Consentable lines, 282–284
Conservation and perpetuity of
    boundary lines, 1
Constructive notice, 50
Contiguity, corner, 71
Conventional line, 79
Conveyance, 66
Conveyancing, 338
Cook v. Babcock, 197
Cook v. Combs, 197
Coppin v. Manson, 412
Cordell v. Sanders, 315, 422
Corlies v. Little, 244
Cornelious v. State, 407
Corner accessories, 253
    accessories to special-purpose
        monuments, 257–258
    fences, 259–262
    memorials, 258–259
Corners, 9, 69. *See also* Lines; Lost corners
    accessory, 71
    center of section, 74–75
    closing, 71
    contiguity, 71
    existent, 70–71
    extinct, 72–73

    half-mile post, 73
    lost, 72
    meander, 73
    missing monument, 76–77
    monument, 74
    monument accessory, 75–76
    never set, 298
    nonexistent, 71
    obliterated, 72
    quarter-section, 71
    reestablished, 72
    section, 71
    sixteenth-section, 72
    standard, 72
    township, 72
    types, 70
    witness, 73
    witness tree, 73
Cornett v. Kentucky River Coal Co., 237
Coterminous owners, 292
County of Yolo v. Nolan, 314
Course, 187, 243–244
Courts, 268
Cowan v. Fauntleroy, 80, 387, 389
Cox v. Finks, 166
Cox v. Hart, 91*fn*, 148, 393, 399
Cragin v. Powell, 26, 30, 98, 136, 137, 288,
    391, 425
Craig v. Russell, 21
Cram v. Ingalls, 198
Crandall method, 225
Craven v. Lesh, 412
Creech v. Johnson, 121, 122, 285
Crockett v. Andrews, 87
Curtis v. Aaronson, 244
Curtis v. Francis, 197
Custom, 44

D

Dagget v. Wiley, 300
Daniel v. Florida Industrial Co., 147
Data collection, errors in, 234
Davenport v. Bass, Com. App, 165
Davies v. Craig, 248
Davis v. Curtis, 291
Dawes v. Prentice, 198
Dean v. Erskine, 197
Declination, 188

Index 459

Dedication, 42
Deed, 41, 58, 329
Delphey v. Savage, 417
Descent, 42
Description errors, 235
Descriptions by adjoiners, 57
Desha v. Erwin, 149
Desideratum of rules, 184
Detective science, 1
Dicus v. Allen, 379–380
Diehl v. Zanger, 3fn, 39, 242, 291, 314, 315, 378, 390, 420
Dimmitt v. Lashbrook, 32
Directions, problems with, 184
 compass, 184–185, 187–188
 course, 187
 declination, 185–186, 188
 directions in reverse order, 206–207
 Dorcas Merrill, 193–196
 GLO surveyor, 192–193
 impact on successful corner location, 188–189
 lands redeemed, 201–203
 light, 191–192
 local attraction, 189–190
 meridian, 187
 needle, 186
 North Carolina case, 190–191
 Parol proof, 198–199
 plaintiff, 196–198
 Sargent's Purchase, 203–206
 tax, 199–200
Discrepancies, 229
 between field notes and plat, 165
Dispute, 268
 as to location of boundary, 274–275
Dissenting opinion, 325–327
Distance in perspective, 243–244
Distances, problems with, 207
 errors in chaining, 210
 errors in taping, 210–211
 length measurement units, 207–208
 measurement allowances, 208–210
 presumption of straight line, 211–218
 slope *vs.* horizontal measurement, 218–219
Dittrich v. Ubl, 26, 30, 312, 391, 411
DiVirgilio v. Ettore, 78
Division

by area, 337–339
by half, 337
line. *See* Partition line
lines, 411
Dobson v. Finley, 301, 302
Doctrine, 268
Dodge vs. Smith, 382
Doe v. Paine, 426
Doe vs. Paine & Sawyer, 382
Dolphin Lane Associates, Ltd. v. Town of Southampton, 48, 359fn, 373
Doolittle v. Bailey, 413
Douglass, Andrew Ellicott, 250
Dowdle v. Cornue, 92fn, 399, 425
DRED v. Dow Sand & Gravel, 267
Duff v. Fordson Coal Co., 131
Duncan v. Hall, 218, 219, 241
Dundalk Holding Co. v. Easter, 48
Dunn v. Ralyea, 27, 235, 397
DuPont v. Randall, 361
Dupont v. Sparring, 291, 314, 315, 420

E

Eagle Gap, 364
Early Court Decision, 101
Edinger v. Woodke, 429
Edmunds v. Griffin, 198
Emery v. Fowler, 112
Eminent domain, 41
Errors, 229
 accidental, 231–232
 in chaining, 210
 of closure, 230, 234
 computational, 232
 in data collection, 234
 description, 235
 index, 230
 in measurements, 229
 natural, 231
 personal, 230
 plat, 235
 random, 230
 in reporting data, 235–240
 systematic, 231
 in taping, 210–211
 total, 233
Escheat, 43
Esmond v. Tarbox, 311, 388

460   *Index*

Establishing, line, 94–95

Estoppel, 42, 272, 273–274

Everett v. Lantz, 381

Evidence, 246

    discovering, 324–330

    recognizing evidence, 245

Evidence and Procedures for Boundary Location, 100

Ewing's Heirs v. Savary, 124

Existent corner, 70–71

Existing perimeter, 129

Extended line, 79

Extinct corner, 72–73

F

Fagan v. Walters, 74

Falby v. Booth, 324*fn*

Farr v. Woolfolk, 268–269, 284

Fay v. Crozer, 106*fn*, 241, 351

Federal District Court, 297

Fellows v. Willett, 149, 395–396

Fence line, 78

Fences, 259–262

Ferch v. Konne, 419

Ferguson v. Bloom, 293, 295

Ferrous metals, 258

Field notes, 143, 253, 380

    admissibility, 146

    ambiguity in, 53–54, 180

    boundary question, 169

    conclusions of law, 176–180

    defendant, 164–165

    discrepancy between field notes and plat, 165

    enlargement of boundary, 170

    field notes of surveyor Job S. Collard, 170–172

    findings of fact, 172–173

    Florida land, 147–149

    government map, 150–151

    Hamblin's notes, 159–162

    issue, 146–147

    language of court, 168

    learning from, 145–146

    misleading evidence in, 167

    north line, 162–164

    principles, 166–167

    private parties, 149–150

    south corner of Riggs survey, 173–175

    southwest boundary of Riggs survey, 175

    subdivision of township, 154–157

    trial line, 152–154

Figures, numbers, and symbols (mathematics of surveys); *see also* Paper survey

    actual survey *vs.* paper survey, 219–224

    adjustments and getting rid of closure errors, 224

    directions, problems with, 184–207

    distances, problems with, 207–219

    mathematics, problems with, 224

    shifting positions, 224–225

    survey types, 181–184

First survey, 89–93

Flagg v. Thurston, 197

Flanders deed, 329

Flynn Group Min. Co. v. Murphy, 394

Foard v. McAnnelly, 280

Footsteps, 99–101, 105, 141–142, 276, 377

    Acampora v. Pearson, 398

    Adams v. Hoover, 390–391

    additional, 139–140

    admissibility of field notes, 146–165

    ambiguity in field notes, 180

    Bayhouse v. Urquides, 383

    Bean v. Bachelder, 387–388

    Bentley v. Jenne, 403

    Billups v. Woolridge, 402

    Cherry v. Slade, 395

    contrary to established rule, 34–36

    Dicus v. Allen, 379–380

    discrepancy between field notes and plat, 165

    Dittrich v. Ubl, 391

    Fellows v. Willett, 395–396

    field notes principles, 166–167

    Gray v. Coykendall, 393–394

    Hall v. Powel, 398

    Henrie v. Hyer, 400–401

    Herbst v. Smith, 385–386

    J.R. Buckwalter Lumber Co. v. Wright, 391–392

    Jury v. Grant, 387

    Kellogg v. Smith, 388–390

Index

461

learning from field notes, 145–146
left, 143–145
Lodge v. Barnett, 397–398
Martz v. Williams, 383
McKinney v. Doane, 392
M'Clintock v. Rogers, 383–384
Millar v. Bowie, 388
misleading evidence in field notes, 167–180
Moody v. Farinholt, 401–402
Moreland v. Page, 386–387
Morris v. Jody, 387
Olson v. Jude, 392–393
Pallas v. Dailey, 393
passing calls, 165–166
Racine v. Emerson, 402–403
Riley v. Griffin, 382–383
Rowell v. Weinemann, 386
Sawyer v. Cox, 384–385
Sellman v. Schaff, 395
Smith v. Dodge, 394
State el rel. Buckson v. Pennsylvania R. Co., 381–382
Strunz v. Hood, 402
of surveyor, 97
Titus v. Chapman, 398–399
U. S. v. Doyle, 381
Van Dusen v. Shively, 396–397
Wacker v. Price, 377–379
Wharton v. Littlefield, 377
Wood v. Starko, 400
wrong set, 142
Yolo v. Nolan, 380–381
Fordson Coal Co. v. Napier, 131, 291
Fordson Coal Co. v. Osborn, 291
Forensic applications, 371
complete and thorough investigation, 374
forensic science of criminal environment, 374
historical knowledge and research, 372–374
modern tools, 374
Forensics, 371
Forensic science, 371, 374
Fore v. Gilliam, 291
Forsaith v. Clark, 198, 200
Foss v. Johnstone, 132
Foster v. Duval County Ranch Co., 111

Frazier v. Bryant, 426
Frederitzie v. Boeker, 314, 419
French arpent system, 333
French Lots, The, 345
French system of longlots, 345, 346
Froscher v. Fuchs, 12, 21, 24, 25, 288, 304
Fuelling v. Fuesse, 410
Fund v. Wetstone, 24
Furbush v. Goodwin, 197

G

Gage v. Gage, 197
Gaines v. City of Sterling, 381, 408
Galbraith v. Lunsford, 279
Galbraith v. Parker, 132, 136, 149, 316, 322, 407
Galpin v. Atwater, 197
Galt v. Willingham, 28, 73, 392, 393
Gammon's surveyor, 361*fn*
Gammon v. Verrill, 359, 361
Garrett v. Cook, 267, 268, 270*fn*
Gazzam v. LesSee of Phillips, 136
Gazzam v. Phillips, 136
Geiger v. Uhl, 410
General Land Office (GLO), 333, 347, 348
George v. Thomas, 426
Gerke v. Lucas, 429
Gibson v. Hjul, 394
Gilbert v. Finberg, 139
Gilmer v. Young, 220
Givens v. U.S. Trust Co., 123, 128*fn*
Glenn v. Whitney, 34, 281
GLO. *See* General Land Office
Goff v. Avent, 66
Goff v. Lowe, 239
Goodman v. Myrick, 33, 396, 412
Goodson v. Fitzgerald, 400
Goodwin v. Garibaldi, 367
Government corner, 96
Government procedure, 296
Government surveyors, 291, 314, 330
Government surveys, 143, 303–304, 414
Government. Yakes v. Williams, 381
Granby Mining Co. v. Davis, 412
Grand Cent. M. Co. v. Mammoth M. Co., 132
Grantees, 351
Grants, 349

Index

Graves v. Amoskeag Co, 198
Gray v. Coykendall, 393–394
Griffith v. Sauls, 379
Grover v. Paddock, 410
Guise v. Shuman, 24
Gulf Oil Corp. v. Marathon Oil Co., 111
Gulf Production Co. v. Spear, 236
Gunter v. Mfg. Co., 301

H

Hagerman, et al. v. Thompson, et al., 6fn
Hagerman v. Thompson, 55, 241
Hagey v. Detweiler, 284
Haggan v. Wood's Heirs, 415
Hagins v. Whitaker, 239
Hale v. Warren, 315, 324
Half-mile post, 73
Hallean lines, 189
Halley v. Harriman, 136
Hall v. Powel, 398
Hall v. Powell, 384
Handheld compass, 187
Haney v. Burgin, 328
Hannah v. Pogue, 267
Hansford v. Chesapeake Coal Co., 129
Hanson v. Township of Red Rock in Minnehaha County, 430–432
Hanstein v. Ferrall, 302
Hardee v. Horton, 138
Harrington v. Boehmer, 132, 381
Harris v. Lavin, 239
Harry v. Graham, 120, 219, 301
Hart v. Gries et al, 400
Harvey v. Mitchell, 196
Hawaii, surveys in, 242–243
Hayes v. Hanson, 202
Helehan v. Ueland, 392
Helm v. Small, 387
Henrie v. Hyer, 35, 36, 210, 400–401
Herbert v. Wise, 80, 387
Herbst v. Smith, 385–386
Herrick's plan, 82
Hess v. Meyer, 149, 210, 413, 418–419, 426, 431
Heybrook v. Index Lumber Co., 433
Hickerson v. Dillard, 393
Hickman creek, 354
Higgins v. Ragsdale, 144, 408

Higuera's Heirs v. United States, 132, 135
Hodges v. Sanderson, 146
Hoekman v. Iowa Civil Township, 92fn, 399, 423
Hoffman v. Beecher, 101fn
Holloway's Unknown Heirs v. Whatley, 62
Home Owners' Loan Corp. v. Dudley, 281
Homer v. Cilley, 200
Hootman v. Hootman, 410, 429
Horizontal measurement, 218–219
Hoskins Heirs v. Boggs, 78, 284
Houston Ice & Brewing Co. v. Murray Oil Co., 415–416
Howell v. Ellis, 236
Howes v. Wells, 237
Humble Oil & Ref. Co. v. State, 105fn

I

Ideal line, 79, 297–298
Idiosyncrasies within system, 340
Imaginary straight line customarily, 9
"Incidental call", 165
Independent surveys, 96, 424
Index errors, 230
Inquiry Notice, 51
Instrumental errors. *See* Index errors
Intention, 61–63
Intent of parties, 60–61
Intent of surveyor, 63–64
Interlocks, 351
Internet services, 374
Investigation, 371, 374
Investigator, 287
Involuntary alienation, 41
Irrigated plots. *See* Suertes
Irvin v. Rotramel, 413
Isogonic lines, 189
Ivalis v. Curtis v. Harding, 142, 289
Ivy v. Parker, 149, 429

J

Jackson Purchase, 333, 343–344
Jackson v. Cole, 384
Jackson v. Douglas, 280
Jackson v. Joy, 383–384
Jackson v. McConnell, 281
Jackson v. Wendell, 384

Index

463

Jackson v. Wilkinson, 384
Jacobs v. Moesley, 291, 313, 314, 419
Jacobs v. Moseley, 280
Jahnke v. McMahon, 280
Jakeway v. Barrett, 206
James v. Drew, 323, 367, 412
Janke v. McMahon, 266
Jarvis v. Swain, 301
Jefferis v. East Omaha Land Co., 137
Jefferis v. Land Co., 429
Jenks v. Morgan, 197
John C. Barnes v. George L. Tolson, 366
Johnson v. M'millan, 222
Johnson v. Westrick, 55
Jones v. Burgett, 57
Jones v. Hamilton, 124
Jones v. Kimble, 5, 367, 412
Jones v. Merrimack Co., 198
Jones v. Pashby, 279
Jordan v. Deaton, 281
J.R. Buckwalter Lumber Co. v. Wright, 292, 391–392
Jurisdictions, 270
Jury v. Grant, 387
Justice v. McCoy, 221
J.W. Robinson v. Elias Mosson et al., 57

K

Kahn v. Delaware, 29, 30
Kaiser v. Dalto, 34, 380
Keesling v. Truitt, 410
Kellogg v. Smith, 197, 243, 388–390
Kelsey v. Lake Childs Co., 149
Kennedy v. Oleson, 146
Kentucky, 333
    case of *Allen v. Blanton,* 339
    Jackson Purchase, 343–344
    land patents, 343
    river, 353–354
Kentucky Coal Lands Co. v. Mineral Development Co, 363
Kentucky Land Office, 344
Kentucky River Timber & Coal Co. v. Morgan, 123
Kerr v. Fee, 82
Keyser v. Sutherland, 132, 393
Kimms v. Libby, 280
Kincaid v. Peterson, 429–430

King v. Carmichael, 432–433
Kip v. Norton, 281
Kirby's Digest, 366, 368
Kirch v. Persinger, 149
Klyce survey, 30
Knapp v. Marlborough, 197
Knight v. Elliott, 313, 314, 367, 410, 419, 426
Knox v. Silloway, 197
Knupp v. Barnard, 292
Krause v. Nolte, 413
"Kuleana" surveys, 242

L

Laflin Borough v. Yatesville Borough, 48
Lampson Lumber Co. v. Caporale, 62
Land, 267, 306. *See also* Survey(s)
    establishment of title to, 40–41
    laws, 366
    rights in, 40–41
    survey. *See Roman centuriatio*
    surveying, 82*fn,* 227–228
    surveyor, 13
Land Ordinance of 1785, 405
Landscapes, 288
Lane v. Darlington, 425
Lane v. James, 202
Langle v. Brauch, 149
Large grants, 219, 348
Larsen v. Richardson, 32, 99, 139, 166, 290, 341
Lawler v. Rice & Goodhue Counties, 149
Lawson v. Viola Tp., 91*fn,* 399
Least Squares adjustment, 225
Ledford patent, 114
Lemmon v. Hartsook, 426
LesSee of Alshire v. J.R. Hulse, 250
LesSee of Brown v. Clements, 136
Lewen v. Smith, 300
Lewis v. Ogram, 280
Lewis v Prien, 407
Liddon v. Hodnett, 149, 281
Liebler v. Abbott, 360
Lilly v. Kitzmiller, 384
Lind v. Hustad, 280
Line(s), 77; *see also* Corner
    agonic, 189
    blazed line, 78
    compromise line, 79

464   *Index*

Line(s) *(Continued)*
  conditional line, 78
  consentable line, 78
  conventional line, 79
  extended line, 79
  fence line, 78
  hallean, 189
  ideal line, 79
  isogonic, 189
  lost line, 79
  lot line, 78
  meander line, 79–80, 429
  never run, 298
  partition line, 80
  property line, 80
  right line, 80–81
  right of way line, 78
  from surveys, 129–130
  true line, 81
  types, 77
Line of Cumberland Mountain, 118
"Line of Occupation", 265, 266
Linscott v. Fernald, 197
Lippincott v. Souder, 244
Lloyd v. Benson, 76, 228
Local government, 143
Locative calls, 165
Lodge v. Barnett, 101*fn*, 397–398
Longlots, 333, 346–348
  French system of, 345
Loring v. Norton, 75
Lost corners, 72. *See also* Corners
  Alabama, 406–407
  application of rules to government
      surveys, 410
  Bailey v. Chamblin, 414
  boundary line agreement, 332
  Carstensen v. Brown, 433
  Case v. Trapp, 417–418
  City of Jacksonville v. Broward, 411–412
  collateral evidence, 312
  corner formerly stood, 310–311
  Delphey v. Savage, 417
  discovering evidence, 324–330
  foregoing problem situation, 331
  Gaines v. City of Sterling, 408
  government surveyor, 330
  *Hanson v. Township of Red Rock in
      Minnehaha County*, 430–432

Hess v. Meyer, 418–419
Higgins v. Ragsdale, 408
*Houston Ice & Brewing Co. v. Murray
      Oil Co.*, 415–416
independent surveys, 424
Kincaid v. Peterson, 429–430
King v. Carmichael, 432–433
Krause v. Nolte, 413
Land Ordinance of 1785, 405
lost v. obliterated corners, 311–312
Nystrom v. Lee, 425–426
original corners, 309
Overton v. Leonard, 426–429
parol evidence and reputation, 320–324
practices, 410–411
procedure, 313–320
Ramsey v. Morrow, 415
restoring lost, 309–313
restoring lost corners, 309–313
Riley v. Griffin, 412
section lines and corners, 313
*State, Bd. of Educational Lands and
      Funds v. Jarchow*, 420–421
Stetson v. Adams, 416–417
"supplement to the Manual", 406
Thomsen v. Keil, 421–422
United States Government, 405
Velasquez v. Cox, 422
Woods v. Johnson, 419–420
Lost line, 79
Lost v. obliterated corners, 311–312
Lot line, 78
Lottings in Georgia, 348
Louisiana, 345–346
Lovelace v. Hightower, 424
Lower precision surveys, 233–234
Lugon v. Closier, 69, 298
Lugon v. Crosier, 71, 75, 315
Lugon v. Crozier, 310*fn*, 311*fn*
Lund v. Parker, 198
Lyon v. Ross, 80, 384, 387

M

Maddox Bros. & Anderson v. Fenner,
      57, 126
Mageehan v. Adam, 384
Mahone pin, 146
Major v. Watson, 291, 309, 313, 314, 419

Index

465

Manual of Instructions for the Survey of the Public Lands of the United States, 409
Manual of Land Surveying, 298
Manual of Surveying Instructions, 296, 406
Manufacturer's National Bank of Detroit v. Erie County Road Commission, 95
Maps, 16, 29, 31, 70, 84, 87, 125, 138, 145, 150, 215, 244, 261, 265, 288, 333, 374, 379–380
  Roman, 334
  unrecorded, 304
Marker, 310
"Marketable title", 45*fn*
Marsh v. Stephenson, 412
Martin v. Carlin, 145
Martin v. Hall, 285
Martin v. Lopes, 267
Martz v. Williams, 383, 413
Maryland Coal and Realty Co. v. Eckhart, 48, 372
Mason v. Braught, 210, 312, 313, 381, 411
Matador Land & Cattle Co. v. Cassidy-Southwestern Commission Co., 57
Mathematical line, 77, 80
Matthews v. Parker, 74
Mayor of Liberty v. Burns, 314, 412, 419
May v. Baskin, 38, 384
MBE. *See* Middle Branch Engineering
McAlpine v. Reicheneker, 429
McClintock v. Rogers, 299, 330, 385, 413
McCourry et al. v. McCourry, 190
McCoy v. Galloway, 300
McCullough v. Absecon Beach Co., 101, 243
McEwen v. Den, 220
McGinley v. Maine Central Railroad Co., 328
McGinley v. Railroad, 330
McIver's LesSee v. Walker, 253
McIver v. Walker, 187
McKinney v. Doane, 314, 392, 419
M'Clintock v. Rogers, 38, 80, 101*fn*, 383–384
Md. Construction Co. v. Kuper, 48
Meander corner, 73
Meander lines, 79–80, 429
Measurement systems, 345

Mechler v. Dehn, 410
Melcher v. Flanders, 198
Mellor v. Walmesley, L. R., 146
Mellow v. Weaver, 267
Melvin v. Marshall, 200
Memorials, 255, 258–259
Mercer v. Bate, 124, 297, 298
Metal detectors, 258
Metal objects, 257
Metes-and-bounds states, 347
Metes, 9, 16, 84
Meyers v. Johnson, 284
Meyer v. Board of Imp. Pav. Dist. No. 3, 407
Middle Branch Engineering (MBE), 110
Miles v. The Pennsylvania Coal Co., 282
Military grants, 348
Millar v. Bowie, 388
Miller v. Topeka Land Co., 429
Miller v. White, 149
Milligan v. Milligan, 76
Mineral Development Co. v. Tuggle, 363
Miner v. Brader, 261
Mining & Smelting Co. v. Davis, 314, 419
Minot v. Brooks, 197
Missing title, 45–50
Missouri v. Iowa, 389
Mitchell v. Hawkins, 393
M'Iver's LesSee v. Walker, 135, 188, 222
M. K. & T. Ry. Co. v. Merrill, 137
M'Nairy v. Hightour, 389
Mock v. Astley, 293
Montana. See Olson v. Jude
Monument, 74
  accessory, 75–76
  missing, 76–77
Monumentation, 73, 307, 338, 340
Moody v. Farinholt, 401–402
Moreland v. Page, 367, 386–387, 408, 410
Morgan v. Renfro, 124
Morrison v. Coghill, 387
Morris v. Jody, 387
Mosher v. Berry, 412
Mounds, 251–252
  stone mounds, 255–256
Mountain surveys, 233
Moyer v. Langton, 432
Multiple markers, 54–56
Myrick v. Peet, 145, 240, 279

Index

N

National City v. California Water and Telephone Company, 62
Neary v. Jones, 425
Neill v. Ward, 208, 343
Nelson v. Hall, 1 McLean, 430
Nesselrood v. Parrish, 92*fn*, 149, 399, 412
Newfound Management v. Sewer, 289
New Hampshire courts, 335
NEW HAMPSHIRE DEPARTMENT OF RESOURCES AND ECONOMIC DEVELOPMENT v. E. Milton DOW and E. Milton Dow d/b/a Dow Sand and Gravel, 267*fn*, 270
New Hampshire, statute law in, 269–270
New Map of Narcoossee, 18
Nichol v. Lytle's Lessee, 280
"Nigh Nicking", 46
Niscia v. Cohen, 112
Nixolin v. Schnerderline, 305
'No-man's land', 50, 353
Nonexistent corner, 71
Nonfederal rectangular surveys, 333; *see also* Paper survey
bounding descriptions, 340–341
division by area, 337–339
division by half, 337
French system of longlots, 345, 346
idiosyncrasies within system, 340
Jackson Purchase, 343–344
longlots in areas, 348
lottings in Georgia, 348
Louisiana, 345–346
military grants, 348
in New World, 334–337
original perimeter survey, 335
proprietors' records, 341–343
special grants, 349
states of rectangular lots, 343
subdivision of Bennington, 336
TennesSee surveyors districts, 344–345
Texas, 347–348
Normanoch Ass'n v. Deiser, 6*fn*
North Carolina, 352
Northern boundary of Carricks Chance, 46
Norwood v. Crawford, 219, 302

Notice, 50–51
Nutting v. Herbert, 197
Nystrom v. Lee, 425–426

O

Obliterated corner, 72
"Office survey", 423
Official plat of United States Government, 367–369
Ogilvie v. Copeland, 92*fn*, 149, 322, 399, 407
Olson v. Jude, 29, 31, 392
Operation of law, 44
Original corners, 291, 309, 314
Original creation and, 287–289
"Original deed writers", 289
Original grant(s), 334, 353–359; *see also* Overlapping grants, resolving
on caveat, 355–356
on chancery, 357–359
Herndon v. Hogan, 354
Hinton v. Stewart's Heir, 356
Zachariah Herndon v. James Hogan, 353
Original grantor, 352
Original line(s), 139, 351
Original lots, 337
Original survey, 89–93, 330
Original surveyor, 9, 40, 93, 126, 287–289
Osborn v. King, 24
Osteen v. Wynn, 111, 268, 283
Outlaw et al. v. Gulf Oil Corporation et al., 400
Outlaw v. Gulf Oil Corporation, 82
Overlapping grants, resolving
absolutely appellees' theory, 364
Caylor v. Luzadder, 367
decisions, 359
Gammon v. Verrill, 360
government surveyors, 366
legal boundary, 361
location of patent, 365
original grants, 353–359
original layout, 362
quarter-section corner, 368–369
ridges, 363
seniority of title, 351–353
Southwestern Improvement Association, 366

Index 467

Overstreet v. Dixon, 82fn
Overton v. Leonard, 426–429
Oxford v. White, 66

P

Pa., 5 Rawle. See Smith v. Moore
Pacheco v. Martinez, 394
Pallas v. Dailey, 393, 394
Pandem Oil Corp. v. Goodrich, 410
Paper survey, 16, 219; *see also* Nonfederal
 rectangular surveys
 appellant, 222
 comparing measurements, 223
 custom of surveyor, 219–220
 distance deficiencies, 222–223
 legal rule, 221
 problems with area recitations,
  223–224
Paramour, The, 189
Parcel, 113–114
Parcels of land, 105, 288
Parent tract, 341
Parker v. Brown, 198
Parkhurst v. Van Cortland, 197
Parol agreement, 272, 274
Parol evidence and reputation,
 320–324
Parol gift, 44
Parran v. Wilson, 74
Partition line, 80
Passing calls, 165–166
Patent from sovereign, 41
Patton on Titles, 372
Peacher v. Strauss, 52, 65, 228
Pearson v. Baker, 118, 240
Peaslee v. Gee, 197
Pennington v. Flock, 66
People of the State of California v.
 Thompson & Gray, Inc., 313
People v. Holmes, 146
Pereles et al. v. Magoon et al., 412
Pereles v. Gross, 95, 403
Pereles v. Magoon, 323, 429
Perkins v. Jacobs, 360
Pernam v. Wead, 430
Perry v. Buswell, 2fn
Perspective, course and distance in,
 243–244

Peterson v. Johnson, 282
Philips v. Shaeffer, 384
Phillips v. Ayres, 245
Pierce v. Richardson, 202
Pike v. Dyke, 312, 388
Pilgrim v. Kuipers, 59
"Pincushions", 107
 corners, 187, 330
Pinkham v. Murray, 202
Piotrowski v. Parks, 111
"Pitched lots", 352
Pitman v. Nunnelly, 124
Pits, 251–252, 256
Plaintiff, 353
 asserts, 316
Plaintiffs' witness and engineer, 423
Plane table-chain, 182
Plane table-intersections, 182
Plane table-stadia, 182
Plat errors, 235
PLSS, *see* Public Land Survey States
Pollard v. Shively, 101fn
Porciones, 347
Porter v. Gaines, 412
Possessor, 263
Post, 250
Potts v. Everhart, 300
Powers Ranch Co., Inc. v, Plum Creek
 Marketing Inc., 269
Practical location, 272
Practical location of boundary,
 275–278
Prescott v. Hawkins, 198
Prescott v. Hayes, 198
Prescription, 43–45
Prescriptive easement, 43
Preston Heirs v. Bowmar, 239, 240
Preston's Heirs v. Bowmar, 119
Preston v. Bowmar, 122
Presumption, 342
Price v. Earl of Torrington, 146
Price v. Mauch, 379–380
Pride v. Lunt, 198
Principles of retracement, 2
 apportionment, 5–6
 example of an ownership, 3
Prior appropriation, 44
"Priority of calls", 99
Private grant, 41

468   *Index*

Probate records, 305–307
Procedure, 287
    block system, 293
    comprehensive review, 289
    corner never set, 298
    about corners and witness trees,
        298–301
    effective and successful retracement
        surveyor, 287
    evidence of relative location, 294
Procedure (*Continued*)
    government procedure, manual of
        surveying instructions, 296
    government surveyors, 291
    government surveys, 303–304, 315
    historical collections, 307–308
    ideal line, 297–298
    line never run, 298
    locating unrecorded maps, plans,
        and sketches, 304
    lost corner, 313
    original corners, 314
    original creation and original
        surveyor, 287–289
    probate records, 305–307
    property owners, 320
    retracement surveyor, 290, 316
    reversing course, 301–303
    texas court, 295
    true control point, 319
    true corner, 292
    unrecorded plat, 304–305
    Wacker v. Price, 317–318
Proctor v. Hinkley, 74
Proper retracement, 352
Property description, 9, 13
Property line, 77, 80
"Proprietors", 372
    records, 335, 341–343
Protracted boundaries
    footsteps, 139–140
    guidance, 130–131
    lines from surveys, 129–130
    protracted lines, 128
    types of protractions, 129
    unmarked line, 130
    West Virginia case, 106
Protracted lines, 107, 128
Protracted surveys, 106

Protraction, 105, 129
    boundary lines, 133
    corners marked in surveys, 134
    deeds of conveyance, 138
    Gazzam case, 136
    Larsen v. Richardson, 139
    M'Iver's LesSee v. Walker, 135
    monumented corners, 137
    prevailing rule, 132
    best-fit approach, 109
    "Blucher State Map", 126
    boundary by parol agreement, 111
    boundary lines, 124
    Bramblet v. Davis of locations, 117
    Coleman County v. Stewart, 12
    court of appeals of Kentucky,
        121–122
    Cumberland Gap, 120
    descriptive matter in field
        notes, 127
    federal court, 122
    Kentucky case, 123
    Ledford patent, 114
    MBE, 110
    original grants and patents, 113
    outside PLSS, 107
    patent calls, 118
    plat in *Bramblet v. Davis*, 115
    in PLSS, 131–139
    protracted lot from known
        perimeter, 108
    protracted subdivided lots within
        known perimeter, 108
    state line, 116
    surveys adjacent, 125
    trial court, 112
    Virginia, 119
Public Lands, 41, 372
Public Land States. *See* Public Land
    Survey States (PLSS)
Public Land Survey States (PLSS), 69,
    102, 131, 288, 406
    outside PLSS, 107–126
    protraction in, 131–139
    witness corner, 73
    witness tree, 73
Public Land Survey System. *See* Public
    Land Survey States (PLSS)
"Public States", 405

Index 469

Puget Mill Co. v. North Seattle Imp. Co., 149
Purchase, The, 343
Pyburn v. Campbell, 380

Q

Quarter-quarter corners, 368
Quarter-section corner, 71, 368
Quesnel v. Woodlief et al., 168
"Quick fix", 263

R

Racine v. Emerson, 26
Radford v. Johnson, 395, 425
Railroad Co. v. Schurmeier, 79
Railroads, 347
Ralston v. Dwiggins, 318
Ralston v. M'Clurg, 76, 124
Ramsey v. Morrow, 407, 415
Randall v. Burk Township, 425
Randall v. Burk Tp., 91*fn*, 399
Randall v. Burk Tp. of Minnehaha County, 252
Randleman v. Taylor, 279
Random errors, 230
Rand v. Dodge, 196, 197, 198
Range lines, 360–361
Rapides Parish Police Jury v. Grant Parish Police Jury, 192
Rautenberg v. Munnis, 327, 328
Re-establishment of Quarter Section Corners on Closing Section Lines between Fractional Sections, 428
Read v. Bartlett, 322, 407
Rectangular grants, 334
Rectangular lots, 106, 333, 343, 345, 348, 352
"Rectangular system of survey", 405
Rectangular systems, 333
Reel v. Walter, 33
Reestablished corner, 72
Reference monuments, 257
Reid v. Dunn, 70, 314*fn*
Relative precision of work, 230
Reliability, 233
Resolving/reconciling errors
  accidental errors, 231–232
  computational errors, 232
  course and distance in perspective, 243–244
  discrepancies, 229
  disposing of lands, 241–242
  errors, 229
  errors in data collection, 234
  errors in measurements, 229
  errors in reporting data, 235–240
  errors of closure, 230
  index errors, 230
  inherent errors, 240
  land surveying as art, 227–228
  lower precision surveys, 233–234
  measurement of precision, 234
  mistakes, 232
  natural errors, 231
  personal errors, 230
  property rights, 240
  random errors, 230
  reliability, 233
  surveys in Hawaii, 242–243
  systematic errors, 231
  total error, 233
Restoration of Lost or Obliterated Corners & Subdivision of Sections, a guide for surveyors, 406
Resultant error, 229
Resurvey, 13, 93–94. *See also* Survey(s)
Retracement, 2, 13, 371. *See also* Survey
  apportionment, 5–6
  Bureau of Land Management, 32–33
  conditions for rule to apply, 6
  diagram from *Tyson v. Edwards*, 21
  establishing internal lines within Rizzo's subdivision, 11
  example of ownership, 3
  footsteps contrary, 34–36
  general rules, 30–31
  people, 55
  plan from *Tyson v. Edwards*, 15
  plat as grid, 21
  plat redrawing, 23
  principles of, 2
  purpose of resurvey, 29
  quarter-quarter section, 7
  retracement surveys, 13
  State Road Department survey, 19
  surveying and boundaries, 28
  surveying principles, 16, 31

470 *Index*

Retracement (*Continued*)
  surveyor, 9, 189, 289, 316
  surveys, 9, 13, 93–94, 227, 340, 341
  true meridian and base line, 10
Retracing lines, 92
Reversing course, 301–303
Revised Statutes of United States, 368
Ricci v. Godin, 77
Richardson v. Bohney, 412
Ridges, 363
Right line, 80–81
Right of way line, 78
Right of way of railroad, 78
Riley v. Griffin, 101*fn*, 382–383, 412
Rittenhouse-type compass, 187
River lots, 347
Rivers case, 25, 141
Rivers v. Lozeau, 7, 93, 290, 381–382
Roadenbaugh v. Egy, 414
Robinson v. Doss, 125
Rock Property Co. v. Hill, 124
Rockwell v. Adams, 281
Rollins v. Davidson, 241, 386
Roman centuriatio, 333
Roman Empire, 333
Roman maps, 334
Rood, 208
Root v. Town of Cincinnati, 386
Roth v. Halberstadt, 63
Routh v. Williams, 29
Rowell v. Weinemann, 241, 386
Rowe v. Kidd, 32, 79, 106, 128, 297
Ruddy v. Rossi, 137
Rule of law, 365
Rules and principles of retracement, 1
Rusha v. Little, 360
Russell v. Dyer, 199

S

Sala v. Crane, 132, 134, 137
Sanborn v. Clough, 197
Sargent vs. Town, 382
Savannah, Florida and Western Railway v. Geiger, 298
Sawyer v. Cox, 384–385
Sayers v. City of Lyons, 386
Scammon v. Scammon, 200, 202
Schraeder Min. Co. v. Packer, 280

Scott v. Thacker Coal Mining, 291
Scribe marking, 248, 249
"Scriveners", 235
Seabrook v. Coos Bay Ice Co., 79, 132
Section corner, 71
Sellman v. Schaaf, 27, 31, 250, 395, 426
Senses, 72
Serrano v. Rawson, 253
Shackelford v. Walker, 351*fn*
Shaw v. Clement, 80, 387
Shelton Logging Co. v. Gosser, 74
Silsby & Co. v. Kinsley, 342, 401
Simmons Creek Coal Co. v. Doran, 304
Simpkins v. Wells, 302
Sitio de granado mayor, 347
Sitio grants, 346, 347
Sixteenth corner, 72
Ski Roundtop, 142
Ski Roundtop Inc. v. Wagerman, 45, 47, 372, 388
Slope measurement, 218–219
Slovensky v. O'Reilly, 393
Smallidge, et al. v. Same, 63
Small v. Chronicle & Gazette Publishing Company, 328
Smart v. Huckins, 62, 64, 65
Smiley v. Fries, 66
Smita v. Young, 305
Smith v. Bradley, 202
Smith v. Dodge, 394
Smith v. Messer, 200, 202
Smith v. Moore, 235, 397
Smith v. Nelson, 281
Sneed v. Osborn, 267
Snider v. Rinehart, 324
South Penn Oil Co. v. Knox, 57
Southwestern Improvement Association, 366
Sparhawk v. Ballard, 198
Square league, 347
Staaf v. Bilder, 29, 31
Stack v. Pepper, 219
Stadin v. Helin, 75, 248
Stafford v. King, 36*fn*, 102, 183, 207, 210, 245, 295*fn*, 296, 400
Stake, 251
  wooden stakes, 250–251
Stake patents, 284
Standard corner, 72

Index

471

Standard measurement, 340
Stangair v. Roads, 322
Stanley v. The County Surveyor of Sheridan County, 414
Stanolind Oil & Gas Co. v. State, 351fn
State, Bd. of Educational Lands and Funds v. Jarchow, 420–421
State el rel. Buckson v. Pennsylvania R. Co., 381–382
State ex rel. Cohen v. Manchin, 87
State of Texas, 347–348
State Road Department survey, 19
State v. Ball, 132, 136
State v. City of Sarasota, 67
State v. Coleman-Fulton Pasture Co., 125
State v. Palacios, 125, 126
State v. Stanolind Oil & Gas Co., 130
Statute law in New Hampshire, 269–270
Statute of frauds, 263, 278–280
    effect, 111, 274
Staub v. Hampton, 27, 30, 31, 102, 400
Steele v. Taylor, 239
Steel v. Co., 403
Steinherz v. Wilson, 107fn
Stetson v. Adams, 416–417
Stewart, S. L., 158–159
Stewart v. Carleton, 38, 40, 260, 291, 314, 315, 378, 379, 420
Stones, 251, 257
    mounds of, 255–256
Stover v. Freeman, 300
St. Paul & P. R. R. Co. v. Schurmeier, 132, 428
St. Paul & P. Ry. Co. v. Schurmier, 135
Straight line, presumption of, 211
    boundary, 215–216
    Enfield v. Day, 217–218
    Hart's Location, 211–214
    law of New Hampshire, 216–217
    New Hampshire townships, 212
    public acts, 214–215
Straight protraction, 129
Straw v. Jones, 198
Strunk v. Geary, 292
Strunz v. Hood, 402
Stump holes, 248–249
Suertes, 347
Suit v. Hershman, 410
Sullivan v. Rhode Island Hospital Trust Co., 62

Sumner v. Sherman, 202
Sumter v. Bracy, 384
"Supplement to the Manual", 406
Surveyor(s), 9, 40, 70, 93, 105, 277, 287, 290, 363
    custom of surveyor, 219–220
    effective and successful retracement, 287
    field notes of surveyor Job S. Collard, 170–172
    footsteps of, 97
    GLO surveyor, 192–193
    government, 291, 314, 330, 366
    intent of, 63–64
    land, 13
    line, 80
    original creation and original surveyor, 287–289
    original surveyor, 9, 40, 93, 126, 287–289
    retracement surveyor, 287, 290, 316
    "sufficiency" test, 67
    Surveyor-General, 347–348
Survey(s), 81, 351. *See also* Paper survey
    accurate survey, 85–89
    actual, 219–224
    boundaries, 84–85
    categories, 85
    establishing, line, 94–95
    Georgia court, 82
    in Hawaii, 242–243
    independent survey, 96
    mountain, 233
    original survey *vs.* first survey, 89–93
    paper, 219–224
    retracement survey *vs.* resurvey, 93–94
    surveying principles, 16, 31, 83
    tracking survey, 96–99
    types, 81, 181–184
Suydam v. Williamson, 122
"Swag", 210
Swanson v. Koeninger, 394
Swift Coal & Timber Co. v. Sturgill, 292, 362
System of land utilization, 334

T

Table of authorities, 437–454
Taping, errors in, 210–211
Tappan v. Tappan, 197, 198

472   *Index*

Tarpenning v. Cannon, 318
Taylor v. French, 202
Taylor v. Higgins Oil & Fuel Co., 400
TennesSee River, 344
TennesSee surveyors districts, 344–345
Terry v. Chandler, 280
Tewksbury v. French, 101*fn*
Texas, 347–348
  court, 295
  treatise, 105
Texas International Petroleum v. Delacroix
    Corp., 193
Thallman v. Thomas, 394
Thatcher v. Matthews, 324*fn*
Thein v. Burrows, 394
Theriault v. Murray, 76
The rule laid down in Haggan v. Wood, 415
The Supreme Court of Wisconsin, 407
Thomas Hinton, heir-at-law of Joseph
    Hinton, deceased, by William
    Morrice, his next friend, v.
    The Heir of William Stewart
    deceased, 354
Thomas v. *Perry*, 168
Thompson v. Darr, 407–408
Thompson v. Hill, 83
Thompson v. McFarland, 389
Thomsen v. Keil, 258, 320
Thornberry v. Churchill, 118, 384, 426
Thurmond v. Espalin, 424
Tiggeman v. Mrzlak, 394
Timber corners, 297
Title, 37, 287. *See also* Apportionment
  ambiguities, 51–54, 64–65
  boundary, 37–38
  boundary line agreements, 60
  combination, 42
  descriptions by abutting tracts, 56–60
  establishment, 38–41
  intention, 61–63
  intent of parties, 60–61
  intent of surveyor, 63–64
  missing title, 45–50
  multiple markers, 54–56
  notice, 50–51
  prescription, 43–45
  property, 11
  seniority of, 351–353
  significance of original survey, 40

unwritten, 42–43
void instrument, 65–67
written, 41
Tituses' survey, 92
Titus v. Chapman, 89, 138, 398–399
Tognazzini v. Morganti, 380–381
Tolson v. Southwestern, 323
Tolson v. Southwestern Improvement
    Association, 366, 429
Tompkins v. American Republics Corp., 62
"Topographical calls", 165
Township corner, 72
Town v. Greer, 149, 150
Tracking survey, 96–99
Transit-stadia, 182
Transit-tape, 182
Treadwell v. Marrs, 393–394
Tree
  bearing objects and, 253–255
  blazed, 247
  names, 246
  rings, 249–250
  witness, 247–248
Triangulating, 194
Trinwith v. Smith, 149, 412
Tripp v. Bagley, 281, 282
Trotter v. Stayton, 317, 403
Trotter v. Town of Stayton, 396–397
True boundary line, 268
True line, 81, 280
Trustees, etc., v. Wagnon, 124
Tucker v. Aiken, 202
Tunisia, 333–334
Turnbow v. Bland, 288, 295*fn*
Tyner v. McDonald, 24, 410
Tyson's case, 24, 25
Tyson v. Edwards, 7, 12, 13, 15, 31, 83,
    381–382

U

Uncertainty, 66*fn*
  error, 230
  meaning, 270–272
United Fuel Gas Co. v. Snyder, 57
United States Forest Service survey, 91
United States Government, 405
United States v. Champion Papers, Inc., 53,
    63, 168, 311

Index

473

United States v. Doyle, 33, 260, 312, 381, 410
United States v. Gallas, 250
United States v. Gratiot, 137
United States v. Hudspeth, 381
United States v. Lane, 381
United States v. State Investment Co., 381
Universal rule of law, 61
Unmarked line, 130
Unrecorded maps, plans, and sketches, 304
Unrecorded plat, 304–305
Unsurveyed public lands, 405
Unwritten agreements, 43
Unwritten title, 42–43
Urban v. Urban, 62

V

Vance v. Fore, 253
Vance v. Gray, 410
Van Dusen v. Shively, 33, 396–397
Varas, 347
Vaughn v. Tate, 367, 410
Vaught v. McClymond, 381, 392, 393, 420
Vermont, 334
Void instrument, 65–67
Voigt v. Hunt, 280
Vosburgh v. Teator, 280
Vowinckel v. N. Clark & Sons, 266

W

Wacker v. Price, 317–318, 377–379
Wade v. McDougle, 353
Walker v. Curtis, 146
Wallace v. Maxwell, 124
Warcynski v. Barnycz, 147
Ward v. Fuller, 198
Ward v. Rodriguez, 424
Washington Rock Co. v. Young, 36, 92fn, 322, 399, 400–401, 407, 413
Wates v. Crandall, 394
Watrous v. Morrison, 149, 280, 381
Weaver v. Howatt, 403
Weaver v. Robinett, 300

Wells v. Burbank, 199, 200
Wells v. Jackson Iron Manufacturing Company, 62, 186, 188, 193
Wells v. New York Mining And Manufacturing Co., 286
Wendell v. Blanchard, 198
Wentzel v. Claussen, 423
Wescott v. Craig, 322
Westcott v. Craig, 408, 410
Westphal v. Schultz, 367
West Tennessee, 343
West Virginia corporation, v. CITY OF WHITE SULPHUR SPRINGS, West Virginia, 85
Wharton v. Littlefield, 377
White v. Amrhien, 425
Whiting v. Gardner, 132
Whitman v. Haywood, 379
Whitmore v. Brown & Gilley, 63
Wightman v. Campbell, 146, 147
Wildeboer v. Hack, 25, 305
Will, 41
William Eagan v. Samuel Hinch heir-at-law of John Hinch, deceased, and John Jack, heir-at law of Samuel Jack deceased, 356
Williamson v. Berry, 122
Williams v. Barnett, 265, 267fn, 330
Williams v. Brush Creek Coal Co., 291
Williams v. Spaulding, 312, 388
Williams v. Tschantz, 291, 414
William White v. Dudley Everest, 341
Willis v. Campbell, 31
Wilson v. DeGenaro, 62
Wilson v. Stork, 312, 411
Wisconsin. See Racine v. Emerson
Wise v. Burton, 252
Wishart v. Crosby, 384
Witness, 252–253
   corner, 73, 257
   house, 253
   points, 257
   tree, 73, 247–248, 298–301
Wooden posts, 250
Wooden stakes, 250–251
Woods v. Banks, 198
Woods v. Johnson, 309fn, 313, 419–420
Woods v. West, 132

474   *Index*

Wood v. Hildebrand, 48
Wood v. Starko, 26, 27, 400
Written title, 41
Wrong correction factor, 340
Wyatt v. Foster & Rafferty, 184

Y

Yocum v. Haskins, 430
Yoder v. Fleming, 384, 398
Yolo County v. Nolan, 149

Yolo v. Nolan, 380–381
Young v. Leiper, 389

Z

Zachariah Herndon v. James Hogan, 353
Zawatsky Constr. Co. v. Feldman
      Development Corp., 48
Zeringue v. Harang, 250
Ziebold v. Foster, 290
Zwakhals v. Senft, 12